ELECTRONIC MATERIALS

ELECTRONIC MATERIALS

Edited by
N. BRUCE HANNAY
Vice President
Research and Patents
Bell Laboratories
Murray Hill, New Jersey

and

UMBERTO COLOMBO
Director, Corporate Research
and Strategic Planning
Montedison
Milan, Italy

PLENUM PRESS • NEW YORK-LONDON

Library of Congress Cataloging in Publication Data

International Conference on Materials Science, 3d, Tremezzo, Italy, 1972.
Electronic materials.

Sponsored by Accademia Nazionale dei Lincei, 13th course of the G. Donegani Foundation.
Includes bibliographical references.
1. Electronics—Materials—Congresses. I. Hannay, Norman Bruce, ed. II. Colombo, Umberto, fl. 1964—ed. III. Accademia Nazionale dei Lincei, Rome. Fondazione Donegani. IV. Title.
TK7871.I57   1972   621.381'028   73-13971
ISBN 978-1-4615-6892-6   ISBN 978-1-4615-6890-2 (eBook)
DOI 10.1007/ 978-1-4615-6890-2

Proceedings of the Third International Conference on Materials Science
held in Tremezzo (Como), Italy, September 4-15, 1972

© 1973 Plenum Press, New York
Softcover reprint of the hardcover 1st edition 1973
A Division of Plenum Publishing Corporation
227 West 17th Street, New York, N.Y. 10011

United Kingdom edition published by Plenum Press, London
A Division of Plenum Publishing Company, Ltd.
Davis House (4th Floor), 8 Scrubs Lane, Harlesden, London, NW10 6SE, England

All rights reserved

No part of this publication may be reproduced in any
form without written permission from the publisher

THE CONTRIBUTORS

A. Ascoli
Centro Informazioni Studi Esperienze
Segrate, Milan, Italy

D. J. Bradley
The Queen's University of Belfast
Belfast BT7 1NN, Northern Ireland

G. Chiarotti
Istituto di Fisica
dell'Università
Rome, Italy

R. L. Colombo
Centro Sperimentale Metallurgico
Rome, Italy

U. Colombo
Montedison
Milan, Italy

A. Eschenfelder
IBM Research Laboratory
San Jose, California

C. Haas
Laboratory of Inorganic Chemistry
Materials Science Center of the University
Zernikelaan, Groningen, The Netherlands

N. B. Hannay
Bell Laboratories
Murray Hill, New Jersey

C. Hilsum
Ministry of Defence, Royal Radar Establishment
Great Malvern Worcestershire, England

G. Marie
Laboratoires d'Electronique et de Physique Appliquée
94, Limeil-Brévannes, France

H. Pötzl
Lehrkanzel für Physikalische Elektronik der Technischen
  Hochschule, and
Ludwig Boltzmann-Institut für Festkörperphysik
Vienna, Austria

H. J. Queisser
Max-Planck Institut für Festkörperforschung
Stuttgart, Germany

K. Seeger
Institut für Angewandte Physik der Universität, and
Ludwig Boltzmann-Institut für Festkörperphysik
Vienna, Austria

D. G. Thomas
Bell Laboratories
Murray Hill, New Jersey

D. Treves
The Weizmann Institute of Science
Rehovot, Israel

P. A. Wolff
Massachusetts Institute of Technology
Cambridge, Massachusetts

PREFACE

This volume constitutes the written proceedings of the Third International Conference on Materials Science, held under the sponsorship of the Accademia Nazionale dei Lincei as the XIII summer course of the G. Donegani Foundation at Tremezzo, Italy, on September 4-15, 1972.

The course of lectures was designed for scientists and engineers with a working knowledge of electronic materials, who sought to extend their knowledge of the newest developments in the field. The rapid pace of research and exploratory development in electronic materials has led to a pressing need for continuing awareness and assessment of new electronic materials, as well as renewal of information in the more traditional areas.

Three classes of electronic materials were selected for the course. Semiconductors provide the foundation for solid state electronics and semiconductor devices represent the most sophisticated and advanced application of materials science and engineering known to modern technology. Yet, the march of progress in semiconductors continues unabated - new semiconductor materials are in the research stage, new process technology is being developed, and new devices are being conceived. The second class of materials dealt with in the course, magnetic alloys and insulators, also has a firm application base; for example, computer performance is often measured in terms of the size of the magnetic memory. The tailoring of materials to provide particular combinations of desired magnetic properties is an integral part of the development of the electronics, just as in the case of semiconductors. The third class of materials covered, optical materials, is by contrast in a quite different stage. There is as yet no large area of practical application. However, it is our strong belief that optical materials and devices have emerged as major opportunity areas for the future. It is too early to say how large this application will be or which materials and devices will be important and for this reason the survey here must necessarily be more speculative. It delineates current trends, but is unable to sort out those that will survive. Laser materials, display materials, and optical communications have been given emphasis in the course because of their particular promise in electronics technology.

We have also designed the course from another point of view. Three aspects of electronic materials are covered. The physics of materials provides the fundamental background for work in the field. Secondly, the materials themselves are considered; thus, solid-state chemistry, crystal growth, film techniques, and new compositions of matter are described. Finally, there is the application of materials in devices; the designs of electronic materials and the corresponding electronic devices in which the materials are used are interacting and inseparable processes. Materials science and engineering furnish an example to other areas of science and technology of the advantages of interdisciplinary research; this volume demonstrates this theme by showing the close relationship between the physics of materials, their chemistry and the device application.

We would like to express our deep appreciation to the many people who helped to make the course a success. Professor Beniamino Segre, Chairman of the Accademia Nazionale dei Lincei and of the "G. Donegani" Foundation and other members of the Foundation provided generous support and assistance. Mr. Ernesto Gianni and his coworkers of the Secretariat of Accademia Nazionale dei Lincei took care of the local organization of the course, while Miss Anna Ingrao acted very ably as Secretary throughout the meeting. Simultaneous translation of a high quality was provided by Mrs. Liana de Pinedo and Mrs. Marina Spani. We are indebted to Miss Nancie Heldt for her careful typing of the manuscripts.

N. Bruce Hannay

Umberto Colombo

March, 1973

CONTENTS

Chapter		Page
	The Contributors	v
	Preface	vii
1	Electronic Structure of Semiconductors P. A. Wolff	1
2	Semiconductor Electroluminescence and Injection Lasers H. J. Queisser	41
3	Compound Semiconductors C. Hilsum	69
4	Hot Electrons in III-V Compound Semiconductors H. Pötzl	89
5	Relaxation Times in Semiconductors K. Seeger	107
6	Diffusion in Compound Semiconductors: A Survey of Recent Results A. Ascoli	127
7	Amorphous Semiconductors C. Hilsum	149
8	Magnetic Semiconductors C. Haas	169
9	Optical Properties of Solids G. Chiarotti	199
10	Nonlinear Optics P. A. Wolff	239

11	Optical Materials N. B. Hannay	261
12	Organic Dyes in Laser Technology D. J. Bradley	285
13	Optical Fiber Waveguides N. B. Hannay	307
14	Large Area Display Materials and Devices G. Marie	317
15	Magnetism C. Haas	371
16	Magnetic Alloys R. L. Colombo	407
17	Microwave Ferrites and Applications D. Treves	451
18	Crystal Growth N. B. Hannay	479
19	Solid State Chemistry N. B. Hannay	505
20	Film Techniques D. G. Thomas	535
21	Semiconductor Integrated Circuit Technology D. G. Thomas	563
22	The Use of Electronic Materials in Computer Memories A. H. Eschenfelder	603
	Index	639

CHAPTER 1

ELECTRONIC STRUCTURE OF SEMICONDUCTORS

P. A. Wolff

Massachusetts Institute of Technology

Cambridge, Massachusetts

INTRODUCTION

The electronic properties of a semiconductor are crucial in device applications. These properties are determined by the crystal structure, the type of bonding in the solid and, ultimately, by the electron wave functions. The aim of these notes is to trace the connection between the underlying quantum mechanics and the macroscopic properties of typical semiconductors. Throughout, we will illustrate the basic concepts with examples drawn from the most important class of semiconductors - those of the diamond type such as silicon, germanium and gallium arsenide. It is hoped that this discussion will give some feel for the band structures of such materials, and demonstrate how this structure determines important characteristics of the crystal. Readers are cautioned, however, that this treatment is necessarily a qualitative one, condensing much semiconductor physics into a few lectures. Those interested in more detailed expositions are referred to the bibliography.

CRYSTAL STRUCTURES

The most important semiconductors, from the point of view of both physics and technology, crystallize in tetrahedral phases. These are structures in which each atom is surrounded by four nearest neighbors which lie at the corners of a tetrahedron. The bond between nearest neighbor atoms is formed by paired electrons of opposite spin. Since each bond requires two electrons, and there are four bonds per atom, the average valence of the atoms in

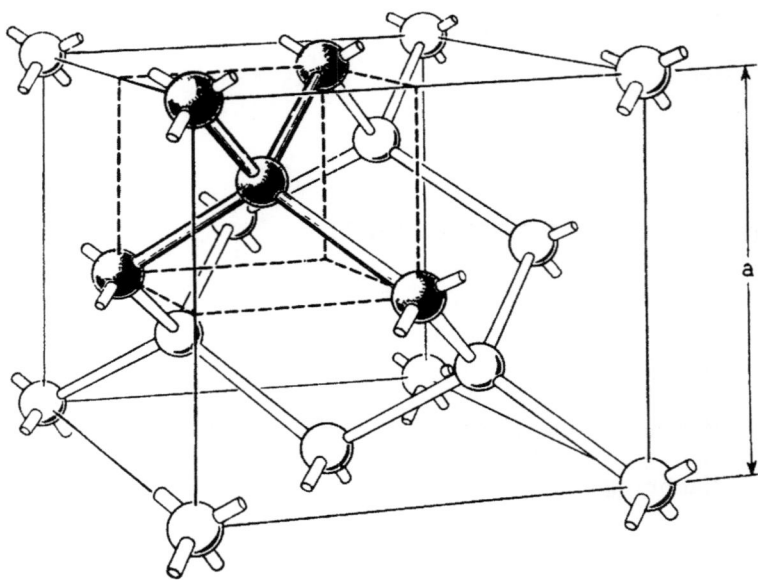

FIGURE 1. Diamond crystal lattice.

a tetrahedrally bonded crystal must be four [1]. This condition is satisfied by the diamond-type crystals (diamond, Si, Ge, grey tin); the III-V compounds, made up from elements of the IIIb (B, Al, Ga, In, Tl) and Vb (N, P, As, Sb, Bi) columns of the periodic table; and the II-VI compounds which involve elements from the IIb column (Zn, Cd) and the VI column (O, S, Se, Te) of the periodic table. There are a few I-VII compounds (CuF, CuCl, CuBr, CuI, AgI) which form tetrahedrally bonded structures, as well as a large number of ternary semiconducting compounds which crystallize in such phases. The latter, in many cases, can be thought of as generalizations of the III-V's. Their study is an increasingly important area in semiconductor physics [1].

The column IV elements (diamond, Si, Ge, α-Sn) crystallize in the _diamond_ structure illustrated in Fig. 1. This lattice is the prototype for almost all cubic, tetrahedrally bonded semiconductors. Its geometry is determined by the covalent bonds between neighboring atoms. Diamond, Si and Ge are ideal examples of covalently bonded crystals. Notice that the diamond lattice is an "open" one compared to ionic crystals such as NaCl, or the closely packed metal structures. This "openness" is made possible by the rigid, directional nature of the covalent bonds.

ELECTRONIC STRUCTURE OF SEMICONDUCTORS

Most of the III-V compounds crystal in the zinc blende structure which is closely related to the diamond lattice. The GaAs lattice, for example, is obtained from the Ge lattice by replacing alternate Ge atoms by Ga and As atoms. Note that Ga and As flank Ge in the periodic table, so one expects Ge and GaAs to be somewhat similar in their properties. The bonding of III-V compounds is partially covalent, partially ionic. Typically, the atoms in these crystals have charges ($\sim \pm 1/2 e$). Though the zinc-blende lattice is closely related to that of diamond, there is a major difference in symmetry between them - the diamond lattice has a center of inversion midway between the two atoms in the primitive cell, whereas the zinc blende lattice does not. This fact has important implications for the nonlinear optical properties of these materials. Only acentric crystals (those lacking a center of inversion) can be strongly nonlinear. The III-V compounds are acentric and have large nonlinear optical coefficients. Column IV semiconductors, on the other hand, have exceedingly small nonlinear coefficients.

Finally, the II-VI compounds (with one or two exceptions) crystallize in the wurtzite lattice. This is a hexagonal structure in which nearest neighbors are tetrahedrally bonded, but in which the further neighbors have a different arrangement from that of the diamond or zinc blende lattices. Since the most important semiconductors (Si, Ge, GaAs) crystallize in the cubic lattices, we will generally concentrate on them. However, many of the concepts we will develop in discussing cubic lattices can be applied to the wurtzite structure as well.

The diamond and zinc blende lattices are made up of two, interpenetrating fcc lattices. The primitive cell, which in either case contains two atoms, is defined by the three vectors;

$$\vec{a} = (1/2, 1/2, 0)\ a_0$$

$$\vec{b} = (1/2, 0, 1/2)\ a_0$$

$$\vec{c} = (0, 1/2, 1/2)\ a_0 \qquad (1)$$

where a_0 is the edge-length of the cubic (non-primitive) unit cell. The reciprocal lattice to a fcc lattice is a bcc lattice, with basis vectors

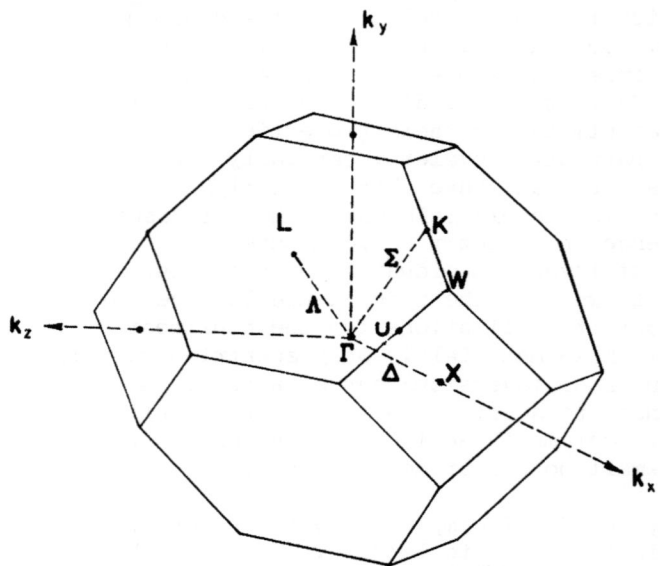

FIGURE 2. Brillouin zone of diamond crystal structure.

$$\vec{A} = \frac{2\pi(\vec{b} \times \vec{c})}{(\vec{a} \cdot \vec{b} \times \vec{c})} = \frac{2\pi}{a_0} (1, 1, -1)$$

$$\vec{B} = \frac{2\pi(\vec{c} \times \vec{a})}{(\vec{a} \cdot \vec{b} \times \vec{c})} = \frac{2\pi}{a_0} (1, -1, 1)$$

$$\vec{C} = \frac{2\pi(\vec{a} \times \vec{b})}{(\vec{a} \cdot \vec{b} \times \vec{c})} = \frac{2\pi}{a_0} (-1, 1, 1). \qquad (2)$$

The Brillouin zone for the reciprocal lattice is obtained by bisecting the lines which join the origin to reciprocal lattice points $\vec{G}_{lmn}$, defined by

$$\vec{G}_{lmn} = l\vec{A} + m\vec{B} + n\vec{C}. \qquad (3)$$

It is illustrated in Fig. 2. Symmetry points are given their standard group-theoretical designations. The most important of these are Γ (zone center), X (100 zone face) and L (111 zone face).

ELECTRONIC STRUCTURE OF SEMICONDUCTORS

ENERGY BANDS

Electronic properties of semiconductors are determined by electron wave functions and energy levels. The relationship between the bulk properties of such crystals, and the underlying quantum mechanics, is extraordinarily close. The device applications of these materials are, in a quite literal sense, "quantum electronics."

The wave functions are solutions of the Schrödinger equation

$$\left[-\frac{\hbar^2 \nabla^2}{2m} + V(\vec{r})\right] \psi = E\psi . \qquad (4)$$

Here $V(\vec{r})$ is the crystal potential which is a periodic function of position:

$$V(\vec{r}) = V(\vec{r} + l\vec{a} + m\vec{b} + n\vec{c}) . \qquad (5)$$

Equation (5) implies that the Hamiltonian of the crystal problem has the translational symmetry of the crystal. From this fact it follows that Eq. (4) has solutions of the Bloch form:

$$\psi_{n,\vec{k}}(\vec{r}) = e^{i\vec{k}\cdot\vec{r}} u_{n,\vec{k}}(\vec{r}), \qquad (6)$$

where $u_{n,\vec{k}}(\vec{r})$ is a periodic function of $\vec{r}$, with lattice periodicity. The quantities $\vec{k}$ and n are the <u>crystal momentum</u> and <u>band index</u> of the state in question. The energy, $E_n(\vec{k})$, of this state is a periodic function of $\vec{k}$ in reciprocal space. The energy levels form a series of <u>energy bands</u>. Within a given band, the energy is a continuous function of $\vec{k}$. In general, however, there are discrete energy gaps, $\hbar\omega_{nn'}(\vec{k}) = E_n(\vec{k}) - E_{n'}(\vec{k})$, between different bands. Each band contains $2N$ states, where N is the number of primitive cells in the lattice. The factor of 2 is a result of spin degeneracy.

During the past decade, an immense calculational effort (hundreds of man years) has been devoted to the band structures of diamond-type semiconductors. Some of these calculations proceed from first principles, seeking to determine the electron wave functions and crystal potential in a rigorous, self-consistent way. Others are semi-empirical in nature, making use of experimental data to fix crucial parameters in the theory. For our present purpose - which is that of surveying the electronic properties of a broad class of semiconductors - the empirical methods are most appropriate. Their relative simplicity allows one to study the properties of many, related crystals, and thus

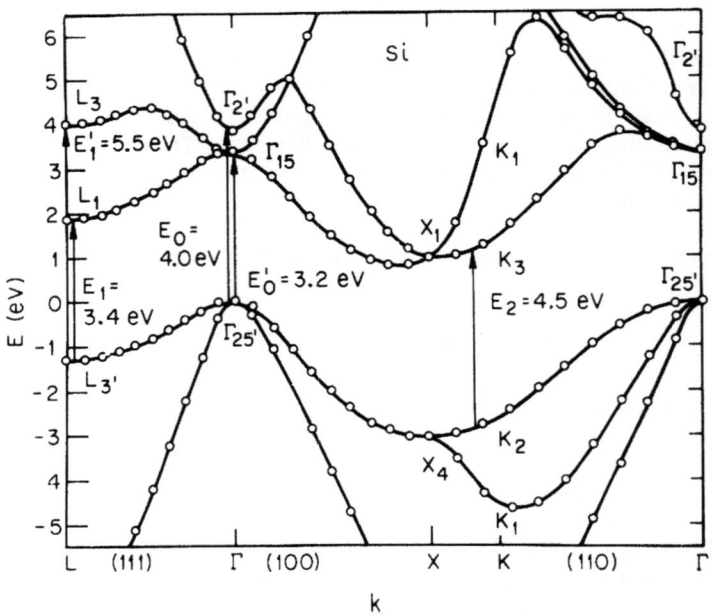

FIGURE 3. Band structure of silicon (after Brust, Cohen and Phillips [3]).

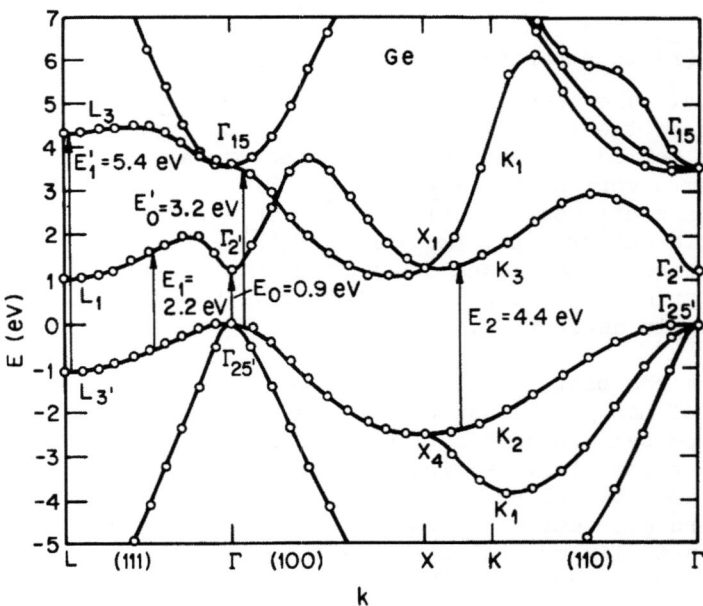

FIGURE 4. Band structure of germanium (after Brust, Phillips and Bassani [3]).

ELECTRONIC STRUCTURE OF SEMICONDUCTORS

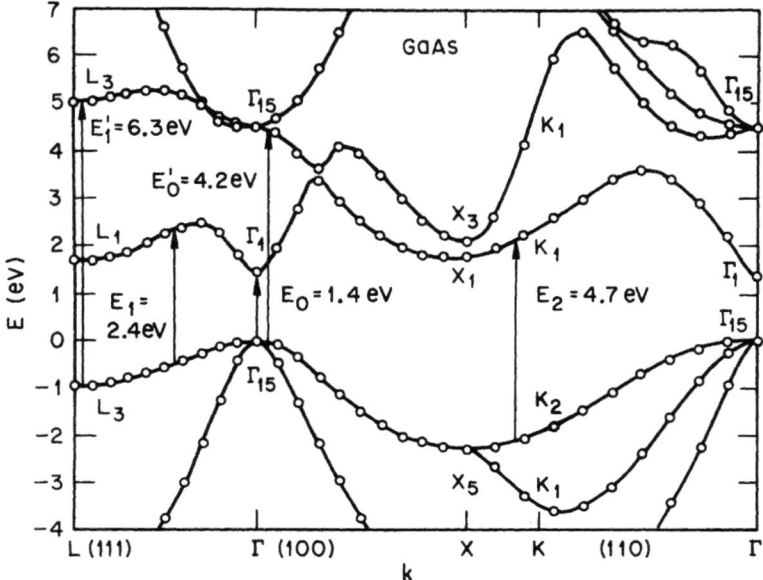

FIGURE 5. Band structure of gallium arsenide (after Cohen and Bergstrasser [3]).

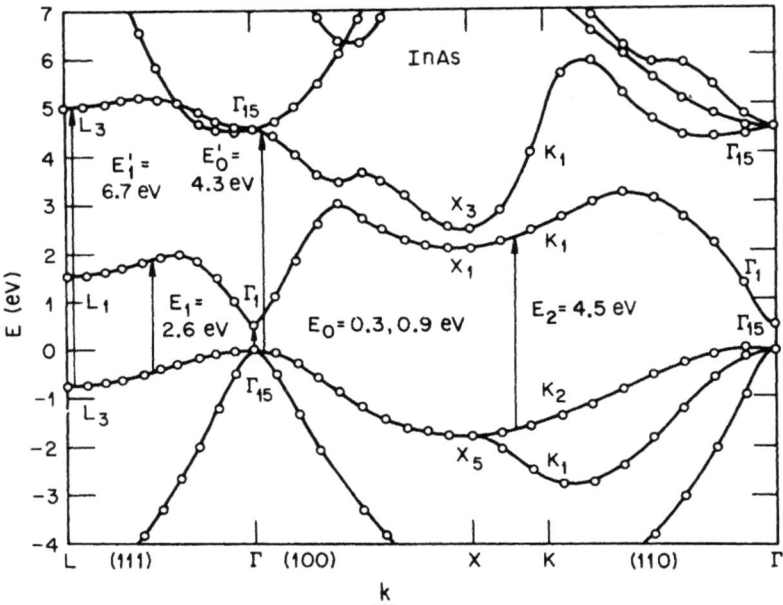

FIGURE 6. Band structure of indium arsenide (after Cohen and Bergstrasser [3]).

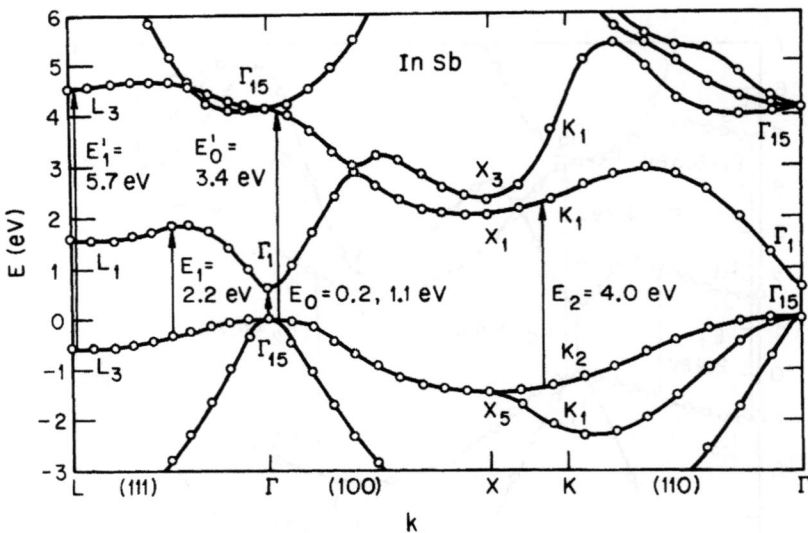

FIGURE 7. Band structure of indium antimonide (after Cohen and Bergstrasser [3]).

learn about chemical trends in such systems. The empirical pseudo-potential method has been particularly widely applied in this way. It has been used to investigate the electronic properties (including band structure, optical properties, charge densities, etc.) of many diamond-type semiconductors [2]. The band structures of several important semiconductors, calculated with this method [3], are illustrated in Figs. 3-7. In these diagrams, the $E_n(\vec{k})$ curves are plotted as a function of crystal momentum $(\vec{k})$ along important symmetry directions in the Brillouin zone. At first glance, the energy bands of these crystals look quite complicated. Fortunately, much of this complexity is not real - but a result of the fact that the bands are folded into the first Brillouin zone. When plotted in an extended zone scheme, they look much more like those of a free electron. We will demonstrate this fact for the semiconductors presently, but first let us consider a simple one-dimensional model which illustrates the essential point.

Imagine an electron moving in one dimension in a weak, periodic potential. The $E(k)$ curve, in an extended zone scheme, is essentially that of a free electron,

$$E(k) = \frac{\hbar^2 k^2}{2m} , \qquad (7)$$

with small gaps at the zone boundaries. At such points the Bragg condition is satisfied,

ELECTRONIC STRUCTURE OF SEMICONDUCTORS

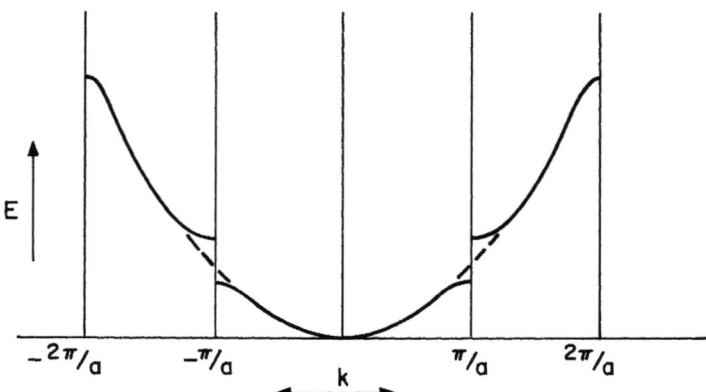

FIGURE 8. Energy bands, for one dimension, in extended zone scheme.

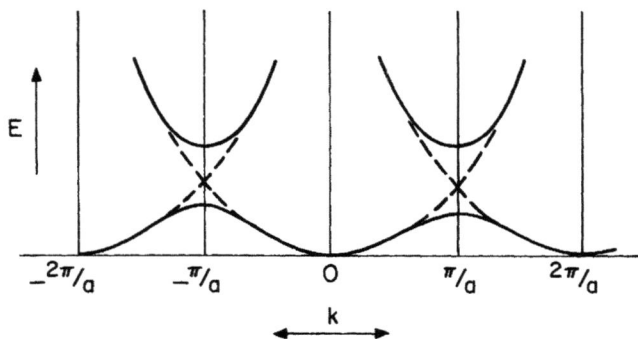

FIGURE 9. Energy bands, for one dimension, in reduced zone scheme.

$$\frac{\hbar^2 k^2}{2m} = \frac{\hbar^2 (k-G)^2}{2m}, \tag{8}$$

implying that the crystal potential can scatter electron waves elastically from state k to (k-G). As a consequence, these degenerate states are completely mixed by the potential - the wave function of an electron at the zone boundary contains equal parts of the plane waves k and (k-G). The mixture can be either a symmetric or an antisymmetric one, with an energy gap between the two states proportional to the Fourier component, V_G, of the crystal potential. This band structure is illustrated in Fig. 8.

Now let us plot the same band structure in the reduced zone scheme. The simplest way to do this is to plot free electron parabolas centered at each reciprocal lattice vector:

$$E(k) = \frac{\hbar^2(k - G)^2}{2m} \qquad (9)$$

where $G = \frac{2\pi n}{a}$ and $n = 0, \pm 1, \pm 2, \ldots$. In the completely free electron picture, the energy bands are as shown above (dotted lines). Most of this structure is unchanged by the crystal potential, but small gaps are created at the zone boundaries. Thus, we have the $E(k)$ curves shown in Fig. 9. This figure is only a slight modification (to take account of the energy gaps produced by Bragg scattering) of the free electron $E(k)$ - yet to the uninitiated eye it may look quite different, because of the folding back procedure which is used to map the free electron $E(k)$ into the first Brillouin zone. The same comments apply to the semiconductor band structures. We will see presently that they too are closely related to the free electron $E(k)$. Parenthetically, it should be noted that the bands we have sketched above are actually a quite good representation of the $E(k)$ curves for Na or K. In these monovalent metals only a couple of bands are required to describe the essential electronic structure of the crystal.

Now let us return to the semiconductors, and apply the same, nearly free electron arguments to them. We see immediately, however, that the problem is more complicated. The tetrahedrally bonded semiconductors have eight valence electrons per primitive cell - enough to fill four bands in the Brillouin zone. Thus, to understand even the valence bands of these materials, we must consider at least four bands. To proceed, let us again draw free electron $E(\vec{k})$ curves centered at each reciprocal lattice point:

$$E(\vec{k}) = \frac{\hbar^2(\vec{k} - \vec{G})^2}{2m} . \qquad (10)$$

The resultant plots of $E(\vec{k})$ along symmetry directions in the first Brillouin zone are shown in Fig. 10. With some imagination, it can be seen that these bands are similar to those calculated for Si or Ge. The circled region, where many bands cross, is the approximate position of the top of the valence band. This degeneracy is partly removed by the crystal potential, which produces an energy gap at $\vec{k} = 0$. As a consequence, the structure of the conduction and valence band edges are quite sensitive to the form of this potential. On the other hand, many features of the band structure are well explained by the free electron model.

ELECTRONIC STRUCTURE OF SEMICONDUCTORS

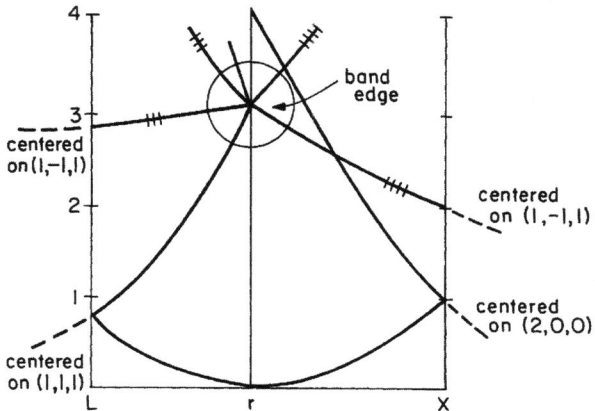

FIGURE 10. Empty lattice (free electron) bands for semiconductors.

For example, all of those illustrated in Figs. 3-7 contain flat conduction and valence bands in the (111) direction. These bands are immediately apparent in Fig. 10. Their flatness results from the fact that when $\vec{k}$ is parallel to (111), it is almost perpendicular to the reciprocal lattice vector $\vec{G}_{1,-1,1}$. Hence the free electron energy

$$E = \frac{\hbar^2(\vec{k} - \vec{G}_{1,-1,1})^2}{2m} \tag{11}$$

is nearly independent of $\vec{k}$, as $\vec{k}$ moves from the Γ to the L point. The steeper valence bands shown in Fig. 10 (centered on $\vec{G}_{1,1,1}$ and $\vec{G}_{2,0,0}$) also have the correct shape. The overall width of the valence band predicted by the free electron model is in good agreement with that determined by soft X-ray experiments. Thus, this model accounts for the gross features of semiconductor band structure - particularly deep in the valence band. However, detailed calculations are required to determine the form of the conduction and valence band edges.

It is also interesting to consider the relation of the energy bands to the states of the free atoms. A well-known approach to the theory of wave function in crystals (the tight binding method) is to expand the Bloch functions in terms of atomic orbitals centered on each atom. In this approximation the wave functions take the form

$$\psi_{\vec{k}} = \sum_i \sum_\alpha \left[c_\alpha e^{i\vec{k}\cdot\vec{R}_i} \varphi_\alpha(\vec{r}-\vec{R}_i) \right], \tag{12}$$

where $\varphi_\alpha(\vec{r}-\vec{R_i})$ is an atomic orbital of the atom at position $\vec{R_i}$. The tight binding method is not useful for quantitative calculation, but gives some feel for the origin of various states in the band structure. For example, when detailed tight binding calculations are carried out for the diamond-type semiconductors, they show that the top of the valence band (which is triply degenerate at $\vec{k} = 0$) is made up of bonding combinations of p-like orbitals on the two atoms in the unit cell. That is, at $\vec{k} = 0$

$$\psi \sim \varphi_A + \varphi_B, \qquad (13)$$

where φ_A and φ_B are p-states centered on atoms A and B in the primitive cell. The three fold degeneracy at $\vec{k} = 0$ in the valence band is thus a result of the three-fold degeneracy (p_x, p_y, p_z) of p-orbitals. Similar comments apply to the III-V semiconductors, but in that case the valence band edge (at $k = 0$) is an asymmetric mixture,

$$\psi \sim \varphi_A + \gamma \varphi_B, \qquad (14)$$

of the (now different) p orbitals on sites A and B. On the other hand, in the conduction band at $\vec{k} = 0$ the states are anti-bonding mixtures of the two orbitals

$$\psi \sim -\varphi_A + \varphi_B \qquad (15)$$

or (in III-V's)

$$\psi \sim -\gamma \varphi_A + \varphi_B. \qquad (16)$$

Similar conclusions can be reached by using group theory to investigate the splitting of the eight-fold plane wave degeneracy at $\vec{k} = 0$ [4] (the circled region in Fig. 10). In the presence of a cubic crystal potential these states separate into four distinct representations of the cubic group - Γ'_{25} (bonding p), Γ_{15} (antibonding p), Γ'_2 (antibonding s), and Γ_1 (bonding s). These representations can be directly identified with the band edge points illustrated in Figs. 4 and 5.

Finally, it should be emphasized that spin-orbit effects have been neglected in the calculations leading to Figs. 3-7. This interaction, which couples the spin of an electron to its orbital motion, is weak except in semiconductors containing atoms of high Z. Nevertheless, it is important because it can remove degeneracies which would otherwise be present in the band structure. We will discuss one or two effects of spin-orbit coupling later. For the time being, we ignore it.

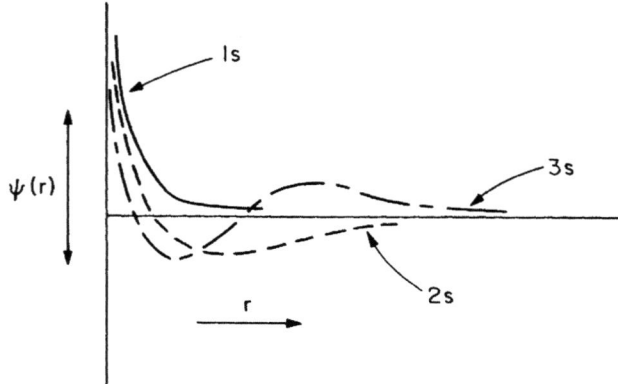

FIGURE 11. Spatial variation of s-like wave functions.

An important question remains - why is the periodic potential weak? In fact, it is not. The true potential can easily be shown to be strong - strong enough, for example, to produce two nodes in the 3s valence electron wave function of silicon as it emerges from the core of the atom. These nodes must be present if the 3s wave function is to be orthogonal to the 1s and 2s core states, as illustrated in Fig. 11. The occurrence of such nodes implies that the valence electron wave function has Fourier components of large k, and that its expansion in terms of plane waves converges slowly [5] (in fact, some 10^6 terms are required). This observation leads us to an apparent contradiction. On the one hand, we have seen that the nearly free electron approximation gives a reasonable description of the energy bands of semiconductors. In its usual form, however, this approximation is based on an expansion of the wave function in terms of a few plane waves. We now see that such an expansion completely fails, implying that the simplest form of the free electron theory is invalid.

This seeming paradox is resolved by using orthogonalized plane waves (OPW's), instead of plane waves, as the basis set [5]. These functions are defined by the formula

$$|OPW, \vec{k}\rangle = e^{i\vec{k}\cdot\vec{r}} - \sum_c (\beta_c \psi_c), \qquad (17)$$

where the coefficients β_c are chosen in such a way as to force $|OPW, \vec{k}\rangle$ to be orthogonal to all core states, ψ_c. The proper choice of β_c is

$$\beta_c = \int \psi_c^* e^{i\vec{k}\cdot\vec{r}} d^3r. \tag{18}$$

In the OPW representation, the matrix elements of the periodic potential are small, at least for many solids. The main function of the strong, close-in part of the potential is to create nodes in the wave function, thereby forcing it to be orthogonal to core states. If this orthogonalization is built into the problem from the start (via the OPW method) the effective potential becomes weak. As a consequence, an expansion of the valence electron wave functions in terms of OPW's,

$$\psi_{\vec{k}} = \sum_{\vec{G}} [a_{\vec{k}+\vec{G}} |OPW, \vec{k}+\vec{G}\rangle], \tag{19}$$

converges quite rapidly. The secular equation which determines the energy levels and expansion coefficients (a's) is formally the same as that of the nearly free electron theory, though the matrix elements which appear in it are those of an effective potential. It is not surprising, therefore, that the energy band structure is similar to that predicted by the nearly free electron theory. However, it should be recognized that the label $\vec{k}$ which indexes the energy bands now refers to OPW's, rather than plane waves.

These ideas are at the heart of the pseudo-potential method, which has been so successful in describing the band structures of a variety of crystals. The pseudo-potential method formally removes the orthogonalization terms (which are responsible for the wiggles near the core) from the wave function. What remains is a relatively smooth, pseudo-wave function,

$$\varphi_{\vec{k}} = \sum_{\vec{G}} [a_{\vec{k}+\vec{G}} e^{i(\vec{k}+\vec{G})\cdot\vec{r}}]. \tag{20}$$

The pseudo-wave function obeys a Schrodinger equation containing an effective potential (the pseudo-potential) which in many cases is weak - and therefore can be treated by perturbation theory. Thus, by a methematical transformation, one recovers a theory which is similar to the nearly free electron theory - and justifies its use in the calculation of energy bands.

ELECTRONIC STRUCTURE OF SEMICONDUCTORS

DISCUSSION OF ENERGY BANDS

Many features of the energy bands illustrated in Figs. 3-7 deserve comment. First of all, it should be noticed that in all cases the valence band maximum is at $\vec{k} = 0$ (Γ point) in the Brillouin zone. The band edge states are triply degenerate with p-character. This degeneracy becomes six-fold when spin is included. However, the degeneracy is partially lifted by the spin-orbit interaction, which splits the band edge states into a four-fold degenerate $p^{3/2}$ manifold, and a two-fold degenerate $p^{1/2}$ manifold. On the other hand, the position of the conduction band edge (lowest state in the conduction band) varies from material to material. In Si and diamond it occurs along (100) directions near the zone face ($\vec{k} \sim 0.8X$). There are six equivalent positions of this type so these crystals have six conduction band minima. Ge has its conduction band edges at the L-points of the Brillouin zone face - there are four such minima. This situation is a bit of an accident, however, because minima at other points in the conduction band (Γ, near X) are only slightly higher in energy than those at L. In GaAs (chemically similar to Ge) the balance shifts slightly, and the conduction band edge is at $\vec{k} = 0$. Such a material, in which the valence band edge and conduction band edge are at the same points of the Brillouin zone (in this case $\vec{k} = 0$) is termed a <u>direct gap</u> semiconductor. Others, for which this is not the case, such as Si or Ge, are called <u>indirect</u>. In the case of GaAs, the conduction band minima near X are next lowest in energy - 0.36 eV above the Γ point. The position of these subsidiary minima is important, since they are responsible for the negative resistance which drives Gunn oscillators. Finally, both InAs and InSb have small, direct gaps at $\vec{k} = 0$.

The reader will have noticed that the band gaps of the materials we have discussed get smaller as one goes lower in the periodic table. This is a fairly general rule and applies to other semiconductor systems as well. For example, among the II-VI crystals CdTe has a much smaller gap than ZnS. In the III-V series the smaller gap crystals (GaAs, InP, InAs, InSb) have direct gaps, whereas the large gap materials (GaP, AlAs, AlP, BP - as well as Ge, Si, SiC and diamond) are indirect. On the other hand, all of the II-VI crystals have direct gaps - even ZnO whose gap is well into the UV.

The curvatures (or <u>effective masses</u>) of valence and conduction band edges are another important feature of the $E_n(\vec{k})$ curves illustrated in Figs. 3-7. These parameters have been measured for most of the semiconductors discussed above. The situation is simplest in cases in which the crystal in question has a single, non-degenerate band edge - examples are the conduction band edge (at $\vec{k} = 0$) of InAs or InSb. Under such circumstances, the $E(\vec{k})$ curve near the band edge point has the form

$$E(\vec{k}) \cong \frac{\hbar^2 k^2}{2m_c^*},$$

characterized by the single parameter, m_c^* - the conduction band effective mass. Values of m_c^* for a few direct gap materials are listed below.

Crystal	Energy Gap	(m_c^*/m)
GaAs	1.52 eV	0.066
InP	1.42	0.077
InAs	0.42	0.023
InSb	0.24	0.014
ZnSe	2.82	0.17
CdTe	1.60	0.096

Note that m_c^* is small in small gap materials (e.g., InSb), and gets bigger as E_G increases. Qualitatively speaking, the masses obey a relation of the form $m_c^* \sim E_G$.

In diamond, Si and Ge the conduction band edges are not at $\vec{k} = 0$, but along the (100) or (111) axes in the Brillouin zone. For such cases, cubic symmetry no longer requires the $E(\vec{k})$ curves to be spherically symmetric. The $E(\vec{k})$ curves have the following form near band edge point (k_0):

$$E(\vec{k}) = \frac{\hbar^2 k_t^2}{2m_t^*} + \frac{\hbar^2 (k_\ell - k_0)^2}{2m_\ell^*}, \qquad (21)$$

where k_t and k_ℓ are the transverse and longitudinal components of the wave vector with respect to the symmetry axis of the ellipsoid - (100) in the case of Si; (111) for Ge. Typical values of the parameters appearing in this expression are listed below.

Crystal	Position of Minima	m_ℓ^*/m	m_t^*/m
Diamond	0.8X		
Si	0.85X	0.92	0.19
Ge	L	1.59	0.082

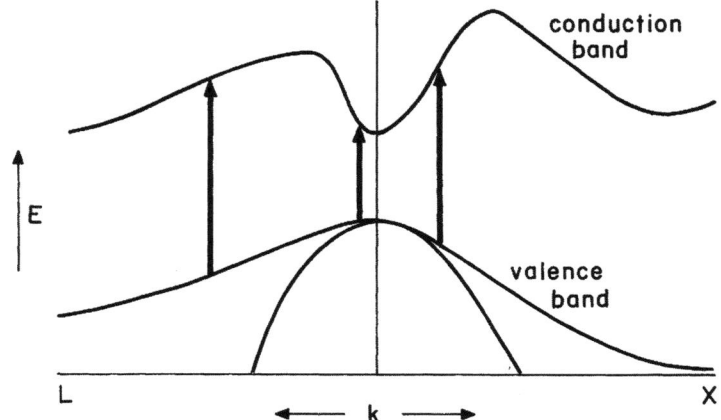

FIGURE 12. Allowed optical transitions.

WHAT USE IS THE BAND STRUCTURE?

In the preceding pages, we have discussed some of the important features of the band structures of common semiconductors. It should be clear that there is an enormous amount of information buried in the $E_n(\vec{k})$ curves. But one may well ask, how much of this is relevant, particularly to practical applications? The answer to this question is that the band structure plays a surprisingly important role in determining the device potential of a given semiconductor material. It is useful, therefore, to summarize at this point some of the ways in which the band structure can affect the properties of semiconducting crystals.

Consider, as one instance, the optical properties of such a material. Most pure semiconductors are relatively transparent to light whose energy is below the band gap ($\hbar\omega < E_G$). As the photon energy approaches the gap, strong absorption sets in, caused by processes in which an electron is excited from the filled valence band to the empty conduction band. So much is common to all semiconductors - but the nature of the absorption edge is entirely different, depending upon whether the material in question is a direct gap or an indirect gap semiconductor. To understand this distinction, we must recall that the wave vector of a light wave is small compared to that of a typical electron in the Brillouin zone (roughly 10^5 cm^{-1} as compared to 10^8 cm^{-1}). Since wave vector (or crystal momentum) is conserved in transitions in crystals, the allowed optical processes are those in which an electron is vertically excited from one band to another, as illustrated in Fig. 12. This picture explains the difference between the absorption edges of direct and indirect gap crystals.

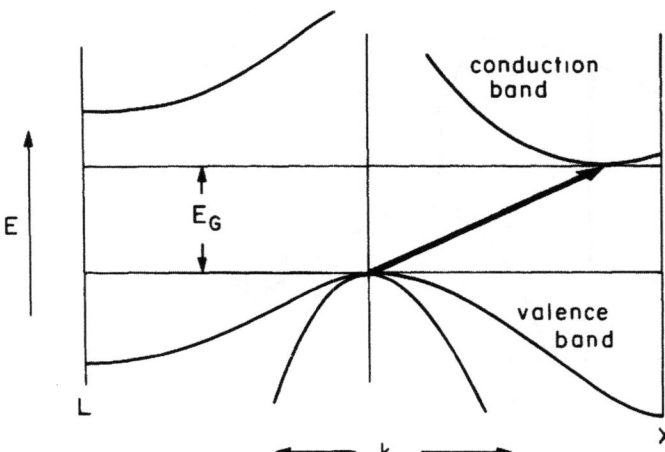

FIGURE 13. Example of an indirect gap semiconductor.

In a direct gap material the minimum of the conduction band is vertically above the maximum of the valence band, and optical transitions between the two states are allowed. As a consequence, the absorption edge is sharp - typically the absorption coefficient rises to values $\alpha \sim 10^5$ cm^{-1} just above the edge. On the other hand, in an indirect gap semiconductor the first transitions which are energetically possible (between valence and conduction bands) are those involving a large change in crystal momentum, as illustrated in Fig. 13. Such processes are optically forbidden, because a light quantum does not have enough crystal momentum to displace the electron, in $\vec{k}$-space, from the valence band edge to the conduction band edge. They can only proceed with the intervention of an added interaction, such as phonon emission or impurity scattering, which provides the necessary momentum. Such processes are considerably weaker than direct, allowed optical transitions. As a consequence, the absorption edges of indirect semiconductors are much less abrupt than those of direct gap materials.

These ideas are central when one considers the use of semiconductors as generators of light - either as lasers or incoherent sources. In such a device, one injects electrons and holes into the crystal, which then recombine with one another. A certain fraction of the recombinations proceed via photon emission - a good device maximizes this optical efficiency. From what has been said above, it is clear that optical recombination is liable to be most efficient in direct gap materials - since the transition

ELECTRONIC STRUCTURE OF SEMICONDUCTORS

between conduction and valence band edges (the states occupied by electrons and holes) is then an allowed one. This statement is confirmed by experiment. To date, all semiconductor lasers (laser action has been produced in about fifty different materials) have employed direct gap crystals. There is even a belief, supported by some calculation, that it is impossible to achieve lasing in indirect materials. This limitation is probably not a fundamental one - forbidden transitions can lase if the background absorption is low - but it is almost certainly true that practical semiconductor lasers will make use of direct gap crystals. Similar considerations apply when one considers the construction of semiconductor lamps; i.e., incoherent light sources. The most efficient emitters, with efficiencies approaching 50%, are direct gap materials, of which GaAs has been most extensively studied. It is only with extreme care (notably in GaP) that reasonable efficiencies have been achieved in indirect materials. Even then, the best efficiency of a practical device is about 5%. The magnitude of the problem can be understood from the fact that the ratio of direct to indirect optical transition rates is about 10^3. Thus, to achieve optical efficiencies in indirect gap semiconductors comparable to those of direct ones, one must do about one thousand times better in suppressing competing, non-radiative recombination. This requirement has led to fascinating studies of the physics and chemistry of electron-hole interactions in crystals such as GaP.

Another set of phenomena, which the band structure controls, is the transport properties of a semiconductor. Here we have in mind the response of carriers to low frequency perturbations - perturbations for which $\hbar\omega \ll E_G$. Examples of such perturbations are low frequency electric or magnetic fields, the fields of impurities, etc. Since the frequencies are low, the states of importance in transport are those near the band edge points - states occupied by carriers, or adjacent to occupied states. For these states, it is the expansion of $E_n(\vec{k})$ about the band edge points that is important - hence our interest in effective masses.

The relation between the band edge $E_n(\vec{k})$ and the dynamics of free carrier motion has been extensively studied. The resulting theory - the famous effective mass theory [6] - gives a precise prescription for using the band-edge $E_n(\vec{k})$ to calculate the electron (or hole) motion. For cases in which the band edge is non-degenerate as in n-type Si or GaAs, this prescription can be stated quite simply - <u>the band edge $E(\vec{k})$ acts as an effective Hamiltonian for the carrier motion</u>. This statement is best illustrated with the simplest example - that of a material such as n-GaAs or n-InSb which has a single, non-degenerate, spherically symmetric conduction band edge for which

$$E \simeq \frac{\hbar^2 k^2}{2m_c^*} . \qquad (22)$$

This is no more than the Hamiltonian of a free particle of mass m_c^*. Electrons in GaAs or InSb behave as particles of this sort. For instance, in a magnetic field, their equation of motion is

$$m_c^* \left(\frac{d\vec{v}}{dt} + \frac{\vec{v}}{\tau} \right) - \frac{e(\vec{v} \times \vec{B})}{c} = e\vec{F}e^{i\omega t}, \qquad (23)$$

where the term $\vec{v}/\tau$ represents the viscous drag due to electron-lattice collisions, $\vec{B}$ is the magnetic field, and $\vec{F}e^{i\omega t}$ a time-varying electric field. If $\omega_c^* \tau \gg 1$, the velocity, $\vec{v}$, has a sharp resonance when $\omega = \omega_c^*$, where

$$\omega_c^* = \left(\frac{eB}{m_c^* c} \right) \qquad (24)$$

is the cyclotron resonance frequency. This resonance is the basis for the cyclotron resonance technique of determining carrier masses. Most of the masses listed in the preceding tables were measured with this method.

The effective mass approximation plays such a central role in semiconductor physics, that it is worth taking some time to consider the rationale behind it [7]. A complete derivation is complicated, but it is fairly easy to discuss the case of a non-degenerate band edge point. The basic problem to be solved is that of the motion of such a carrier as it is perturbed by a slowly varying field. Under these circumstances, we expect the complete wave function to be a mixture of near band edge states, <u>from a single band</u>, as follows:

$$\psi \simeq \sum_{\vec{k}'} [c_{\vec{k}'} \psi_{n,\vec{k}'}(\vec{r})]. \qquad (25)$$

We now choose the coefficients $c_{\vec{k}'}$ in such a way as to make ψ an approximate solution of the Schrodinger equation:

$$H\psi = (H_o + \mathcal{H}) = E\psi, \qquad (26)$$

ELECTRONIC STRUCTURE OF SEMICONDUCTORS

where $H_o = -\frac{\hbar^2 \nabla^2}{2m} + V$ [Eq. (4)] and $\mathcal{H}$ is the slowly varying perturbation. Now multiply this equation by $\psi^*_{n,\vec{k}}$ and integrate. The result is

$$[E_n(\vec{k}) - E]c_{\vec{k}} + \sum_{\vec{k}'}[\mathcal{H}_{\vec{k}\vec{k}'} c_{\vec{k}'}] = 0 \quad (27)$$

where

$$\mathcal{H}_{\vec{k}\vec{k}'} = \int \psi^*_{n,\vec{k}} \mathcal{H} \psi_{n,\vec{k}'} d^3r. \quad (28)$$

If $\mathcal{H}$ is a slowly varying function in space, one may approximate

$$\mathcal{H}_{\vec{k}\vec{k}'} \simeq \int e^{-i\vec{k}\cdot\vec{r}} \mathcal{H} e^{i\vec{k}\cdot\vec{r}} d^3r \equiv \mathcal{H}_{\vec{k}-\vec{k}'}. \quad (29)$$

Furthermore, near the band edge point one can expand $E_n(\vec{k})$ in a power series. To further simplify matters, we consider a case in which this edge is at $\vec{k} = 0$, where

$$E_n(\vec{k}) \simeq E_n(o) + \frac{\hbar^2 k^2}{2m^*_c}. \quad (30)$$

The formula for $c_{\vec{k}}$ then becomes

$$\frac{\hbar^2 k^2}{2m^*_c} c_{\vec{k}} + \sum_{\vec{k}'} (\mathcal{H}_{\vec{k}\vec{k}'} c_{\vec{k}'}) = \Delta E c_{\vec{k}} \quad (31)$$

with $\Delta E = E - E_n(0)$. This equation takes on a more familiar form after a Fourier transformation. We define

$$\varphi(\vec{r}) = \int c_{\vec{k}} e^{i\vec{k}\cdot\vec{r}} d^3k \quad (32)$$

and find

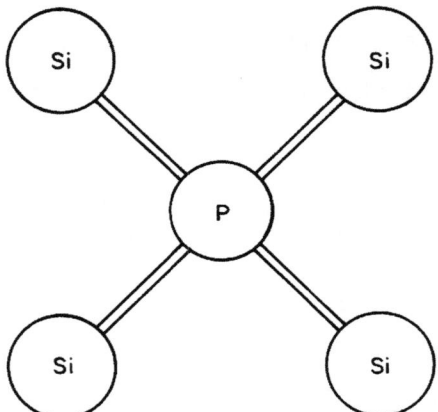

FIGURE 14. Schematic picture of a donor (P) in a silicon crystal.

$$\left[-\frac{\hbar^2 \nabla^2}{2m_c^*} + H(\vec{r}) \right] \varphi(\vec{r}) = \Delta E \varphi(\vec{r}), \qquad (33)$$

which is the Schrodinger equation of a particle with mass m_c^* moving in the perturbing potential $H(\vec{r})$. This derivation is a much simplified one, which does not apply to some important cases (notably those with degenerate bands), and gives no clue as to how one might systematically improve the approximation. Nevertheless, it contains the essential truth of the matter, and indicates why, in many situations, the motion of carriers in semiconductors can be discussed as if they were free particles, rather than electrons moving in a strong, complicated periodic potential. So long as the perturbations are slowly varying, which they almost invariably are in device applications, the effects of the periodic potential can be rigorously included by treating the electron as a free particle, with a Hamiltonian given by $E_n(\vec{k})$.

To conclude this section, we consider an important application of the effective mass approximation - the calculation of the energy levels of donor states [7]. Imagine a Si crystal, into which is substituted (doped) a few P atoms. We will assume that the concentration of P atoms is low; perhaps 1 in 10^5. This is a typical doping level in a useful semiconductor crystal. In the vicinity of an impurity the crystal appears, schematically, as shown in Fig. 14. Note that phosphorus has one higher valence than silicon. Thus, after four of the phosphorus atom's valence electrons are utilized in bonding to neighboring Si atoms, there is a fifth electron left over. This electron is much less firmly bound. Often, it will leave the impurity and wander through the crystal in the conduction band. However, it can also be weakly

ELECTRONIC STRUCTURE OF SEMICONDUCTORS

bound to the impurity site (because of the net positive charge of P relative to Si) in an orbit somewhat like that of a hydrogen atom, but much larger in size. The problem at hand is to calculate the wave functions and energies of such states.

For this purpose, we use the effective mass approximation. Let us first examine the simplest case - in which there is a single conduction band minimum at $\vec{k} = 0$. The effective mass Hamiltonian is then

$$\left[-\frac{\hbar^2 \nabla^2}{2m_c^*} + \mathcal{H}(\vec{r}) \right], \qquad (34)$$

where $\mathcal{H}(\vec{r})$ is the potential due to the impurity. Such a Hamiltonian would describe, for example, the donor levels of a GaAs crystal when doped with Se or Te. The perturbation $\mathcal{H}(\vec{r})$ is the Coulomb potential of the impurity reduced by the dielectric constant (ε) of the pure crystal; i.e.,

$$\mathcal{H}(\vec{r}) = -\frac{e^2}{\varepsilon r}. \qquad (35)$$

In GaAs, $\varepsilon = 11.6$. This value is typical of the diamond-type and III-V semiconductors. The effective mass Schrodinger equation becomes

$$\left[-\frac{\hbar^2 \nabla^2}{2m_c^*} - \frac{e^2}{\varepsilon r} \right] = E\varphi. \qquad (36)$$

This is the wave equation of a hydrogen atom, with the electron mass replaced by m_c^* and the Coulomb potential reduced by the dielectric constant. The wave functions and energy levels for this problem are well known. In particular,

$$E_n = \frac{m_c^* e^4}{2\varepsilon^2 \hbar^2 n^2} = \left(\frac{m_c^*}{m_o}\right) \frac{1}{\varepsilon^2 n^2} (13.5 \text{ eV}) \qquad (37)$$

where $n = 1,2,3...$ is the principal quantum number. In a material such as GaAs where $(m_c^*/m_o) \simeq 0.07$ and $\varepsilon = 11.6$, the lowest donor state has an energy of about 0.006 eV. This energy is well below kT at room temperature so we expect almost all donors in GaAs to be ionized under such conditions - and to considerably lower temperatures. This is an important conclusion which applies to many covalently bonded semiconductors (Si, Ge, GaAs, InAs, InSb, etc.). It is also interesting to estimate the radius of the donor orbit, which is given by the usual hydrogen atom formula expressed in terms of m_c^* and e^2/ε. The <u>effective Bohr radius</u> for the donor is

$$a_o^* = \frac{\hbar^2 \varepsilon}{m_c^* e^2} = \varepsilon \left(\frac{m_o}{m_c^*}\right) (5 \times 10^{-9} \text{ cm}) \qquad (38)$$

For GaAs, $a_o^* \simeq 100$ Å - a huge value. The orbit contains more than 10,000 primitive cells. It is the immense size of this orbit which is responsible for the success of the effective mass approximation.

A slightly more complicated case is that of a donor in Si or Ge. These materials have multiple conduction band minima displaced from $\vec{k} = 0$. Moreover, their band edges are not spherically symmetric. One now expects the donor wave function to be a mixture of Bloch states from the vicinity of each conduction band minimum [7]. Thus, the wave function has the form

$$\psi \simeq \sum_{\vec{k}',\alpha} [c_{\vec{k}',\alpha} \psi_{n,\vec{k}',\alpha}(\vec{r})], \qquad (39)$$

where the subscript α refers to a particular band edge point. In Si, for example, the donor wave function is built up of wave packets centered at each of the six conduction band minima. It can be shown that the separate $c_{\vec{k},\alpha}$'s satisfy an effective mass equation of the form

$$\left[-\frac{\hbar^2 \partial^2}{2m_t \partial x^2} - \frac{\hbar^2 \partial^2}{2m_t \partial y^2} - \frac{\hbar^2 \partial^2}{2m_\ell \partial_z^2} - \frac{e^2}{\varepsilon r}\right] \varphi_\alpha = E \varphi_\alpha \qquad (40)$$

where

$$\varphi_\alpha(\vec{r}) = \int c_{\vec{k},\alpha} e^{i(\vec{k}-\vec{k}_\alpha) \cdot \vec{r}} d^3k, \qquad (41)$$

$\vec{k}_\alpha$ is the position of the αth band edge point in the Brillouin zone, and we have chosen the z-axis along the heavy mass direction of the ellipsoidal energy surface. This Schrodinger equation is that of a hydrogen atom with different effective mass in different directions - an amusing problem in one-particle quantum mechanics.

ELECTRONIC STRUCTURE OF SEMICONDUCTORS

The equation does not separate, as that for the ordinary hydrogen atom does, and thus must be solved approximately. For the ground state, the variational principle works quite well. A good approximation to the normalized wave function is

$$\varphi_\alpha(\vec{r}) = \frac{1}{\sqrt{\pi a^2 b}} e^{-\sqrt{(x^2+y^2)/a^2 + z^2/b}}. \tag{42}$$

This functional form is exact when $m_\ell = m_t$, and is also accurate in the limit $m_\ell/m_t \to \infty$. For the actual values of m_t and m_ℓ observed in Si or Ge, the parameters a and b are determined by minimizing the energy,

$$E = \int \varphi_\alpha^* H \varphi_\alpha d^3 r. \tag{43}$$

This calculation gives the results:

$$\text{Si} \begin{cases} a = 25.0 \times 10^{-8} \text{ cm.} \\ b = 14.2 \times 10^{-8} \text{ cm.} \end{cases}$$

$$\text{Ge} \begin{cases} a = 64.5 \times 10^{-8} \text{ cm.} \\ b = 22.7 \times 10^{-8} \text{ cm.} \end{cases}$$

For Ge the calculated energy of the 1s state is 0.0092 eV. Experimental values for various donors lie in the range 0.0096 - 0.0127 eV. In Si, the agreement is poorer. The calculated value is 0.029 eV; experimental values range from 0.045 to 0.067 eV. These discrepancies are believed to be due to failure of the effective mass approximation near r = 0. One expects such central cell corrections to become larger as the orbit size decreases - exactly what is observed in going from GaAs to Ge to Si.

Finally, we look briefly at the form of the wave functions for the donor states. In the case of a semiconductor with a single band edge point in the conduction band

$$\psi = \sum_{\vec{k}'} [c_{\vec{k}'} \psi_{n,\vec{k}'}(\vec{r})]. \tag{44}$$

With the aid of the Bloch form for $\psi_{n,\vec{k}'}(\vec{r})$, this equation can be rewritten as follows:

$$\psi = \sum_{\vec{k}'} [c_{\vec{k}'} e^{i\vec{k}'\cdot\vec{r}} u_{n,\vec{k}'}(\vec{r})]. \quad (45)$$

If the donor state is spread out in space, the range of $\vec{k}$'s that contribute to this sum is quite small, and one may approximate $u_{n,\vec{k}'}(\vec{r})$ by $u_{n,o}(\vec{r})$ - the Bloch function at the band edge point. With this replacement, the expression for ψ takes the form

$$\psi \simeq \sum_{\vec{k}'} [c_{\vec{k}'} e^{i\vec{k}'\cdot\vec{r}} u_{n,o}(\vec{r})] = \varphi(\vec{r}) u_{n,o}(\vec{r}). \quad (46)$$

This result has a nice physical interpretation. It shows that the electron moves locally in the Bloch state $u_{n,o}(\vec{r})$. However, if the electron is followed over distances of the order of a_o^*, its wave function is slowly modulated by the function $\varphi(\vec{r})$ as a result of the weak Coulomb attraction of the donor ion. In short, ψ can be visualized as the short wave length pattern, $u_{n,o}(\vec{r})$, modulated by the long wave length envelope function $\varphi(\vec{r})$.

CARRIER CONCENTRATION AND DOPING OF SEMICONDUCTORS

In a pure semiconductor at low temperatures, the valence band states are completely filled and the conduction bands are entirely empty. The energy gap between the two sets of states is finite (no band overlap), but not too large. Most of the semiconductors we will consider have band gaps less than 2.5 eV. This value should be contrasted with the 5 eV, or greater, gap which is typical of insulating materials (NaCl, diamond, quartz, etc.). By this criterion, all of the materials whose band structures are illustrated in Figs. 3-7 are true semiconductors.

A crystal whose bands are completely full, or completely empty, is insulating, since the currents carried by electrons in states $\vec{k}$ and $-\vec{k}$ exactly cancel. Therefore, pure semiconductors at low temperatures are insulators. However, as such a crystal is heated or doped with impurities it develops a finite concentration of mobile carriers and a finite conductivity. In this section we will study how the mobile carrier concentration varies with temperature and doping.

ELECTRONIC STRUCTURE OF SEMICONDUCTORS

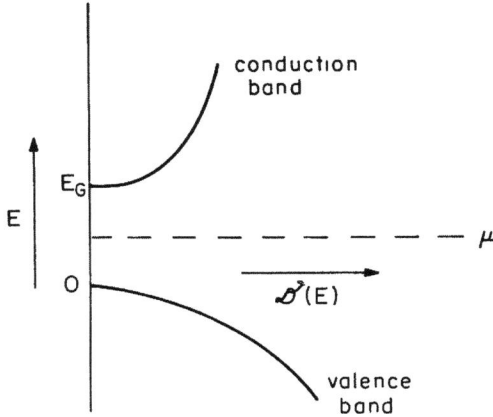

FIGURE 15. Density of states versus energy for a typical semiconductor.

Let us first consider the case of a pure semiconductor crystal at finite temperature. Such a material, in which the carrier concentration is determined by temperature rather than impurity concentration, is termed an <u>intrinsic</u> semiconductor. As usual in Fermi systems, the occupation probability of the various electronic states is given by the Fermi factor

$$f(E) = \left[\frac{1}{e^{\beta(E-\mu)} + 1}\right], \qquad (47)$$

where E is the energy of the state in question, μ the chemical potential, and $\beta = 1/kT$. To use this function to calculate the carrier density, we must also know the density of states versus energy, $\mathcal{S}(E)$, for the semiconductor. This function is plotted schematically in Fig. 15. We have chosen the zero of energy at the top of the valence band, and indicated the chemical potential somewhere near the center of the gap. For an intrinsic semiconductor this choice is about correct, as we will see presently.

The number of electrons in the conduction band is given by the integral

$$N = n(\text{Vol.}) = \int_{E_G}^{\infty} \frac{\mathcal{S}(E) dE}{[e^{\beta(E-\mu)} + 1]} \qquad (48)$$

which, in general, is fairly complicated to evaluate. However, the calculation greatly simplifies if the Fermi level is positioned as shown above - since then $e^{\beta(E-u)} \gg 1$ for all E in the range $E_G \leq E < \infty$. For the time being we will assume that this is the case, and later show that the choice is a consistent one. The expression for n becomes

$$N = n(\text{Vol.}) \simeq e^{\beta\mu} \int_{E_G}^{\infty} \mathcal{N}(E) e^{-\beta E} dE \tag{49}$$

i.e., just the formula one gets from Maxwell-Boltzmann statistics.

In a similar way, the number of holes in the valence band is given by the integral

$$p = p(\text{Vol.}) = \int_{-\infty}^{0} \mathcal{N}(E) \left[1 - \frac{1}{e^{\beta(E-\mu)} + 1} \right] dE. \tag{50}$$

In the valence band $\beta(E-\mu)$ is large and negative (again assuming μ near the center of the gap) and the opposite approximation to that used in the conduction band applies;

$$e^{\beta(E-\mu)} \ll 1. \tag{51}$$

Hence

$$p(\text{Vol.}) \simeq e^{-\beta\mu} \int_{\infty}^{0} \mathcal{N}(E) e^{\beta E} dE$$

$$= e^{-\beta\mu} \int_{0}^{\infty} e^{-\beta E} \mathcal{N}(-E) dE. \tag{52}$$

Finally, to keep the crystal neutral, we require that the number of electrons and holes be equal, n = p. This condition determines μ.

To use these formulas, we must have available forms for the density of states in the conduction and valence bands [8]. If there is a single parabolic conduction band minimum $\left(E = \frac{\hbar^2 k^2}{2m_c^*} \right)$,

$$\mathcal{N}(E) = \frac{8\pi}{(2\pi)^3} \frac{(m_c^*)^{3/2}}{\hbar^3} \sqrt{2(E-E_G)} \quad (\text{Vol.}). \tag{53}$$

The integral which determines n is now

$$n = \frac{8\pi}{(2\pi)^3} e^{\beta\mu} \frac{(m_c^*)^{3/2}}{\hbar^3} \sqrt{2} \int_{E_G}^{\infty} e^{-\beta E} \sqrt{E-E_G}\, dE$$

$$= \frac{2e^{\beta(\mu-E_G)}}{\hbar^3} (2\pi k T m_c^*)^{3/2}. \tag{54}$$

Similarly, if there is a single maximum in the valence band, one finds

$$p = \frac{2e^{-\beta\mu}}{\hbar^3} (2\pi k T m_v^*)^{3/2}. \tag{55}$$

The condition $n = p$ determines the Fermi level. Hence

$$\mu = \frac{E_G}{2} + \frac{3kT}{2} \ln\left(\frac{m_v^*}{m_c^*}\right). \tag{56}$$

Note that μ is exactly at the center of the gap if $m_c^* = m_v^*$, and nearly there even when the masses are different. It is important to realize, however, that this formula only applies to an undoped crystal. It must be modified when impurities are present.

The formulas for n and p are often written in the form

$$n = N_c(T) e^{-\beta(E_G-\mu)}$$

$$p = N_v(T) e^{-\beta\mu} \tag{57}$$

where

$$N_c(T) \equiv \frac{2}{\hbar^3} (2\pi kTm_c^*)^{3/2}$$

$$N_v(T) \equiv \frac{2}{\hbar^3} (2\pi kTM_v^*)^{3/2} \tag{58}$$

are *effective* densities of states for the conduction and valence bands.

To determine the <u>intrinsic carrier concentration</u>, it is most convenient to multiply the equations for n and p to obtain the result

$$np = n_i^2 = N_c N_v e^{-\beta E_G} \tag{59}$$

or

$$n_i \simeq 5 \times 10^{15} \left(\frac{m_c^* m_v^*}{m^2}\right)^{3/4} T^{3/2} e^{-\frac{\beta E_G}{2}}. \tag{60}$$

Typical values at room temperature are

Crystal	$n_i(T = 273°K)$	$E_G(eV)$
Ge	10^{12}	0.8
Si	2×10^9	1.1
GaAs	$< 10^6$	1.4

Note that n_i drops exceedingly rapid as E_G increases from 0.8 to 1.4 eV. In a crystal with large band gap ($E_G \sim 5$ eV) essentially <u>no</u> carriers are generated by thermal excitation. Many of the observed, qualitative differences between semiconductors and true insulators have their origin in this rapid variation of n_i with E_G. Because of it, materials whose band gaps differ by only a factor of two or three can have completely different electrical properties.

So far, we have discussed carrier concentrations in intrinsic semiconductors - those without appreciable doping. Next, we consider the (technologically far more important) situation in which electrically active impurities are deliberately introduced into the crystal. Such materials are termed <u>extrinsic</u> semi-conductors if the carrier concentration is determined by doping rather than thermal activation. To be specific, we will treat the case of an n-type material - a semiconductor which contains excess electrons in the conduction band. Such carriers are produced by substituting specific types of impurity atoms (donors) into the crystal - usually atoms of valence one higher than that which they replace in the host material. For example, P, As and Sb are donors in Si or Ge; S, Se and Te are donors in GaAs (where they substitute for As). Under ordinary circumstances, the weakly bound donor states are ionized. The extra electron provided by each donor wanders, relatively freely, in the conduction band. p-type materials (those containing excess holes) can be produced in a similar way. In this case the dopant is an acceptor, an impurity which removes electrons from the valence band. B, Al, Ga and In are acceptors in Si or Ge; Mg, Zn and Cd are acceptors in GaAs. Notice that most acceptors, as one might expect, have a valence one lower than the atom they replace. Such impurities have one-too-few electrons to complete the tetrahedral bonding. The hole is nothing more than this incomplete bond wandering through the crystal. In discussing carrier concentration, there is complete symmetry between n-type and p-type cases - thus, we have chosen to concentrate on the former.

In a doped semiconductor, the equations for n and p remain the same as those derived above:

$$n = N_c e^{-\beta(E_G - \mu)}$$

$$p = N_v e^{-\beta \mu}. \tag{61}$$

Moreover,

$$np = n_i^2 = N_c N_v e^{-\beta E_G}, \tag{62}$$

as before. The condition which changes is that of charge neutrality since, in an n-type material, ionized donors will be present. The neutrality condition becomes

$$n = p + N_d^+, \tag{63}$$

where N_d^+ is the density of ionized donors. This quantity is determined by the formula

$$N_d^+ = N_i \left\{ 1 - \frac{1}{[1/2 e^{\beta(E_i - \mu)} + 1]} \right\}, \tag{64}$$

where N_i is the total donor density (ionized and unionized). The factor of 1/2 that appears in the Fermi function results from the fact that the donor state can be occupied in two ways, by spin-up or spin-down electrons, but not both simultaneously.

One can now use the charge neutrality condition

$$n = N_c e^{-\beta(E_G - \mu)} = p + N_d^+$$

$$= N_v e^{-\beta\mu} + N_i \left[1 - \frac{1}{1/2 e^{\beta(E_i - \mu)} + 1} \right] \quad (65)$$

to determine μ, and from it n and p. Unfortunately, this equation is relatively complicated. We will not attempt to solve it in detail, but merely study important limiting cases to extract the essential physics.

The simplest case is that in which $N_i \ll n_i$. In this limit, it can easily be seen that $n \simeq p \simeq n_i$; the N_d^+ term plays practically no role in the charge neutrality equation. Though simple, this case is important since it sets a limit on the operation of many semiconductor devices. Imagine, for example, a semiconductor at room temperature in which $n_i \ll N_i \ll N_c$. This is the typical situation in most materials used in devices. If the crystal is heated n_i increases towards $\sqrt{N_v N_c}$, in so doing becoming larger than N_i. Once the condition $n_i \gg N_i$ is achieved the carrier concentration in the crystal, which now consists of almost equal numbers of electrons and holes, is produced by thermal excitation across the gap rather than the impurities. As a consequence, any variations of carrier concentration built into the crystal by specific doping profiles, are wiped out once $n_i > N_i$. The whole crystal becomes intrinsic, and usually the device ceases to operate. Thus, the condition $n_i \simeq N_i$ sets an upper temperature limit on the operation of most semiconductor devices. It is this condition, in part, which makes Si a preferable material to Ge for many applications.

Now we return to the formula for the Fermi level, and consider the more interesting case $n_i \ll N_i \ll N_c$. Under these conditions, the hole density in the crystal is small ($p \ll n_i \ll n$) and we may use the approximate formula $n = N_d^+$ to determine μ. The resulting equation,

$$N_c e^{-(E_G-\mu)} = N_i \left[1 - \frac{1}{1/2 e^{\beta(E_i-\mu)} + 1} \right] \tag{66}$$

is quadratic in the quantity $x = e^{\beta(E_G-\mu)}$, with the solution

$$x = \frac{1}{2}\left(\frac{N_c}{N_i}\right) + \frac{1}{2}\sqrt{\left(\frac{N_c}{N_i}\right)^2 + \frac{4N_c}{N_i \lambda}}$$

where $\lambda = 1/2 e^{-\beta(E_G-E_i)}$. Thus

$$N \simeq \frac{2N_i}{\left[1 + \sqrt{1 + \frac{4N_i}{N_c} e^{\beta(E_G-E_i)}} \right]}. \tag{67}$$

So long as $\beta(E_G-E_i) \lesssim 1$, we see that $n \simeq N_i$ and is temperature independent. In this range, nearly all donors are ionized. Ultimately, however, as temperature is reduced one reaches the condition $\beta(E_G-E_i) \gg 1$. The carrier density then drops rapidly as the electrons "freeze out" onto donor sites.

We may summarize this discussion as follows. With regard to carrier concentration there are three important ranges of temperature in covalently bonded smiconductors. At high temperatures the material is intrinsic $(n_i \gg N_i)$; the carrier concentration is strongly temperature dependent and largely unaffected by doping. The semiconductors of importance in device work are, for the most part, those in which this transition to intrinsic behavior occurs well above room temperature. Below the intrinsic region, there is a considerable temperature range in which carrier concentration is determined by doping, and is nearly temperature independent. Here $n = N_i$, and is under the control of the device engineer. This range (sometimes called the saturation range) is that of major practical importance. Finally, at fairly low temperatures, there is a freeze-out range in which most of the electrons are bound to donors.

One final observation is in order in this section on carrier concentration. The carrier concentrations of interest for devices usually lie in the range 10^{14}-10^{18}/cc, i.e. some four to eight orders of magnitude lower than the atom density in the crystal.

To achieve such control of the carrier concentration, it is necessary to start with semiconductor crystals that are pure (at least of electrically active impurities) to about 1 part in 10^9. This is a severe chemical problem which, to date, has only been surmounted in a few materials - most notably Si, Ge, GaAs, and InSb. The problem of purity and chemical control, which re-occurs again and again in device work, has forced a collaboration between physicists and engineers, on the one hand, and chemists and metallurgists on the other. Successful device groups are generally those which can bring all these disciplines to bear on the problem at hand.

CARRIER TRANSPORT: MOBILITY AND DIFFUSION

In the preceding section we have seen how free carriers occur in semiconductors. These carriers are responsible, in most cases, for the operation of semiconducting devices. Thus, it is important to understand how the carriers respond to perturbing forces, such as electric and magnetic fields, or density gradients. As a first step in this direction, we consider the behavior of a free carrier in an electric field. This discussion will lead us to the concept of <u>carrier mobility</u>.

For simplicity, we study the case of a semiconductor having a single, non-degenerate, isotropic band edge (e.g., n-GaAs). In a perfect crystal of such a material, the carrier motion is determined by the effective mass Hamiltonian

$$H^* = \frac{p^2}{2m^*} - e\vec{F}(t)\cdot\vec{r}, \qquad (68)$$

where $\vec{F}(t)$ is a time varying electric field. In using the effective mass approach, we are tacitly assuming that the frequencies involved in $\vec{F}(t)$ are small compared to the band gap ($\hbar\omega \ll E_G$). This condition is well satisfied in most devices. However, it is never true that the crystal is perfect. Scattering centers - defects, impurities, lattice vibrations (phonons) - are invariably present. As a consequence, the equations of motion must be modified to include their influence:

$$\vec{v} = \dot{\vec{r}} = \frac{\partial H}{\partial \vec{p}} = \frac{\vec{p}}{m^*}$$

$$\dot{\vec{p}} = -\frac{\partial H}{\partial \vec{r}} + \dot{\vec{p}}\bigg|_{collision} = e\vec{F} + \dot{\vec{p}}\bigg|_{collision}. \qquad (69)$$

ELECTRONIC STRUCTURE OF SEMICONDUCTORS

In general, a detailed solution of the Boltzmann operation is required to determine $\dot{\vec{p}}\big|_{\text{collision}}$. We will use a simple, phenomenological method for describing collisions, which is to replace $\dot{\vec{p}}\big|_{\text{collision}}$ by $-\vec{p}/\tau$, where τ is a collision time for the carriers. In effect, this assumption implies that carriers collide (on the average) every τ seconds with complete destruction of the field-induced momentum at each collision; i.e., that carrier momentum is random after collision. This method of treating collisions should be handled with care, but is essentially correct for many scattering problems in semiconductors.

With the relaxation time approximation, the equation of motion takes the form

$$\dot{\vec{p}} + \frac{\vec{p}}{\tau} = e\vec{F}(t). \tag{70}$$

If the field is sinusoidal in time, one finds

$$\vec{v} = \frac{\vec{p}}{m^*} = \frac{e\vec{F}e^{-i\omega t}}{m^*\left(-i\omega + \frac{1}{\tau}\right)}. \tag{71}$$

In the limit $\omega \to 0$, $\vec{v} = \left(\frac{e\vec{F}\tau}{m^*}\right)$. The carrier drifts at a velocity which varies linearly with applied field; the coefficient in this relation is the <u>mobility</u>,

$$\mu = \left(\frac{e\tau}{m^*}\right), \tag{72}$$

usually measured in units cm^2/volt sec. Note that when $\omega = 0$, $\vec{v}$ is in phase with the electric field. As a consequence, there is a net flow of power from the field to the electron.

The processes responsible for collisions in crystals are basically of two sorts - phonon scattering and impurity scattering. In either case, the perturbation destroys the periodicity of the lattice and thus makes possible a transition (scattering) from one Bloch state to another. Phonon scattering occurs via phonon absorption or phonon emission. Crystal momentum is conserved in these processes. Thus

$$\vec{k}_f = \vec{k}_o \pm \vec{K}, \tag{73}$$

where $\vec{k}_o$ and $\vec{k}_f$ are the initial and final wave vectors of the electron, and $\vec{K}$ is the wave vector of the phonon. Phonon collisions also conserve energy;

$$\frac{\hbar^2 k_f^2}{2m^*} = \frac{\hbar^2 k_o^2}{2m^*} \pm \hbar\omega(\vec{K}), \qquad (74)$$

where $\omega(\vec{K})$ is the frequency of the phonon with momentum $\vec{K}$. Phonon scattering can be caused by both acoustic and optical phonons. In nonpolar semiconductors (such as Si or Ge), acoustic scattering is the more important. For scatterings in which the carrier remains within a given conduction band minimum (intra-valley scattering) the $\vec{K}$ vectors of the phonons involved are small, and the acoustic phonon dispersion relation is nearly linear: $\omega(\vec{K}) \simeq v_s|\vec{K}|$. Here v_s is the sound velocity (about 10^5 cm/sec. in Si and Ge). This speed is much lower than that of a thermal electron at ordinary temperatures. From this fact, and the conservation laws, one can easily show that electron-acoustic phonon scattering is nearly elastic. The energy change, as a function of the scattering angle, is given by the approximate formula

$$\Delta E = \pm \frac{4v_s}{v_o}\left(\frac{\hbar^2 k_o^2}{2m^*}\right) \sin(\theta/2), \qquad (75)$$

where $v_o = \frac{\hbar k_o}{m^*}$. Since $v_s/v_o \ll 1$, the percentage change in energy is always small. One may also ask how the probability of phonon scattering varies with scattering angle. Bardeen and Shockley [9] have shown that acoustic phonon scattering cross-sections are independent of angle. The net result is that acoustic phonon scattering is isotropic and nearly elastic; i.e., essentially randomizes the carrier velocity. This is precisely the sort of scattering required for the relaxation time approximation to hold. It is not surprising, therefore, that this method works well in describing acoustic scattering - though we have certainly not proved the point. By contrast, optical phonon scattering is highly inelastic and usually must be treated by more sophisticated methods [10]. It is particularly important in polar materials, such as GaAs.

To conclude this discussion of phonon scattering, we briefly consider the temperature dependence of the scattering rate. The Bardeen-Shockley [9] theory shows that the mean free path for acoustic phonon scattering is independent of carrier energy, and varies inversely with lattice temperature. As a consequence, the collision rates varies as $T^{3/2}$:

$$\frac{1}{\tau} = \frac{v_{electron}}{\ell} \propto T^{3/2}$$

and

$$\mu \propto T^{-3/2}. \qquad (76)$$

In n-Ge, the temperature dependence of μ is $T^{-1.66}$ - in reasonable agreement with the acoustic phonon prediction. On the other hand, in p-type Ge μ varies as $T^{-2.33}$. The variations for n- and p-type Si are $T^{-2.5}$ and $T^{-2.7}$. Discrepancies between the observed temperature variations and those predicted by the acoustic phonon theory are believed to be due to other scattering processes - optical phonons, intervalley scattering, and interband scattering among the valence bands. The problem is a complicated one, and we will not pursue it further. One general feature is clear, however. As temperature is increased, the number of phonons in the lattice increases, and the mobility of carriers decreases rapidly. This behavior is common to most semiconductors.

Since electron-phonon scattering rates increase rapidly with temperature, one might hope to achieve large carrier mobilities in semiconductors by cooling the crystals. Such a stratagem works over a certain range of temperature, but in most cases ultimately fails because a new scattering mechanism, ionized impurity scattering, takes over at low temperature. Impurity scattering is the solid state version of Rutherford scattering - just as the donor state is the solid state version of the hydrogen atom. The essential feature of such scattering that we need to recognize is the fact that the Rutherford scattering cross-section varies inversely with the square of the carrier energy:

$$\frac{d\sigma}{d\Omega} = \left[\frac{e^2}{4\varepsilon E \sin^2(\theta/2)} \right]^2. \qquad (77)$$

Here E is the carrier energy. From this result it follows that the scattering rate varies as $E^{-3/2}$, or $T^{-3/2}$ after averaging over a Maxwellian velocity distribution. The rapid rise of impurity scattering at low temperatures is a direct manifestation of the energy dependence of the Rutherford cross-section.

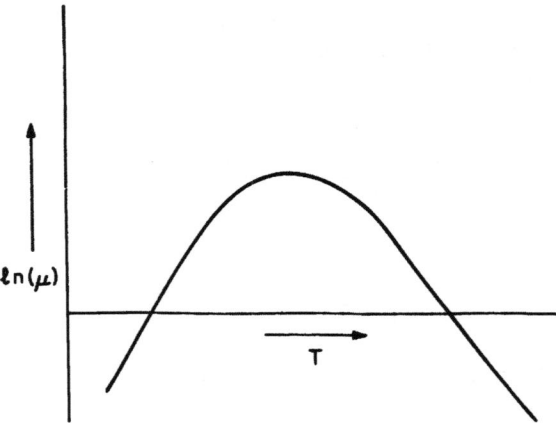

FIGURE 16. Temperature variation of mobility for a typical diamond-type semiconductor.

In most semiconductors, carriers scattering results from a combination of phonon and impurity scattering. The former dominates at high temperatures; the latter at low temperatures. Between, there is a temperature range in which neither scattering mechanism is particularly strong, and the sample has its maximum mobility. Thus, a plot of $\ln\mu$ vs. T generally has the form shown in Fig. 16.

So far in this discussion of carrier transport, we have considered the response of electrons (or holes) to applied electric fields; i.e., the problem of carrier mobility. An equally important form of carrier motion is caused by carrier density gradients. This is <u>diffusion</u>. Diffusion currents play a vital role in many devices since the doping level is often deliberately made non-uniform. The p-n junction is a well known example of such a situation.

The diffusion current is proportional to the particle density gradient

$$\vec{j}_{\text{diffusion}} = -D\vec{\nabla}n. \tag{78}$$

Here $\vec{j}$ is a particle (rather than an electrical) current and D is the diffusion coefficient. In general, D must be calculated from a transport theory similar to that used to determine the mobility. However, in near equilibrium situations there is an important relation between D and μ, namely

$$D = \frac{\mu k T}{e}. \tag{79}$$

Equation (79) is known as the **Einstein relation**, and is valid when currents are induced in a Maxwellian distribution. Fortunately, this is the situation of primary importance in device applications.

The Einstein relation can be simply derived by noting that, in exact equilibrium, mobility and diffusion currents must cancel. Imagine a set of carriers in a semiconductor which are subject to an electrostatic potential, $\varphi(\vec{r})$. In thermal equilibrium, the distribution of carriers in space is determined by the Boltzmann factor;

$$n(\vec{r}) = n_o e^{-e\varphi(\vec{r})/kT}. \tag{80}$$

Note that under these circumstances the carriers feel an electric field ($-\vec{\nabla}\varphi$) and are also subject to a carrier density gradient. Thus, both mobility and diffusion currents flow, but the total current vanishes. That is

$$\vec{j}_{total} = 0 = \vec{j}_{diff} + \vec{j}_{mobility}$$

$$= -n(\vec{r})\mu\vec{\nabla}\varphi(\vec{r}) - D\vec{\nabla}n(\vec{r}). \tag{81}$$

With the aid of Eq. (80) this result takes the form

$$\left(\frac{De}{kT} - \mu\right) n(\vec{r}) \vec{\nabla}\varphi(\vec{r}) = 0, \tag{82}$$

and Eq. (79) follows immediately.

REFERENCES

[1] Valence rules for the formation of tetrahedrally bonded semiconductors are discussed by N. A. Goryunova, <u>Chemistry of Diamond-like Semiconductors</u>, M.I.T. Press, Cambridge, Mass., 1965.

[2] An extensive discussion of the empirical pseudopotential method and its applications is given by Marvin L. Cohen and Volker Heine, <u>Solid State Physics</u>, Vol. 24, Academic Press, New York (1970).

[3] D. Brust, J. C. Phillips, and F. Bassani, Phys. Rev. Letters <u>9</u>, 94 (1962); D. Brust, M. L. Cohen, and J. C. Phillips, Phys. Rev. Letters <u>9</u>, 389 (1962); M. L. Cohen and T. K. Bergstrasser, Phys. Rev. <u>141</u>, 789 (1966).

[4] C. Kittel, *Quantum Theory of Solids*, John Wiley and Sons, Inc., New York (1963), p. 213.
[5] The OPW method, and its relation to the pseudopotential method are discussed in an article by Volker Heine, *Solid State Physics*, Vol. 24, Academic Press, New York, 1970.
[6] J. M. Luttinger and W. Kohn, Phys. Rev. **97**, 869 (1955).
[7] A detailed discussion of the effective mass approximation, and its application to donor and acceptor state problems, is given by W. Kohn, *Solid State Physics*, Vol. 5, Academic Press, New York.
[8] Charles Kittel, *Introduction to Solid State Physics*, John Wiley and Sons, New York, 1971, Chap. 7.
[9] J. Bardeen and W. Shockley, Phys. Rev. **80**, 72 (1950).
[10] H. Ehrenreich, Phys. Rev. **120**, 1951 (1960).

CHAPTER 2

SEMICONDUCTOR ELECTROLUMINESCENCE AND INJECTION LASERS

H. J. Queisser

Max-Planck Institut fur Festkorperforschung

Stuttgart, Germany

This chapter consists of three sections. Semiconductor fundamentals are described first, followed by sections on recombination and light-emitting devices.

SEMICONDUCTOR FUNDAMENTALS

This section serves as an introduction to the fundamental aspects of p-n junction devices. The basic description of a junction at equilibrium without bias is presented first. Next we treat the situation with externally applied voltage. A survey of presently used junction structures and basic technological methods follows. The section closes with a portion briefly summarizing junction devices where diodes rather than transistors are emphasized because diode structures are of particular significance for light-emitting devices.

The p-n Junction

The significance of ambipolar conduction by holes and electrons and the feasibility of doping has been pointed out in the preceding chapter. We now investigate what happens when regions of p-type and n-type conductivity are immediately adjacent in a semiconductor crystal; this is the p-n junction.

Figure 1 gives a simple approach. The top part shows the fixed donors ⊕ and acceptors ⊖ and the resulting free charge

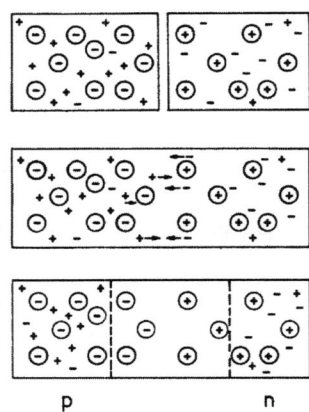

FIGURE 1. Schematic representation of the formation of a p-n junction. Top part: separate pieces of p-type and n-type material; encircled charges are the fixed positive donor ions and negative acceptors. Plus and minus signs represent the mobile electrons and holes. Middle part: schematic for the transient after contact, diffusion sets in as indicated by arrows. Lower part: junction in equilibrium with space charge formed which prevents further diffusion.

---

carriers - and + without circles. In addition to the extrinsic carriers from donors and acceptors we have symbolically included hole-electron pairs from thermal generation. The middle portion of Fig. 1 is a schematic attempt to show what happens immediately after bringing the pieces into intimate contact. The most important feature is diffusion. There are more holes, of course, in the p-type crystal, therefore there will result a tendency to diffuse into the adjacent n-type piece, where holes are minority carriers. The same is true for the electrons, which tend to diffuse towards the p-type side. The diffusion of neutral particles would only come to an end after there is no longer any gradient in the concentrations. However, we deal here with charged particles: positive holes and negative electrons. They experience forces exerted on them by electrical potentials.

The diffusion of electrons and holes in our case becomes balanced by the potential of a space charge which introduces an equal but opposite flow. The bottom portion of Fig. 1 shows this situation. In the middle we see unbalanced charges of negative acceptors and positive donors. This is the space charge region. We can think of it as resulting from recombinations of the majority carriers with the minorities that have diffused in from the other side. Equilibrium is reached when the repulsive space charge potential exactly balances the diffusion tendency.

SEMICONDUCTOR ELECTROLUMINESCENCE AND INJECTION LASERS

After this intuitive reasoning we consider the situation mathematically and follow Shockley's treatment [1].

The p-type region is characterized by a hole density p_p and a much smaller electron density n_p. They obey the equilibrium condition

$$p_p \cdot n_p = n_i^2, \qquad (1)$$

which is a law of mass action; n_i is called the intrinsic density.

The doping fixes the Fermi level E_F closer to the valence band than to the conduction band. We write for convenience $E_F = -q\varphi$, where $(-q)$ is the charge of the electron. The hole density is

$$p = n_i \exp[q(\varphi-\psi)/kT] \qquad (2)$$

This Eq. (2) is written in a slightly different form than that presented in Chapter 1, in which N_v is the effective density of states in the valence band. We have used Boltzmann statistics as a good approximation for not too high a doping. ψ is the electrostatic potential; its zero is chosen such that $\psi = \varphi$ when the material is intrinsic, which makes Eq. (2) rather convenient.

For the electrons we have in analogy

$$n = n_i \exp[q(\psi-\varphi)/kT] \qquad (3)$$

Multiplying (2) and (3) confirms Eq. (1). As shown in the earlier lectures, n_i can be expressed as a product of the effective densities and an exponential in which the band gap $E_g = E_c - E_v$ enters. This was treated in Chapter 1. The relations (2) and (3) hold for majorities and minorities in the n-regions as well as the p-region, where of course ψ and φ attain different values, as we will see.

The Fermi potential φ has the characteristics of a "chemical potential" in the sense of thermodynamics. Chemical equilibrium means that the potential must be equal everywhere. Thus the requirement in joining pieces of semiconductors with differing doping is to line up the Fermi levels rather than the band edges, as is shown in Fig. 2. We see that the potential for an electron in the conduction band is higher in the p-side than in the n-side. A potential difference V_{bi} ("built-in voltage") has developed. It is this potential wall resulting from the space charge that balances the diffusion tendency.

If we again use Boltzmann statistics we can calculate the ratio of electrons at the higher potential in the p-side to that in the n-side as

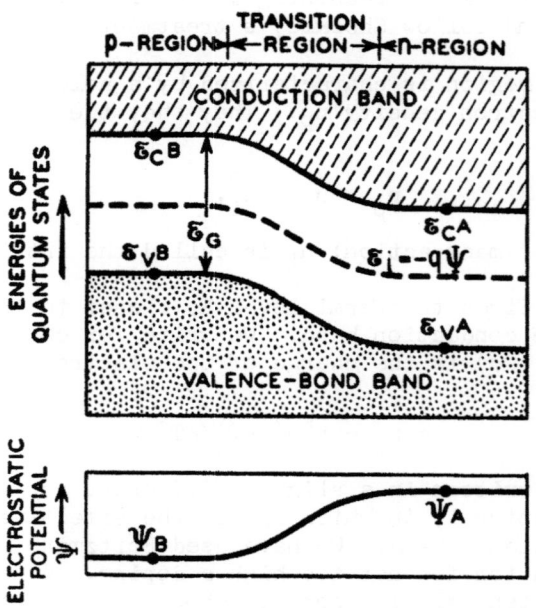

FIGURE 2. Energy levels and electrostatic potentials in a p-n junction after Shockley [1].

$$n_p/n_n = \exp(-qV_{bi}/kT), \qquad (4)$$

similarly

$$p_n/p_p = \exp(-qV_{bi}/kT) \qquad (5)$$

which gives us the built-in voltage in terms of the carrier densities resulting from the doping of the two sides

$$V_{bi} = (kT/q) \ln (n_n/n_p) \qquad (6a)$$

$$= (kT/q) \ln (n_n p_p/n_i^2) \qquad (6b)$$

We call

$$kT/q = V_{th} \qquad (7)$$

the "thermal voltage", it is about 25 mV at room temperature.

We can see from Eqs. (6) that the potential barrier V_{bi} must be greater if the doping levels n_n, p_p increase, because the diffusion tendencies will be accordingly stronger.

SEMICONDUCTOR ELECTROLUMINESCENCE AND INJECTION LASERS

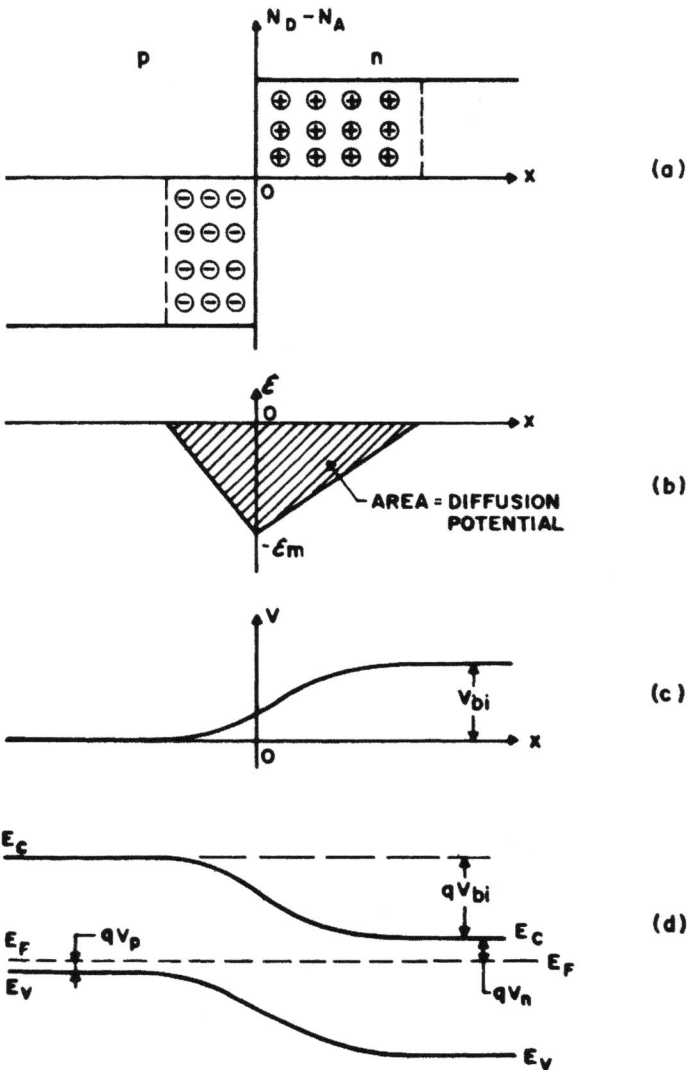

FIGURE 3. Abrupt p-n junction without applied bias, (taken from Sze [2]) (a) shows model for impurity distribution in the space charge layer, (b) gives electric field, (c) potential and (d) the energy bands, all as a function of distance x.

How can we understand the origin of V_{bi} in terms of the space charge, which we know to originate in the fixed donor and acceptor ions no longer neutralized by free carriers? The Poisson equation

$$\nabla^2 \psi = -4\pi \rho / \kappa \qquad (8)$$

gives us this relation of potential ψ to space charge density ρ, where κ is the static dielectric constant of the material. For the simple case of two rectangular charge distributions with densities N_D and N_A of Fig. 3 we can easily integrate (8) and get the field $\mathcal{E}$ in the junction

$$\mathcal{E}(x) = - \frac{qN_A(x+x_p)}{\kappa} \quad \text{for} \quad -x_p \leq x < 0 \tag{9a}$$

$$\mathcal{E}(x) = \frac{qN_D}{\kappa}(x-x_n) \quad \text{for} \quad 0 < x \leq x_n \tag{9b}$$

where x_n and x_p are the coordinates of the ends of the space charge layer. The two total amounts of opposing space charges must be equal: $N_D x_n = N_A x_p$, which means that the space charge layer always extends more deeply into the material of lower doping level.

The potential can be obtained by one further integration yielding

$$\varphi(x) = \mathcal{E}_m \left(x - \frac{x^2}{2W} \right) \tag{10}$$

where $\mathcal{E}_m$ is the maximal field, as calculated from (9) and shown in Fig. 3(b) and $W = (x_n + x_p)$ is the total width of the space charge region. Here we cannot go into more detail for the calculations, which are described completely in Shockley's treatment [1] or in standard texts of semiconductor principles such as the book by Sze [2].

We now have determined the properties of the junction, since we can for example express the space charge width W as dependent on the doping levels if we assume an "abrupt" p-n junction [2]:

$$qW^2 = 2 \left(\frac{N_A + N_D}{N_A N_D} \right) V_{bi} \tag{11}$$

This width can be measured directly through the capacitance of the junction [2].

We have covered the essential idea of the p-n junction at equilibrium, which means without applied bias. We must now continue by treating the case of nonequilibrium with bias applied and currents flowing.

Current-Voltage Characteristics of a Junction

Nonequilibrium means that the chemical potentials of holes and electrons will not be equal and can no longer be described by a common potential φ. We therefore define quasi-Fermi levels φ_n and φ_p for electrons and holes by

$$\varphi_n = \psi - V_{th} \cdot \ln(n/n_i) \qquad (12a)$$

$$\varphi_p = \psi + V_{th} \cdot \ln(p/n_i) \qquad (12b)$$

in generalization of (2) and (3).

We therefore get

$$n_n p_n = n_p p_p = n_i^2 \exp[(\varphi_p - \varphi_n)/V_{th}] \qquad (13)$$

A difference of the chemical potential of the two species can be attained by an externally applied bias V_a

$$(\varphi_p - \varphi_n) = V_a \qquad (14)$$

If the bias $V_a = 0$, of course we return to equilibrium, and Eq. (13) reduces to Eq. (1).

The quasi-Fermi levels φ_n, φ_p will now become functions of position. Far away from the junction there is equilibrium again, but in the junction itself φ_n and φ_r must differ. Application of bias V_a can strongly deplete or enhance carriers, we see this from Eq. (13) and (14). The sign of V_a is decisive now.

Positive V_a means "forward" bias, the plus pole attached to the p-side. Then $np > n_i^2$ from Eq. (13) holds. It means that current will flow. Forward bias reduces the potential barrier V_{bi}, thus diffusion dominates and charge is transported, current flows. "Reverse bias" means minus to the p-side and negative V_a. Equation (13) predicts for this case a carrier deficiency, thus no current by diffusion occurs. We have enhanced the potential barrier V_{bi} and overcompensated diffusion.

The difference $np - n_i^2$ determines the current flow. From Eq. (13) and (14) and (7) this is seen to be proportional to $[\exp(V_a/V_{th}) - 1]$. In this way one can derive [1,2] the "ideal rectifier equation" which links current I to bias V_a at a temperature indicated by V_{th}:

$$I = I_s(\exp(V_a/V_{th}) - 1); \qquad (15)$$

there I_s is a constant into which enter doping levels and lifetimes of the carriers. The significance of I_s is seen for negative bias exceeding $-V_{th}$. Then the current approaches a constant value of $-I_s$; we get a "reverse saturation current", independent of voltage.

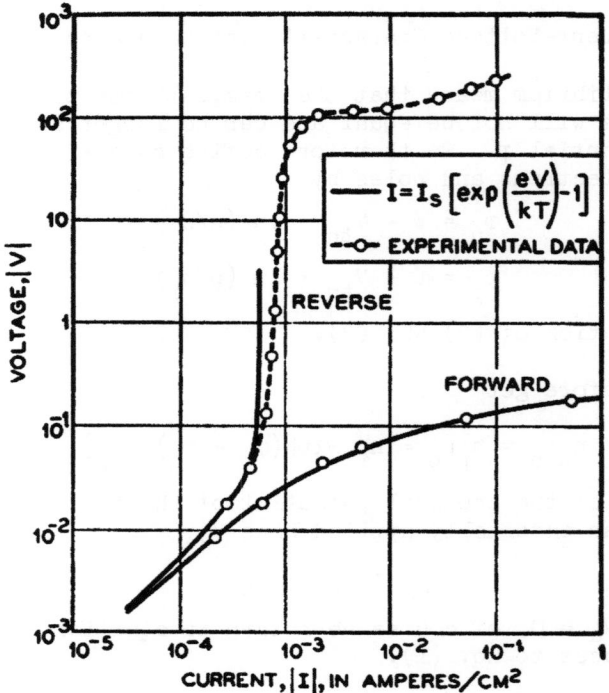

FIGURE 4. Ideal rectifier, theory and experimental points for a germanium junction [1].

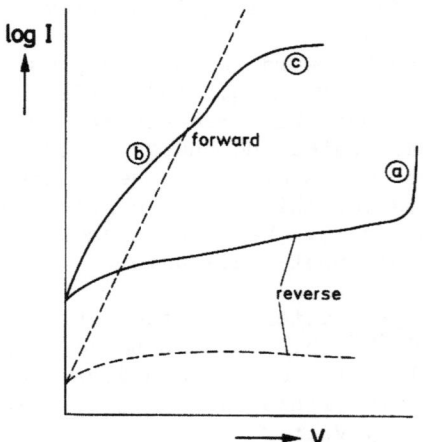

FIGURE 5. Ideal rectifier behavior (dashed line) and actual characteristics (solid line) of junctions deviating from the simple model, as often encountered in practice: at (a) avalanche effect, at (b) recombination in space charge layer, at (c) series resistance. Notice logarithmic scale for the current I versus linear bias scale.

SEMICONDUCTOR ELECTROLUMINESCENCE AND INJECTION LASERS

In the forward direction the current increases rapidly with voltage. For room temperature and $V_a > V_{th}$ there is an increase of a factor e in current when we increase the voltage by 25 mV (or by a decade for increases by 59 mV). Figure 4 shows a well-known graph depicting this rectifier behavior.

Actual junctions deviate from Eq. (15) in three important features, as shown schematically in Fig. 5. At (a) we get avalanche effects, the field is so high that one carrier can gain enough energy in the field of the space charge layer to create additional hole-electron pairs by breaking valence bonds. At (b) the current behaves as

$$I \sim \exp(V_a/nV_{th}) \qquad (16)$$

where $1 \leq n \leq 2$. This results from carrier recombination at traps in the space-charge layer [3]. Occasionally $n > 2$ is found which is usually an indication of poor quality of the junction and field inhomogeneities such as resulting from precipitated metal impurities. Finally at (c) the series resistance limits current flow.

Junction Structures

Many ways have been explored to realize p-n junctions. The junction must be situated in good crystalline material and cannot be formed by simply pressing n- and p-type material together. Grown junctions can be made by changing conductivity type during pulling from the melt; this is, however, impractical.

Epitaxy is a good way to make junctions in some cases, especially for the compound semiconductors emphasized here. This is discussed in Chapter 20.

Junctions can be fabricated by allowing, say, acceptors into n-type materials. Such alloying techniques were often employed for germanium devices and lead to inexpensive and quick processes. Control of the junction properties is, however, not very good.

Diffusion is the most important present technology for today's junction devices. Donor or acceptor impurities are added to a stream of an inert carrier gas which flows over a heated piece of semiconductor. The dopant atoms enter the crystal and diffuse into it. Diffusion time, temperature, and the diffusion coefficient of the impurity inside the host crystal determine the impurity profile. Junctions can thus be made with typical depths of a few microns from the exposed surface. Good control is achieved of the essential parameters, which are junction depth and doping level.

Oxide layers on the semiconductor can be used to prevent diffusive entry of the impurity. Patterns of oxide-protected areas and oxide-free windows enable us to control doping spatially. This technique led to the present integrated circuits in silicon. The stability of SiO_2 is of great importance here.

Junctions are very sensitive to surface effects. They must therefore be protected wherever the reach surfaces. Often this is done by etching the surface just before the device is encapsulated in an evacuated package or some protective plastic. In silicon the junctions are well secured by oxide layers, covering the junction.

Bombardment with high-energy dopant ions from an accelerator is a recent technology for junction formation. One can to a certain extent control the depth of "implantation" by choice of particle energy. An advantage is the relatively low process temperature which reduces unwanted contamination.

Junction Devices

The p-n junction is the essence of many semiconductor devices. Solid state electronics depends heavily on this concept. In this lecture we cannot possibly cover the entire field, the readers are referred to textbooks [2]. Here we can only attempt a broad general survey.

The name "ideal rectifier equation" is indicative of one of the most important uses of p-n junctions. The non-linear and non-symmetric behavior of the current in its dependence upon voltage are obviously useful to obtain a rectifier that passes current when biased forward and essentially blocks current flow when reverse biased. Both the rectification of alternating current to direct current for power applications and the unidirectional flow of current for logic applications in a computer benefit from this important property of a p-n junction.

The forward characteristic of a junction implies recombination of electrons with holes near or within the junction. We have seen that holes diffuse from the p-side and electrons from the n-side towards each other. Since the current far away from the junction is carried mainly by the majority carriers, there must be recombination taking place, which means that the other type of carrier takes over the transport of charge. This act of recombination becomes important in itself as soon as one wants to utilize radiative recombination processes. A hole can recombine with an electron by emitting a photon with an energy close to that of the band gap of the crystal. This process of light generation is the topic of the following sections.

A diode based on the rectifying action of a p-n junction can also be used to detect light and to transform it into electrical power. This is the exact inverse of creating light with an applied bias. The solar cell is one example for the conversion of light to electricity. Imagine that light with energy above the band gap is absorbed near a p-n junction in a semiconductor. Visible light from the sun is absorbed in silicon, which is the most commonly used material for solar cells. The absorption process is equivalent to creation of a hole-electron pair. This carrier pair is separated in the electric field of the junction. Equation (9) describes this field. Thus the electrons drift in the field towards the n-side, holes drift to the p-side because of this built-in field. We have a generator which delivers a direct current output with the positive pole being at the p-side of the junction. Satellites in space rely heavily on this means of electricity generation from the available sunlight.

The same principle as just described holds for many other devices in which photons must be detected. Photon-counting diodes and light sensitive elements are made according to the same principle. Often one can use avalanche multiplication, as discussed in conjunction with Fig. 5 at (a), to have a built-in amplification in the device. Each absorbed photon then created not just one pair, but an avalanche of carriers. Nuclear particle detectors can also be made with this basic idea, we have a solid-state analogon of the ionization chamber.

The transistor is of course the most famous of all junction devices. It consists of two junctions separating three regions. Let us take a n-p-n transistor for example. The starting material for such a device as a piece of n-type semiconductor, say silicon. Into this crystal one diffuses acceptors to form a p-n junction, which is called the base-collector junction. Into most of this p-type region a second diffusion is directed with the aid of masking oxide layers, as outlined before. This diffusion drives donors into the crystal and reconverts the upper portions close to the surface to n-type conductivity. Thus we have a three-layer sequence n-p-n. The top portion of heavily doped n-type is called the emitter.

Transistor action depends on the fact that one can control the current in the base-collector junction by electrons emitted from the emitter region. The base-collector junction is reverse biased, thus passes little current and represents a high impedance. The emitter-base junction is forward biased and great changes in current can be attained by small changes of the applied forward bias. Since the electrons have a finite lifetime they diffuse through the narrow base layer and reach the collector, where they change the current drastically. Many different circuits and applications utilize this basic feature of two junctions in close vicinity.

Four layers separated by three p-n junctions are also in use today; these are the "four layer-diodes". They have a high-impedance OFF state, which switches into a low-impedance ON state after a certain threshold voltage is exceeded. Four-layer devices can also be used as controlled rectifiers, where the rectification is controlled by an externally applied current.

This brief summary is too short to be a real introduction into the extensive and fascinating field of solid state devices, of which the semiconductor junction devices form the most important and most widely used family. The junction devices rely on the possibility to establish both n-type and p-type conduction within the same semiconductor. Sometimes different semiconductors are used, in this case one speaks of "hetero-junctions". This added degree of freedom in device design and materials selection is useful for laser construction. Nonzero lifetime of the minority carriers plays an important role, as we have seen for the transistor because the emitted electrons must traverse the base to reach the collector. Nonzero lifetime is also important to extract energy out of a solar cell. Otherwise the electron could not reach the n-side and deliver voltage and current to the outer load. Lifetime is therefore an important parameter. Most of the next section deals with process that determines the lifetime, namely recombination. Recombination is the act of removal of an excess minority carrier by disappearing together with a majority carrier and leaving behind their energy.

RECOMBINATION

This section deals with the process of electron-hole recombination, especially the mechanism of radiative recombination. In this type of recombination the excess energy of the carrier pair is removed from the crystal by the emission of a photon. Before we start with a treatment of the radiative recombination of a hole-electron pair, we must treat the unwanted process by which pairs are lost without utilization of their excess energy towards production of electromagnetic radiation. The excess energy is here absorbed by the lattice as heat.

Nonradiative Processes

Nonradiative processes usually dominate the recombination. Only a few percent of all transitions are at best radiative when we are dealing with low levels of excess carriers. Therefore the nonradiative processes play an important role. Nevertheless they are not all understood today nor really under control in most cases. The problem of nonradiative transitions is indeed one which presents formidable difficulties in both theory and experiment.

Experimentally it seems clear that one cannot easily study something that does not leave a clear reaction product such as a photon but rather results in transfer of heat to the crystal lattice. One can only infer indirectly to nonradiative recombination by looking at optical emission spectra. The theory is quite complicated too, because it cannot treat carriers and lattice separately but must base all predictions on the electron-phonon interaction, which is difficult.

A hole-electron pair cannot simply create at once the rather large number of phonons which are needed to conserve the pair energy, which is approximately the gap energy, being of the order of one eV. This simultaneous process is improbable.

Impurities can aid in the nonradiative recombination. One type of carrier is captured first at an impurity. The other carrier might then be captured from its free band-state into a bound state of high energy and then successively lose energy to the lattice by dropping into closer and less energetic orbits around the impurity. This is called a "cascade process". There is a yet insufficient experimental evidence for this process. One suspects that impurities will be very important, however, to foster recombination. Beside bringing together the two carriers into large overlap, these centers can also accomodate excess crystal momentum. This is particularly important for indirect semiconductors where the band extrema are located at different k-values, as described by Professor Wolff. "Deep" impurities [3] in the middle of the forbidden energy gap are of particular importance, but those deep centers are the least understood.

There is another important nonradiative mechanism, called "Auger effect", in close analogy to the same effect in atomic X-ray spectra. This is a three-particle phenomenon. One pair recombines and delivers the resulting energy to a third particle, a majority carrier near the recombining pair. This carrier is therefore ejected deep into its band and then gradually slides back down in energy its $E(k)$ curve by emission of lattice phonons, in particular optical phonons. The probability of this reaction is proportional to the square of the majority carrier density, because two of these particles participate. Therefore the Auger-effect is of particular significance for heavily doped materials. For a quantitative consideration one has to take into account the condition of conservation of crystal momentum and evaluate the density of states that are available to the energy-receiving partner. Therefore the details of the band structure enter explicitly, and one has to evaluate each material individually for this recombination mechanism. In general it seems, however, that this Auger-effect does not satisfactorily account for the carrier lifetimes in lowly-doped materials.

There is another important variant of the Auger-effect. It is the recombination of a pair close to a neutral donor or acceptor. Such a neutral dopant atom has a bound electron or hole around it, therefore we again have a three-particle affair. The pair recombines and delivers the energy to the bound carrier. This carrier is then excited to higher-lying bound states or completely ionized. Eventually it returns to thermal equilibrium by phonon emission. Here we have all three particles located closely to each other, and the probability of this process to happen is not necessarily proportional to the square of the carrier density. Therefore it can be an important mechanism in crystals of moderate doping. This process has first been clearly evaluated in GaP doped with sulfur donors [4]. Characteristic sideband emission of reduced energy has been seen, where the energy differences correspond to the excitations of the third particle. The Auger transitions are thus seen indirectly, because their characteristic energies are now found as missing in the radiative transitions.

With this review of qualitative features of nonradiative processes we conclude our consideration of nonradiative reaction channels. All nonradiative mechanisms present losses for the conversion of electrical power into light. It is therefore an important task to understand what is going on, but we have seen it to present principal difficulties both in theory and experiment. It is easier, as we will see, to study radiative transitions with spectroscopy and identify the processes. This topic will now be discussed.

Radiative Transitions

Electrons and holes can recombine and give off a photon of energy $h\nu$, this process is called "radiative recombination". Carriers at the band edges represent an excess of

$$E_g = E_c - E_v \approx h\nu \tag{17}$$

Conservation of momentum requires the transitions to involve no change in k-vector, i.e. to be "direct". The momentum of a photon can be neglected, because it is

$$\hbar k_{photon} = h\nu/c, \qquad (18)$$

which is a negligible fraction of the Brillouin zone, as discussed in Chapter 9. In semiconductors with indirect transitions, such as silicon, optical absorption or emission is only possible with an accompanying phonon emission to establish conservation of the k-vector.

We will now classify the types of transitions and discuss each class separately. First we consider free carriers which recombine with free or bound partners. Next we treat excitons, both the case of free excitons and those bound to attractive centers. A special case of great practical importance is the binding of excitons to isoelectronic impurities. Finally we consider the very important problem of pair spectra, which result from recombination of localized electrons and holes.

(i) <u>Free carriers</u>. It may seem surprising that the simplest process of recombination from free electrons in the conduction band with free holes in the valence band is rarely seen, unless we go to highly doped crystals and strong excitation, such as in laser diodes. For low excitation rates of minority carriers in pure crystals these carriers end up in the band extrema where the density of states is smallest and other processes present better chances for recombinations. Photo-excited energetic carriers in the bands usually undergo lossy scattering with phonons much more rapidly than recombination. Therefore one can in most cases think of "thermalized" charge carriers which first lose all their excess energy within their bands and attain a temperature only slightly above lattice temperature. After this process is finished we get radiative (and nonradiative) recombinations across the band gap.

The recombination of a free carrier with a bound carrier is often seen. Figure 6 shows one example for GaAs. Lines "A" and "Cu$^-$" result from electrons in the conduction band recombining with a hole bound to an acceptor close to the valence band. The presently used nomenclature to describe this is: (e, A^o), which means that the free electron e recombines at an acceptor A which is neutral - indicated by the superscript o - because the hole p surrounds the negative acceptor ion A^-. In Fig. 6 we have two acceptors, line A resulting from carbon and "Cu" from a deep copper-acceptor.

From the energy of these (e, A^o) lines one can deduce the ionization energy E_a of a hole at the acceptor when the band gap E_g is known, since $E_a = E_g - h\nu (e, A^o)$. In general this optical determination of ionization energies is in agreement with Hall effect data. (See upper abscissa in Fig. 6).

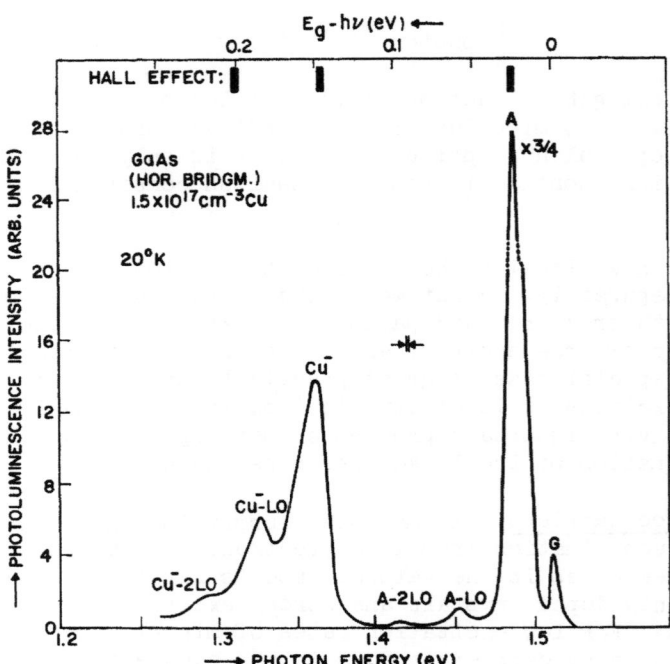

FIGURE 6. Example of an emission spectrum from a GaAs crystal after excitation of hole-electron pairs by illumination with light whose energy exceeds the band gap. ("Photoluminescence").

The recombination lines very often have satellite lines at lower photon energies than the main line. In Fig. 6 we see examples. In most cases such satellites occur at energies lower by the energy of one or several longitudinal optical phonons. This means that we have recombination with simultaneous generation of phonons. The emitted photon must of course be reduced in energy by the amounts needed for the phonon generation. The strength of the satellites is a measure for the coupling of the electron-hole system to the lattice. One can show that the longitudinal phonons couple most strongly to the electrons; the satellites result mostly from emission of phonons with zero wave-vector k. The situation is different from recombinations in indirect materials, where finite k are involved in order to preserve crystal momentum for the transition, therefore also acoustic phonons with large momenta play an important part in the recombination.

(ii) <u>Excitons</u>. Chapter 9 treats the exciton in some detail. We will not repeat the discussion here. Exciton recombination is important in many semiconductors. Rarely does one see the free exciton recombination. It takes very pure crystals, very low temperatures, and gentle excitation to detect recombination from the exciton ground state and the excited states. Otherwise "bound excitons" dominate the emission spectra.

Impurities in the crystal easily capture excitons, where the excitons then recombine. Therefore emission resulting from these "bound excitons" greatly exceeds that from free excitons. After the exciton is generated it can diffuse inside the crystal until it is captured. Mobile excitons of small reduced masses and large excitonic radii are particularly susceptible to capture and decay as bound excitons. The semiconductors with low effective masses and large static dielectric constants favor the bound exciton recombination.

Bound exciton emission results in very sharp lines if one has good and pure crystals without strain. The lines are actually narrower in energy than the thermal energy kT, because they are bound particles and not moving ones with thermal velocities. The sharpness of the lines makes feasible many spectroscopic experiments, especially the Zeeman effect in a magnetic field. From the splitting of the lines one can deduce what kind of a center is binding the exciton, this helps greatly in the interpretation of spectra and also serves to identify impurities. The binding energy of the exciton can be estimated according to the "Haynes rule". It states that the binding energy is approximately 10% of the binding energy for the corresponding hole or electron if the center is an acceptor or donor. The problem of what centers can bind excitons at all is not an easy one. Many calculations have been made, most of which use the ratio of the effective masses for electrons and holes as the critical parameter for the existence of bound states. Neutral impurities seem to more easily bind excitons than do ionized ones.

A new and important type of binding centers are the isoelectronic impurities. These are impurities which have the same number of outer electrons as the host atom they replace. Nitrogen replacing phosphorus in GaP is an important representative. Both elements are in the same column of the periodic table, they are therefore isoelectronic. Such impurities differ from ordinary donors or acceptors which are by definition not isoelectronic. The potential of an isoelectronic impurity is weaker than that of the charged donor or acceptor. A single particle therefore cannot be bound in general. However, an exciton may in many cases be bound. One of the two particles is usually more tightly bound while the other one is on a more extended orbit. An example is of N in GaP, which is called an "isoelectronic acceptor" because the nitrogen atom plus the localized electron around it appears similar to the negative charge of an ordinary acceptor binding a hole on an extended less localized orbit.

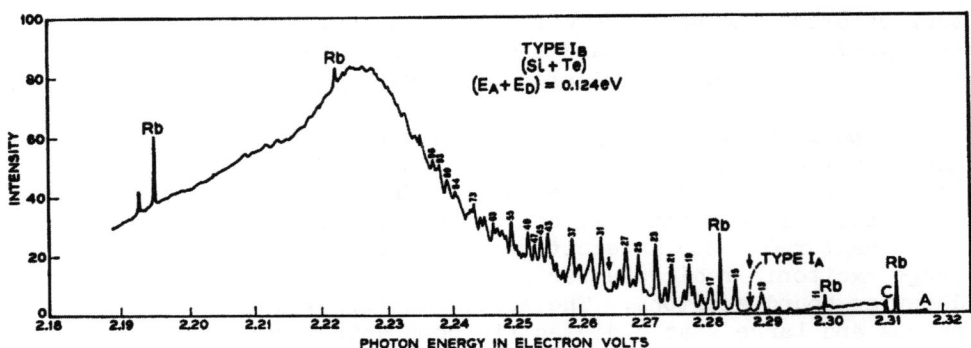

FIGURE 7. Photoluminescence "pair spectra" of GaP, doped with Si-acceptors (on P-site) and Te-donors. A and C are bound-exciton lines, Rb a calibration line. The numbers designate the shells [6].

Recombination of excitons bound to isoelectronic nitrogen is an efficient source of green recombination radiation in GaP [5]. There are only two particles + and - bound to an isoelectronic trap - not three as in the case of neutral donors or acceptors. Thus we cannot have a non-radiative Auger effect and the radiative efficiency is increased.

(iii) <u>Pair spectra</u>. A hole bound by an acceptor can recombine with an electron bound by donor. If acceptor and donor pair are separated by a distance r, the emitting photon has an energy

$$h\nu = E_g - (E_A + E_D) + e^2/\kappa r, \qquad (19)$$

where E_A and E_D are the ionization energies of the acceptor and the donor. The last term is included because we must consider the Coulomb interaction $e^2/\kappa r$, where κ is the dielectric constant. Equation (19) means that we must get a series of discrete emission lines because r can only take discrete values since the atoms are not continuously distributed but fixed on discrete lattice positions. Figure 7 shows one example of a series of pair lines resulting from tellurium donors and silicon acceptors in GaP. For very large pair separations, $1/r$ becomes small and gives a continuum which is seen to result in a broad band of "distant pairs" near 2.22 eV in Fig. 7. The number of available recombination partners increases as the radius r increases since the shells of atoms at that distance r increase in area and thus in the number of possible sites for the partner. But on the other hand the probability for recombination decreases with increasing distance, this causes the emission-intensity to decrease.

Pair spectra with sharp discrete lines have been found in a number of semiconductors, mostly in indirect materials. In GaAs and InP one can only find the broad continuum band and does not see discrete lines. In these direct gap semiconductors the ionization energies of donors and acceptors are too small, therefore the energy of close pairs with its large Coulomb contribution (see Eq. 19) exceeds the band gap energy E_g. The transitions from donor to acceptor are nevertheless quite important in these materials, too. Such transitions can be distinguished from band-to-impurity transitions by a number of criteria, such as the dependence of emission intensity on temperature and excitation rate. A detailed analysis of these radiative transitions gives much information about the impurity distributions.

Pair spectra, especially their discrete lines, are an exceptionally beautiful and useful object of solid-state investigations. Many experiments in this field have increased our knowledge of electronic processes in semiconductors to great details. GaP has become a model substance for that reason. For example, time-resolved spectroscopy has shown that indeed the closer pairs recombine more rapidly than distant pairs [7]. Recently the effect of strain in the lattice due to foreign atoms of different atomic radii has been investigated by means of pair spectra [8]. The scope of these lectures does not permit us to go into more details. Our next topic concerns the utilization of recombination processes for generation of light in devices.

LIGHT-EMITTING DEVICES

This section treats the two types of junction devices which are presently used for generation of visible and infrared light: electroluminescent diode and semiconductor injection laser. These two devices involve an unusual amount of materials research. These light-emitting devices employ compound semiconductors with complicated doping. Further it is necessary for room-temperature operation of lasers to utilize heterojunctions of mixed-crystal semiconductor systems. The materials technology thus goes far beyond the comparatively simple requirements of doped elemental semiconductors, such as the silicon technology. We must now treat multicomponent materials, where the phase diagrams alone present a complicated topic for research and development. On the other hand, we will see that once one masters this technology there are many ingenious ways to employ the properties of solids. We will first give a survey of light-emitting diodes, then treat the semiconductor injection laser.

Light-Emitting Diodes

The principle of electroluminescence from semiconductor p-n junctions follows immediately from what we have covered in the previous section. A forward biased junction generates excess holes and electrons in the junction region, which must be made to recombine radiatively. There are presently two major lines of development. One uses the direct radiative recombination in a mixed crystal of gallium arsenide-phosphide, the other one uses excitonic recombination at impurities in the indirect material gallium phosphide.

Gallium arsenide is a semiconductor with direct allowed transitions from the valence band to the conduction band, as was discussed in Chapter 1. Both band extrema are located at the center of the Brillouin zone; no phonon assistance is required for optical transitions. This means that there is a strong coupling between an electromagnetic field and the electron-hole system of this crystal. We can therefore expect large radiative recombination rates. But unfortunately a p-n diode of GaAs emits only invisible infrared light since the band gap at room temperature is about 1.4 eV which corresponds to a wavelength of almost 9000 Å, invisible to the human eye, but useful for optoelectronic purposes in connection with silicon detectors.

Very efficient luminescence-diodes have been fabricated with GaAs by a special doping technique. Amphoteric doping with elements from the fourth group of the periodic table is employed. Silicon is the most commonly used dopant for this purpose. A silicon impurity replacing a gallium atom acts as a donor. When Si replaces an arsenic atom it is an acceptor. Silicon is therefore called "amphoteric" in its doping behavior. Closely compensated structures can be made by silicon doping, which means that almost equal donor and acceptor concentrations are achieved. Such structures are particularly effective for emission of radiation. Compensated material emits at lower photon energies than heavily doped crystals, this fact causes less self-absorption and increases the external efficiency of photon emission.

One semiconductor with direct optical transitions at much higher photon energies is GaN, but this material is extremely difficult to grow. No junctions have thus far been produced by in GaN doping, only metal-semiconductor structures have been fabricated which emit ultraviolet and visible light but require undesirably high voltages.

Visible light from direct semiconductors can be obtained if one resorts to mixed crystals. This subject is treated in detail by Dr. Hilsum. Figure 8 shows the energy positions of the

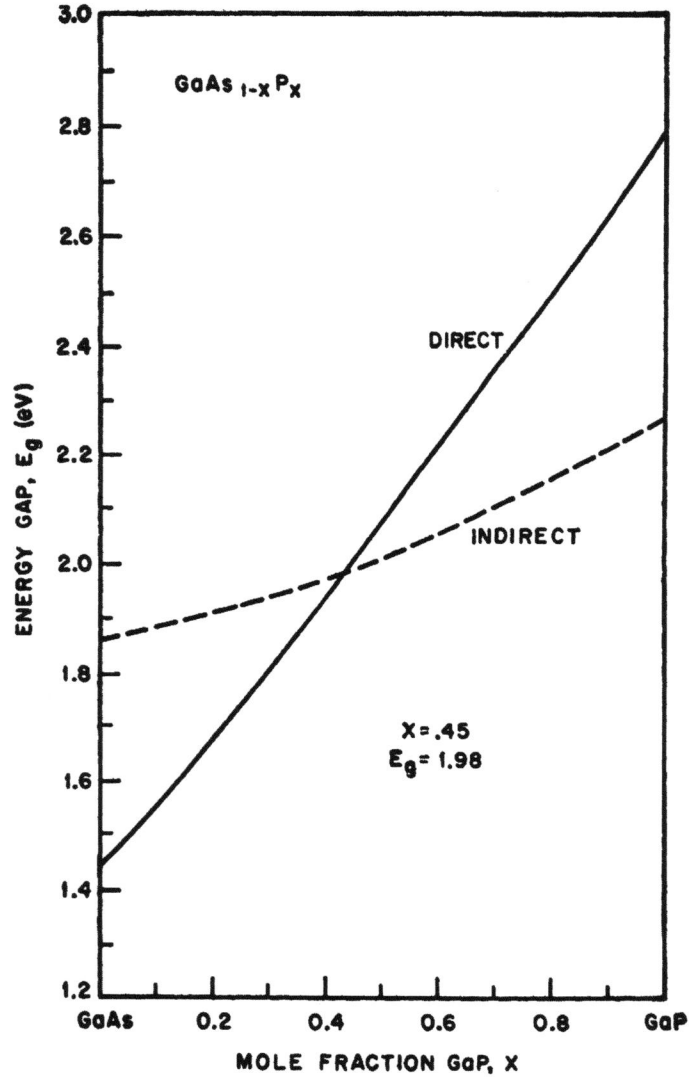

FIGURE 8. Gap energy of the mixed-crystal system GaAs-GaP. At 45% GaP the mixed crystal changes from direct optical transitions to indirect ones, since from there on towards higher GaP concentrations the lowest conduction band lies no longer in the center of the Brillouin zone (broken line, marked "indirect") [9].

conduction band minima for the system gallium arsenide - gallium phosphide. One obtains approximately linear variation in this system, as shown by the lines which connect the band minima of the pure constituents. The most important feature for our purpose is the crossover point at roughly 45% gallium phosphide. Up to this composition the mixed crystal maintains the nature of direct transitions for the lowest energy, for higher phosphorus content the lowest conduction band minimum is located near the Brillouin zone boundary and no longer in the center of the zone. The composition of 45% GaP corresponds to the emission of red light with 1.98 eV photon energy. The use of a small, efficient lamp with such mixed crystal material is already widespread; luminescent displays and signal lamps are being produced in large quantities.

Gallium arsenide single crystals are used as substrates for vapor-phase epitaxy of the mixed crystal. It is necessary to slowly vary the composition; otherwise the abrupt change in the lattice constant introduces too much strain. The strain would be relieved by the formation of a dislocation network which in turn leads to inhomogeneities and precipitation effects. These lattice defects enhance nonradiative transitions which are of course most undesirable in a lamp structure. Therefore the epitaxy must be guided such that initially pure GaAs is deposited and gradually more phosphorus is introduced into the gas stream for the vapor phase epitaxy. The process of epitaxy is described in detail in Chapter 20. Doping can also be introduced by impurity admixtures into the gas. Thus the entire structure of the p-n junction in the mixed crystal of Ga (As,P) can be achieved in one continuous and automated process. Large-area slices result, which can then be processed into individual diodes or into more complicated structures, such as segments for alphanumeric displays. The technology for these optoelectronic devices is very interesting, but the scope of these lectures does not allow us to go into more detail.

Gallium phosphide represents the other approach to produce visible light in p-n junction devices. We have seen that GaP is an "indirect material", (which expression is an abbreviated jargon to indicate that the optical transition of lowest energy is indirect). Therefore one must utilize the action of impurities in order to get good efficiencies for optical transitions. Electrons and holes are swept by forward bias into the junction region in their respective bands. The carriers then lose some of their energy (which was supplied by the bias) by capture into localized impurity states. Excitons can also be formed in the junction region by the forward bias action. These excitons are quickly captured by attractive impurities. Finally these bound excitons decay radiatively. This mechanism of recombination via localized states can be very efficient, because bound excitons have a rather high probability of decaying radiatively. Although we are dealing with an indirect material and we are sacrificing some of the carriers' energies, the overall process of light emission competes favorably with the direct transitions in the mixed crystals we described before. GaP offers an added advantage: by proper choice of doping we can change the color of the emission. A very efficient red emission can be achieved, it results from excitonic decay at close pairs of zinc and oxygen in GaP. Green emission is obtained by nitrogen doping, we had earlier discussed that the isoelectronic impurity N in GaP can bind excitons. The efficiency here is smaller than in the case of the red emission, but the human eye is so much more sensitive in the green portion of the spectrum that the green solid-state lamps appear very useful.

The technology for GaP electroluminescent devices is more complicated than that of the Ga(As,P) epitaxy. It is difficult to grow large GaP crystals of high perfection as easily as GaAs. The high vapor pressure of GaP near the melting point is troublesome. A new type of growth technique is needed for large crystals. The volatile melt is protected by a liquid layer of an oxide to prevent decomposition. The added foreign material limits crystal perfection. Liquid phase epitaxy is then used to generate good GaP material on top of low-quality substrates. It should be clear from even this very superficial discussion that we are here facing a very interesting and demanding problem of complex solid-state technology. The reader is referred to a thorough and detailed discussion of all these questions, which can be found in a recent authoritative review by Bergh and Dean [9].

Junction Lasers

Coherent light can be generated if we achieve population inversion and have an optical resonator of sufficient quality. This is the basic idea of the laser. The semiconductor laser is one of the many forms of realizing this idea. Semiconductor lasers are small and can be pumped most easily, merely by application of a very low voltage ($\sim 2V$) in order to forward-bias a junction. The optical resonator is easily achieved also. Cleaving the crystal often suffices to produce well-defined end planes which act as mirrors to the light because of the high index of refraction of GaAs. Figure 9 gives a sketch of the semiconductor laser structure. Into n-type GaAs a zinc diffusion is made, which generates the junction. Forward bias causes radiative recombination as discussed before. The light travels along the directions indicated by the arrows in the figure. If there is population inversion this light will be amplified because it can stimulate further transitions resulting in more photons which are coherently added to the existing light. One speaks here of optical "gain". The details of these considerations can, for example, be found in the review article by M. H. Pilkuhn [10].

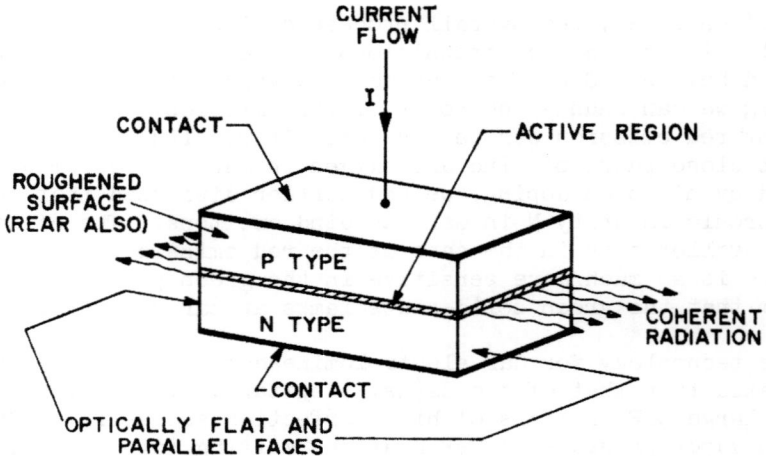

FIGURE 9. Schematic of the diode laser structure. Forward bias is applied as indicated. The arrows indicate the direction of the emitted coherent light.

If we use band-to-band transitions for laser action by large forward currents, then the laser condition of population inversion implies that the quasi-Fermi levels must be separated by more than the energy of the emitted light:

$$qV_a = q(\varphi_p - \varphi_n) > h\nu, \qquad (20)$$

which really means that the battery must at least provide the energy for the photons. From Eq. (20) we deduce that at least one of the sides of the junction must be degenerate: the doping must be so heavy that the Fermi level lies within the band. Only then will there be sufficiently many excess pairs to insure stimulated band-to-band optical transitions at $h\nu \approx E_g$.

Lasers of this sort have first been operated at low temperatures or only with short pulses of the high currents needed to generate coherent light. The optical gain must be large enough to overcome all loss mechanisms for the photons. There is therefore a threshold current density J_{th} above which laser action is seen, below the threshold only noncoherent spontaneous emission is observed. The threshold density is given by the formula of Lasher and Stern [1], with reflectance losses at the ends neglected:

$$J_{th} \propto \alpha N^2 \gamma d_o \Delta E / \eta \lambda^2 \qquad (21)$$

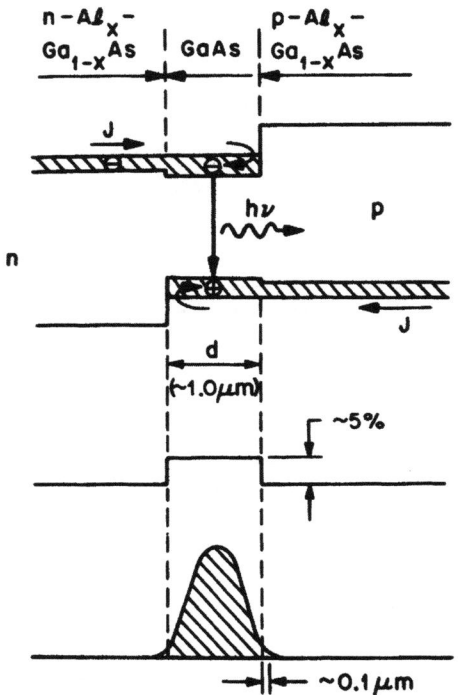

FIGURE 10. Double-hetero structure of the semiconductor laser (from Hayashi, Panish and coworkers [12]). Shows are: composition (x is typically 0.2 to 0.4); band scheme of junction at forward bias, refractive index steps at the heterojunctions; optical mode confinement. (top to bottom).

---

where N is the refractive index, d_0 the width of the active region where light is generated and amplified, ΔE is the spectral linewidth of the spontaneous emission, α is the internal loss of photons in the laser, η the internal quantum efficiency of converting a hole-electron pair to a photon, λ the wavelength of the radiation, and finally γ is a factor which describes the spread in the energy distribution of the injected carriers and effectively depends on d_0 again. Equation (21) implies a strong increase of the threshold with temperature. Mostly this is caused by the increasing spread of the carrier distribution with increasing temperature. The hotter the carriers are the farther they diffuse into the other part of the junction into which they are injected. This dilutes the carrier density and increases the lasing threshold. For a while it seemed therefore very difficult, if not hopeless, to obtain a semiconductor laser that operated with constant current - not pulsed - at room temperature.

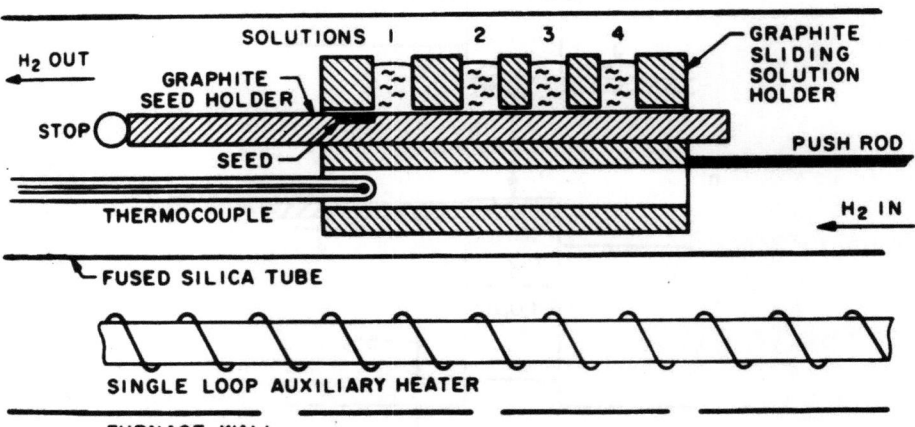

FIGURE 11. Sketch of the equipment used by Panish et al. for the liquid-phase epitaxy method to produce multi-layer hetero-structures for room temperature lasers [12].

Panish, Hayashi and their coworkers [12,13] overcame the problem by making a "double hetero-structure" as indicated by the schematic of Fig. 10. The main idea is to confine the injected electrons into as narrow a region d_o as possible. Heteroepitaxy from the liquid phase (see Fig. 11) is used to produce junctions between GaAs and (Ga,Al)As. The mixed crystal has a higher band gap than the pure GaAs. This fact helps to confine the radiation in the active zone, but even more important is the confinement of the injected carriers by the resulting potential barrier, which can be seen in Fig. 10. The efficiency is enhanced because less injected carriers are lost and the threshold density becomes so low that - with proper precautions to cool the device - one can make lasers that operate continuously at room temperature. This is a great achievement with obvious implications for opto-electronic uses. At present the most severe problem is still the lifetime of these junction lasers. The large currents and the high photon densities required apparently stress the structures so strongly that the devices quickly degrade. Much research is presently done in order to understand the physics and chemistry behind these degration effects.

One of the most promising uses for the junction lasers is in optoelectronic communications. One thinks of coupling junction lasers with glass fiber cables to obtain a communication link with very high bandwidth. The laser can be modulated to carry large amounts of information. Pulse-code modulation techniques appear most promising at present. The light of the GaAs emission is ideally suited to silicon detectors. Materials research is urgently needed in many areas connected with this project (see Chapters 11 and 13).

CONCLUSION

Our brief survey is an introduction into a fascinating field of interdisciplinary efforts in solid state research. The materials needed for the more and more complex principles of solid state devices have already reached a high level of sophistication. On the other hand the physical means of analyzing materials, especially by optical techniques, have increased tremendously. This interplay will prove to continue as a powerful approach in the fields of materials research and semiconductor devices.

REFERENCES

[1] W. Shockley, "Electrons and Holes in Semiconductors", D. van Nostrand, Princeton, 1950.
[2] S. M. Sze, "Physics of Semiconductor Devices", Wiley, New York, 1969.
[3] See the tutorial survey: H. J. Queisser in "Festkörperprobleme", Volume XI; O. Madelung, editor Pergamon-Vieweg, 1971, p. 45.
[4] D. F. Nelson et al., Phys. Rev. Lett. $\underline{17}$, 1262 (1966).
[5] For an introduction to the problem of isoelectronic traps, see D. G. Thomas in Proceedings of the Eighth Internat. Conf. on the Physics of Semiconductors, Kyoto (1966); J. Phys. Soc. Japan $\underline{21}$, Suppl. (1966), p. 265. A more recent review is by W. Czaja, same as [3], p. 65.
[6] D. G. Thomas et al., Phys. Rev. $\underline{133}$, A 269 (1964).
[7] D. G. Thomas, J. J. Hopfield, and W. M. Augustyniak, Phys. Rev. $\underline{140}$, A 202 (1965).
[8] T. N. Morgan and H. Maier, Phys. Rev. Lett. $\underline{27}$, 1200 (1971).
[9] A. A. Bergh and P. S. Dean, Proc. IEEE $\underline{60}$, 156 (1972).
[10] M. H. Pilkuhn, physica status solidi $\underline{25}$, 9 (1968).
[11] G. Lasher and F. Stern, Phys. Rev. $\underline{133}$, A 553 (1964).
[12] I. Hayashi, M. B. Panish, P. W. Foy, and S. Sumski, Appl. Phys. Letters $\underline{17}$, 109 (1970).
[13] A good introduction is the article "A new class of diode lasers" by M. B. Panish and I. Hayashi, Scientific American, July 1971, p. 32.

CHAPTER 3

COMPOUND SEMICONDUCTORS

C. Hilsum

Ministry of Defence, Royal Radar Establishment

Great Malvern Worcestershire, England

PHYSICS

Silicon is a remarkable material with many interesting properties. It can fulfil the requirements for most semiconductor devices, and is the material selected for the overwhelming majority of them. But, in fact, silicon is a rather poor semiconductor if judged on its physical properties alone, comparing unfavorably with germanium and gallium arsenide, as shown in Table 1.

TABLE 1. A Comparison of Ge, Si, and GaAs.

	Energy Gap (300°K)	Electron Mobility
Germanium	0.67 eV	4000 cm^2/vs
Silicon	1.12	1500
Gallium Arsenide	1.40	9000

Silicon owes its predominant position to its developed technology. Many millions of dollars have been spent on devising methods of producing silicon slices and silicon devices, and it is more economic to try variations of this technology to produce new devices rather than to exploit new materials. Fortunately for those who find silicon a rather dull semiconductor there are a number of functions which it cannot fulfil. The exploitation of new semiconductors is always fascinating, but it is rewarding economically as well if we seek out these fields in which silicon operates poorly, if at all. With this philosophy in mind, we begin our study of compound semiconductors by analyzing the ways in which they differ from silicon.

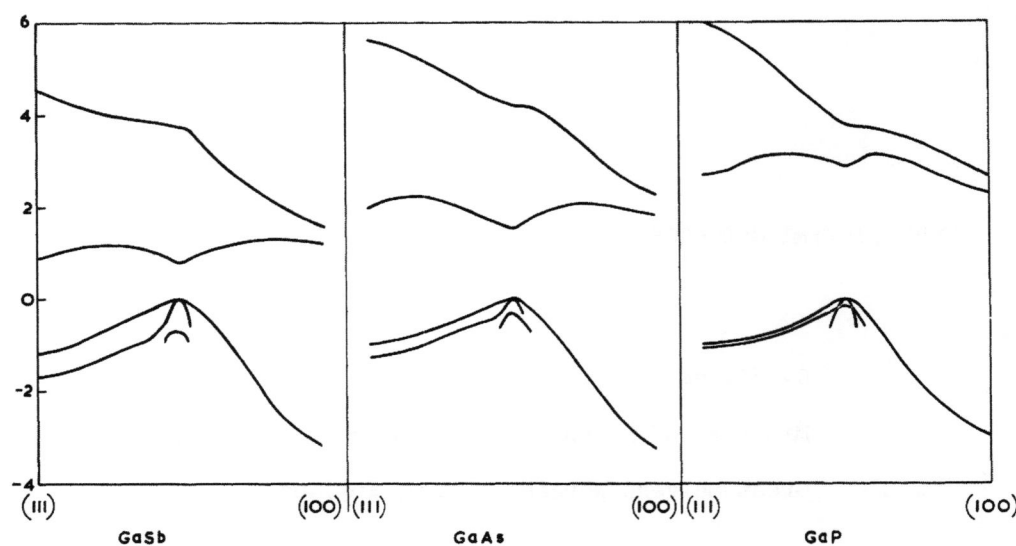

FIGURE 1. Band structure of gallium compounds.

Band Structure

We will restrict ourselves, for the moment, to semiconductors with the zinc-blende lattice, since their symmetry has many points in common with silicon. The first point to note is that the valence-band is the same shape for all zinc-blende materials, with a maximum at the center of the zone. The energy drop to the 111 edge, at the L point, is 1.1 eV, and to the 100 edge, at the X point is 3.3 eV. Band structure theory teaches us that the effective mass is roughly proportional to the <u>direct</u> gap for the particular minima we are considering [1]. It <u>follows</u> that in compounds with the conduction band minimum at the center of the zone the conduction band effective mass will be proportional to the energy gap. In indirect gap materials, and particularly those where the X minimum is the lowest, the mass will not vary much, since the direct gap which determines the mass will be so much larger than the indirect gap which characterizes the material.

As the mass of the atoms gets lighter, so the melting points rise, as do all characteristic energies. The conduction band moves away from the valence band, but the movement is different in different points of the Brillouin zone. The X points move least, the L points more, and the Γ points, at the center of the zone, move most. When all the separations are small the Γ is lowest, so the small gap compounds, made from heavy atoms, have direct transitions. The reduction in atomic number gives larger gaps, and an increasing tendency for the compound to become indirect (Fig. 1).

The choice in energy gaps and effective masses available to us is illustrated in Table 2. We should note the anomalously high mass of InP compared with GaAs.

TABLE 2. Energy Gaps and Effective Masses for III-V Compounds.

Compound	Energy Gap (eV)	Type of Transition	Effective Mass
InSb	0.18	Direct	0.014
InAs	0.34	Direct	0.022
GaSb	0.72	Direct	0.044
InP	1.33	Direct	0.078
GaAs	1.40	Direct	0.065
AlSb	1.6	Indirect	0.4
AlAs	2.2	Indirect	-
GaP	2.3	Indirect	0.35
AlP	2.4	Indirect	-

Scattering Mechanisms

In a perfect lattice the carrier motion is determined by its mass and the carrier-lattice interaction. In compounds this interaction is peculiarly effective because the lattice is made up from charged dipoles. For example, the free gallium atom normally has three electrons associated with it, and the arsenic atom five. When these atoms are brought together to form GaAs, there is electron exchange (Fig. 2). The electron cloud surrounding the molecule moves so that negative charge is passed to the gallium atom. The GaAs assembly now behaves as a dipole. We say that the compound shows "polar" scattering, and can express the strength of the scattering by an effective charge, the fraction of an electron which is transferred [2]. It is not wise to take this number too literally; there are eight atoms shared between the two atoms, and the precise ratio in which they are shared depends on the position chosen for an arbitrary boundary drawn between these atoms. There are at least three values for effective charge, which depend on the position of this boundary, and they are each defined in a mathematical relationship for one of the electrical or optical properties of the semiconductor. In Table 3 we tabulate the Callen effective charge,

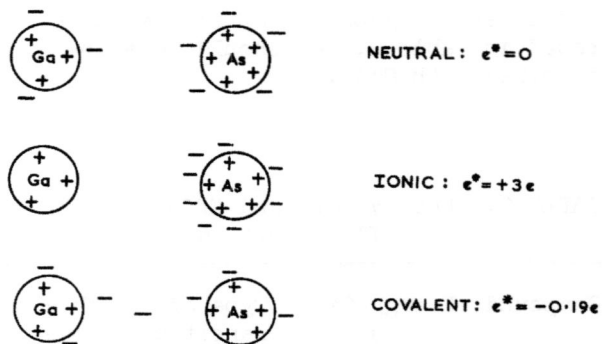

FIGURE 2. Schematic view of charge transfer and bonding.

which is the charge used to calculate electrical properties. It is possible from a knowledge of e* to calculate the polar mobility, using the effective masses from Table 2, and these are also shown in Table 3. Though polar scattering is dominant in most compounds, other scattering mechanisms cannot be completely forgotten. In Table 3 the effect of these mechanisms is included, and the resultant electron mobility tabulated [3].

TABLE 3. Predicted and Observed Electron Mobilities for III-V Compounds.

Compound	e^*_c	Calculated Mobility cm^2/vs	Observed Mobility cm^2/vs
InSb	0.16e	71,000	78,000
InAs	0.22	32,000	28,000
InP	0.27	5,500	5,200
GaSb	0.13	20,000	4,000
GaAs	0.19	8,500	8,900
GaP	0.24	180	200

We should emphasize at this point the differences between an elemental semiconductor like Si and a compound like InP or GaAs. They can be summarized in Table 4.

TABLE 4. Summary of Features of GaAs and Si.

	Mass	Transition	Scattering
Si	Large	Indirect	Acoustic and Optical
GaAs	Small	Direct	Polar

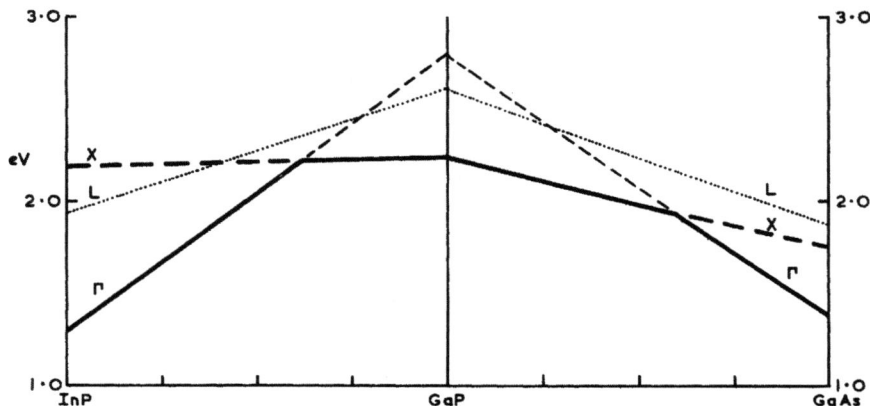

FIGURE 3. Variation of band structure with composition for In-GaP and GaAs-P.

It is of course possible to take other compounds where the differences are not as marked, but the Si-GaAs comparison is a good one, because it emphasizes the contrast. One point that cannot be made in such a comparison is the versatility of compounds. Energy gaps can be selected in the range from several electron volts down to a few tenths of an eV.

Mixed Crystals

The choice of properties made available by the use of compound semiconductors was broadened still further by the discovery that many compounds can be mixed in pairs to give properties intermediate between those of the two primary compounds. For example, a uniform material of composition $GaAs_xP_{1-x}$, with x controlled anywhere between 0 and 1, can be synthesized from the vapor phase. Some mixed crystals can be made by liquid phase epitaxy, and a few pulled from the melt. In general the properties of the mixed crystal can be derived from those of the compounds by assuming that <u>all</u> energy gaps vary linearly with composition. We must be cautious in our interpretation of this statement. Taking the system InGa-P, with an energy gap of 1.3 eV for InP and 2.3 eV for GaP, we do <u>not</u> deduce that the compound $In_{0.5}Ga_{0.5}P$ has a gap of 1.8 eV. Since all gaps vary linearly, the Γ gap changes from 1.3 eV to 2.8 eV, the X gap from 2.2 eV to 2.3 eV and the L gap from 1.9 eV to 2.7 eV. At the 50% point the Γ gap would be 2.05 eV, and the X gap 2.25 eV. The mixed crystal would have a direct energy gap, unlike GaP (Fig. 3).

This ability to mix crystals is of particular advantage in optical applications where one may wish to specify a wavelength accurately, as for example in a filter with a defined cut-off. The spot wavelength available from InSb, InAs, GaSb, InP and GaAs would not give us an efficient coverage of the infra-red region of the spectrum. Mixed crystals enable us to bridge the gaps. Similarly for electroluminescent devices we often prefer a direct-gap semiconductor, and for emission in the visible we need a gap larger than 1.9 eV. The compound with the largest direct gap is GaAs, but its energy gap is too small for this purpose, only 1.4 eV. Mixed crystals of either GaAs-P or InGa-P offer direct gaps of well over 1.9 eV.

There are two further points which should be noted about mixed crystals. First, not every combination of compounds mixes easily. Guidance in this is given by the phase diagram, with the separation of liquidus from solidus. The smaller the area enclosed by these two lines, the easier the system is to make. Examples of easy systems are InAs-P, GaAl-As and GaAs-P. Difficult systems are InSb-As and GaSb-P.

The second point is a caveat we should enter on the statement that gaps vary linearly with composition. This is rarely accurately so, in that there is usually some small curvature in the relationship. Occasionally there is a marked departure from linearity. For InAs-Sb the energy gap goes through a minimum at a composition near 50% [4].

II-VI, IV-VI and Other Compounds

Though the III-V compounds and mixed crystals offer a wealth of properties, there is little available at very low and very high energy gaps. The aluminum compounds, which have the largest energy gaps are difficult to make, and they decompose easily. II-VI compounds, such as ZnS, have been known for longer than any other semiconductor, because of their luminescent properties. We can learn much about their physics by analogy with the III-V compounds. Not all of the compounds have a zinc-blend structure, some being hexagonal. All have a similar valence-band structure to the III-V family, the maximum being at the zone center. The conduction band for all the well known materials has also a minimum at the zone center, and so this family as a whole tends to be direct gap. This is true even for compounds with a gap of 2.5 eV or more, unlike the III-V family. Mercury selenide and mercury telluride have very small gaps - in fact they are semi-metals. In themselves they offer little, but as mixed crystals with CdSe and CdTe they give systems with an energy gap readily controllable from a few tenths of an eV down to zero. It is in fact possible by varying the composition of Cd-HgTe to go smoothly down to zero and up again, with the conduction and valence-bands

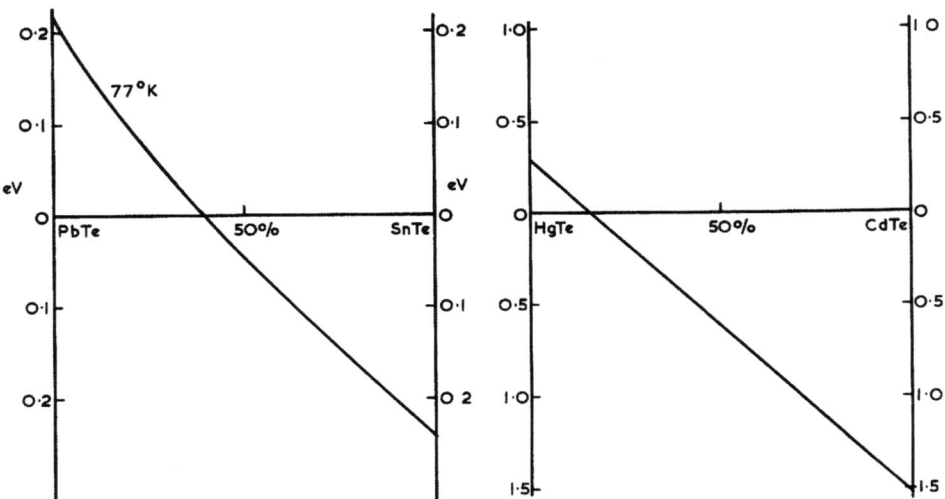

FIGURE 4. Variation of energy gap with composition for CdHgTe and PbSnTe.

crossing [5]. The II-VI compounds are more polar than the III-V materials. This tends to reduce their mobility, but makes them more interesting for studies of acousto-electric effects (Fig. 4).

IV-VI compounds, particularly the lead salts, were of greater importance about 20-30 years ago. Galena, PbS, was the best catswhisker rectifier for detectors in early radio receivers. Later, during and after the war, the lead chalcogenides found use as the polycrystalline layer in infra-red photocells. The lead salts are cubic, and their valence band is quite different from zinc-blend and hexagonal crystals. In fact they are direct-gap materials, but the gap is at the edge of the zone, with the result that the conduction band effective mass is large. This reacts in undesirable ways on their ability to compete in applications with materials like InSb and InAs, and they are studied little today. On the other hand the mixed crystal systems Pb-SnSe and Pb-SnTe are becoming more important, since like CdHgTe, their energy varies smoothly with composition from a few tenths of an eV to zero and up again [6].

Though gallium and aluminum nitride are III-V compounds, in many respects they behave more like II-VI compounds. They have a hexagonal crystal structure, and very large direct energy gaps. Both are difficult to make, and still wait full exploitation. Their known properties reflect the state of technology today rather than their full potentiality.

Silicon carbide [7] is an unusual compound which does not fit easily into the framework we have devised, since both elements come from the same column of the Periodic Table. It has a complex hexagonal structure with many polytypes, and is noticeable for its stability at high temperatures, and its hardness. SiC was the original material used for semiconductor lamps, and is still studied for this purpose in the Soviet Union. Two other uses are as high temperature rectifiers and cold cathodes, but neither offers a very exciting future for the material. It has an indirect energy gap near 2.5 eV, and its effective mass is high. Hence the mobility is only 500 cm^2/vs.

Ternary Compounds

Our list of compounds that have proved useful is now complete, but there still remains room for further exploration. The search has proved most useful in branching out symmetrically from the 4th Group of the Periodic Table, and we would like to follow this system to find more materials. The I-VII compounds are not semiconductors, and we now turn to ternary compounds.

Certain rules may be formulated which fit our choice of the III-V and II-VI families, and they may readily be broadened to cover ternaries [8]. The two rules are

1. The average number of valence electrons per atom should be 4.

2. The system should have normal valence ie the electrons given up by the cation to form a tetrahedral bond should be equal to the number of electrons required to form an octet on the anion.

From a binary compound $A_{(1-x)}B_x$ this gives two equations

$$A(1-x) + Bx = 4 \qquad (1)$$

$$A(1-x) = (8-B)x \qquad (2)$$

The first equation alone can be satisfied by compounds such as I-V_3, the second by compounds such as II_2-IV. Only I-VII, II-VI, and III-V satisfy both equations.

We broaden this by considering two classes of ternaries, those in which two elements share the role of cation, and those in which two act as anion. For both cases the first rule, on the average number of valence electrons, is the same.

We then have for $A_{1-x-y}B_xC_y$ the rules

$$A(1-x-y) + Bx + Cy = 4 \qquad (3)$$

__Dicationic__ $\qquad A(1-x-y) + Bx = (8-C)y \qquad (4a)$

__Dianionic__ $\qquad A(1-x-y) = (8-B)x + (8-C)y \qquad (4b)$

and deduce a series of ten families in Table 5.

TABLE 5. Ternary Analogues of 4th Group Elements.

Dicationic	Dianionic
I - III - VI$_2$	II$_2$ - V - VII
I - IV$_2$ - V$_3$	III$_3$ - IV$_2$ - VII
I$_2$ - IV - VI$_3$	II$_3$ - IV - VII$_2$
I$_3$ - V - VI$_4$	II$_4$ - III - VII$_3$
II - IV - V$_2$	III$_2$ - IV - VI

Very little is known about the dianionic materials, and they present obvious problems in manufacture. The I-III-VI$_2$ compounds, exemplified by silver thiogallate, mostly have the chalcopyrite structure, which is similar to zinc blend. A number are known to be semiconductors with quite interesting electrical properties, some of which are given in Table 6. However it is as optical materials that they are likely to prove most useful since they show pronounced non-linear optical effects. The I-IV$_2$-V$_3$ series, such as CuSi$_2$P$_3$, are mostly of zinc blend structure. This is also true of the I$_2$-IV-VI$_3$ compounds, such as Cu$_2$SiTe$_3$. Little more is known about them. Some I$_3$-V-VI$_4$ compounds occur naturally as minerals, for example Cu$_3$AsS$_4$ (enargite) and Cu$_3$SbS$_4$ (famatinate). The family more widely studied are the II-IV-V$_2$ compounds, the immediate analogues of III-V compounds. One general point is worthy of comment. As we synthesize compounds by moving away symmetrically from the IV Group, the energy gap increases. Ge has a gap of .8 eV, GaAs of 1.6 eV, ZnSe of 2.4 eV. However, the ternary analogue of GaAs, ZnGeAs$_2$ has a gap of only 0.8 eV, and the compound corresponding to ZnSe, CuGaSe$_2$ a gap of 1.7 eV.

TABLE 6. Electrical Properties of Some Ternary Compounds.

Compound	Energy Gap (eV)	Conductivity Type	Mobility cm^2/vs
$CdGeAs_2$	0.5	p,n	1000(n)
$CdSnP_2$	1.2	p,n	-
$ZnGeP_2$	2.2	-	-
$ZnSiP_2$	2.3	-	-
$AgInSe_2$	1.2	n	750
$AgInSe_2$	1.5	p,n	200(n)
$AgInS_2$	2.0	n	150
$AgGaS_2$	2.7	-	-
$CuAlS_2$	3.5	p	-

We shall see in the section dealing with applications that no economic outlet has yet been found for these semiconductors. It is unlikely that a random widening of the search will prove any more fruitful.

Doping Anomalies

No account of compound semiconductors would be complete without some mention of the problems of doping. The III-V compounds behave normally on the introduction of impurities, with one exception that will be mentioned later. Elements from the IInd group dope p type, and those from the VI group act as donors. The IV group elements behave in a more complex fashion since they can go on either lattice site. A good example of this is silicon which if introduced into the crystal at low temperature acts as an acceptor, while a high temperature doping gives donors. It is in fact possible to make a p-n junction in GaAs using Si as the sole dopant. There is no evidence in III-V compounds that departures from stoichiometry have pronounced electrical effects, or, to be more exact, that they can act as donors or acceptors. There is still perhaps a mystery about GaSb, which has never been made with less than 5×10^{16} acceptors/cm^3, even though it is reported to be much purer than this. The other compound families

can show large electrical effects due to excess or deficiency of
one component, and annealing in vacuum or a gaseous ambient is an
accepted treatment for many compounds. It is this readiness to
form defects which is probably responsible for our inability to
dope many II-VI compounds both n and p type. There appears to be
little problem if the energy gap is less than 1.6 eV, and CdTe
gives good p-n junctions. The larger gap II-VI compounds are
always n type, except for ZnTe, which is always p type. Any
attempt to introduce acceptors into ZnS, for instance, gives high-
resistivity material. It is as though the crystal automatically
forms defects which act as donors, and these compensate exactly.
The process is called "automatic compensation". A similar effect
occurs in GaN and in the ternary compounds. This effect has
hindered exploitation of II-VI compounds, because the p-n junction
is a most versatile device.

A different anomaly occurs in the doping of some III-V
compounds, but this has proved extremely useful. A GaAs crystal
that has less than about 5×10^{16} donors/cm^3 is readily converted to
a high-resistivity condition by doping with oxygen or chromium.
This material, though it contains over 10^{16} impurity centers/cm^3,
behaves in many ways like intrinsic GaAs, with a resistivity of
over 10^6 ohm-cm, and a carrier concentration less than 10^7 cm^{-3}.
In this form it is called "semi-insulating" GaAs. It is used as
a substrate for the deposition of epitaxial layers, and finds
application in certain device structures and for the characterization of material deposited simultaneously on n$^+$ substrates.
Similar effects occur in InP.

APPLICATION

The scene is now furnished for our requirements, and we are
able to define the fields in which it would be appropriate to
search for applications for compounds. We shall exploit the
properties which are lacking in silicon. The applications can be
conveniently classified under two headings, optical applications
and electrical applications.

Optical Applications

The main optical applications are as filters, photocells,
photo-emitters, modulators, and light emitting devices. In all of
these we exploit the wide range of energy gaps which compound
semiconductors make available to us, and an equally important
property, the direct nature of the conduction-valence band
transition. In other chapters light emitting devices are dealt
with at length - here we shall concentrate on photocells.

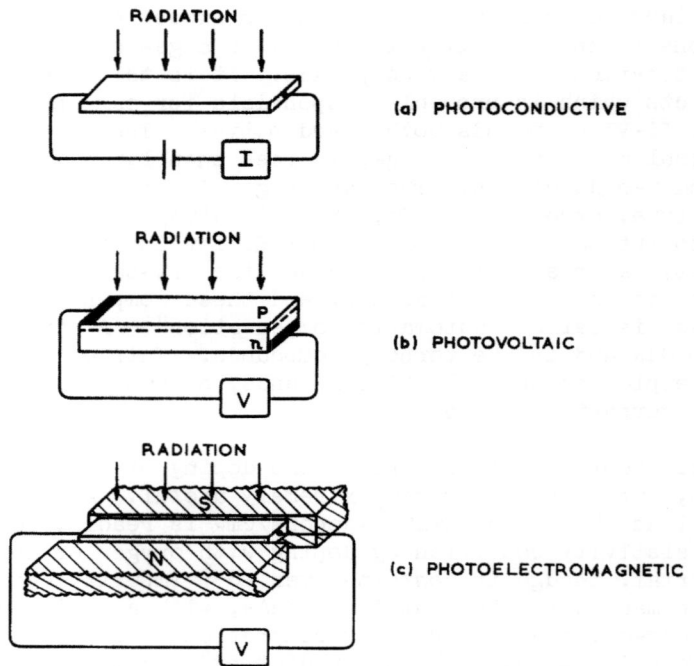

FIGURE 5. Photocell Modes of Operation.

(i) <u>Photo cells-modes of operation</u>. Semiconductors are sensitive to radiation because light creates electron-hole pairs, and these modify the electrical properties of the material. For intrinsic photoelectric effects, the radiation must be at a short enough wavelength so that the photons have energy greater than the energy-gap. A photocell will thus have a spectral response from very short wavelengths up to a cut-off wavelength, and beyond this cut-off its response drops rapidly to zero. At wavelengths much shorter than this the response will also suffer, but the decrease with wavelength decrease will be rather gradual. At energies much greater than the energy gap, E_G, the absorption coefficient becomes very high, and the radiation is absorbed in a very thin surface layer. Surface recombination will then act to reduce the sensitivity.

The simplest mode of operation of a photo-sensitive device is a photoconductive cell (Fig. 5a). This is a strip of uniformly doped material of arbitrary width and length chosen to suit the optical system, and of thickness not greater than the diffusion length of minority carriers. Ohmic contacts are fixed to the two ends, the exposed surfaces are treated to reduce surface recombination and, if ultimate sensitivity is needed, an anti-reflection coating is applied. There are some general principles which can be deduced for giving guidance in the choice of semiconductor.

There are three important parameters which express the sensitivity of a photocell. The first, the spectral response, has been mentioned above. The second, the responsivity, expressed in volts/watt, is a useful measure of the performance under large-signal conditions. The third, the detectivity, brings in the signal-noise ratio, and describes the ability of the cell to detect small signals. It is simple to deduce from first principles that the responsivity, R, for an n-type semiconductor is proportional to τ/n_o, τ being the carrier lifetime and n_o the equilibrium electron concentration. R is maximized by reducing n_o, but clearly it is not possible to reduce n_o below the intrinsic concentration n_i. It is obvious that R can be increased more if n_i is small. Since n_i increases rapidly as E_G falls, a good design will keep E_G high. There is then a direct relationship between cut-off wavelength and responsivity - long wavelength response brings poor sensitivity. The semiconductor must be chosen to limit the response to the shortest wavelength tolerable. Having done this, and established the maximum E_G, we can look to other ways of reducing n_i. Our knowledge of semiconductors teaches us that n_i is proportional to $m^{*3/2}$. We need to keep m^* small, while remembering that the energy gap is already defined. A direct gap at the zone center is helpful here, particularly in small gap compounds. The last thing we can do to reduce n_i is to cool the cell, though this is only worthwhile if $N_D < n_i$. A few more points are worth noting. The sensitivity of a photoconductive cell can be increased by increasing the voltage across it. For optimum performance the drive should be pushed up until excessive heating occurs. Naturally the cell is designed with an efficient heat sink, often made of anodized aluminum or copper, but nevertheless there remains a disadvantage in using small energy gaps, since the tolerable temperature rise is thereby reduced. Further, the simple discussion above considers only the direct effects of reducing n_o. In practice τ will usually decrease with n_o, so the effects of using a pure crystal will be still larger. The optimum sensitivity in most semiconductors is not found in intrinsic material. Provided the hole mobility is lower than the electron mobility, crystals with a slight excess of acceptors are the best.

Similar remarks may be made about detectivity, though the square-root dependence of noise voltage on n_o reduces the need to keep n_i low.

The second type of photocell used is a p-n junction, either grown or diffused (Fig. 5b). Such cells have the advantage that they operate with no bias voltage, and may be easily coupled into amplifiers. The most widely used photocell is the solar battery, always so far made from silicon. Silicon is not the optimum semiconductor, since its conversion efficiency is usually only 12 or 13%. Higher figures are theoretically possible using CdTe or GaAs, which match the sun's emission spectrum better, and quite recently a cell of GaAlAs was reported with an efficiency of 18% [9]. The exploitation of compounds here will be difficult, since nearly all solar cells are used on satellites, and weight is at a premium. The low density of silicon can be matched by only a few compounds, and these give a lower efficiency.

Aside from this particular application, the photo-voltaic mode merits comparison with the photoconductive. The principles governing the choice of material are similar. The responsivity now is proportional to the zero-voltage resistance of the junction, which in turn depends on the saturation current I_s in the current-voltage equation

$$I = I_s [\exp(qV/kT) - 1] \quad (5)$$

To improve the responsivity I_s must be kept low. We have

$$I_s \propto \left[\frac{1}{n_o} \sqrt{\frac{\mu_p}{\tau_p}} + \frac{1}{p_o} \sqrt{\frac{\mu_n}{\tau_n}} \right] n_i^2 \quad (6)$$

the first term depending on electron concentration, hole mobility and lifetime in the n region, and the second term corresponding for the p-region. The doping levels must be high in any case to give a good junction, but over-doping to reduce I_s may reduce the carrier lifetimes as well. Again there is a premium on keeping n_i low.

A third mode of operation of photocell is the photoelectro-magnetic mode (Fig. 5c). Here the crystal strip is mounted in the gap of a small permanent magnet, with a field near 10,000 oersted. Carriers created by the radiation in the front surface diffuse away towards the back, and they are separated by the magnetic field so that a voltage appears across the terminals. The PEM cell is useful at room temperature in materials with low energy gap, small carrier lifetime and high mobility, InSb being a particular example. Material doped slightly p-type is optimum [10], and in suitable semiconductors the sensitivity can be rather higher than the other two modes.

(ii) <u>Photocell materials</u>. For many years the lead chalcogenides, PbS, PbSe and PbTe have been widely used as infra-red photocells. They are not single crystal devices, and their preparation owes much to an empirical technology, designed to deposit a polycrystalline layer with photosensitive boundaries between the crystallites. Though these are photoconductive cells, in that their resistance changes on illumination, their mode of operation is probably more akin to a series-parallel connection of a series of biased photo-voltaic cells. The III-V compounds are now replacing the lead salts, and in particular InAs and InSb cover the infra-red to 5 microns. Room-temperature operation is possible, but the cells are often cooled to give higher sensitivity.

For still longer wavelength response mixed crystals of CdHgTe or PbSnTe are used. Similar photocells can be made from PbSnSe, but the system is metallurgically more difficult to synthesize. The energy gap of these two systems varies linearly with composition, since the end point compounds are direct gap. However the symmetry of the end points is different, and we can think of HgTe and SnTe as having "negative" energy gaps. In both systems the gap decreases steadily from the value characteristic of CdTe (1.6 eV at 77°K) or PbTe (0.18 eV at 77°K) through zero and up again. Both therefore are suitable for photocells sensitive in the far-infra-red, and both have been made in compositions suitable for response out to at least 30 microns [11].

Sensitive photocells which detect visible light are in great demand, and the best material for this appears to be CdS. In general one is interested in responsivity rather than detectivity, but in the larger gap materials there is no question of reducing N_D below n_i. Here one is concerned with all kinds of deep levels which act as traps for the carriers. Careful exploitation of these can raise the lifetime, and hence the sensitivity, though at the expense of response time. Single crystal CdS photocells are available, but more often polycrystalline layers are used.

This section has been devoted to intrinsic photosensitivity. However, it is a general result that any process whereby radiation disturbs the electron population can be used for detection. For detection of wavelengths beyond 50μ impurity absorption or free carrier absorption is feasible. The hosts most often used are Germanium and InSb, but the recent provision of extremely pure GaAs epitaxial layers has led to the developments of photocells with good infra-red response [12].

Electrical Applications

There are so many potential electrical applications for compounds that it is not sensible to list them. Instead we will select certain basic properties and show how they can be exploited. One striking feature of a number of compounds is their

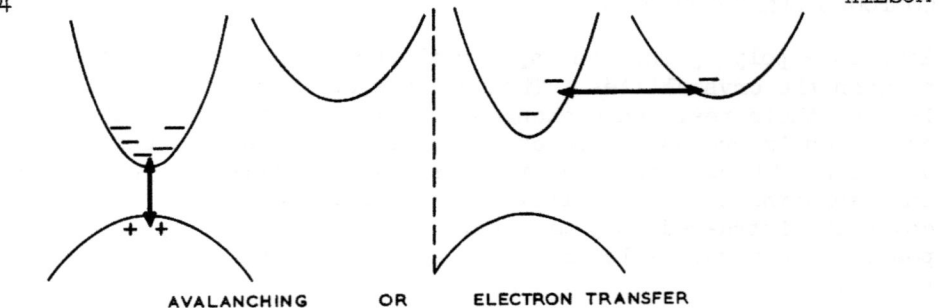

AVALANCHING   OR   ELECTRON TRANSFER

FIGURE 6. High electric field effects.

high electron mobility. This is used to good purpose in galvanomagnetic devices, based on either the Hall effect or magnetoresistance. The Hall effect is proportional to μ and magnetoresistance to μ^2. InAs and InAs-P are the preferred choice for Hall voltage generators because they combine a high mobility with a small dependence of carrier concentration on temperature. InSb is used for magnetoresistance devices, usually as a eutectic with NiSb. The eutectic crystallizes with long conducting needles in it, and these increase the magnetoresistive effect [13].

(i) <u>Microwave devices</u>. It might be thought that high mobility could be exploited directly in transistors intended for operation at high frequency. A unipolar transistor (or JFET) has a certain spacing between source and drain, and the frequency response is dominated by the transit-time of carriers. Suitable materials are Si and GaAs and one might think that since carriers in GaAs are five or six times more mobile, the gap could be scaled accordingly. The assumption in this argument is that the carrier velocity is proportional to the electric field, and this is only true for velocities up to about 10^6 cm/sec. When we attempt to drive carriers in Si faster than this they meet extra resistance from the lattice, and their incremental mobility falls. In direct gap compounds stranger things occur.

Since the conduction band mass at the center of the zone is low, the carriers can be accelerated to high velocities by easily attainable electric fields. As the field increases the carrier at first can give up the energy acquired by the field by interacting with the lattice, giving up a polar phonon. If the field continues to increase the energy acquired becomes more than the carrier can easily lose, so it retains some energy. It finds a new equilibrium at a higher potential energy - in other words it has risen in the conduction band. It is now a "hot" electron. Further field increases cause it to rise further and further, and one of two things happens. The carriers can ionize additional electron hole pairs, and the semiconductor breaks down, or the carriers transfer to other minima in the conduction band (Fig. 6). We must expand

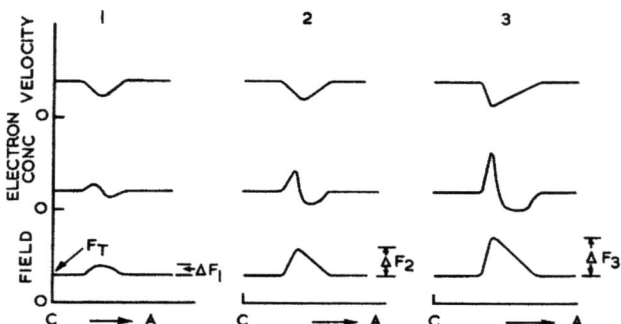

FIGURE 7. Three sequences in domain formation (C = cathode, A = anode).

this point. There are three main sets of minima in zinc-blend compounds, the Γ minimum, which is the lowest in energy in these direct-gap materials we are dealing with, the three X minima at the 100 edge of the zone, and the four L minima, at the 111 edge. Both these sets have higher effective mass than the Γ minimum. If we assume the X minima are lower than the L, and are energy Δ_{IV} above Γ, the material will readily avalanche if Δ_{IV} is greater than the energy gap, E_G. Should E_G be the greater the carrier can acquire an energy Δ_{IV} without avalanching occurring. The electrons now have available to them X states as well as Γ states. They will share their time between the available states, and the density of states in the three X minima is much greater than that in the single Γ minimum. At high fields nearly all the carriers will therefore transfer to the X minima. Here their mass is high, their mobility low, and their velocity low. The electron system has therefore moved from a situation where the velocity was high at low fields to one where it is low at higher fields. The current, which is proportional to the velocity, has therefore decreased as the voltage increased. This is a negative differential resistance (NDR).

In the prototype material for electron transfer GaAs, E_G is 1.4 eV and Δ_{IV} is 0.36 eV. The critical field at which negative resitance occurs is a little over 3000 volts/cm, a quite modest field [14]. The NDR reveals itself in a number of ways, depending on the quality of the semiconductor, the type of contacts, and the circuit in which the sample is placed. If the sample has ohmic contacts, random doping fluctuations and is in a resistive circuit, a dipole domain of high field will form at the largest doping fluctuation and then migrates to the anode. Subsequently the domain will nucleate at the cathode and move to the anode. The domain is first formed at a position where there is a slight accumulation of carriers adjacent to, and on the cathode side of a slight depletion (Fig. 7). By Poisson's Law the field in this

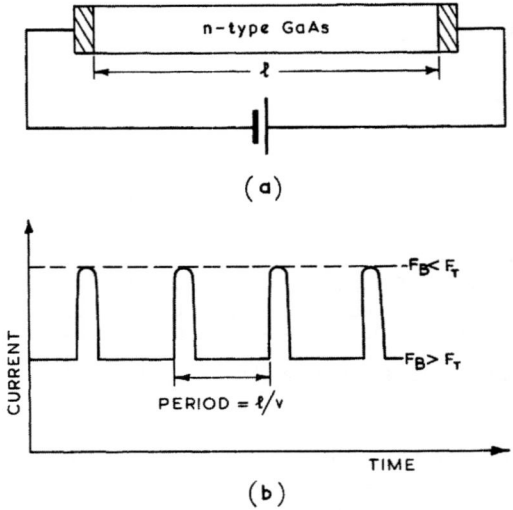

FIGURE 8. The GUNN effect.

region is slightly higher than that outside. If the whole sample is raised to the threshold field F_T, this region will be at a slightly higher field, $F_T + \Delta F$. The carriers in it therefore move more slowly than those in front and behind. The carriers behind will overtake, and build up the accumulation. Those in front will move away stretching and deepening the depletion. ΔF therefore increases and the domain grows. This growth requires more volts, and so the field outside drops well below threshold. The current drops too. An equilibrium is reached with the field in the domain F_D much higher than the field outside, F_R, with the detailed relationship between them determined by the precise form of the negative-resistance ie of the electron velocity-electric field curve. The fall in current due to domain formation can easily be a factor of two in samples a few mm long. The current-time curve is therefore a series of equally spaced spikes, the current being high momentarily when the domain is quenched at the anode and reforms at the cathode, but falling rapidly to its lower equilibrium value as the domain grows to its full height. The lower value is held constant as the domain moves from cathode to anode, so the spikes are separated by the electron transit time. This effect, the series of spikes appearing under DC drive, is called the GUNN effect [15] (Fig. 8). Since the velocity is about 10^7 cm/sec, a separation of 10 microns yields a transit time of 10^{-10} secs or a frequency in the microwave region. In fact if samples as short as this are used the situation becomes more complex, because few circuits are purely resistive at 10 GHz. Any reactive components will cause rf fields to be built up, because the rapid current spikes will interact with the circuit. In

COMPOUND SEMICONDUCTORS

addition other modes of instability are possible, pure accumulation layers or LSA modes [16]. This last mode is rarely, if ever, observed, but it is worth mentioning. If a sample is operated in a high Q circuit of a resonant frequency five to ten times higher than that corresponding to the transit time, the rapid RF swings of field can prevent domains ever building up. The RF field is sufficiently high to bring the resultant of DC±RF below F_T, so that any incipient domains collapse. The field across the sample swings up and down at the resonant frequency, and power is generated.

(ii) <u>Microwave materials</u>. The NDR in GaAs has led to simple microwave oscillators and amplifiers [17]. A source which is to work at X-band (10 GHz) would be 10 microns thick, and this should need only about 3 volts to work. This is exactly what happens though optimum power is generated by overdriving the voltage, using about 12V. The simplest way to construct the device is to deposit an epitaxial layer of appropriate thickness with about 10^{15} electrons/cm^3 on to an n$^+$ substrate. A lattice of silver-tin dots 100μ in diameter is evaporated on, through an exposed photoresist, and dice with one dot separated by scribing and breaking. Each dice is then encapsulated in a suitable microwave package. The diodes are variously called Gunn Diodes or transferred electron oscillators (TEO), and now form an integral part of many microwave systems. Their simplicity and cheapness makes them ideal for local oscillators at frequencies up to Q band (35 GHz), and they also are the primary source for low power portable radar sets, such as intruder alarms, which operate on the Doppler principle. CW Power levels of about 500 mW can be obtained from TEO's, and pulse powers of 100W-1000W, depending on the frequency. It has recently been suggested [18] that InP might prove a superior material to GaAs for TEO's, but the research on this is still at an early stage.

The avalanching process can also be used to generate microwave power, because the avalanching current acts as an inductive current component, which is in parallel with the junction capacitance. Silicon can be used, and is the material exploited commercially at present, but compounds may be more efficient. GaAs is being studied for this purpose, and it does show some advantages but most of the devices have a limited life.

GaAs is also used now for varactor diodes, mixers and detectors. It is quite feasible to build a microwave integrated circuit on semi-insulating GaAs, which incorporates several of the devices mentioned above. It can certainly be said that compound semiconductors are proving vital for obtaining the best performance from solid state microwave systems.

CONCLUSION

Exploitation of compound semiconductors has now proved so successful that few laboratories would need to justify the existence of a group specializing in this topic. We should, however, not forget that as more functions can be fulfilled by silicon and the growing number of established compounds, so the need for further exploration becomes less. There are still devices to be invented, but it is unlikely that a random search among exotic ternaries and quaternaries will yield results. The key to the wider use of more materials lies in the basic physics, and our ability to understand the processes which occur in semiconductors In this way we can appreciate the limitations of the compounds we have, and can point the way to the materials we need.

REFERENCES

[1] E. O. Kane, J. Phys. Chem. Solids 1, 249, 1957.
[2] C. Hilsum, Semiconductors and Semimetals, Vol. 1, Academic Press, New York, 1966, p. 1.
[3] D. L. Rode, Phys. Rev. B2, 1012, 1970, B3, 3287, 1971, Physica Stat. Sol. 53(b), 245, 1972.
[4] J. C. Woolley and J. Warner, Canad. J. Phys. 42, 1879, 1964.
[5] D. Long and J. C. Schmit, Semiconductors and Semimetals, Vol. 5, Academic Press, New York, 1970, p. 175.
[6] I. Melngailis and T. C. Harman, Semiconductors and Semimetals, Vol. 5, Academic Press, New York, 1970, p. 111.
[7] R. B. Campbell and Hung-Chi Chang, Semiconductors and Semimetals, Vol. 7B, Academic Press, New York, 1971, p. 625.
[8] A. S. Borshchevskii, N. A. Goryunova, F. P. Kesamly and D. N. Nasledov, Phys. Stat. Solidi. 21, 9, 1967.
[9] J. M. Woodall and H. J. Hoval, App. Phys. Letters 21, 1972.
[10] P. W. Kruse, Semiconductors and Semimetals, Vol. 5, Academic Press, New York, 1970, p. 15.
[11] C. Verie and J. Ayas, App. Phys. Lett. 10, 241, 1967.
[12] G. E. Stillman, C. M. Wolfe and J. O. Dimmock, GaAs and related compounds, Institute of Physics Conference Series No. 9, 1970, p. 212.
[13] H. Weiss, Semiconductors and Semimetals, Vol. 1, Academic Press, New York, 1966, p. 315.
[14] C. Hilsum, Proc. IRE 50, 185, 1962.
[15] J. B. Gunn, IBM, J. Res. Devel. 8, 141, 1964.
[16] J. A. Copeland, J. App. Phys. 38, 3096, 1967.
[17] Bulk-Effect Devices - IEEE Trans. on Electron Devices ED-13, January 1966, and ED-14, September 1967.
[18] C. Hilsum and H. D. Rees, Electronics Lett. 6, 277, 1970.

CHAPTER 4

HOT ELECTRONS IN III-V COMPOUND SEMICONDUCTORS

H. Pötzl

Lehrkanzel für Physikalische Elektronik der Technischen
Hochschule, and
Ludwig Boltzmann-Institut für Festkorperphysik
Vienna, Austria

INTRODUCTION

Hot electron effects in compound semiconductors are of great interest not only because of the physics involved but also because of the attractive applications in semiconductor devices. These applications are extensively reviewed by Dr. Hilsum (this volume). The present paper is intended to give an outline of the methods of numerical calculation which have led to the great progress in our knowledge of hot electron distribution functions during the last years.

First the relevant features of the electronic band structure and the electron-phonon interaction in III-V compounds will be briefly discussed. Then the Boltzmann transport equation will be presented. The Monte Carlo method and the iterative procedure for a numerical solution of this equation will be described in detail. Finally two examples will be given of the resulting hot electron distribution functions which refer to n-InSb at 77°K lattice temperature and to n-GaAs at 300°K lattice temperature.

Band Structure

The most important features of the band structure of several III-V compounds are shown in Fig. 1. The plots indicate the maximum of the valence band at Γ (center of Brillouin zone) and the three lowest minima (valleys) of the conduction band at Γ, X (boundary of Brillouin zone in 100-direction of k-space) and L (boundary of Brillouin zone in 111-direction). With the

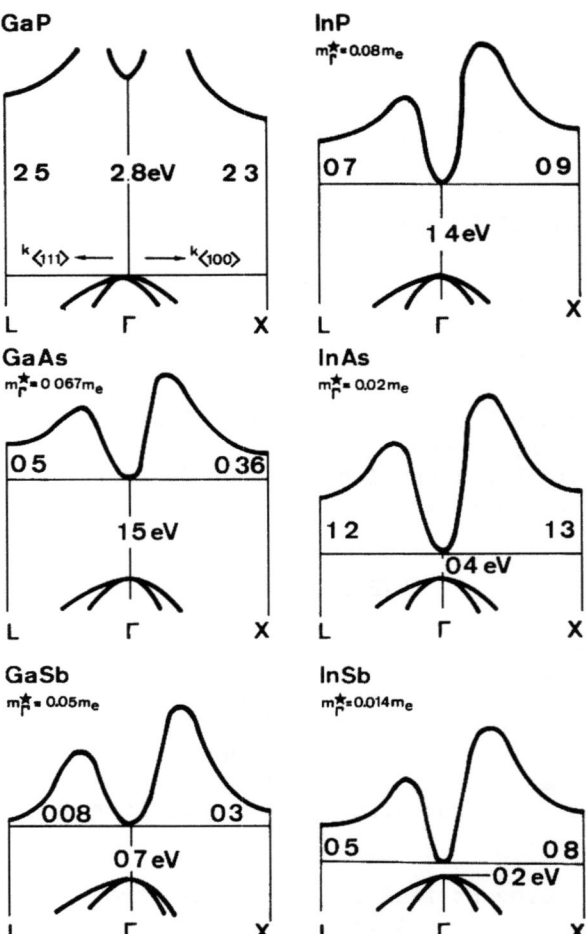

FIGURE 1. Energy band structure of several III-V compounds (the numbers indicate approximate energy differences in eV).

exception of GaP these compounds are direct semiconductors with the lowest valley at Γ. The large curvature of the band structure in the Γ - valley results in a low effective mass and corresponding high mobility of the conduction electrons. Hence the electrons acquire a high drift momentum in an electric field and there is strong energy transfer to the electron ensemble leading to pronounced hot electron effects. The Γ valley is spherically symmetrical. However, there is still an essential deviation from the simple relation

$$E = \frac{\hbar^2 k^2}{2m^*_\Gamma} \qquad (1)$$

between the energy E and the electronic crystal momentum $\hbar\vec{k}$ (m^*_Γ effective mass in the Γ valley): because of the quantum mechanical interaction between conduction and valence band states a non-parabolic band structure results which is well described by the Kane theory [1]. To a first approximation the energy-momentum relation is given by

$$E\left(1 + \frac{E}{\Delta E}\right) = \frac{\hbar^2 k^2}{2m^*_\Gamma} \qquad (2)$$

where ΔE is the width of the forbidden gap. This relation implies increasing effective mass at increasing energy. The overall effective mass therefore becomes dependent both on the lattice temperature and on the electric field strength. Furthermore it should be noted that the electronic wave functions are also affected by non-parabolicity the electronic states being represented by a mixture of "pure conduction band states" and "pure valence band states".

The minima or satellite valleys of L-type and X-type are characterized by a low effective mass and a high density of states. They are anisotropic, which fact in most investigations, is not taken into account because of the lack of knowledge of the band parameters.

In InAs and InSb the energy separations between the central and the upper valleys of the conduction band are larger than the width of the forbidden band. Therefore the energy which a hot electron needs for creating an electron-hole pair is smaller than the energy necessary for a transfer from the central to the satellite valleys. Hence carrier multiplication (avalanching) will occur at fields too low for any significant intervalley transfer.

The compound GaSb is characterized by an extremely small Γ-L energy separation. Because of the larger density of states in the L-valleys approximately half of the conduction electrons will populate these valleys at room temperature and two-band conduction can be observed even without heating the electrons. In electric fields of a few kV/cm additional electron transfer causes a variation of the slope of the current-voltage characteristic.

The Gunn effect takes place in GaAs and InP. By heating the electrons in a strong electric field a rapid transfer to the

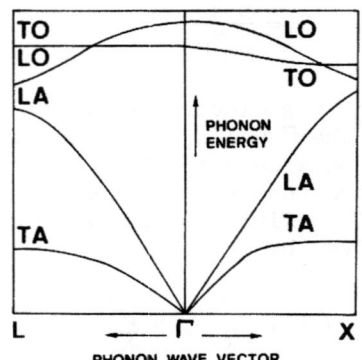

FIGURE 2. Schematic phonon dispersion diagram for GaAs.

satellite valleys is initiated above a certain threshold. The resulting increase of the overall effective mass leads to a decrease of the current at increasing electric field strength thus giving rise to a negative differential conductivity. This implies the Gunn instability which is used for microwave generation.

Scattering

Hot electron phenomena are determined by the interaction of the electrons with phonons. During the heating process energy is gained by the electrons from the electric field and transferred to the lattice by phonon emission. Intervalley transfer can only take place by emission or absorption of an "intervalley phonon" providing the corresponding change of the electronic crystal momentum $\hbar\vec{k}$. Therefore we have to discuss electron scattering by phonons.

A typical phonon dispersion diagram is shown schematically in Fig. 2. The branches denoted by L and T belong to longitudinal and transverse lattice vibrations, respectively. The transverse vibrations are doubly degenerate along the directions of high symmetry shown in the figure. Thus there are 6 branches three of which are called optical modes (LO, TO) and the others acoustic modes (LA, TA).

If the electron is scattered between states belonging to the same conduction band minimum the process is called intravalley scattering. Because of momentum conservation the corresponding intravalley phonons are characterized by a small wave number in the vicinity of the Γ-point in Fig. 2. There the acoustic phonons

have small energy and therefore acoustic intravalley scattering
is almost elastic. The energy of the optical phonons in this
region can be considered constant (independent of the wave number)
and given by $k_B\Theta$ where k_B is the Boltzmann constant and Θ the
characteristic temperature (Debye temperature). These temperatures
are in the range from 260°K to 500°K and therefore intravalley
scattering by optical phonons is very inelastic and, in fact,
provides the fundamental energy transfer mechanism between hot
Γ-valley electrons and the lattice.

The interaction between electrons and phonons can occur in
two ways. The change of the electronic band structure caused by
the lattice vibrations is the origin of the deformation potential
scattering. This is the only scattering mechanism in semiconductors like Ge and Si. In compound semiconductors, however,
the lattice exhibits an electric dipole moment which oscillates
due to the lattice vibrations. The oscillating dipole moment acts
on the electrons resulting in the absorption or emission of
phonons. This polar scattering is called piezoelectric if acoustic
phonons are involved. Polar scattering by longitudinal optical
phonons (po scattering), on the other hand, is the dominant
intravalley scattering mechanism in the Γ-valley of III-V compounds
at room temperature and therefore shall be discussed in detail
below.

Intervalley scattering occurs by the interaction with large
wave number phonons. The intervalley transfer from Γ to X for
instance which causes the Gunn effect in GaAs can only occur by
emission or absorption of a phonon which belongs to the vicinity
of the X-point (Fig. 2). It is seen that the characteristic
temperature of the intervalley phonons is in the same order of
magnitude as for intravalley optical phonons.

Scattering mechanisms can be conveniently discussed in terms
of the corresponding scattering rates $\lambda(k)$ which denote the
probability per unit time that a carrier with crystal momentum
$\hbar\vec{k}$ is scattered out of $\vec{k}$ to any final position in k-space by the
mechanism under consideration. For actual transport calculations,
however, it is also necessary to know the probability distribution
of the electron after the scattering event and therefore the
transition probability $S(\vec{k},\vec{k}')$ is used to which λ is simply
related by

$$\lambda(k) = \int S(\vec{k},\vec{k}')\, d^3k'. \qquad (3)$$

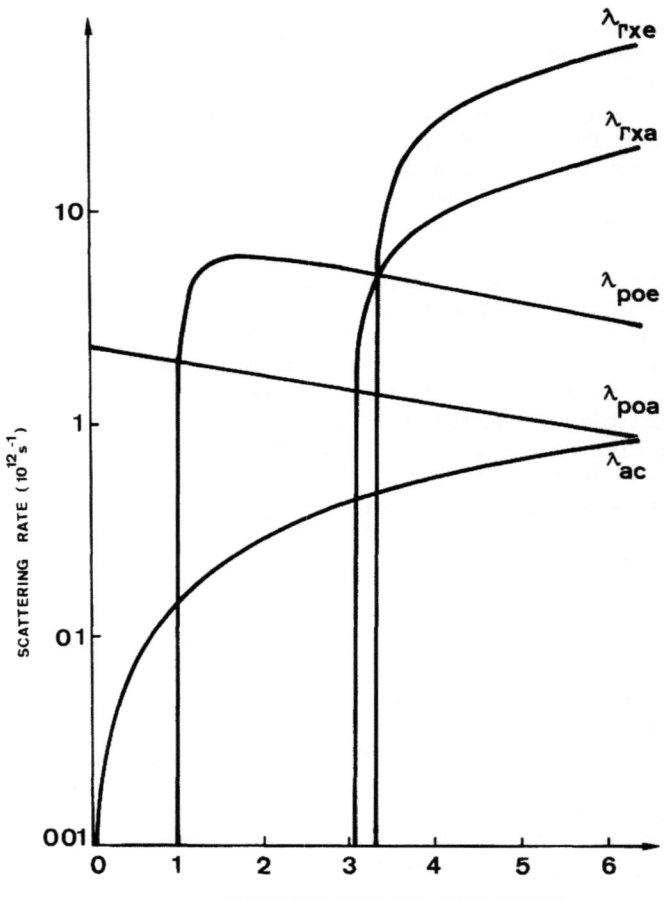

FIGURE 3. Schematic plot of scattering rates for the Γ-valley of GaAs as functions of normalized electron wave number - poe: polar optical emission; poa: polar optical absorption; ac: acoustic deformation potential; ΓXe: emission of intervalley phonons; ΓXa: absorption of intervalley phonons.

Figure 3 shows as a schematic example scattering rates for the Γ-valley of GaAs as functions of the normalized electron wave number

$$\sqrt{E/k_B\Theta} = \hbar k / \sqrt{2m^*\Gamma\, k_B\Theta}\ .$$

Intravalley scattering by polar emission (poe) and absorption (poa) of longitudinal optical phonons and by acoustic deformation potential scattering (ac) and intervalley scattering from Γ to X by emission (ΓXe) and absorption (ΓXa) of intervalley phonons are indicated.

At an electron energy below $k_B\Theta$ phonon emission is impossible and λ_{poe} vanishes. Absorption is strongly dependent on the lattice temperature but only weakly on the carrier energy in this range. Above $k_B\Theta$ there is first a steep increase of poe followed by a gradual decrease of both poe and poa. The latter leads to a very important consequence: if po scattering is the only effective mechanism there will exist a maximum electric field strength beyond which a stable electron distribution cannot exist. In reality there will be other mechanisms confining the carrier energy, primarily avalanche multiplication and intervalley transfer. It can be concluded that po scattering favours a rapid increase of the carrier energy beyond a certain threshold field and therefore assists the appearance of a pronounced Gunn effect.

Furthermore po scattering has a focusing effect on the electron momentum both at low and at high electric field strength. In the first case the electrons which are only weakly scattered by poa at low energies will take up energy from the field until they enter the region $E > k_B\Theta$. Then they will be rapidly scattered by poe and lose nearly all their energy and hence also their momentum. In taking up energy again they only gain momentum in the direction parallel to the field the electron distribution being "cooled down" in the transverse direction. In the high field case the carriers can enter the high energy region $E > k_B\Theta$. A strong anisotropy of po scattering becomes apparent in this region because of the form of the transition probability:

$$S(\vec{k},\vec{k}') \propto |\vec{k}-\vec{k}'|^{-2}. \tag{4}$$

Thus small angle scattering is strongly favored over large angle scattering and an electron can undergo many scattering events without significantly changing its direction in k-space. Between successive collisions the electric field will focus the electron more and more along the field direction [2].

This development is only interrupted by the few large angle scattering events which can bring the electrons into states in which they are decelerated rather than accelerated by the field leading to the almost complete loss of their energy.

From this discussion it becomes apparent that po scattering has a tendency to cause "streaming" electron distributions in k-space which are strongly peaked in the vicinity of the axis of cylindrical symmetry parallel to the electric field. This tendency contributes further to a rapid energy gain of the carriers beyond a certain threshold field. In GaAs these hot carriers will according to Fig. 3 enter a region where strong intervalley scattering sets in. The respective threshold energy is only slightly different for absorption and emission of an intervalley phonon because it is primarily due to the energy separation between the Γ-valley and the X-valleys ($\Delta_{\Gamma X}$).

Finally scattering by ionized impurities shall be briefly considered which is most effective in limiting the carrier mobility at low lattice temperatures. It is elastic and very anisotropic in that small angle scattering is strongly dominant. Because of this it is very ineffective in changing the electron momentum. It causes considerable difficulties in hot electron calculations because it gives rise to high scattering rates with small effect which raise the possibility of numerical errors. Fortunately its influence decreases at increasing carrier energy and therefore it is of minor importance at high electric fields.

The Boltzmann Transport Equation

The interaction of electrons and phonons in an electric field is described by the Boltzmann transport equation which, under the assumption of spatial uniformity, can be considered as a continuity equation in k-space. The electron density is described by a distribution function $f(\vec{k},t)$ which varies with time t because of the electric field $\vec{F}$ and the collision processes. The equation can be written as

$$\frac{\partial f}{\partial t} + \frac{e}{\hbar} \vec{F} \cdot \nabla_k f + \lambda(k)f - \int f(\vec{k}',t)S(\vec{k}',\vec{k})d^3k' = 0 \qquad (5)$$

The second term accounts for the change of the electron momentum by the electric field which according to the force equation is given by

$$\frac{d\vec{k}}{dt} = \frac{e}{\hbar} \vec{F}. \qquad (6)$$

This relation describes the collision-free trajectories of the electrons in k-space which are straight lines parallel to the electric field direction along which the electron moves at constant rate. The third term arises from collisions which scatter the electron out of $\vec{k}$ with the total scattering rate λ. The final term describes the rate of carriers scattered into k

HOT ELECTRONS IN III-V COMPOUND SEMICONDUCTORS

from any other part of k-space. It is this term which causes the greatest difficulties in the solution of the transport equation because it can only be evaluated if the distribution function (df) is already known in the whole k-space. Stated otherwise, by this term the transport equation becomes of integro-differential type rather than being a partial-differential equation.

In principle the df can be expanded in a series of Legendre polynomials of cos $(\sphericalangle (\vec{k},\vec{F}))$. If the df is not very anisotropic it will be sufficient to keep the first two terms. This procedure is known as diffusion approximation which is extensively reviewed in [3]. The "streaming" distributions of hot electrons in III-V compounds, however, lead to a poor convergence of this series. Therefore other methods have to be applied for the determination of the df. Different other approximations to the actual form of the df have been used some of which are reviewed in [4]. A decisive progress, however, is due to the development of two numerical methods for the solution of the Boltzmann equation.

The first one is a Monte Carlo procedure [5,6,7] which consists of a simulation of the electron motion in k-space on a computer. This motion consists of the collision-free trajectories according to (6) and of the scattering processes which cause a discontinuous change of the electron k-vector governed by the probability distribution $S(\vec{k},\vec{k}')$.

In order to carry out the simulation random numbers must be generated obeying the probability distribution of the times of free flight (without scattering). Furthermore, it is necessary to determine the scattering process which terminates the free flight. This is done by a second set of random numbers which select one of the scattering mechanisms according to their scattering rates $\lambda(k)$. Finally, random numbers are required which yield the position $\vec{k}'$ of the electron after the scattering event. The magnitude of $\vec{k}'$ is given by the energy conservation because the energy after the scattering is a unique function of the initial energy if the scattering process is specified. Hence it is sufficient to generate random numbers yielding the angle $\sphericalangle (\vec{k},\vec{k}')$.

The procedure is essentially simplified by the concept of self scattering which was first introduced by Rees [8]. This is an artificial non-physical scattering process by which the electron is not scattered at all but remains at its position in k-space. The rate of this process is now chosen in such a way that the overall scattering rate $\lambda(k)$ becomes a constant, say Γ, over the whole k-space. This implies the tremendous advantage that the times of free flight now obey the simple probability distribution

$$P(t) = \Gamma e^{-\Gamma t} \tag{7}$$

instead of being a complicated function of k.

To obtain a stationary df by the Monte Carlo technique the k-space is divided into discrete cells, and the time is recorded which the electron spends in each cell during the total simulation. Usually some ten thousand scattering processes must be simulated to obtain statistical convergence to the static distribution function. Any macroscopic mean value like drift velocity and mean energy can be obtained from this procedure.

The second successful numerical method for the solution of the Boltzmann equation is the Rees' iterative method [8,9] which is closely related to the path variable method by Budd [10]. The basis of the iteration procedure is the calculation of the carriers scattered into the region under consideration - as given by the last term in (5) - from the df obtained by the preceding step of the iteration. Stated otherwise: If we have arrived at $f_n(\vec{k})$ by n iterations aimed at the determination of a stationary df we use $f_n(\vec{k})$ to calculate the last term in (5):

$$\int f_n(\vec{k}') \, S(\vec{k}',\vec{k}) \, d^3k' = g_{n+1}(\vec{k}) \tag{8}$$

If this function is known the Boltzmann equation reduces to a pure differential equation which can be solved in a straight forward manner to obtain the n+1 iteration $f_{n+1}(\vec{k})$. Thus the integro-differential form of the Boltzmann equation is avoided.

Again an essential simplification is achieved by the introduction of self scattering with a scattering rate $\Gamma - \lambda(k)$. In this way the whole iteration procedure can be expressed by the following pair of formulas

$$g_{n+1}(\vec{k}) = \int f_n(\vec{k}')S(\vec{k}',\vec{k})d^3k' + \{\Gamma - \lambda(k)\} f_n(\vec{k}), \tag{9}$$

$$f_{n+1}(\vec{k}) = \int g_{n+1}(\vec{k} - e\vec{F}t'/\hbar)e^{-\Gamma t'} dt'. \tag{10}$$

Two different coordinate systems are best suited for the two iteration steps. Spherical polar coordinates are adequate for (9) because of the simple form of energy conservation in this system relating the magnitudes of $\vec{k}$ and $\vec{k}'$. On the other hand, cylindrical polar coordinates are best suited for (10) because the collision free motion according to (6) yields straight lines parallel to the axis of this coordinate system. Therefore spherical polar to cylindrical polar coordinate changes and vice versa will be carried out between any two steps.

A great advantage of the iterative method is the following: If Γ is chosen sufficiently large the time response of the df to a step in the electric field can be directly obtained. Assume that at the time $t_o = 0$ the electric field undergoes a sudden change from F_1 to F_2. Then the stationary df belonging to F_1 is taken as the initial df $f_o(\vec{k})$ and the successive iteration steps yield $f_1(\vec{k})$, $f_2(\vec{k})$,..., $f_n(\vec{k})$,... It can be shown that these functions are identical with the df at $t_1=1/\Gamma$, $t_2=2/\Gamma$,..., $t_n=n/\Gamma$,.... So the whole development of the df with time caused by the field step is obtained. Furthermore, the response to a small field step can be Laplace-transformed to yield the small-signal frequency response of the electron ensemble to a small a.c. field superimposed on a large d.c. field. Results of this type can be directly compared to experiments in which the small signal a.c. conductivity of electrons heated by a large d.c. field is measured [14,16].

Hot Electron Distribution Functions

A full account of the tremendous amount of information obtained by use of the above methods is beyond the scope of this introductory lecture. The advanced reader is referred to [6,8,9,11,12]. An excellent review has been presented recently by Fawcett [13]. We shall restrict ourselves here to examples showing some features of hot electron distribution functions.

A convenient test material for hot electron investigations is n-InSb. The high electron mobility implies considerable heating at comparatively low electric field. Figure 4 shows a set of df along the axis of cylindrical symmetry calculated by the iterative method [14]. At zero electric field strength the well known Maxwell-Boltzmann distribution appears. At 25 V/cm the maximum of the distribution is higher as a consequence of transverse "cooling" and a sharp edge appears caused by the onset of polar optical phonon emission. The change of slope indicates that the df at lower energy is "heated" more than at higher energies because the emitted optical phonons provide an effective energy transfer to the lattice. The edge becomes even more pronounced at higher fields indicating the increasing influence of emission processes. It is evident from these curves that a Maxwell-Boltzmann distribution shifted in momentum space according to the drift velocity is a poor approximation to the actual shape of the distribution function. It also indicates that a "two temperature model" seems reasonable describing the df by two Maxwellians with differing electron temperatures which

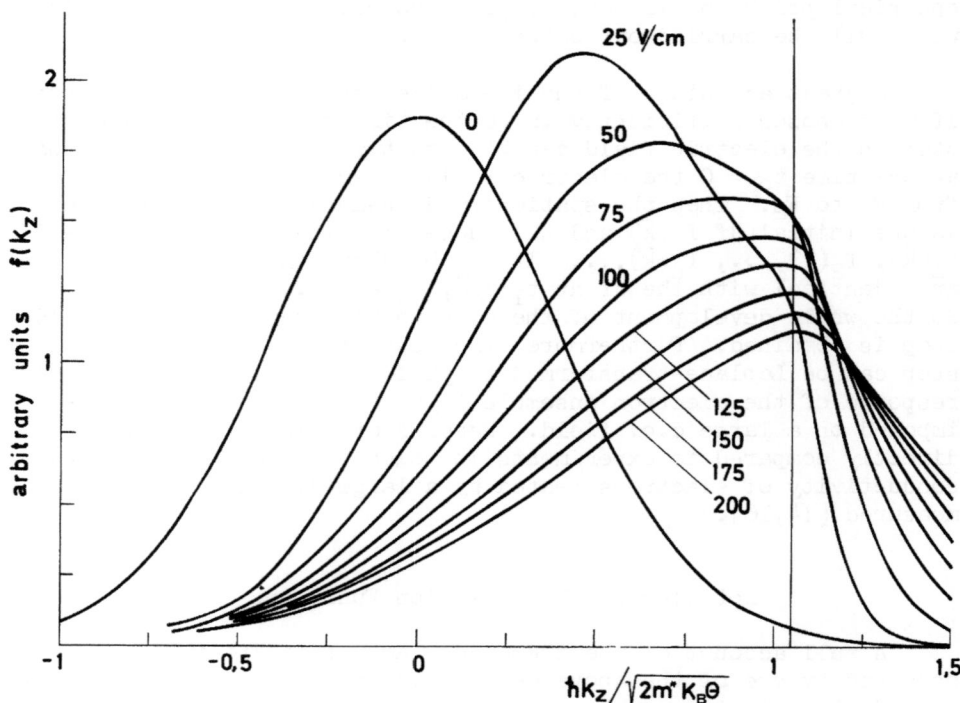

FIGURE 4. Distribution functions at different values of the electric field strength plotted along the axis of cylindrical symmetry as a function of $\hbar k_z/\sqrt{2m^* k_B \Theta}$ (n-InSb, 77°K lattice temperature, parabolic band structure, $\Theta = 260°K$).

---

intersect at the optical phonon energy [15]. Above 80 V/cm the maximum stays at the same place at which phonon emission sets in. It is typical that the maximum of the df is found at a boundary in k-space at which there is a transition from weak to strong scattering. At increasing electric field strength an increasing number of electrons is seen to escape the polar scattering mechanism and to reach high energies. These will cause electron-hole pair production initiating avalanche breakdown which actually is known to occur above 250 V/cm (the simple parabolic band structure used in these calculations overestimates the high energy tail of the df).

From the stationary df the drift velocity is readily obtained as a function of electric field strength as shown in Fig. 5. Curve (a) corresponds to the simplified model neglecting the non-parabolicity of the band structure which was included in the calculation of curve (b). Non-parabolicity is seen to have

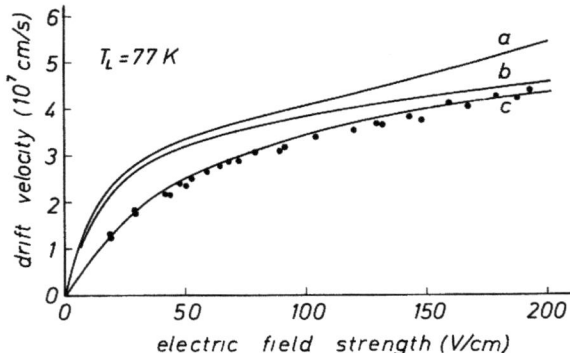

FIGURE 5. Drift velocity vs. electric field strength (n-InSb, 77 K): a) $m^* = 0.014\ m_o$, parabolic band structure, no impurity scattering; b) $m^* = 0.012\ m_o$, non-parabolic band structure, no impurity scattering; c) $m^* = 0.012\ m_o$, non-parabolic band structure, $n = 10^{14}\ cm^{-3}$ (electron concentration), $N_I = 4 \times 10^{14}\ cm^{-3}$ (ionized impurity concentration) full circles: Experimental data by Bonek [16].

---

increasing influence with increasing electric field strength. Curve (c) includes also impurity scattering and compares nicely with experimental data [16]. The general shape of the curves is characterized by a decrease of the differential mobility (slope of curves) with increasing electric field indicating the increasing influence of optical phonon emission.

For practical applications hot electron effects in InSb are of minor importance. Interest concentrates on the Gunn-effect materials GaAs and InP because the velocity-field characteristics exhibit negative differential conductivity. Therefore we shall conclude with a discussion of hot electron distribution functions in the Γ-valley of n-GaAs at 300°K lattice temperature as obtained by Hillbrand [17] from Monte Carlo calculations assuming the simple parabolic band structure.

In Fig. 6 the df is plotted along an axis which is parallel to the electric field direction and lies in the vicinity of the cylindrical symmetry axis. At an electric field strength 2 kV/cm the only structure of the df is a very slight discontinuity at the onset of poe. It is much less pronounced than in Fig. 5 because the Debye temperature $\Theta = 420°K$ (36 meV) is not so much different from the lattice temperature as in the former case and therefore poa is not weak compared to poe. At higher fields the maximum of the df stays close to the position where poe sets in. However, a pronounced edge becomes apparent which is caused by

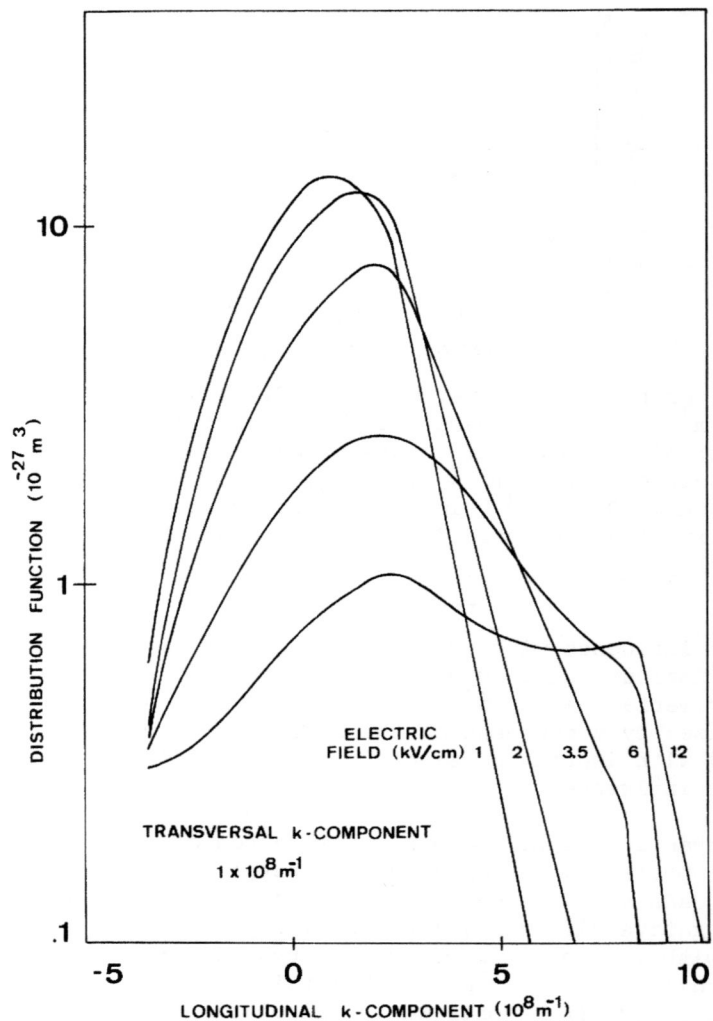

FIGURE 6. Electron distribution functions in the Γ-valley of GaAs plotted along an axis parallel to the electric field direction at constant transverse k-vector component 1×10^{-8} m^{-1}. Lattice temperature T=300°K, non-equivalent intervalley coupling constant 0.5×10^9 eV/cm, parabolic band structure.

---

the onset of intervalley transfer. Because of this strong scattering mechanism the df is cooled down beyond this edge which corresponds to the Γ-X energy separation $\Delta_{\Gamma X}$ = 0.36 eV. The electrons in this high energy tail are frequently scattered between Γ and X. In the X valleys the electrons are not

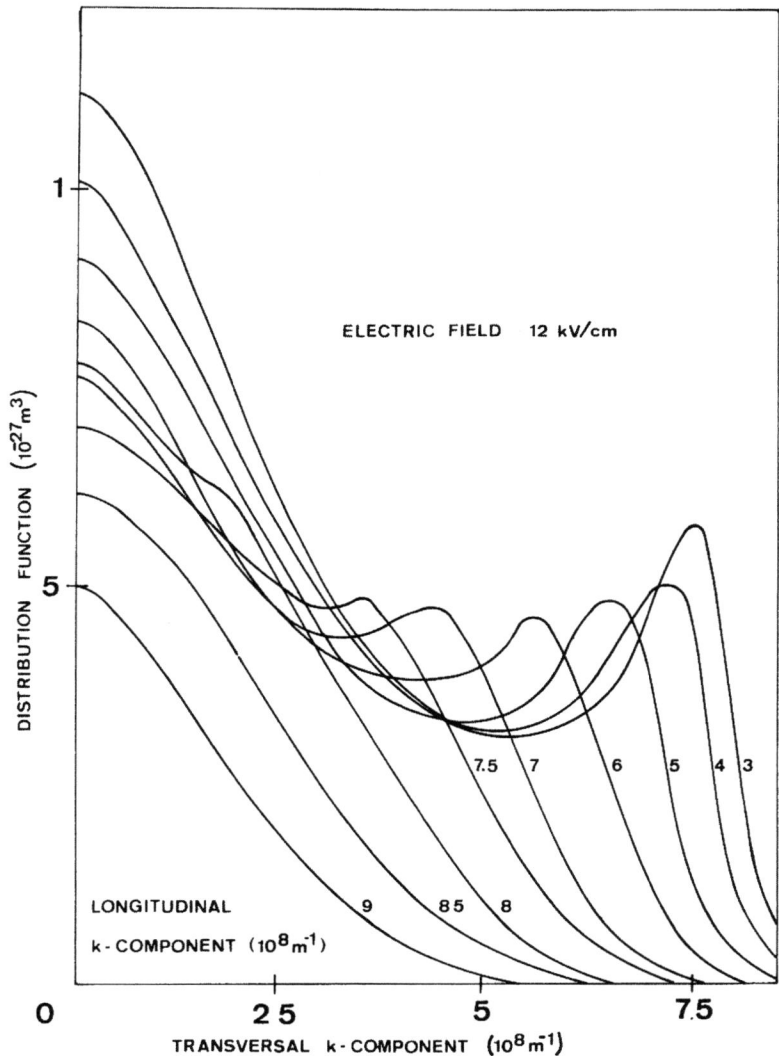

FIGURE 7. Electron distribution functions in the Γ-valley of GaAs plotted along different straight lines transverse to the field direction at a constant field strength of 12 kV/cm.

noticeably heated by the electric field because of their large effective mass. Therefore the X→Γ transitions cause a "cool" tail of the df in the Γ valley. Although there is some similarity between this intervalley scattering and the poe some important differences have to be noted. At an X→Γ transition the electrons change their energy only by a small fraction (roughly 1/10).

Their momentum, however, is completely randomized. Therefore they are homogeneously distributed over a spherical shell in k-space in the neighborhood of the sphere defined by $E = \Delta_{\Gamma X}$. This results in an increase of the electron population with an energy close to $\Delta_{\Gamma X}$ which leads to a "population inversion" i.e. a larger population at higher energies than at lower energies. The subsidiary maximum at 12 kV/cm in Fig. 6 arises in this way (this maximum might, however, be due to the assumption of the parabolic band structure c.f. [6]).

In Fig. 7 the df is plotted along different straight lines in momentum space which are perpendicular to the electric field direction. The electric field strength is kept constant at 12 kV/cm. The transverse population inversion which appears very pronounced in these figures is also confirmed by the rigorous non-parabolic calculations [6]. Again these regions of high population are fed by the $X \rightarrow \Gamma$ intervalley transfer. The streaming of the carriers parallel to the electric field after the transfer assists the build up the inversion in those parts of the sphere $E = \Delta_{\Gamma X}$ where the field direction is tangential to the sphere. Therefore the inversion is most pronounced in the curves with small longitudinal electron momentum (for instance 3×10^{-8} m^{-1} longitudinal k-vector component in Fig. 7).

The rather complicated structures of hot electron df demonstrated in this section are essential for a quantitative treatment of the Gunn effect. It has been discussed in order to show that simplified assumptions concerning the df cannot lead to satisfactory results and therefore the powerful numerical methods have to be used which were outlined in section 3, beginning on p. 96.

ACKNOWLEDGMENT

Thanks are due to Dr. H. Hillbrand for providing his unpublished results.

REFERENCES

[1] E. O. Kane, J. Phys. Chem. Solids 1, 249 (1957).
[2] W. P. Dumke, Phys. Rev. 167, 783 (1968).
[3] M. Asche and O. G. Sarbei, Phys. Stat. Sol. 33, 9 (1969).
[4] E. Conwell, High Field Transport in Semiconductors, Academic Press 1967.
[5] T. Kurosawa, Proc. Int. Conf. Phys. Semiconductors, Kyoto, 424, (1966).
[6] W. Fawcett, A. D. Boardman and S. Swain, J. Phys. Chem. Solids 31, 1963 (1970).

[7]   A. D. Boardman, W. Fawcett and J. G. Ruch, Phys. Stat. Sol. (a) 4, 133 (1971).
[8]   H. D. Rees, J. Phys. Chem. Solids 30, 643 (1969).
[9]   H. D. Rees, IBM J. Res. Develop. 13, 537 (1969).
[10] H. Budd, Proc. Int. Conf. Phys. Semiconductors, Kyoto, 420, (1966).
[11] A. Alberigi Quaranta, C. Jacoboni and G. Ottoviani, Rivista del Nuovo Cimento, Serie 2, 1, 445 (1971).
[12] C. Hilsum, Hot Electrons in Compound Semiconductors, Proc. Int. Conf. Phys. Semiconductors, Warsaw, Vol. 1, 585 (1972).
[13] W. Fawcett, Non-Ohmic Transport in Semiconductors, Winter College on Electrons in Crystalline Solids 1972, International Centre for Theoretical Physics, Miramare P.O.B. 586, 34100 Trieste (Italy).
[14] D. Kranzer, H. Hillbrand, H. Pötzl and O. Zimmerl, Acta. Phys. Austr. 35, 110 (1972).
[15] G. Persky and D. J. Bartelink, IBM J. Res. Develop. 13, 607 (1969).
[16] E. Bonek, J. Appl. Phys. 43, 5101 (1972).
[17] H. Hillbrand, unpublished results.

CHAPTER 5

RELAXATION TIMES IN SEMICONDUCTORS

K. Seeger

Institut für Angewandte Physik der Universität
and
Ludwig Boltzmann-Institut für Festkörperphysik
Vienna, Austria

There are various relaxation processes in semiconductors, and probably the most important one is the relaxation of drift momentum, mv_d, which is the product of the effective mass m and the drift velocity v_d. Its change with time is due to the electric force eE_z (we will not consider here magnetic forces or temperature gradients) and collision processes with a probability $1/\bar{\tau}_m$:

$$\frac{d(mv_d)}{dt} = eE_z - \frac{mv_d}{\bar{\tau}_m} \tag{1}$$

We have introduced a Cartesian coordinate system with the field in the z direction. At equilibrium the equation of motion [1] is solved for v_d:

$$v_d = \frac{1}{n}\int_{-\infty}^{\infty} v_z f(v) d^3v = \frac{e}{m}\bar{\tau}_m E_z = \mu E_z \tag{2}$$

where $\mu = (e/m)\bar{\tau}_m$ is the mobility; $\bar{\tau}_m$ is the momentum relaxation time, averaged over the distribution of the carriers:

$$\frac{1}{\bar{\tau}_m} = \frac{1}{n \cdot v_d}\int_{-\infty}^{\infty} \frac{1}{\tau_m(v)} v_z f(v) d^3v \tag{3}$$

n is the total carrier concentration given by

$$n = \int_{-\infty}^{\infty} f(v) d^3v \qquad (4)$$

Equation (3) can be shown to derive from the Boltzmann equation in the relaxation time approximation for the simple model of the band structure [1].

Similarly the energy balance equation

$$\frac{d\langle\varepsilon\rangle}{dt} = eE_z v_d - \frac{\langle\varepsilon\rangle - \varepsilon_L}{\tau_\varepsilon}$$

is obtained for the average kinetic energy of a carrier

$$\langle\varepsilon\rangle = \frac{1}{n} \int_{-\infty}^{\infty} \varepsilon(v) f(v) d^3v; \quad \varepsilon(v) = \frac{m}{2}(v_x^2 + v_y^2 + v_z^2)$$

where ε_L is the energy in thermal equilibrium with the crystal lattice, which for a nondegenerate electron gas is $\frac{3}{2} k_B T$, and τ_ε is the "energy relaxation time" given by

$$-\left(\frac{\partial\langle\varepsilon\rangle}{\partial t}\right)_{coll} = -\frac{\langle\varepsilon\rangle - \varepsilon_L}{\tau_\varepsilon} = -\frac{1}{n}\int_{-\infty}^{\infty} \varepsilon(v) \frac{f(v) - f_o(v)}{\tau_m(v)} d^3v \qquad (7)$$

f_o is the zero-field velocity distribution.

In order to illustrate the physical meaning of the two relaxation times let us consider the case that a dc field applied to the sample for a long time is switched off. This is illustrated by Fig. 1 for the case that $\tau_\varepsilon > \tau_m$. We notice that the current continues to flow for a time τ_m which is typically of the order of magnitude 10^{-13} sec. By then, collisions with lattice vibrations or other crystal imperfections such as impurities have randomly distributed momentum. But still the carrier gas is "hot". We could introduce an "electron temperature" T_e and a distribution function

$$f(v) \propto \exp(-\varepsilon(v)/k_B T_e) \qquad (8)$$

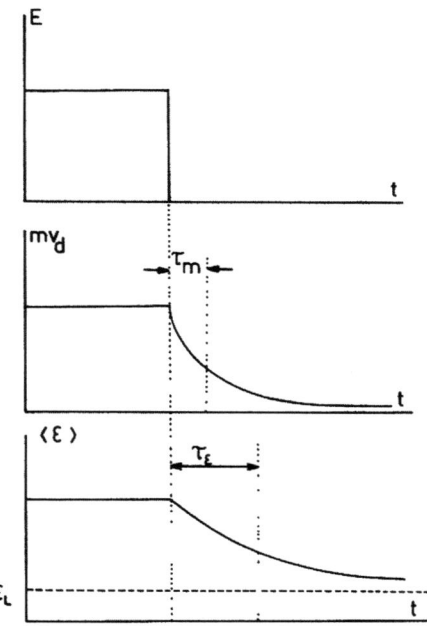

FIGURE 1. Decrease of drift momentum and average kinetic energy of a carrier after switching off the electric field which had been applied for a long while, illustrating the momentum and energy relaxation times, τ_m and τ_ε, respectively.

---

where $T_e > T$, the lattice temperature. Not until after a time τ_ε which is typically of the order of 10^{-12} sec will the average carrier energy $\langle \varepsilon \rangle = \frac{3}{2} k_B T_e$ return to its equilibrium value $\varepsilon_L = \frac{3}{2} k_B T$. We notice that τ_ε is of interest to hot-carrier phenomena [2,3] such as the Gunn effect and its high-frequency limit [4].

In this respect we also have to deal with another relaxation time which in the many-valley model of the band structure required for Gunn-effect semiconductors describes the intervalley transitions. This is the "intervalley relaxation time" τ_i. Its reciprocal, $1/\tau_i$, is the probability for a carrier to be scattered out of one valley into another by either absorption or emission of a phonon of large momentum $\sim (\pi/a)\hbar$ called an "intervalley phonon" (a = dimension of the unit cell). In n-type Ge at 85°K calculated values of τ_i are of the same order of magnitude as τ_ε [5] as shown in Fig. 2.

Still another relaxation time which is of importance in position dependent phenomena such as Gunn domains is the

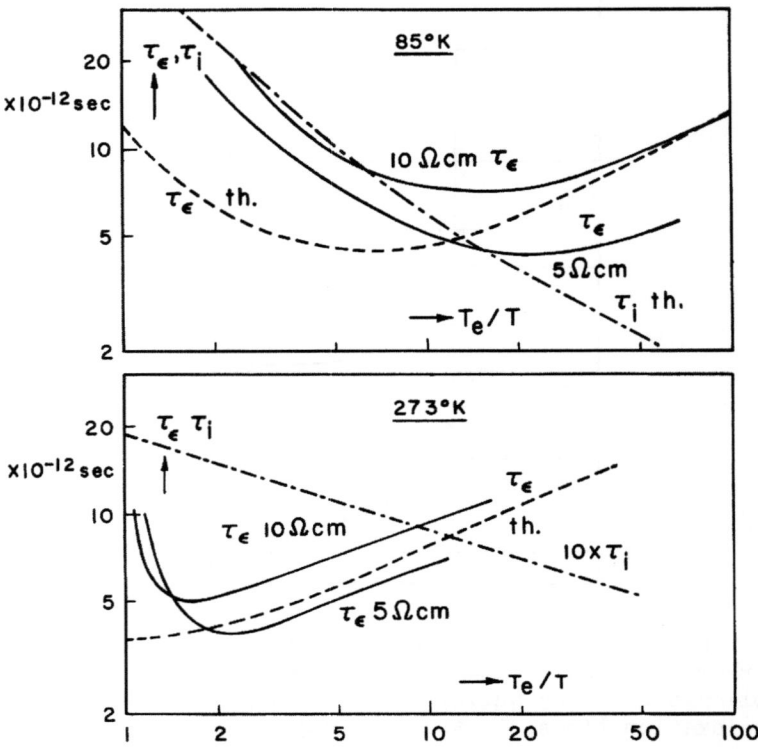

FIGURE 2. Calculated energy relaxation time τ_ϵ and intervalley relaxation time τ_i for transitions between equivalent valleys of the conduction band of germanium, vs electron temperature in units of the lattice temperature, for lattice temperatures of 85 and 273°K (th). The full lines have been obtained from experimental results for n-type Ge samples of 5 ohm-cm and 10 ohm-cm room temperature resistivity (ref. 5).

"dielectric relaxation time" τ_d given by

$$\frac{1}{\tau_d} = \frac{n_0 e}{\kappa \kappa_0} \cdot \frac{\partial v_d}{\partial E} \tag{9}$$

where n_0 is the average carrier concentration, κ is the relative dielectric constant, and κ_0 is the permittivity of free space. By a low-frequency small-signal theory where the space charge $n - n_0 \propto \exp(i\omega t)$, one can show that the imaginary part of ω equals $1/\tau_d$, neglecting diffusion, and since in the negative-differential-mobility region τ_d is negative, the wave is amplified [6].

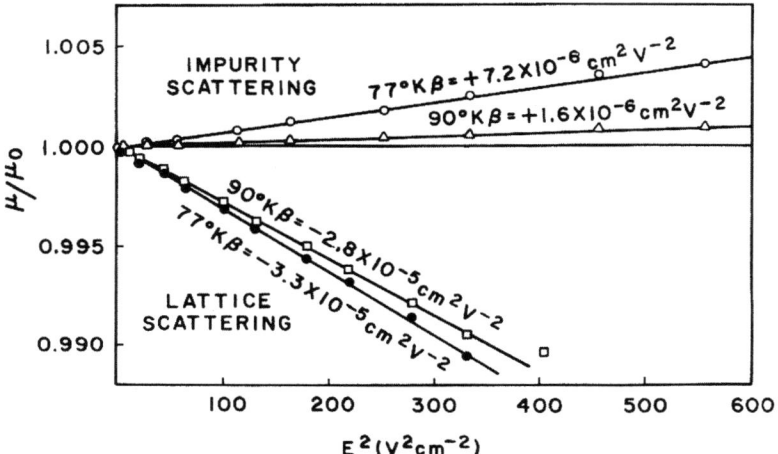

FIGURE 3. Dependence of the mobility ratio μ/μ_o on the square of the electric field strength for two samples of different impurity concentration, each at two different temperatures (after J. B. Gunn, Progr. in Semicond. 2 (1957) 213 (A. F. Gibson, ed.) London).

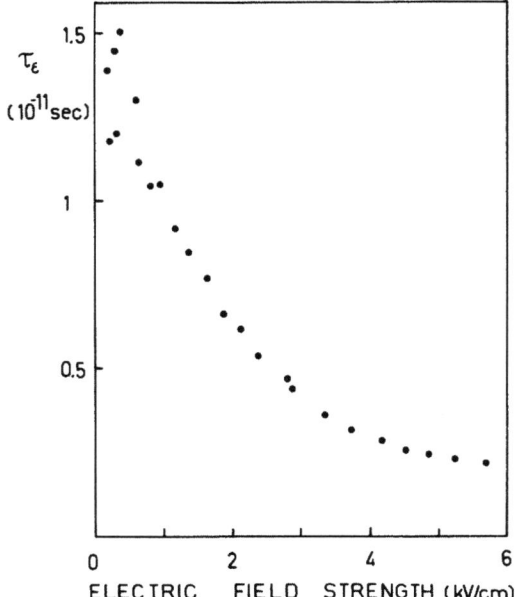

FIGURE 4. Energy relaxation time of n-type GaSb at 77° K obtained from a comparison of the j(E) characteristics with theory assuming two types of valleys for the conduction band [6].

τ_d is also of importance if it is positive, in semiconductors where it is larger than the momentum relaxation time $\bar{\tau}_m$. This is the case in low-mobility high-resistivity materials such as amorphous semiconductors and organic semiconductors. The validity of normal mobility calculations from the Boltzmann equation has been questioned [7], and the unusual behavior of these materials is far from being understood [8].

The remaining part of this paper shall be devoted to experimental methods of determination mainly of the energy relaxation time τ_ε. For both $\bar{\tau}_m$ and τ_ε there are indirect and direct methods available. Consider first $\bar{\tau}_m$: An indirect method consists of a measurement of the Hall mobility and an effective-mass determination by magneto-optics and a calculation of $\bar{\tau}_m$ from $\bar{\tau}_m = \mu \cdot m/e$. A direct method involves the measurement of the ac conductivity as a function of frequency,

$$\sigma_{ac} \propto \frac{\tau_m}{1 + \omega^2 \tau_m^2} \qquad (10)$$

assuming monoenergetic carriers for simplicity. Notice, however, that not always a relaxation time τ_m can be defined, and then the frequency response is different as has been shown by Pötzl et al. [3].

Similarly, τ_ε can be obtained from the nonlinear current-vs-field characteristics. Consider e.g. the case of "warm electrons" where the mobility μ depends on the field intensity E according to

$$\mu = \mu_0 (1 + \beta E^2) \qquad (11)$$

and β is a coefficient. Figure 3 shows experimental data of μ/μ_0 plotted vs E^2 for two samples of different impurity concentrations where either lattice scattering with $\beta < 0$ or impurity scattering with $\beta > 0$ are predominant. Assuming for the mobility a dependence on electron temperature T_e given by a function $\mu = \mu(T_e)$ one finds for the energy relaxation time

$$\tau_\varepsilon = \frac{3}{2} \cdot \frac{k_B T}{e} \cdot \frac{\beta}{\mu_0} \cdot \left(\frac{d \ln \mu}{d \ln T_e}\right)^{-1}_{T_e = T} \qquad (12)$$

The last factor is a number of the order of magnitude 1, and hence τ_ε can roughly be calculated from β, i.e. from the current-vs-field characteristics. Figure 4 shows data for hot

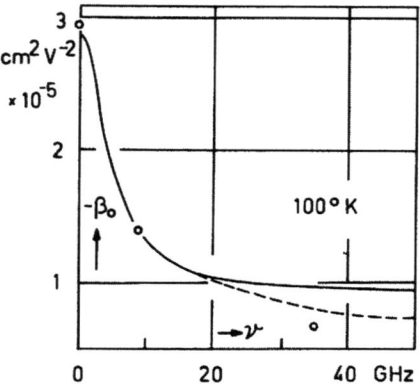

FIGURE 5. β_o as given by Eq. (14) as a function of frequency. Curves are calculated from Eq. (15).

---

electrons in n-type GaSb at 77°K obtained in a similar way [9]. The field dependence suggests a dependence of τ_ε on the average carrier energy $\langle\varepsilon\rangle$, and the decrease of $\langle\varepsilon\rangle$ with time indicated in Fig. 1 must therefore be non-exponential, i.e. τ_ε is a relaxation time only in a less strict sense of the word.

Direct determinations of τ_ε for warm carriers have been made in a field configuration where combined ac and dc fields in parallel are applied to the sample:

$$E = E_o + E_1 \cos \omega t \qquad (13)$$

The change in dc current density due to the addition of the ac field to the dc field E_o,

$$\langle\Delta j\rangle_\| = \langle j\rangle_\| - j_o = \tfrac{3}{2} \beta_o E_o \sigma_o E_1^2 \qquad (14)$$

where

$$\beta_o = \frac{\beta}{1 + \omega^2 \tau_m^2} \cdot \frac{1}{3}\left(1 + \frac{2 + \omega^2 \tau_m \tau_\varepsilon}{1 + \omega^2 \tau_\varepsilon^2}\right) \qquad (15)$$

is observed as a function of the frequency $\nu = \omega/2\pi$ ($j_o = j$ for $E_1 = 0$). Figure 5 shows results obtained in n-type Ge at 100°K together with 2 curves calculated from Eq. (15) with $\tau_\varepsilon = 33.6$ psec for both of them and $\tau_m = 2$ psec (dashed) and $\tau_m = 0$ (full curve). The value of 2 psec for τ_m taken from mobility data is

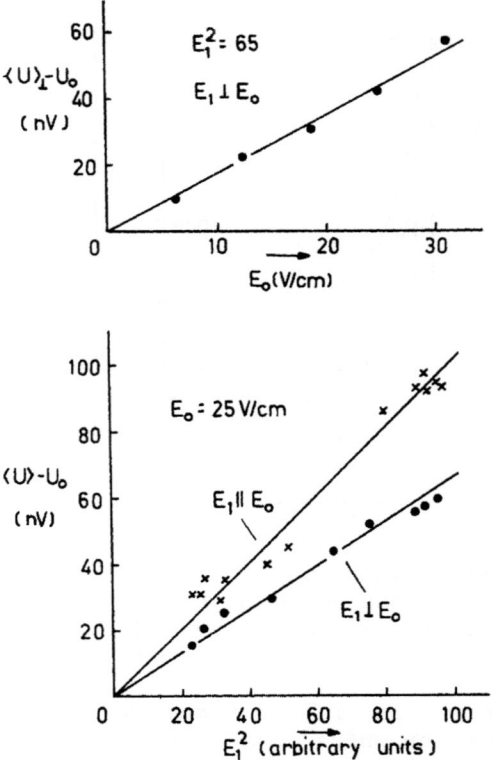

FIGURE 6. Change in voltage between the ends of a n-GaAs sample due to the application of a 337 μm radiation of a few mW/cm^2 power polarized either perpendicular or parallel to the applied dc field E_o: (a) plotted vs E_o, (b) plotted vs the laser power $\propto E_1^2$ [10].

in good agreement with the experimental values. All frequencies are in the microwave range.

If the ac field is perpendicular to the dc field, the change in dc current density

$$\langle \Delta j \rangle_\perp = \langle j \rangle_\perp - j_o = \frac{1}{2} \cdot \frac{\beta E_o \sigma_o E_1^2}{1 + \omega^2 \tau_m^2} \qquad (16)$$

is independent of τ_ε. The ratio of both experimental results,

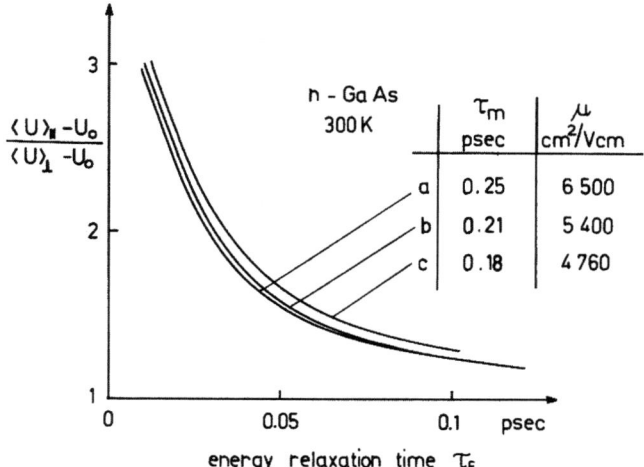

FIGURE 7. Plot of the ratio $\langle\Delta U\rangle_\parallel/\langle\Delta U\rangle_\perp$ given by the right side of Eq. (17), vs τ_ε with τ_m as a parameter (after Seeger et al., l.c.).

$$\frac{\langle\Delta j\rangle_\parallel}{\langle\Delta j\rangle_\perp} = 1 + \frac{2 + \omega^2 \tau_m \tau_\varepsilon}{1 + \omega^2 \tau_\varepsilon^2} > 1 \qquad (17)$$

has the practical advantage of depending only on $\omega\tau_\varepsilon$ and $\omega\tau_m$ and being independent of β, E_1 etc. Using for the ac field the linearly polarized light of a HCN laser with a wavelength of 337 μm, measurements have been made with 3 n-type GaAs samples at room temperature [10]. Figure 6a and b show $\langle\Delta U\rangle \propto \langle\Delta j\rangle^2$ plotted vs E_0 and vs E_1^2, respectively, for one sample. The straight lines indicate the verification of Eq. (16). Figure 7 shows a plot of the right side of Eq. (17) vs τ_ε for various values of τ_m obtained from the mobilities of the samples. The observed values of $\langle\Delta j\rangle_\parallel/\langle\Delta j\rangle_\perp$ are about 1.5 which yields $\tau_\varepsilon = 0.5$ psec. Notice that in n-GaAs at 300°K where polar optical scattering predominates τ_ε is two orders of magnitude and τ_m one order of magnitude smaller than in n-Ge at 100°K where nonpolar scattering predominates. What is the temperature dependence in both cases?

Figures 8 and 9 show plots of τ_ε and τ_m for polar and nonpolar optical scattering vs temperature in units of the Debye temperature θ, calculated from the quantum mechanical transition probabilities. At room temperature in n-GaAs, $\tau_\varepsilon > \tau_m$ but at low temperatures $\tau_\varepsilon \approx \tau_m$ if τ_m is not dominated by impurity scattering. In pure n-Ge at 100°K τ_m is due mainly to acoustic

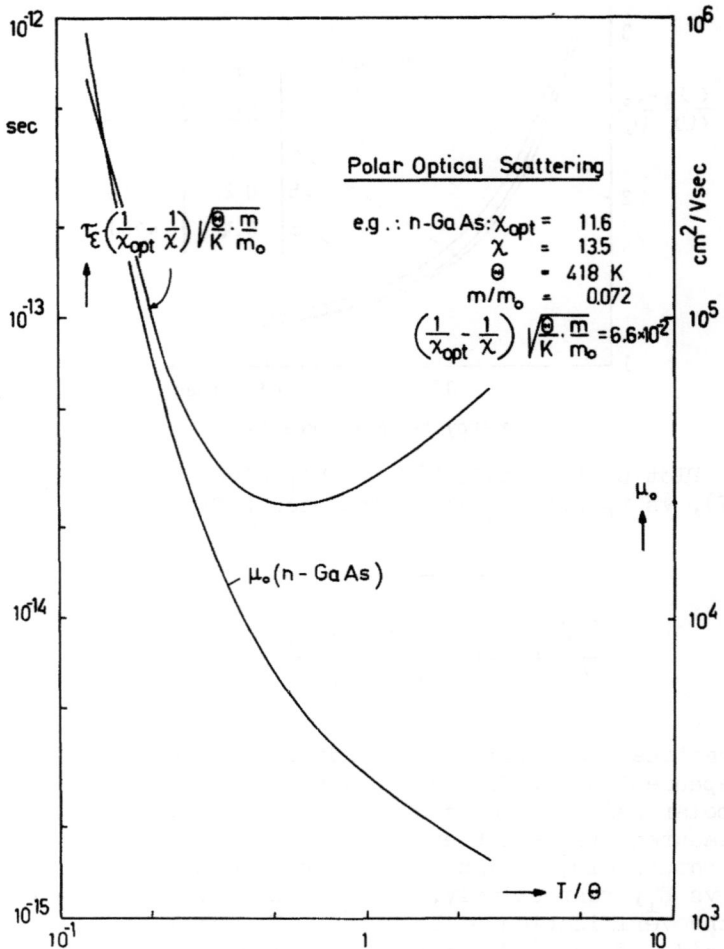

FIGURE 8. τ_ε and τ_m for polar optical scattering, as a function of lattice temperature in units of the Debye temperature, for warm electrons. κ_{opt} and κ are the optical and the static dielectric constants, respectively. For n-type GaAs the factor behind the relaxation times has a value of 6.6×10^{-2}.

---

phonon scattering, and this value of τ_m is obviously $\ll \tau_\varepsilon$ considering the ratio of deformation potential constants, $\varepsilon_{opt}^2 / \varepsilon_{ac}^2 = 0.4$.

Let us now deal with a more accurate method for a determination of τ_ε called "harmonic mixing" or "optical rectification"

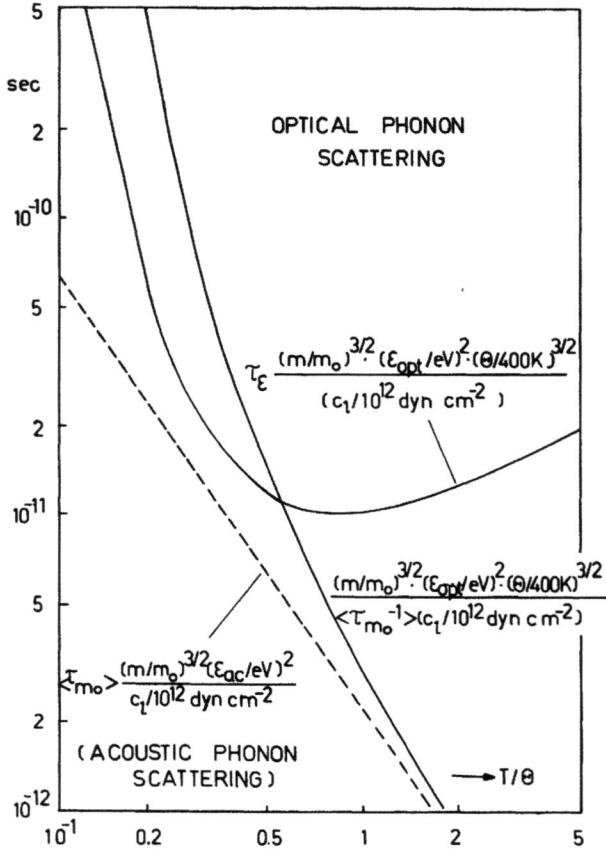

FIGURE 9. τ_ε and τ_m for nonpolar optical scattering, and τ_m for nonpolar acoustic scattering, as a function of lattice temperature in units of the Debye temperature θ, for warm electrons (for acoustic scattering: $\theta = 400$ K assumed). c_l is the longitudinal elastic constant (normally denoted by c_{44}).

[11]. We apply to a sample a fundamental wave and its second harmonic in parallel. The total field intensity in the sample is given by

$$\vec{E} = \vec{E}_1 \cos(\omega t + \varphi) + \vec{E}_2 \cos 2\omega t; \quad \vec{E}_1 \| \vec{E}_2 \qquad (18)$$

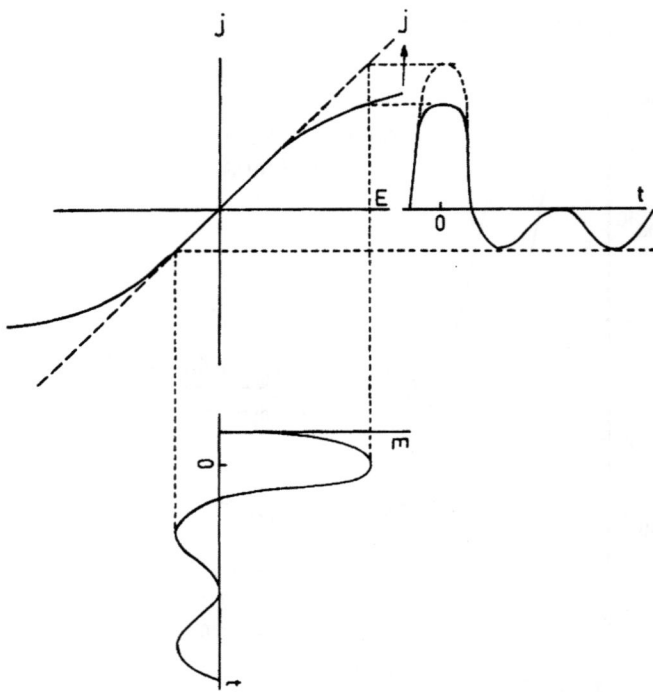

FIGURE 10. The time dependent field intensity obtained by the superposition of a fundamental wave and its second harmonic, yields a distorted current $j(t)$ and therefore a nonvanishing time average $\langle j \rangle$ when subjected to a nonlinear j-E characteristics.

---

where φ is a phase difference arbitrarily chosen. E_1 and E_2 are the field amplitudes. The lower part of Fig. 10 shows $E(t)$ for $E_1 = E_2$ and $\varphi = 0°$. For any value of the phase the time average of E vanishes. The right side of the same figure shows $j(t)$ obtained with a nonlinear $j(E)$ characteristics. Clearly, $\langle j \rangle \neq 0$, slightly negative in this case. In fact,

$$\langle j \rangle \propto \cos 2(\varphi + \psi) \tag{19}$$

where for the simple case of $\tau_m \ll \tau_\varepsilon$

$$\tan 2\psi = + \frac{2\omega^3 \tau_\varepsilon^3}{1 + 3\omega^2 \tau_\varepsilon^2} \tag{20}$$

and for the case $\tau_m = \tau_\varepsilon$ which we denote simply by τ

$$\tan 2\psi = - \frac{\omega^3 \tau^3}{1 + 2\omega^2 \tau^2} \tag{21}$$

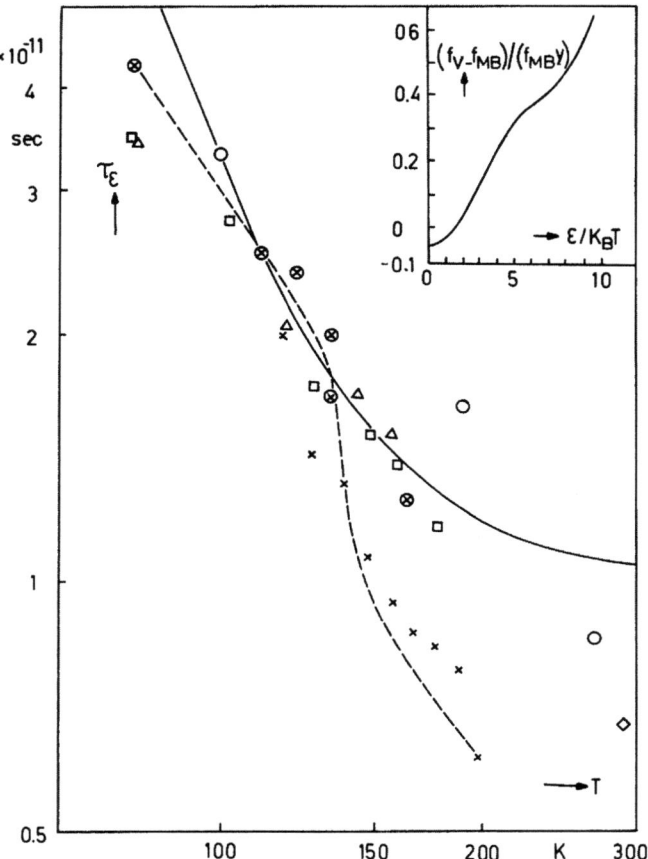

FIGURE 11. Observed and calculated values of τ_ε for n-type Ge plotted vs temperature. Open circles, diamond, and crosses: data obtained by combined dc and ac fields by K. Seeger, Z. Physik 172 (1963) 68; A. F. Gibson, J. W. Granville, and E. G. S. Paige, J. Phys. Chem. Solids 19 (1961) 198, and T. N. Morgan and C. E. Kelly, Phys. Rev. 137 (1965) A1573, respectively. Other data for samples of room temperature resistivities between 2 and 30 ohm-cm by harmonic mixing (ref. 7). Calculations with a Maxwell Boltzmann distribution f_{MB} (K. Seeger, l.c.) and a variational method distribution f_V (K. Hess, unpublished).

For a null method, $\langle j \rangle = 0$ or $\psi = -\varphi + \pi/4$ and therefore ψ is obtained from the setting of a phase shifter and τ_ε calculated from ψ.

Results of τ_ε for n-type Ge samples of various impurity concentrations obtained by this method are plotted vs temperature in Fig. 11. The open circles have been obtained by the method of

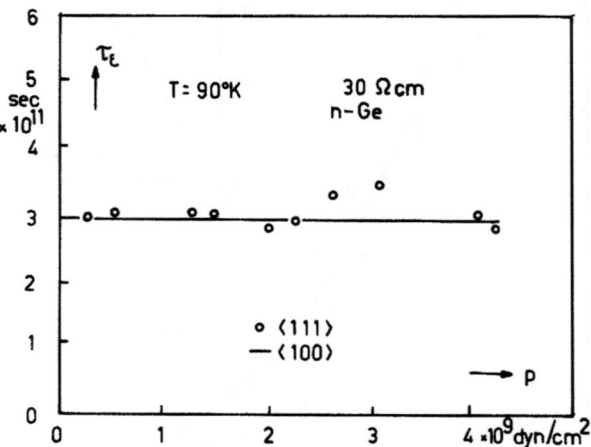

FIGURE 12. τ_ε in n-type Ge under uniaxial pressure.

combined ac and dc fields. The curves are theoretical for a
Maxwell Boltzmann distribution, f_{MB}, (full line) and a distribution
obtained by variational methods, f_v. The inset shows the
relative deviation of f_v from f_{MB}, divided by an expression $\propto E^2$.
The experiment cannot decide uniquely between the two distributions,
unfortunately.

The application of a uniaxial pressure does not change τ_ε in
n-type Ge shown by Fig. 12. This is not true in p-type Si shown
in Fig. 13. The dashed curve has been calculated [12]. In this
material τ_ε is even slightly anisotropic at low temperatures shown
in Fig. 14. Such an anisotropy persists to higher temperatures in
p-type Ge (Fig. 15). For p-type tellurium results shown in
Fig. 16 have been obtained [13]. The experimental data seem to
favor polar optical energy loss. Figure 17 shows experimental
results obtained for n-type GaAs and GaSb and curves calculated
also for polar optical energy loss. The inset shows τ_ε vs the
electron temperature T_e for n-type GaSb obtained from Shubnikov-de
Haas measurements. This method works only at very low temperatures
but - in contrast to previous methods - for a degenerate electron
gas where in most cases an ohmic j-E characteristics is observed.
We will now investigate the Shubnikov-de Haas effect in more
detail.

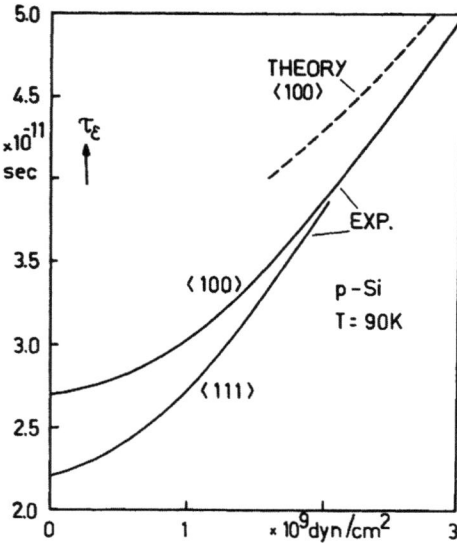

FIGURE 13. τ_ε in p-type Si under uniaxial pressure. Dashed curve: theoretical.

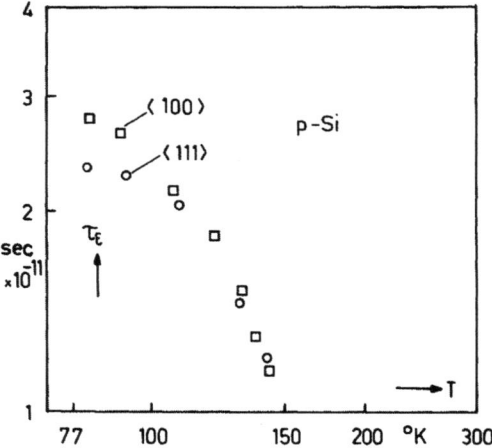

FIGURE 14. τ_ε in p-type Si vs temperature. Field directions indicated.

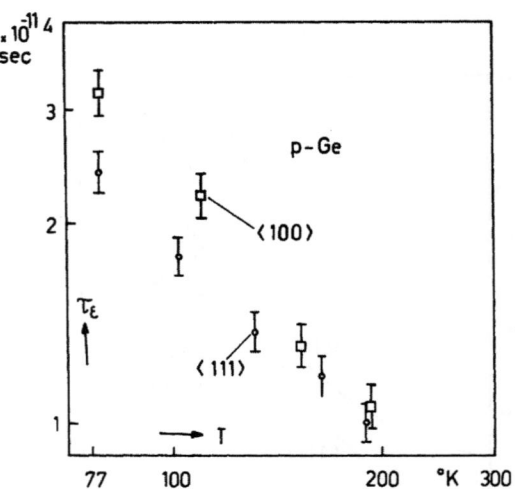

FIGURE 15. τ_ε in p-type Ge vs temperature. Field directions indicated.

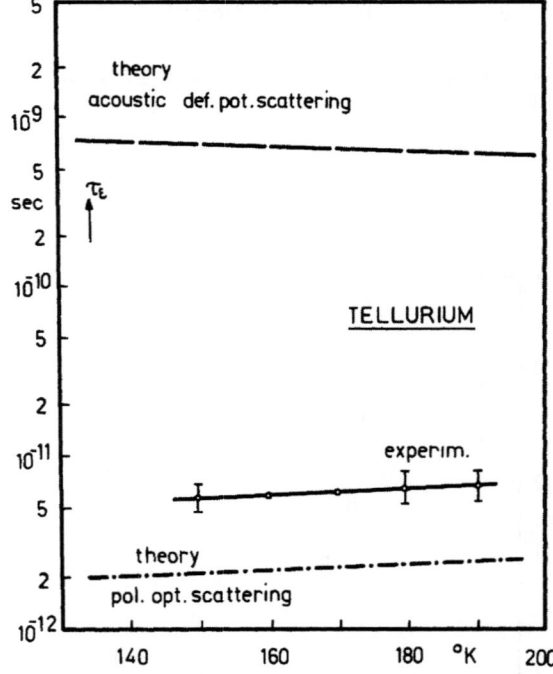

FIGURE 16. τ_ε in p-type Te vs temperature for $\vec{E}\|c$-axis; circles: experimental data; dash-dotted curve: theoretical assuming energy loss via polar optical scattering; dashed curve: theoretical assuming energy loss via deformation potential scattering with a deformation potential constant of 5 eV.

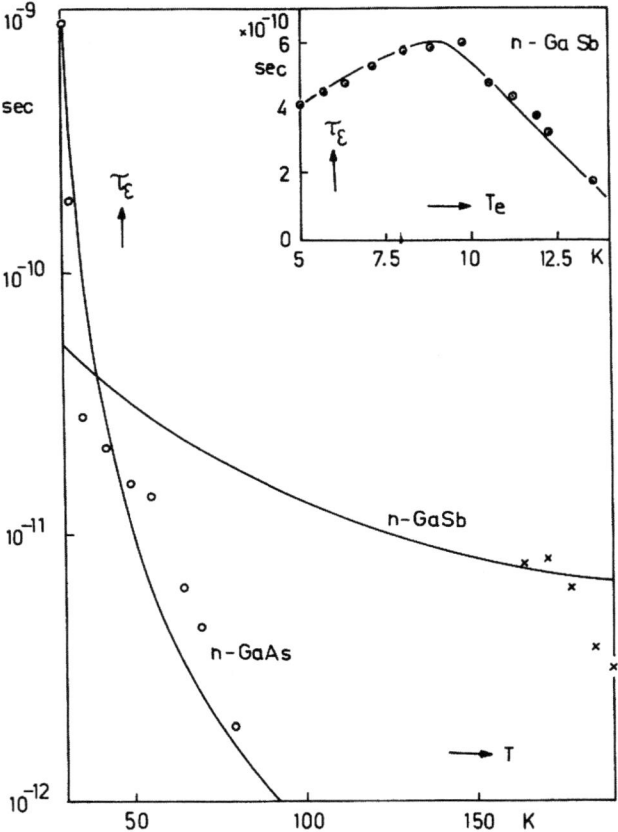

FIGURE 17. τ_ε in n-type GaAs and n-type GaSb vs T. Inset: τ_ε in n-type GaSb vs T_e.

Figure 18 shows the longitudinal magnetoresistance of n-type InAs plotted vs the magnetic induction B for various lattice temperatures (left side) and at the lowest temperature for various values of the applied dc field intensity (right side) [14]. The two sets of curves are very similar. One may relate corresponding curves and e.g. say that at a field intensity of 320 mV/cm the electron temperature is 12.5°K which is 3 times the lattice temperature, i.e. the electrons are hot, not warm. Also one can thus determine the energy loss rate as a function of the electron temperature plotted in Fig. 19. The kink in the curve can be explained as a transition from acoustic to optical phonon energy loss.

Finally let us consider a method for degenerate direct semiconductors applicable also at high lattice temperatures.

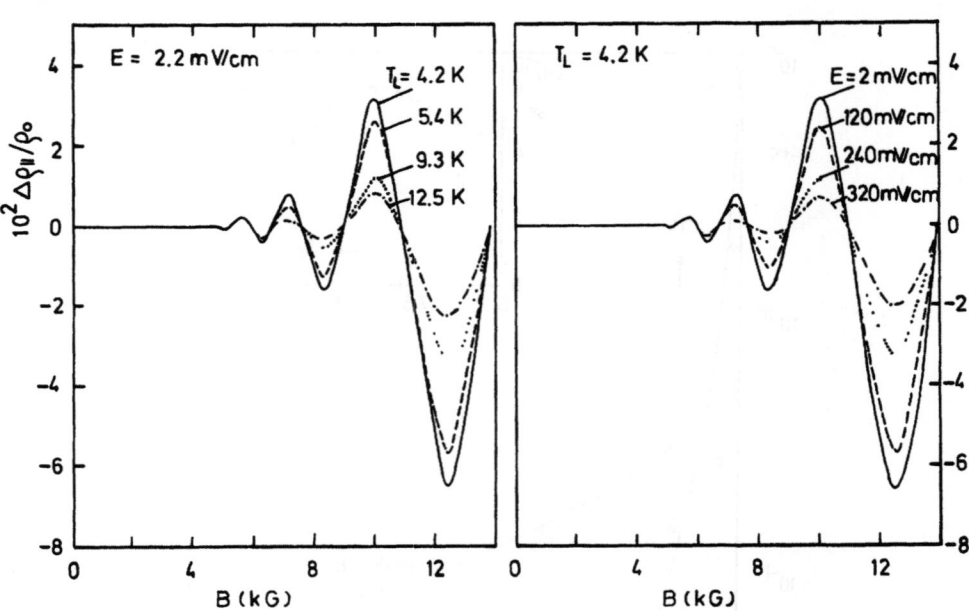

FIGURE 18. Oscillatory component of the longitudinal magneto-resistance of n-type InAs. Left side: Low electric field intensity, various lattice temperatures. Right side: Constant lattice temperature, various electric field intensities.

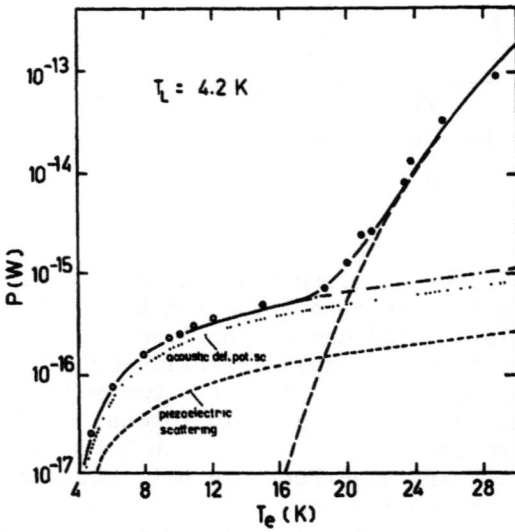

FIGURE 19. Energy loss rate vs electron temperature for n-type InAs at 4.2° K lattice temperature; dashed curve: calculated for polar optical energy loss; dashed-dotted curve: calculated for combined screened piezoelectric and screened nonpolar acoustic energy loss; full curve: all three mechanisms combined; open circles; Shubnikov-de Haas data; full circles: data from j-E characteristics.

RELAXATION TIMES IN SEMICONDUCTORS

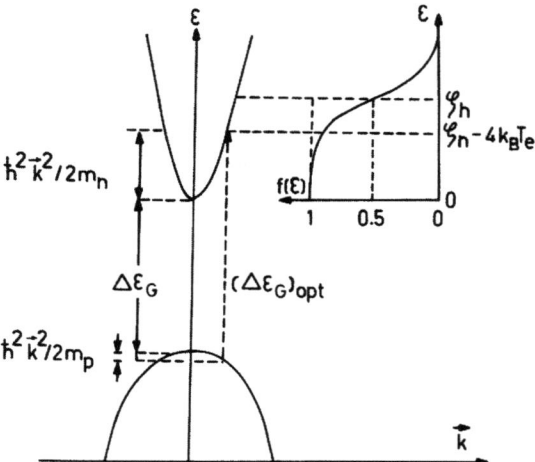

FIGURE 20. Burstein-Moss shift in the energy band model of a direct degenerate semiconductor. In the diagram on the right-hand side the Fermi Dirac distribution is illustrated.

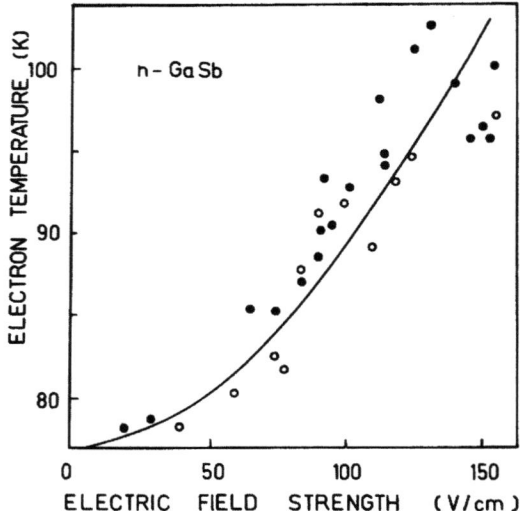

FIGURE 21. Electron temperature vs dc electric field strength obtained in n-type GaSb with $3.5 \times 10^{17}/cm^3$ carriers from the Burstein-Moss shift.

Finally let us consider a method for degenerate direct semiconductors applicable also at high lattice temperatures.
Figure 20 shows an optical transition from the valence band to the Fermi level ζ_n in the conduction band (actually to $4 kT_e$ below ζ_n). Since all states below this level are occupied, this is the

lowest possible direct transition energy, i.e. the absorption edge is shifted to a higher energy than the gap energy. This is called the "Burstein-Moss shift". If the carriers are heated by a dc electric field the degeneracy is lifted and the absorption edge moves. Again comparing this set of curves with one obtained at increased lattice temperatures yields the relation between electron temperature and electric field strength shown for n-type GaSb in Fig. 21 [15]. A combination of these data with the j-E characteristics or with time-dependent measurements would give the energy loss rate; however, for time-dependent measurements fast infrared detectors are required and therefore at present this method is limited to low-temperature measurements where the relaxation times are long, just as it is the case for the Shubnikov-de Haas method.

REFERENCES

[1] L. Spitzer, "Physics of Fully Ionized Gases", Interscience Publ., New York 1967, Appendix.
[2] C. Hilsum, this conference.
[3] H. W. Pötzl, this conference.
[4] Relaxation phenomena of importance in transferred-electron devices have been investigated by D. Jones and H. D. Rees (to be published) and by H. A. Hillbrand, thesis, T. H. Wien/Austria 1972.
[5] K. Seeger and D. Schweitzer, Proc. Int. Conf. Phys. Semic. Kyoto 1966, p. 415.
[6] See e.g. P. N. Butcher, Rep. Progr. Phys. XXX Pt. 1 (1967) 1.
[7] H. J. Queisser, H. C. Casey, and W. van Roosbroeck, Phys. Rev. Lett. 26 (1971) 551; W. van Roosbroeck and H. C. Casey, Phys. Rev. B 5 (1972) 2154.
[8] C. Hilsum, this conference.
[9] H. Heinrich, K. Hess, W. Jantsch, and W. Pfeiler, J. Phys. Chem. Solids 33 (1972) 425.
[10] F. Kuchar, A. Philipp, and K. Seeger, Verh. Deutsch. Phys. Ges. 6/1972, p. 586.
[11] K. Hess and K. Seeger, Z. Physik 218 (1969) 431; K. Seeger and K. Hess, Z. Physik 237 (1970) 252.
[12] K. Hess, thesis, Univ. Vienna/Austria 1970.
[13] H. Kahlert, K. Hess, and K. Seeger, Solid State Commun. 7 (1969) 1149.
[14] G. Bauer and H. Kahlert, Phys. Rev. B 5 (1972) 566; Proc. Internat. Conf. Phys. Semic. Cambridge (1970) 65; H. Kahlert and G. Bauer, phys. stat. sol. (b) 46 (1971) 535.
[15] H. Heinrich and W. Jantsch, Phys. Rev. B 4 (1971) 2504.

CHAPTER 6

DIFFUSION IN COMPOUND SEMICONDUCTORS: A SURVEY OF RECENT RESULTS

A. Ascoli

Centro Informazioni Studi Esperienze
Segrate
Milan, Italy

INTRODUCTION

Diffusion in semiconductors comes in for a lot of attention because of its fundamental and technological interest. So much experimental and theoretical work has been devoted to this subject that its collection has already provided material for several review articles and two books have been entirely devoted to it: one, by Boltaks, in 1961 [1], the other, by Sharma, in 1970 [2]. Since the latter gives excellent coverage of the literature until 1969, we have made a survey here of the results published from 1970 till now (mid-1972). Late 1969 references are occasionally quoted and appear in some of the tables of this chapter.*

A variety of sensitive techniques can be employed to detect the presence and the migration of impurities in semiconductors. Some of these techniques are typical of semiconductors. This fact contributes to making diffusion in semiconductors a subject in itself.

This chapter deals only with compound semiconductors. We shall classify the subject-matter in three parts: diffusion in III-V compounds, in II-VI compounds, in IV-VI and other compound semiconductors.

---

*Note added in proof: a third book [104] just appeared on the subject, but it was not available when this lecture was given.

DIFFUSION IN III-V COMPOUNDS

Diffusion in III-V compounds can take place via vacancies, via vacancy pairs (a metal vacancy plus an anion vacancy), and, at not too high temperatures, via interstitials. Interstitial-substitutional equilibria, like those obtainable by continuous reactions between interstitials and vacancies, are also likely in these compounds, in which impurities can even diffuse by simultaneous mechanisms. Lattice distortion due to the presence of impurities favors self-diffusion and the diffusion of other impurities. This effect may be enhanced or hindered by charge interactions between pre-existing impurities and diffusing atoms.

Table I offers a representative set of results published in the last three years on diffusion in III-V semiconductors. It can be seen that nearly half the work in this period has been devoted to the diffusion of group-II elements. Samples mentioned in the tables are single crystalline unless otherwise stated.

A few observations can be formulated on the basis of the contents of Table I.

A large variety of methods has been employed and this calls for some caution when comparing independent results. Self-diffusion seems to be attributable to interstitials. Interstitial-substitutional equilibrium plays an important role in Zn diffusion; occasionally dislocations have been cited at the lower temperatures, divacancies at the higher ones. Formation of complexes is assumed in heavily doped samples. Cd seems to diffuse in GaAs through Ga vacancies. The irradiation increases the concentration of vacancies and, by vacancy capture, decreases the diffusion coefficient of Zn. Si in GaAs takes substitutional positions in the Ga or in the As sublattices - the alternative depending on the concentration of impurities. Gallium divacancies are thought to be responsible for the diffusion of sulphur in GaP and GaAs. The formation of a complex consisting of a gallium vacancy and three sulphur atoms is also considered.

The formation of a chemical compound has been suspected to favor the diffusion of S in InSb. The diffusion coefficient for Te in InSb takes larger values than the diffusion coefficient for Se in the same compound but they approach the same value as the melting temperature of the host substance is approached. Turnbull's dissociative mechanism has been tentatively supposed to explain the diffusion of inert gases in ion-bombarded GaAs.

Figure 1 shows data by Ting and Pearson [17] on the concentration- and annealing-time-dependence of the effective diffusion coefficient of Zn in GaAs at 600°C as obtained by the

Table I

Diffusion in III-V Semiconducting Compounds.

Host Substance	Diffusing Species	Interpretation Model or Results (when model is not stated)	Ref.
GaAs	Ga	interstitials	3
GaAs	Ga and As	$D \simeq 10^6$ at 47at. % As; $D \simeq 1.7$ at 52at. % As	4
InSb	surf. self-d.	$Q \simeq 0.61$-0.74 eV	5
GaAs	Cu	differences in the solubilities and diffusion coefficients of different impurity centres	6,7
GaAs	Ag	higher conc. of surf. vacancies on B than on A polar faces; different surf. interactions	7,8
GaAs	Au	nonlinear penetration plots on A polar faces, linear penetration plots on B polar faces, $D_A > D_B$ — rapid saturation of A polar faces with impurities; differences in surf. D and vol. D due to polarity of {111} faces	7
GaAs (Te-doped)	Cu	subsurface region; Cu-vacancy complexes; deep region: Cu-Te complexes	9
InSb	Cu	higher conc. of surf. vacancies on B than on A faces; rapid saturation of A faces with impurities; differences in surf. D and vol. D due to polarity of {111} faces	7
InSb	Ag		7
InSb	Au	different number of unpaired electrons on different polar faces	8
GaP	Be	interstitial-substitutional diffusion model; D enhanced by stacking faults and dislocation clusters and strongly conc.-dependent; non-gaussian penetration profiles	10
GaP_xAs_{1-x}	Zn	substitutional for t<8h d. annealing, interstitial by induced dislocations for t>8h d. annealing	11
GaAs	Zn	As divacancies	12-14
		Q depends on polarity of {111} faces	7
		interstitial-substitutional d. for high Zn conc.	15
		nonequilibrium interstitial-substitutional d. (see Fig. 1)	17,20,22,23
		nonequilibrium vacancy	19
		20-35% of Zn precipitates in heavily damaged diffused region	24

d.=diffusion; surf.= surface; vol.= volume; conc.= concentration.

Table I
(continued)

Diffusion in III-V Semiconducting Compounds.

Host Substance	Diffusing Species	Interpretation Model or Results (when model is not stated)	Ref.
GaAs(epitaxial)	Zn	$D_o = 6.3 \cdot 10^{-4}$ cm^{2}s^{-1}, Q=1.7 eV	18
GaAs(Te-doped)	Zn	interstitials (see Fig. 2) formation of substitutional, neutral Zn$^-$Te$^+$ complexes in heavily Te-doped samples with E_f=0.2 eV	3 16
GaAs; InAs (fast n irradiated)	Zn	at low temp. and low Zn conc., vacancy; at high temp. and high Zn conc., dissociative mechanism	21
GaAs	Cd	Ga vacancies	12-14,25,26
In$_{0.18}$Ga$_{0.82}$As	Zn	$D_o = 2.1 \cdot 10^{-3}$ cm^{2}s^{-1}, Q=1.7 eV	18
GaSb	Zn	simultaneous vacancy and interstitial mechanisms	27
InAs	Zn	dissociative mechanism	18,28
InAs	Hg	$D_o = 1.45 \cdot 10^{-5}$ cm^{2}s^{-1}, Q=1.32 eV	29
GaAs	Al$^+$	$D_{Zn} \gg D_{Al}$, Ga$_{1-x}$Al$_x$As conversion	30
GaAs	Si	at low conc., Si goes substitutional in the Ga sublattice as a singly ionized donor, at high conc., Si goes substitutional in the As sublattice as a singly ionized acceptor	31
GaAs	Ge	$D_o = 10^{-6}$ cm^{2}s^{-1}, Q=1.8 eV (see Fig. 3) amphoteric impurity	32 33
GaAs(Zn-doped)	Sn	$D = 3.5 \cdot 10^{-14}$ cm^{2}s^{-1} at 850°C, $D = 5 \cdot 10^{-13}$ cm^{2}s^{-1} at 1050°C; high carrier conc. achieved	34
GaAs(SiO$_2$ masked)	Sn	D enhanced at mask edges due to thermal strain or p-impurity coupling	35
GaP	S	$V_{Ga} V_P$ divacancy	36,39

Table I
(continued)

Diffusion in III-V Semiconducting Compounds.

Host Substance	Diffusing Species	Interpretation Model or Results (when model is not stated)	Ref.
GaP	Zn	anomalous distribution with steep conc. gradient near the junction	37
GaAs	O	$D_o = 2 \cdot 10^{-3} \text{cm}^2 \text{s}^{-1}$, $Q = 1.1$ eV	38
GaAs	S	V_{Ga} V_{Ga} divacancy - $V_{Ga}S_3$ complex	36
GaAs (epitaxial)	Te	D increases because of defects accumulated at the film-substrate interface	40
InSb	S	two simultaneous diffusion fluxes; substitutional (along Sb vacancies) + interstitial	41
InSb	Te	substitutional (Sb sublattice) and interstitial simultaneous mechanisms	42
GaAs	Fe	$D_o = 1.2 \cdot 10^{-1} \text{cm}^2 \text{s}^{-1}$, $Q = 2.64$ eV	43
GaAs	Co	$D_o = 2.2 \cdot 10^{-3} \text{cm}^2 \text{s}^{-1}$, $Q = 2.32$ eV	43
GaAs	Cr, Fe, Co	formation of a high-resistivity, p-type compensated layer	44
InSb	Co	interstitials (fast) and dislocations (structure sensitive); chemical compounds for Co conc. larger than solubility limit	45
GaAs	Kr and Xe	Turnbull dissociative mechanism?	46
GaP-GaAs InP-InAs (hetero-junctions)	P and As	$D_P \simeq 5 \cdot 10^{-11} \text{cm}^2 \text{s}^{-1}$, $D_{As} \simeq 3 \cdot 10^{-11} \text{cm}^2 \text{s}^{-1}$ at 1000°C and 40 atm	47

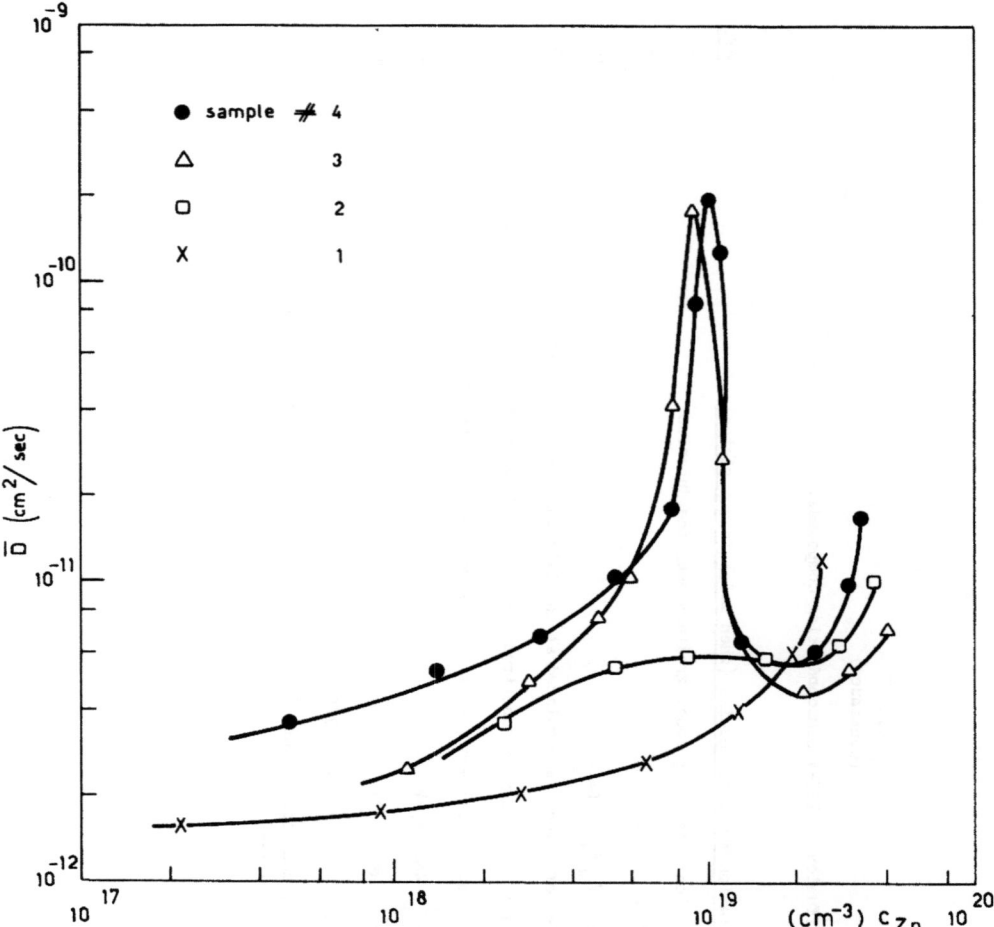

FIGURE 1. The effective diffusion coefficient of Zn in GaAs vs. Zn concentration at 600°C as obtained from Boltzmann-Matano analysis [17].

Boltzmann-Matano method. Over long diffusion-times, a peak appears at intermediate concentrations. This may denote an inversion of prevalence between the influence of lattice distortion and charge interactions on the diffusion coefficient.

Data by Winteler [3] on the dependence of the chemical diffusivity of Zn in GaAs at 400°C as a function of Zn concentration are shown in Fig. 2. Maxima of D for intermediate Zn concentrations are evident here too, although less pronounced than in Fig. 1.

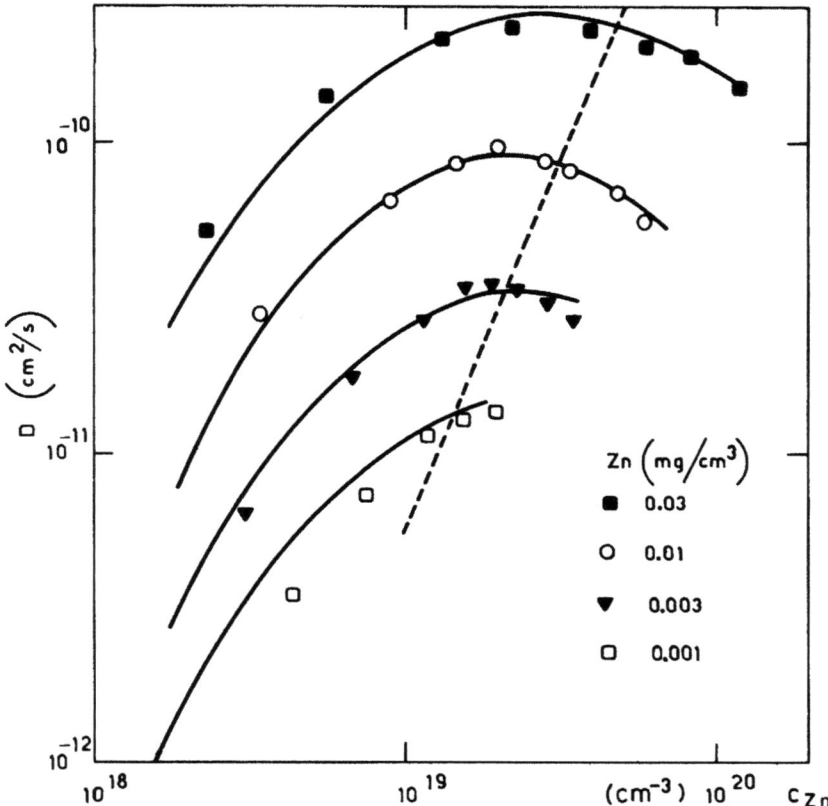

FIGURE 2. Dependence of the chemical Zn diffusivity in GaAs at 1100°C as a function of Zn concentration in GaAs. As vapor: 5 mg cm^{-3}; Zn vapor: as indicated. The dotted line represents the calculated dependence of D_{Zn}^{iso} on concentration. D_{Zn}^{iso} is the diffusion coefficient of Zn when the concentration of all point defects in the sample is constant [3].

Figure 3 shows results by Lavrishchev et al. [32] on the diffusion of Ge in GaAs crystals at an excess pressure (2 atm) of As. D and Q data displayed in Table I are provided by the dotted straight line.

DIFFUSION IN II-VI COMPOUNDS

Diffusion in II-VI compounds is often "anomalous", i.e. concentration-dependent, native-defect dependent, etc., probably because of ready impurity-impurity and impurity-defect

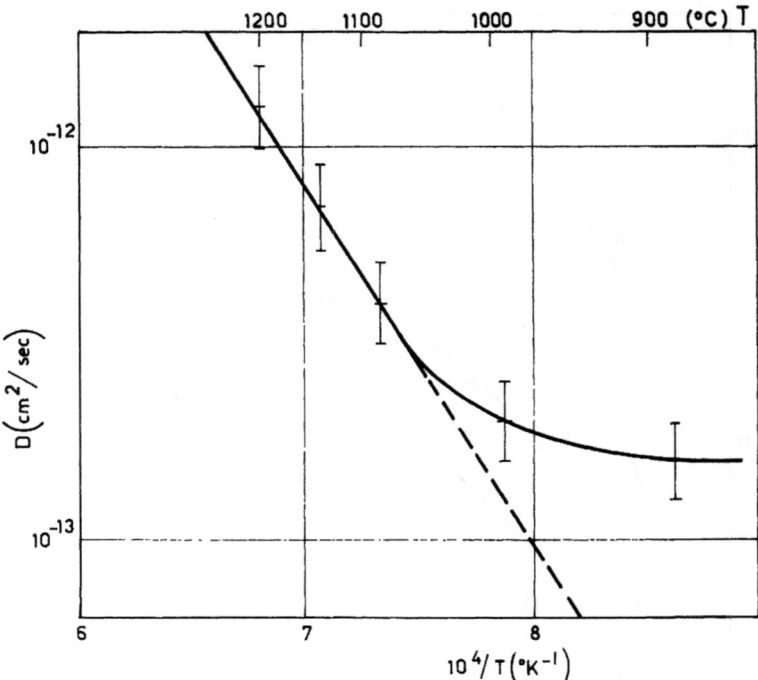

FIGURE 3. Temperature dependence of the diffusion coefficient of Ge in GaAs crystals [32].

interactions. The concentration of native defects is strongly influenced by the vapor pressure of the components when the sample is heated below melting point - and thus when it undergoes diffusion annealings. Therefore, the measured diffusion coefficient depends not only on the temperature but also on the partial vapor pressures of the components. This dependence of D on various interdependent parameters is no doubt responsible for scarce reproducibility of the data; and giving significance to systematic classification involves not only careful control of experimental conditions but also and essentially a scrupulous characterization of sample material. By comparing previous data collections on diffusion in II-VI compounds with those collected in Table II referring to the last three years, an intensification of researches becomes very evident, probably because improved techniques for precise characterization of materials have begun to reduce the scatter of results - to the advantage of significant comparisons. Remarkable sets of independent results are piling up on self-diffusion and on the diffusion of noble metals. However, more complex models are now turning up, such as doubly ionized Frenkel defects for Zn in ZnSe or neutral Se interstitials plus doubly ionized Se vacancies for Se in ZnSe. Instead, the diffusion of Zn in ZnTe still seems to be entrusted to simple vacancies. The results and the interpretation models for diffusion of S in CdS strongly depend on the component pressure on the sample; doubly ionized vacancies (double donors) and neutral interstitials have been assumed as models. A common pattern has been observed for chalcogen self-diffusion in Zn and Cd chalcogenides, and similar models appear to explain this process in all II-VI compounds. The same similarity has not been observed for the metal self-diffusion, which in the tellurides is different from the other compounds: not influenced by the metal partial pressure in CdTe and ZnTe, it increases with metal vapor pressure in CdS, ZnS and CdSe. The presence of O_2 lowers the D values for Cd in CdS by an order of magnitude, whereas the effects of surface cold-work, due to pre-anneal preparation, are within standard error. No significant difference is brought out if Cd diffusion in CdS is performed parallel or perpendicular to c axis [57]. γ and n irradiations unquestionably lower the activation energy for diffusion, as can be noticed both in S diffusion in CdS and in Hg and Cd diffusion in HgTe. Cd seems to diffuse by singly-ionized interstitials in CdS and by doubly-ionized interstitials in CdSe; Se vacancies are confirmed as contributing to diffusion in CdSe; Hg vacancies are held responsible for self-diffusion in HgS_xSe_{1-x} and S and Te vacancies for self-diffusion in HgS_xTe_{1-x}.

Figure 4 shows the effects of heating in component vapor and n-irradiation preliminary treatments on the γ-radiation enhancement of diffusion of S in CdS, as obtained by Mannanova and Niyazov [55]. The possibility of strongly enhancing diffusion effects by different types of irradiation confirms that technological importance can be attributed to irradiation treatments.

Figure 5 [56] gives an example of the agreement that can be expected between theoretical and experimental results: the points indicated are on the chemical diffusivity in CdS at high Cd activity, the solid line has been calculated from self-diffusion data.

Figure 6 [64] shows results on the S (curve a), Se (curve b) and Hg (curve c) self-diffusion in HgS_xSe_{1-x} at 300°C for $0<x<40\%$. Some scatter of the experimental data is evident; a maximum may (curves b and c) or may not (curve a) appear in the dependence of the diffusion coefficient on concentration.

It can also be noticed in Table II that a systematic series of researches starts to bring out the superimposition of slow (interchange) and fast (interstitial) mechanisms on diffusion of both noble metals and Zn in II-VI compounds. Once again, diffusion coefficients of Au in CdS under a Cd atmosphere are higher than those under a S atmosphere.

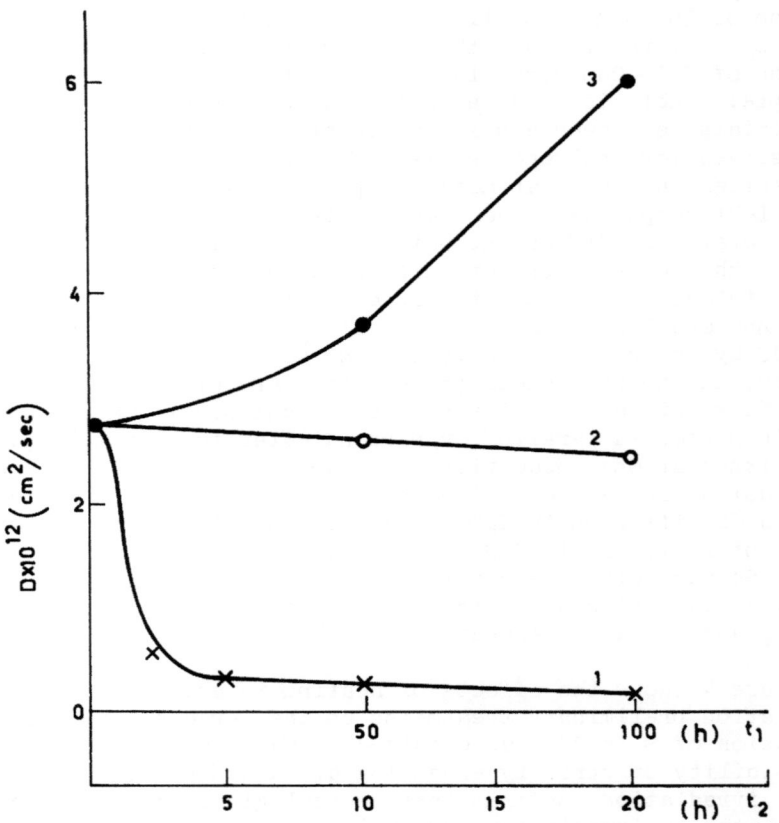

FIGURE 4. Influence of heating in S vapor (1) or in Cd vapor (2) and of prolonged irradiation with reactor neutrons (3) on the diffusion coefficient of S in CdS. Each point corresponds to a γ-ray dose of $1.8 \cdot 10^8$ R. The time scale t_1 represents the duration of neutron irradiation (in hours) and t_2 represents the duration of heating in S or Cd vapor (in hours) [55].

Slow and deep components for Yb^{3+} in CdS are also reported in Table II, while Al (see the same table) seems to go substitutionally into the Zn sublattice in either ZnSe, ZnTe and their mixed crystals.

Sn in HgTe and Te in CdS show two-slope penetration profiles: a front and a deep penetration section.

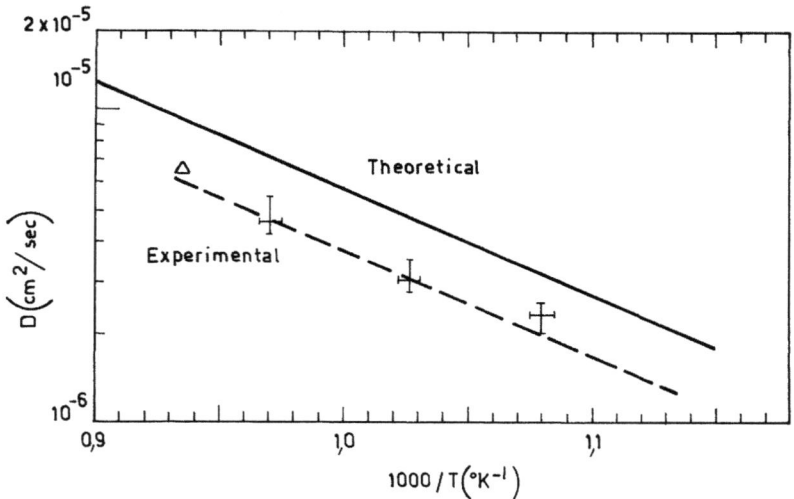

FIGURE 5. Temperature dependence of the chemical diffusivity in CdS at high Cd activity: △ from ref. 103, + from ref. 56. The solid line has been calculated from self-diffusion data of ref. 53.

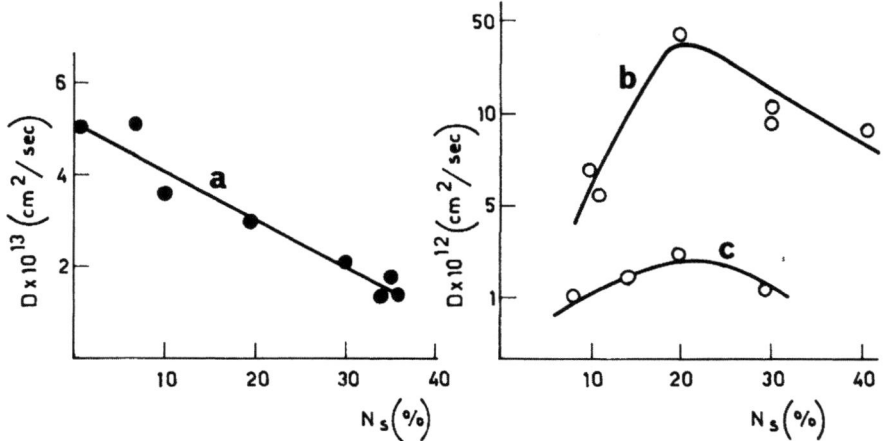

FIGURE 6. S (curve a), Se (curve b) and Hg (curve c) self-diffusion in $HgSe_{1-x}S_x$ single crystals ($0 < x < 0.4$) at 300°C [64].

Table II

Diffusion in II-VI Semiconducting Compounds.

Host Substance	Diffusing Species	Interpretation Model or Results (when model is not stated)	Ref.
ZnS	S	various kinetics, according to firing atmospheres (vacuum, He, H_2S, HCl, O): oxidation, exchange reaction, gas-solid interface sublimation, sintering of ZnS particles; $D_o = 2.16 \cdot 10^4 \text{ cm}^2\text{s}^{-1}$, $Q = 3.15$ eV	48
ZnSe	Zn	doubly ionized Frenkel defects	49
ZnSe	Se	neutral Se interstitials + doubly ionized Se vacancies	50
ZnTe	Zn	vacancy	50
CdS	S	S vacancies, 3 times faster than Cd diffusion	51
		doubly ionized vacancy, double donors, at $P_{Cd} = 4$ atm; neutral interstitials (Cd doubly ionized vacancy) at $P_{S_2} = 2$ atm vacancy (see Fig. 4)	52, 53
			54, 55
CdS	chemical diff.	good agreement with self-diffusion (see Fig. 5)	56
CdS	Cd	singly ionized interstitials, few donors compared to S vacancies small amount of fast diffusing atoms appears in excess Cd vapour D independent of annealing in S vapour	53, 57, 52
CdS_xSe_{1-x}	growth of mixed crystals	evidence of a single diffusion process	58
CdSe	Cd	doubly ionized Cd interstitials and Se vacancies interstitial Cd	59, 60
CdTe	Cd	interstitial	61
CdTe	chemical diff.	$\tilde{D} = (5 \pm 2) \, 10^{-5} \text{cm}^2\text{s}^{-1}$ at 900°C; $Q = 1.1$ eV	62
CdS	Se, Te	two components: vacancy (slow) + interstitial (fast)	63
CdSe	S, Te		
CdTe	S, Se		

Table II
(continued)

Diffusion in II-VI Semiconducting Compounds.

Host Substance	Diffusing Species	Interpretation Model or Results (when model is not stated)	Ref.
$HgS_{0.2}Se_{0.8}$	S,Se,Hg	Hg vacancies (see Fig. 6)	64,65
$HgS_{0.2}Te_{0.8}$	S,Hg	S and Te vacancies	64,65
HgSe	Se	$D_o = 2 \cdot 10^{-7} cm^2 s^{-1}$, $Q = 0.76$ eV	65
HgSe	Hg	$D_o = 5.6 \cdot 10^{-9} cm^2 s^{-1}$, $Q = 0.50$ eV	65
HgTe	Hg	$D_o = 1.8 \cdot 10^{-8} cm^2 s^{-1}$, $Q = 0.58$ eV	65
HgTe(γ irrad)	Hg	$D = 3 \cdot 10^{-13} cm^2 s^{-1}$ at r.t., value of D at 250°C without irradiation	66
ZnS	Au	$D_o = 1.75 \cdot 10^{-4} cm^2 s^{-1}$, $Q = 1.16$ eV	49
CdS	Cu,Ag,Au	measured D values are in agreement with tracer results	67
CdS	Cu,Ag,Au	interchange (slow) + interstitial (fast)	68
CdS	Cu,Ag,Au	diffusion attended by dislocation generation; dislocation density increased by a factor roughly proportional to the square of the ionic radius of the diffusing element	69
CdS	Ag	$D_o = 7 \cdot 10^{-3} cm^2 s^{-1}$, $Q = 0.63$ eV	70
CdS	Au	fast diffusion due to vacancy deficiency in the substitutional sublattice in saturated Cd or CdS vapour; dissociative mechanism in S_2 vapour; positively charged interstitial in Ar atmosphere	71
CdSe	Cu	high conductance due to $Cu_{2-x}Se$, low conductance due to Cu compensation of CdSe	72
CdSe	Cu,Ag,Au	measured D values are in agreement with tracer results	67
CdTe	Cu	$D_o = (8.2 \pm 0.8) \cdot 10^{-8} cm^2 s^{-1}$, $Q = (0.64 \pm 0.04)$ eV; surface related effects	73
ZnS	Cd	$D = 10^{-10} cm^2 s^{-1}$ at 1100°C; linear dependence of lnD on composition of $Zn_xCd_{1-x}S$ mixed crystals	74
CdS	Zn	$D = 5 \cdot 10^{-12} cm^2 s^{-1}$ at 1100°C; mixed crystals	74

Table II
(continued)

Diffusion in II-VI Semiconducting Compounds.

Host Substance	Diffusing Species	Interpretation Model or Results (when model is not stated)	Ref.
CdS	Zn	substitutional (slow) + interstitial (fast)	75
CdS	Te	dissociative model (slow, surface) + dislocations (fast, deep)	76
HgTe (γ irrad)	Cd	$D = 1.8 \cdot 10^{-12} \text{ cm}^2 \text{s}^{-1}$ at r.t., value of D at 210°C without irradiation	66
CdS	In	displacement of Cd by In	77
CdS	Yb^{3+}		78
CdTe	In	Cd ion vacancies (slow, deep) + saturated surface component	79
		at low Cd vapor pressure, substitutional through Cd sublattice, at high Cd vapor pressure, anomalously fast diffusion gives evidence of other mechanism	
ZnSe $ZnSe_{0.5}Te_{0.5}$ ZnTe	Al	substitutional Al in Zn sublattice	80
HgTe	Sn	surface region: slow d. due to formation of a (HgTe+SnTe) phase; deep region: fast d., positive charge of the diffusing atom	81
ZnS	O	model application of d. with simultaneous oxidation reaction	82
CdTe	O	interstitial nonionized atoms, binding of Cd to O substitutional in the Te sublattice?	83
ZnS	Cl	effects of firing atmospheres and impurity addition studied	84
ZnS	Mn	$D_o = 2.3 \cdot 10^3 \text{ cm}^2 \text{s}^{-1}$, $Q = 2.46$ eV	49

Table III

Diffusion in IV-VI Semiconducting Compounds.

Host Substance	Diffusing Species	Interpretation Model or Results (when model is not stated)	Ref.
PbSe (various dopings)	Se	neutral interstitial in high Se excess; singly ionized interstitial in slight Se or Pb excess; vacancy pair or some other defect pair in high Pb excess; anions move as rapidly as Pb	85
PbSe	interdiffusion	different D values in p and n material are explained by a Schottky-Wagner and a Frenkel mechanism	86
PbTe	Te	interstitialcy mechanism (dumbell Te split interstitial along a (111) direction)	87
PbTe	Pb	Frenkel disorder in the Pb sublattice: interstitialcy in the Pb-rich side, vacancy in the Te-rich side	87
PbTe		predominant cation defect is of the Frenkel type, and is singly ionized	88
PbS (polycr. film)	O	grain boundary d. + fast chemisorption	89
PbS	Se	$D_o = 1.41 \cdot 10^{-6} \text{cm}^2\text{s}^{-1}$, $Q = (0.93 \pm 0.08)$ eV	85
PbSe	Ni	formation of a $Ni_3Pb_2Se_2$ compound by interdiffusion	90
PbTe	Se	$D \simeq (1-2)10^{-10} \text{cm}^2\text{s}^{-1}$ at 700°C	85
PbTe	Ni	mostly a slow (front) and a fast (deep) penetration sections; d. parameters strongly dependent on crystal composition, cation vacancy concentration, time	91
GeTe-SnTe (heterojunction)	interdiffusion	$D_o = 1.3 \cdot 10^{-5} \text{cm}^2\text{s}^{-1}$, $Q = 0.60$ eV	92

Table IV

Diffusion in Semiconducting Compounds Other Than III-V, II-VI and IV-VI.

Host Substance	Diffusing Species	Interpretation Model or Results (when model is not stated)	Ref.
CdSb	Fe	$D_0 = 1.3 \cdot 10^{-6} cm^2 s^{-1}$, $Q = 0.74$ eV in single crystals; evidence of grain boundary d. in polycrystals	93
CuS	Cu	Cu vacancies	94
Cu_2S	Cu	ionic diffusivity of Cu^+	95
$Cu_{1.8}Se$	Cu, Ag	positive ions	96
CuS and FeS	Cu and Fe	mathematical treatment of kinetic reaction and migration of ions in nonstoichiometric thick layers with "smooth phase boundaries"	97
αAg_2S	Ag	collinear interstitial	98
βAg_2Se, βAg_2Te	Ag	interstitials	99
βAg_2Se	Ag	interstitial Ag ion and Ag ion vacancy	100
$Ti-Ti_5Si_3$	Si	epitaxial growth of Ti_5Si_3 fibers by d. of Si	101
$Ti-Ti_5Ge_3$	Ge	epitaxial growth of Ti_5Ge_3 fibers by d. of Ge	102

DIFFUSION IN IV-VI AND OTHER COMPOUNDS

Table III collects the results on diffusion in IV-VI compounds. Here too, self-diffusion has so far caught most of the attention, with interstitials or Frenkel disorder being the predominant explanations. Grain boundaries and chemisorption, of course, play an important role in O diffusion in polycrystalline films of PbS. In certain conditions, also Ni in PbTe shows two-slope (front and deep) penetration profiles.

Finally, a variety of interpretations is brought out by Table IV, which collects a miscellanea of host substances and diffusing species. Of course, diffusion of Fe in CdSb polycrystals takes place along grain boundaries. The proposal of collinear interstitial to explain the results on the cationic conductivity of Ag in αAg_2S is interesting.

Also interesting, for mechanical applications, is the possibility of using epitaxial growth of Ti_5Ge_3 or Ti_5Si_3 fibers in a Ti matrix by diffusion of Ge or Si to obtain fiber-reinforced eutectic materials with grown-in fibers.

CONCLUDING REMARKS

As a conclusion, we may say that a systematic comparison of independent results certainly helps us to understand mechanisms but so large is the variety of modes for diffusion in compound semiconductors that the published results are still unsufficient for satisfactory in-group comparisons. A certain delay in publishing results due to industrial secrecy limits the availability of already-obtained data and so hinders wide-span comparisons.

ACKNOWLEDGMENT

Thanks are due to Miss R. Sarno for invaluable help in checking the References and for tireless patience in critical typing of the tables.

REFERENCES

[1] B. I. Boltaks, Diffuziya v Poluprovodnikakh, Moskwa, 1961; English translation by J. I. Carasso, Diffusion in Semiconductors, Edited by H. J. Goldsmid, Infosearch Ltd., London, 1963.

[2] B. L. Sharma, Diffusion in Semiconductors, Trans Tech Publications, Clausthal-Zellerfeld, 1970.

[3] H. R. Winteler, Helv. Phys. Acta 44, 451 (1971).
[4] B. D. Lainer, V. V. Rakov, M. G. Milvidskii and
     I. A. Magidson, Dokl. Akad. Nauk SSSR 185, 142 (1969).
[5] A. A. Galaev and M. I. Ryaboi in Poverkh. Diffuz. Rastekanie,
     Mater. Konf. (1967, publ. 1969) p. 78 (Ed. Ya. E. Geguzin,
     Izd. Nauka, Moskwa).
[6] V. I. Safarov, V. E. Sedov and T. G. Yugova, Fiz. Tekh.
     Poluprov. 4, 150 (1970).
[7] B. I. Boltaks, T. D. Dzhafarov, G. S. Kulikov and
     V. M. Mikhelashvili, Fiz. Tverd. Tela 12, 3256 (1970);
     Sov. Phys. Solid State 12, 2631 (1971).
[8] T. D. Dzhafarov and G. S. Kulikov, Fiz. Tverd. Tela 12, 1564
     (1970); Sov. Phys. Solid State 12, 1237 (1970).
[9] T. D. Dzhafarov, Fiz. Tverd. Tela 12, 2801 (1970); Sov. Phys.
     Solid State 12, 2259 (1971).
[10] M. Ilegems and W. C. O'Mara, J. Appl. Phys. 43, 1190 (1972).
[11] Y. Ono and K. Kurata, Jap. J. Appl. Phys. 11, 55 (1972).
[12] M. Fujimoto, Y. Sato, M. Ikeda, K. Kudo and N. Hishinuma,
     Denki Tsushin Kenkyujo Kenkyu Jitsuyoka Hokoku 18, 1655
     (1969).
[13] M. Fujimoto, Rev. Phys. Chem. Jap. 40, 21 (1970).
[14] M. Fujimoto, Y. Sato, M. Ikeda and K. Kudo, Rev. Elec.
     Commun. Lab. 18, 624 (1970).
[15] C. H. Ting and G. L. Pearson, J. Appl. Phys. 42, 2247 (1971).
[16] A. I. Blasku, B. I. Boltaks, T. D. Dzhafarov and
     F. P. Kesamanly, Fiz. Tekh. Poluprov. 5, 755 (1971);
     Sov. Phys. Semicond. 5, 664 (1971).
[17] C. H. Ting and G. L. Pearson, J. Electrochem. Soc. 118,
     1454 (1971).
[18] T. T. Lavrishchev, B. G. Abramov and S. S. Khludkov,
     Izv. Akad. Nauk SSSR, Neorgan. Mater. 7, 2081 (1971).
[19] B. Tuck, Phys. Stat. Solidi B45, K157 (1971).
[20] B. I. Boltaks and F. S. Shishiyanu in Poluprov. Soedin. Ikh
     Tverd. Rastvory (1970) p. 98 (Ed. S. I. Radautsan, Red.-Izd.
     Otd. Akad. Nauk Mold. SSR, Kishinev).
[21] E. P. Savin and B. I. Boltaks, Fiz. Tekh. Poluprov. 5, 1331
     (1971); Sov. Phys. Semicond. 5, 1173 (1972).
[22] M. A. H. Kadhim and B. Tuck, J. Mater. Sci. 7, 68 (1972).
[23] B. Tuck and M. A. H. Kadhim, J. Mater. Sci. 7, 585 (1972).
[24] C. H. Ting and G. L. Pearson, J. Electrochem. Soc. 119, 96
     (1972).
[25] M. Fujimoto, K. Kudo and N. Hishinuma, Jap. J. Appl. Phys.
     8, 725 (1969).
[26] K. Gamo, K. Aoki, K. Masuda and S. Namba, Jap. J. Appl. Phys.
     10, 1118 (1971).
[27] A. S. Kyuregyan and V. M. Stuchebnikov, Fiz. Tekh. Poluprov.
     4, 1591 (1970); Sov. Phys. Semicond. 4, 1365 (1971).
[28] B. I. Boltaks and E. P. Savin, Fiz. Tekh. Poluprov. 4, 567
     (1970); Sov. Phys. Semicond. 4, 470 (1970).

[29] B. L. Sharma, R. K. Purohit and S. N. Mukerjee, J. Phys. Chem. Solids 32, 1397 (1971).
[30] R. G. Hunsperger and O. J. Marsh, Appl. Phys. Lett. 19, 327 (1971).
[31] T. T. Lavrishchev and S. S. Khludkov, Izv. Akad. Nauk SSSR, Neorgan. Mater. 7, 2079 (1971).
[32] T. T. Lavrishchev, L. P. Vasileva, R. K. Zayatinov and S. S. Khludkov, Arsenid Galliya 2, 129 (1969).
[33] T. T. Lavrishchev and S. S. Khludkov, Izv. Akad. Nauk SSSR, Neorgan. Mater. 7, 310 (1971).
[34] C. F. Gibbon and D. R. Ketchow, J. Electrochem. Soc. 118, 975 (1971).
[35] C. F. Gibbon, E. I. Povilonis and D. R. Ketchow, J. Electrochem. Soc. 119, 767 (1972).
[36] A. B. Y. Young and G. L. Pearson, J. Phys. Chem. Solids 31, 517 (1970).
[37] A. E. Widmer and R. Fehlmann, Solid State Electron. 14, 423 (1971).
[38] J. Rachmann und R. Biermann, Solid State Commun. 7, 1771 (1969).
[39] T. T. Lavrishchev and A. P. Vyatkin, Arsenid Galliya 2, 205 (1969).
[40] L. N. Aleksandrov, A. V. Kamenskaya and V. M. Zaletin, Arsenid Galliya 3, 29 (1970).
[41] G. I. Rekalova, U. Kebe and L. A. Mezrina, Fiz. Tekh. Poluprov. 5, 776 (1971); Sov. Phys. Semicond. 5, 685 (1971).
[42] G. I. Rekalova, U. Kebe, T. V. Persiyanov, V. M. Krymov and E. D. Krymova, Fiz. Tekh. Poluprov. 5, 158 (1971); Sov. Phys. Semicond. 5, 134 (1971).
[43] V. A. Uskov and V. P. Sorvina, Izv. Akad. Nauk SSSR, Neorgan. Mater. 8, 758 (1972).
[44] S. S. Khludkov, G. L. Prikhodko and T. A. Karchina, Izv. Akad. Nauk SSSR, Neorgan. Mater. 8, 1044 (1972).
[45] D. N. Nasledov, Yu. S. Smetannikova, K. I. Vinogradova and V. K. Yarmarkin, Phys. Lett. A40, 224 (1972).
[46] Hj. Matzke, Radiat. Eff. 3, 93 (1970).
[47] T. D. Dzhafarov, T. T. Dedegkaev and L. M. Dolginov in Fiz. Elektron.-Dyrochnykh Perekhodov Poluprov. Prib.(1969) p. 188 (Ed. S. M. Ryvkin, Izd. Nauka Leningrad. Otdel., Lenningrad); and in Physics of p-n Junctions and Semiconductor Devices (1971) p. 209 (Eds. S. M. Ryvkin and Yu. V. Shmartsev, Consultants Bureau, New York).
[48] M. Sakaguchi, T. Hirabayashi and K. Oguro, Nippon Kagaku Kaishi 1, 34 (1972).
[49] V. A. Williams, J. Mater. Sci. 7, 807 (1972).
[50] M. M. Henneberg and D. A. Stevenson, Phys. Stat. Solidi B48, 255 (1971).
[51] E. A. Secco and R. Swee-Chye Yeo, Can. J. Chem. 49, 1953 (1971).

[52] L. A. Sysoev, A. Ya. Gelfman, A. D. Kovaleva and N. G. Kravchenko, Izv. Akad. Nauk SSSR, Neorgan. Mater. $\underline{5}$, 2208 (1969).
[53] V. Kumar and F. A. Kroeger, J. Solid State Chem. $\underline{3}$, 387 (1971).
[54] Kh. Kh. Mannanova and Kh. R. Niyazov, Izv. Akad. Nauk Uzb. SSR, Ser. Fiz.-Mat. Nauk $\underline{15}$, 75 (1971).
[55] Kh. Kh. Mannanova and Kh. R. Niyazov, Fiz. Tekh. Poluprov. $\underline{5}$, 785 (1971); Sov. Phys. Semicond. $\underline{5}$, 695 (1971).
[56] V. Kumar and F. A. Kroeger, J. Solid State Chem. $\underline{3}$, 406 (1971).
[57] E. D. Jones, J. Phys. Chem. Solids $\underline{33}$, 2063 (1972).
[58] H. F. Taylor, V. N. Smiley, W. E. Martin and S. S. Pawka, Phys. Rev. B$\underline{5}$, 1467 (1972).
[59] W. D. Callister Jr., C. F. Varotto and D. A. Stevenson, Phys. Stat. Solidi $\underline{38}$, K45 (1970).
[60] N. V. Dmitrieva, A. V. Vanyukov and S. G. Yakovlev, Elektron. Tekh. Nauk-Tekh. Sb. Mater. $\underline{5}$, 150 (1970).
[61] K. Zanio, J. Appl. Phys. $\underline{41}$, 1935 (1970).
[62] Yu. V. Rud' and K. V. Sanin, Fiz. Tekh. Poluprov. $\underline{6}$, 886 (1972).
[63] J. Żmija, Acta Phys. Polon. A$\underline{40}$, 435 (1971).
[64] F. F. Kharakhorin, D. A. Gambarova, F. A. Zaitov and R. V. Lutsiv, Izv. Akad. Nauk SSSR, Neorgan. Mater. $\underline{6}$, 564 (1970).
[65] F. F. Kharakhorin et al., Izv. Akad. Nauk SSSR, Neorgan. Mater. $\underline{5}$, 2212 (1969).
[66] F. A. Zaitov, N. V. Alekseev and G. E. Popovyan, Izv. Akad. Nauk SSSR, Neorgan. Mater. $\underline{7}$, 1624 (1971).
[67] J. Żmija, Acta Phys. Polon. A$\underline{39}$, 531 (1971).
[68] J. Żmija and M. Demianiuk, Acta Phys. Polon. A$\underline{39}$, 539 (1971).
[69] J. Pielaszek and J. Żmija, Phys. Stat. Solidi A$\underline{4}$, K123 (1971).
[70] S. N. Baranovskii and M. V. Demidenko. Izv. Vyssh. Ucheb. Zaved. Fiz. $\underline{13}$, No. 7, 12 (1970).
[71] G. K. Malysheva, Fiz. Tekh. Poluprov. $\underline{5}$, 481 (1971); Sov. Phys. Semicond. $\underline{5}$, 420 (1971).
[72] A. Matsuda, H. Okushi, M. Saito and M. Kikuchi, Solid State Commun. $\underline{9}$, 2241 (1971).
[73] H. Mann, G. Linker and O. Meyer, Solid State Commun. $\underline{11}$, 475 (1972).
[74] W. J. Biter and F. Williams, J. Luminescence $\underline{3}$, 395 (1971).
[75] J. Żmijia and L. Sados, Biul. Wojsk. Akad. Tech. $\underline{20}$, No. 4, 105 (1971).
[76] E. Nebauer and J. Lautenbach, Phys. Stat. Solidi B$\underline{48}$, 657 (1971).
[77] S. S. O'Tuama and J. Richter, J. Appl. Phys. $\underline{41}$, 1861 (1970).
[78] W. W. Anderson and H. J. Chang, J. Electrochem. Soc. $\underline{118}$, 1451 (1971).

[79] L. V. Maslova, O. A. Matveev, Yu. V. Rud' and K. V. Sanin in Fiz. Elektron.-Dyrochnykh Perekhodov Poluprov. Prib.(1969) p. 211 (Ed. S. M. Ryvkin, Izd. Nauka Leningrad. Otdel., Leningrad); and in Physics of p-n Junctions and Semiconductor Devices (1971) p. 234 (Eds. S. M. Ryvkin and Yu. V. Shmartsev, Consultants Bureau, New York).
[80] M. Aven and E. L. Kreiger, J. Appl. Phys. 41, 1930 (1970).
[81] F. A. Zaitov, Fiz. Tverd. Tela 13, 278 (1971); Sov. Phys. Solid State 13, 219 (1971).
[82] E. Mendoza, R. E. Cunningham and J. J. Ronco, J. Catal. 17, 277 (1970).
[83] F. F. Vodovatov, G. V. Indenbaum and A. V. Vanyukov, Fiz. Tverd. Tela 12, 22 (1970); Sov. Phys. Solid State 12, 17 (1970).
[84] M. Sakaguchi, K. Oguro and T. Hirabayashi, Nippon Kagaku Zasshi 92, 313 (1971).
[85] Y. Ban and J. B. Wagner Jr., J. Appl. Phys. 41, 2818 (1970).
[86] R. W. Brodersen, J. N. Walpole and A. R. Calawa, J. Appl. Phys. 41, 1484 (1970).
[87] M. P. Gomez, D. A. Stevenson and R. A. Huggins, J. Phys. Chem. Solids 32, 335 (1971).
[88] T. D. George and J. B. Wagner Jr., J. Appl. Phys. 42, 220 (1971).
[89] V. I. Il'in, Fiz. Tekh. Poluprov. 4, 229 (1970).
[90] M. L. Kantor, N. N. Myuller, L. M. Ostrovskaya and L. I. Sotnikova, Izv. Akad. Nauk SSSR, Neorgan. Mater. 7, 580 (1971).
[91] T. D. George and J. B. Wagner Jr., J. Phys. Chem. Solids 30, 2459 (1969).
[92] A. V. Agafonova, O. G. Vasilkova, V. E. Lebedeva and N. N. Myuller, Zavod. Lab. 36, 1091 (1970).
[93] Sh. Mavlonov, A. Radzhabov, A. Sadiev and A. A. Kuliev, Diffuz. Poluprov. (1967, Publ. 1969) p. 272; Ref. Zh., Fiz. E abstr. n. 5E820 (1970).
[94] I. Bartkowicz, E. Fryt and S. Mrowec, Zesz. Nauk Akad. Gorn.-Hutn. Krakovie, Ceram. 14, 19 (1969).
[95] A. Etienne, J. Electrochem. Soc. 117, 870 (1970).
[96] J. Kur and A. Scuczyszym, Biul. Wojsk. Akad. Tech. 19, No. 12, 37 (1970).
[97] S. Traikov and N. Filipovska, Tehnika (Beograd) 25, 62 (1970).
[98] I. Bartkowicz and S. Mrowec, Phys. Stat. Solidi B49, 101 (1972).
[99] N. Valverde, Z. Phys. Chem. N. F. 70, 128 (1970).
[100] T. Takahashi and O. Yamamoto, J. Electrochem. Soc. 118, 1051 (1971).
[101] F. W. Crossman and A. A. Yue, Met. Trans. 2, 1545 (1971).
[102] A. S. Yue and F. W. Crossman, Met. Trans. 1, 322 (1970).
[103] R. C. Whelan and D. Shaw, Phys. Stat. Solidi 36, 705 (1969).
[104] D. Shaw, Ed., Atomic Diffusion in Semiconductors, Plenum, New York, 1972.

CHAPTER 7

AMORPHOUS SEMICONDUCTORS

C. Hilsum

Ministry of Defence, Royal Radar Establishment

Great Malvern Worcestershire, England

In this book so far we have restricted our studies to crystalline materials, both because it is relatively simple to apply our physics to single crystals, and because of the number of versatile electronic devices which can be made from them. Nevertheless semiconducting properties can be found in non-crystalline media, and it is also possible that useful devices can be produced which would not be available from the conventional materials. This chapter is therefore devoted to amorphous semiconductors. We must stress at the outset that this is intended only as a brief survey, in contrast to the thorough study on crystalline semiconductors. We include it not just for completeness, but because this is one of the growth areas in the field, with development of the new physical theories which must be invoked and of the new devices which are being invented.

THE GLASSY AND AMORPHOUS STATES

It is commonly said that a glass is a supercooled liquid. This simple statement is not a complete definition, but it does introduce an important concept, the link between the glassy and liquid states. A rigorous definition is that a glass is a material which has been formed by cooling a liquid, and it has become solid by a steady increase in viscosity without any discontinuous changes in volume, heat content, or entropy. Like a liquid, then, a glass has short-range order, but no long range order. Obviously there are no regularly spaced atomic sites, but nevertheless there is a form of structure, revealed by the radial distribution function, which is the probability of finding an atom within a certain distance of another atom. This shows a series of peaks which become more blurred and ill-defined as the distance increases.

Some glass-like materials can be formed by evaporation, sputtering, electrolytic deposition or chemical vapor deposition. These materials can be distinguished in several ways from polycrystalline films, and they are similar to glasses in having no long-range order. However they cannot be called glasses since if the appropriate melt of their constituents were cooled, it would crystallize. They are accurately described as amorphous.

The glassy state is metastable, and in theory there will always be a tendency to devitrify. In many glasses this tendency is so minimal that it can be ignored - few people will have seen a window-pane become cloudy. The instability of other glasses is readily demonstrated - the glassy form of selenium will transform into one of the crystalline forms if the material is held for an hour or so at 120°C, well below the melting point. Materials which cannot be vitrified by quenching, but are available as amorphous films, are generally less stable than glasses. It is also often difficult to ensure that a defined amorphous state has been secured, for the properties of the layer vary markedly with the method of preparation. However these uncertainties are associated more with the thermodynamics of the materials than with the techniques of film formation. Materials which form true glasses can also be successfully made as films by evaporation, sputtering or in other ways. These methods are adopted when it is not convenient to use melt quenching. For example, many device configurations require the glass to be in the form of a thin layer.

AMORPHOUS SEMICONDUCTOR MATERIALS

The standard glasses we meet in everyday life are based mostly on silicon oxide, boric oxide or lead oxide. Such glasses generally conduct by ion movement, and have no relevance to our work. Our interest is shared between two classes of amorphous materials, the amorphous form of the standard semiconductors like Ge and Si, and certain true glasses which have semiconducting properties.

The interest in amorphous Ge, Si and III-V compounds arises because we hope that we can learn something about the physics of the amorphous state by similarities between and differences from the properties of single crystals of the same materials. This process of extrapolation has had a limited success, but it is difficult to get data which is truly representative of amorphous semiconductors because the properties depend markedly on the method of preparation. Selenium is perhaps the one semiconductor which can be readily studied in both the amorphous and crystalline form, and indeed has device applications in both forms, amorphous Se being used in xerography, and crystalline Se for photocells. Glassy selenium is readily devitrified by heating, but its stability is improved by the addition of either As or Ge. The effect is however reversed if the As proportion exceeds 60% or the Ge proportion 30%. These binary systems are true semiconductor glasses, and have many points in common with the more complex systems which are now the object of detailed study because of their device potential. Hundreds of combinations of over 20 elements have semiconducting properties, but most promise appears to be shown by glasses based on elements from the VI Group, the chalcogenide glasses.

Arsenic trisulphide has been known for some years as a very stable infra-red transmitting glass. Good glassy properties are retained for sulphur proportions increased up to 90%, or arsenic up to 70%. Arsenic tellurides are difficult to make as glasses, but amorphous films can readily be prepared by deposition on to cold substrates. These two systems and the arsenic selenides have proved good hosts for the study of semiconductor properties, but have not yet demonstrated device potential. Most compositions in the Ge-Te system crystallize when quenched, but a small region near the eutectic, $Ge_{15}Te_{85}$, gives glasses, and this has proved to be important for devices, both in its own right, and as a phase which separates from more complex systems.

Two well-studied ternary systems, Ge-As-Te and Si-As-Te have important glass forming regions [1], which are shown in Fig. 1. It should be remembered that the boundaries given on these diagrams are not unique, but depend on the thermal history. A material which crystallizes when quenched to room temperature may form a glass when quenched with liquid nitrogen. Another important chalcogenide glass is the quaternary system formed by combining the two ternaries given above. A particularly useful composition for devices, one of the prototype materials, is $Si_{12}Te_{48}As_{30}Ge_{10}$, though the precise proportions are probably not significant. This glass is often known as STAG.

Though the chalcogenides are thought by most laboratories to have the greatest device possibility, they do have the disadvantage of a low maximum temperature for device operation. This is probably linked with the low transition temperature, T_G, the temperature at which there is a second order phase transition, an increase in heat capacity and a change in expansion coefficient. There is therefore some interest in glasses made with different glass formers, and of these the metal-oxides have received most

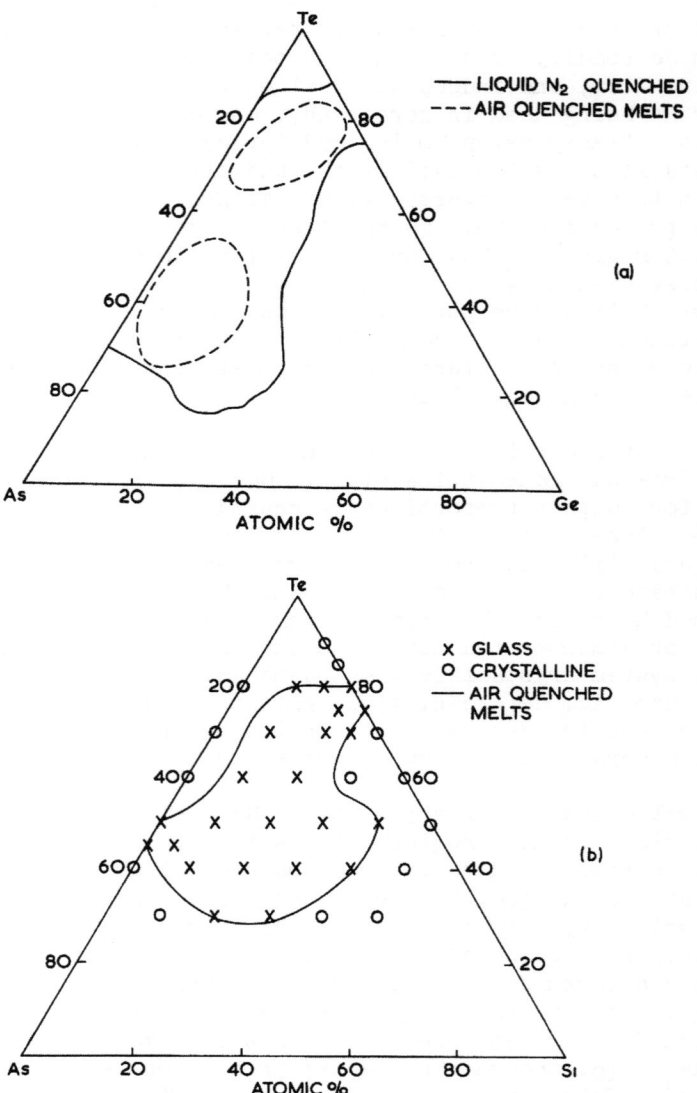

FIGURE 1. Glass forming regions of (a) Ge-As-Te (b) Si-As-Te

attention. Among these are the iron phosphates, the copper phosphates and the vanadium phosphates. Vanadium oxide alone can be prepared as a glass, and in addition can be mixed with tellurium oxide or barium oxide. The chemistry of the oxide systems is more complex than that of the chalcogenides. Ions

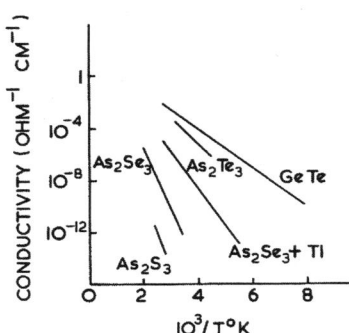

FIGURE 2. Temperature dependence of conductivity for chalcogenides.

can exist in several valence states, and the thermal treatments and melt atmosphere must be controlled accurately to achieve reproducible glasses.

The more complex chalcogenide and oxide glasses are difficult to prepare as homogeneous uniform materials. Often there is a two-phase structure, and sometimes one of the phases is crystalline. Some glasses show a sub-structure of tiny drops a micron or so in diameter. It is not then sufficient to know the composition of the glass under examination - one must also know the degree of crystallinity and of homogeneity.

SEMICONDUCTING PROPERTIES

Glasses have a high resistivity compared with the standard semiconductors. At room temperature the resistivity can be as high as 10^{16} ohm-cm, for As_2S_5, or as low as 100 ohm cm for GeTe. Small quantities of impurity have little effect on the conductivity of most amorphous materials. Thin films sputtered from highly doped germanium are highly resistive, and this type of behavior is observed in general. However silver added to As_2S_3 in quantities of 0.1 to 1% causes a dramatic reduction in resistivity [2], and copper, gold, tin and lead have a similar though less marked effect. It is likely that the impurity here does not create impurity levels, but reduces the disorder in the glass.

All glasses show an increase in conductivity with temperature, and in chalcogenides there is a simple exponential variation with $1/T$ for an appreciable temperature range near room temperature (Fig. 2). The behavior has some similarities to an intrinsic semiconductor. In some other amorphous materials there are two or three activation energies over a temperature range of several hundred degrees.

Mobilities are universally low, and often it is difficult to establish even the sign of the Hall coefficient. Where it can be measured, it is negative. On the other hand the thermo-electric power is nearly always positive. The mobility deduced from the application of normal semiconductor formulae is about 10^{-1} cm^2/v.s. for chalcogenides, and rather lower than this for amorphous germanium.

Most amorphous materials show a well defined optical absorption edge lying between 0.3 and 2.5eV, the more stable glasses in general having their edges at the higher energies. The shape of the edge follows an exponential law, and therefore does not correspond to either direct or indirect transitions. This edge shape is said to obey Urbach's rule, and was observed some time ago for alkali halides, and some of the II-VI compounds. A number of models have been proposed to explain it, including field broadening of exciton lines or of direct allowed transitions. Since no model has received universal acceptance, it is not possible to define an optical activation energy. An arbitrary scale can be established by noting the energy value E_1 at which the absorption coefficient reaches 10^4 cm^{-1}. Above the exponential tail a number of glasses have absorption coefficients varying as $[\hbar\omega - E_0]^2/\hbar\omega$, a similar relationship to that for indirect transitions in crystals. E_0 then represents an energy gap, but there is no certainty that there is a zero in the density of states. E_0 is slightly less than E_1, and for chalcogenides corresponds to an absorption coefficient smaller than 10^3 cm^{-1}. The values of E_0 and E_1 are roughly twice the electrical activation energy E, as would be expected of an intrinsic semiconductor, with a Fermi level near the centre of the gap. A more accurate comparison for the chalcogenides shows that 2E is 10-20% less than E_0.

Glasses are non-ohmic. The current increases more rapidly than the field for fields greater than about 10^4 volts/cm. There is the possibility, at least in thin films, of electrode effects, giving carrier injection, space-charge growth, and finally space-charge limited currents. In many experiments electrode effects are eliminated, but large non-ohmic effects are still observed. It is generally accepted that these effects are due to the Poole-Frenkel mechanism, a lowering of a coulombic barrier by the electric field. The field F lowers the ionization energy of the centre by $2\sqrt{Fe^3/K}$. This amount of barrier reduction would occur in the absence of tunnelling. At low temperatures tunnel currents will give an additional field dependent contribution.

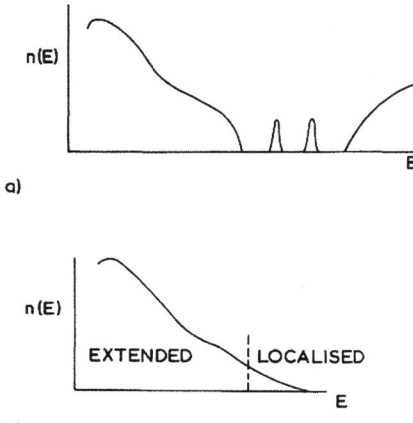

FIGURE 3. Band structure in (a) ordered imperfect crystal (b) disordered material.

BAND MODELS

In the past three years there has been a coalescence of views on the band structure of amorphous semiconductors. Different authors however arrive at similar end-points after following a variety of paths. We consider here the three main approaches. Cohen, Fritsche, and Ovshinsky [3] develop their model by considering an imperfect crystal as an intermediate state between the perfect crystal and the disordered solid. For a perfect crystal the density of states has band edges which are sharp, square root singularities within the bands, and distinct energy separations between the upper boundary of one band and the lower boundary of the next band. Each localised imperfection which gives a large change in crystal potential creates a localized bound state in the gap between the two bands. Additional imperfections give extra bound states, and if there are a large number of imperfections the bound states become a subsidiary band within the energy gap (Fig. 3a). The square root singularities within the main bands are by now rounded off, but the sharp edges remain. As the crystal perfection decreases still more the subsidiary bands within the energy gap merge with the main bands, giving tails of localized states continuous with the bands of extended states (Fig. 3b). This could well represent the band structure of amorphous germanium. A more complex material, such as a chalcogenide, can show compositional disorder as well as structural disorder. The tail states will now be much more pronounced, and the upper energy tail from one band may overlap with the lower energy tail of the next band,

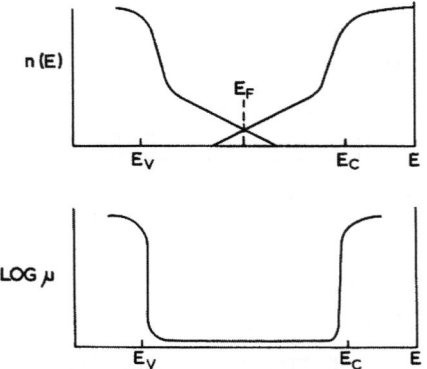

FIGURE 4. The mobility gap.

giving localized states throughout the "forbidden" gap. The Fermi level will be within the region of overlap, and the valence band states above the Fermi level will empty and operate as electron traps. Similarly the conduction band states below the Fermi level will become hole traps.

Though there are now allowed states at all energies, the nature of the states changes with energy. A carrier in the localized states moves by phonon-assisted hopping, and its mobility is therefore low. In the extended states the carriers will be subject to a number of scattering processes, but their mobility will be 2 or 3 orders of magnitude greater. The mobility is therefore a function of energy, showing a steep change at energies corresponding to the previous conduction and valence band edges. A mobility gap therefore replaces the density-of states gap (Fig. 4).

Mott [4,5] developed his model from some work by Anderson [6] on electron transport in a tight-binding impurity band. In a compensated semiconductor there will be variations in local fields due to the random arrangement of charged acceptors. A regular lattice of square potential wells gives a band of width in energy J. A potential energy U added to the wells at random causes little change while U/J is small. When U/J is larger than about 0.5 the mean free path is comparable with the lattice spacing, and the phase of the wave functions now varies randomly from atom to atom. At a critical value of U/J, thought by Anderson to be near 5, the wave functions become localized. Mott suggests that the Anderson criterion is satisfied in the tail states of bands, giving a range of localized states, and there is a discontinuous change in conductivity at the band edges.

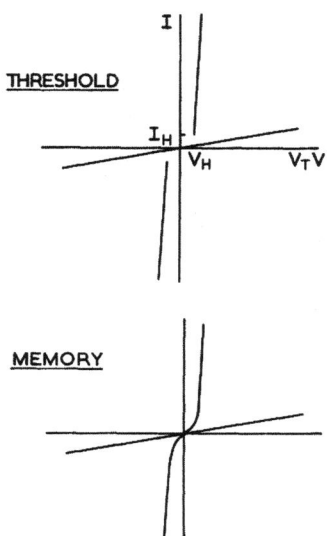

FIGURE 5. Electrical characteristics of glass switches.

Gubanov [7] treats the glassy solid as a "frozen" liquid, and the liquid as a crystal that has melted. A one-dimensional chain of atoms in the crystal has its periodicity destroyed when the solid melts. The potential energy now has maxima and minima displaced in real space, and wells in which the depth is altered by a random amount. The spatial distortions give the more important effects, and they lead to a broadening of the energy bands, and a diffuseness of the edges. Large fluctuations in short-range order give rise to localized states.

All three models predict tail states extending to several tenths of an eV, and localized state densities of 10^{18}-10^{19} cm^{-3}. They differ in their explanations for the cause of the localized states. Cohen finds they arise from density fluctuations in a random atomic structure, Mott from random well depths in a periodic well structure, and Gubanov from random spacing of wells of uniform depth, with a small effect due to well depth variations.

APPLICATIONS

Amorphous semiconductors, in the form of selenium or arsenic-selenium glasses, have been used for many years as the photosensitive layer in document copying machines. This important

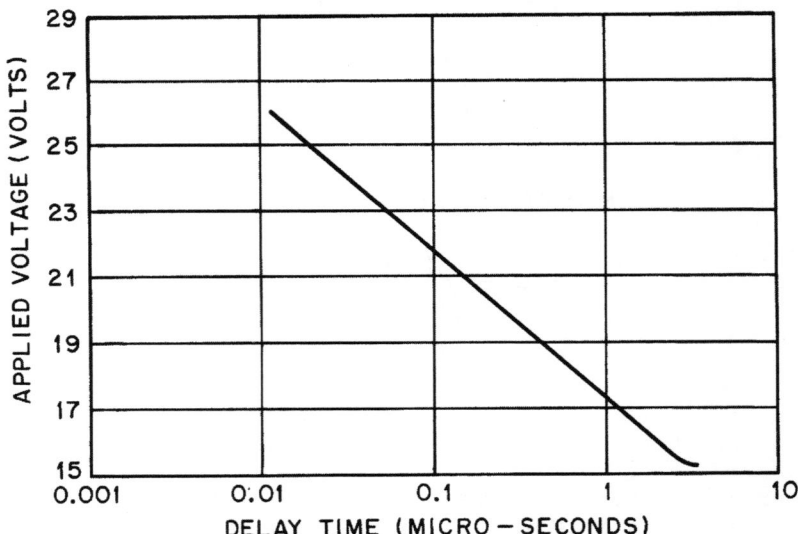

FIGURE 6. The delay time as function of applied voltage V_T = 15 volts [8].

application however did not generate the international interest in this topic which it probably deserved. This waited until the discovery and exploitation of glass switching devices.

Glass Switches

The two main types of glass switch, with characteristics illustrated in Fig. 5, are the threshold switch and the memory switch - sometimes called monostable and bistable switches respectively. The threshold switch shows a high resistance at voltages below a critical value, V_T, at which it changes to a conducting state. This threshold voltage is almost proportional to the electrode separation, but varies with glass composition. In the conducting state the current can be varied without appreciably changing the voltage drop across the device, provided it is not reduced below a holding current I_H. At lower currents the device switches back to the resistive state.

AMORPHOUS SEMICONDUCTORS

The memory switch similarly has a resistive state and a threshold voltage, but the conducting state is maintained even when the bias is removed. The device is returned to the resistive state by the passage of a short high current pulse. Both threshold and memory switches have symmetrical current-voltage characteristics. Threshold switching is actually more complex than the simple description above would indicate. Switching does not occur immediately V_T is exceeded. The device remains in the high resistive state for a period of time before switching, and this delay decreases rapidly with voltage increases above V_T [8] (Fig. 6). Also there is a clear link between threshold and memory operation. A memory switch must be held in the conducting state for a certain minimum time to establish a stable low resistance. Otherwise the device performs like a threshold switch.

Device construction has much in common with conventional semiconductor devices. Two designs may be used, the sandwich or "through" switch (Fig. 7), and the gap cell [9] (Fig. 8). Care must be taken in the choice of electrode materials. Molybdenum appears to be satisfactory, if a thick film of a good thermal conductor is deposited on top of it, and pyrolytic carbon has been used for gap cells.

A large number of glass compositions have been tried for each type of switch, and obviously the more successful materials are proprietary. However it is known that the $Si_{12}Te_{48}As_{30}Ge_{10}$ glass is good for threshold devices, and materials approximating to the $Ge_{15}Te_{85}$ eutectic are suitable for memories [10]. For either application additions of a few per cent of sulphur, selenium, phosphorus and antimony are common.

Oxide glasses show similar switching effects, but less work has been done on determining optimum materials and compositions. The copper phosphate glasses were originally thought to be promising [11], but recently attention has turned away from these towards vanadates and niobates. The technique used for the deposition of the glass as a thin film is critical to the successful operation of the devices. The simplest method is evaporation, but without care there is risk of contamination by the crucible material, and fractional distillation of the components. Electron beam heating overcomes the first problem, and flash evaporation the second. In general the method is slow, and the layers have defects and often pinholes. Sputtering is more often used, particularly with R.F. excitation. This gives good uniform films of predicted composition, but the composition of the cathode must be modified to allow for the different sputtering yields of the various elements in the glass. Chemical vapor deposition has not been widely used, because the yields are high only at substrate temperatures where there is a risk of devitrification.

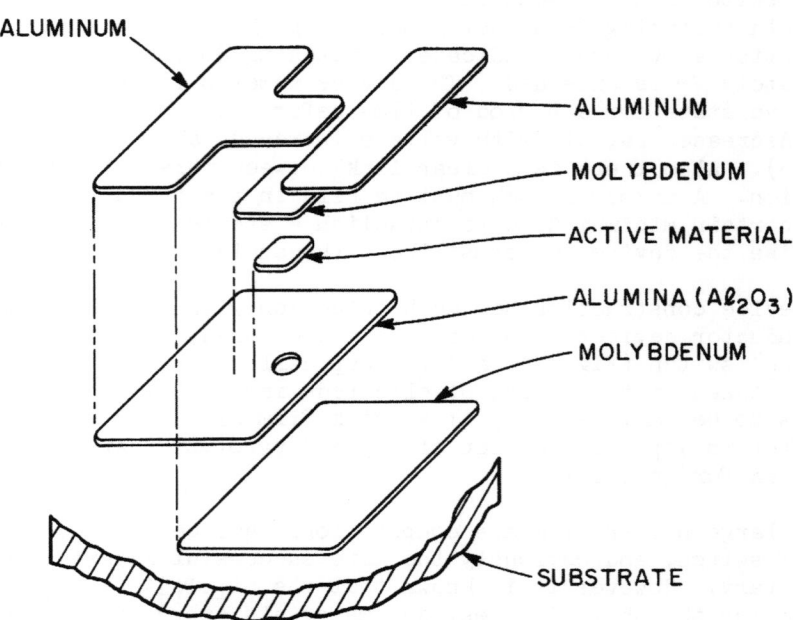

FIGURE 7. Exploded view of thin film sandwich device [9].

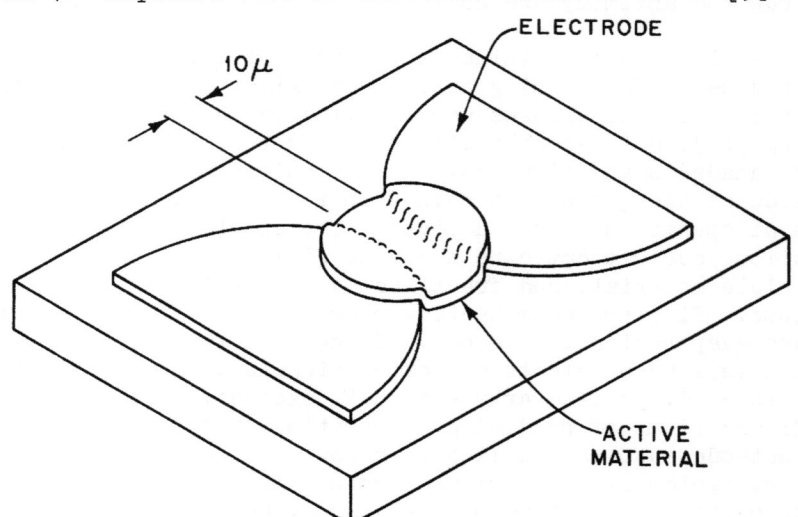

FIGURE 8. Thin film gap device [9].

Since all the preferred techniques of glass deposition are slow, there is a tendency to keep the glass thickness small, and sandwich devices are generally in the thickness range 0.5 - 2 microns. The threshold voltage is then in the range 10-30 volts. It should be noted that most devices show a much higher V_T when first tested, in the range 50-100V, but V_T falls rapidly during the initial switching cycles. A plateau is reached after about 50 operations. Nominally identical devices show a small spread in V_T, but this is being reduced as manufacturing techniques improve. The holding current is an important parameter in threshold switches, and it is generally near 1 mA, with a holding voltage between 1 and 2 volts. A number of attempts have been made to reduce I_H, but these have had only limited success.

For many purposes a threshold voltage of over 50 volts is desirable, and it is not easy to achieve this in a sandwich construction. The gap cell is more suitable for this purpose. The anode-cathode separation can be set during the photo-lithographic processing, and a gap of 10-20 microns is common. With such structures V_T is between 50 and 250 volts, depending on the glass used. The spread in V_T for a defined structure is larger for gap cells than for sandwich devices.

The operating mechanism of threshold devices is still a subject of controversy. The current-voltage characteristic shows a negative-resistance with an S-shape, and it is well known that if a device with such a characteristic is biased in the negative resistance region it will switch along the load line to a higher current, with the formation of a conducting filament. This may be contrasted with the behavior of devices with an N shaped negative resistance. In this case high field domains form, which move towards the anode, as in the Gunn Effect. Though there is agreement that a conducting filament develops in the threshold switch, there are two main theories on its nature.

We can well appreciate that the conductivity of a glass increases rapidly with temperature. A relatively low electric power dissipation in a glass can cause a sufficient local temperature rise to reduce the resistance locally. As a result more current passes in that region, the temperature rises further and there is a thermal runaway. There can be no doubt that this mechanism is adequate to explain threshold switching in configurations where there is poor thermal contact to a heat sink. This is the case in devices made on bulk glass, where there is an anode-cathode separation of 100 microns or more. For thinner devices there is more opportunity of heat loss through the electrodes, and a thermal theory alone cannot then explain the switching process. If a term for the field dependence of conductivity is included in the analysis, most of the parameters

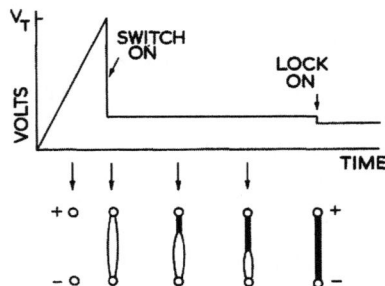

FIGURE 9. Filamentary conduction and crystallization in memory switch.

of switching can be explained [12,13,14]. The device temperature is about 30°K above ambient at V_T, but after switching rises 400-500°K in a filament a micron or so in diameter.

It cannot be denied that there are temperature rises in switches, but some authors maintain that these are incidental to the operation of the device. They suggest that an electronic mechanism must be involved, with injection of space-charge and modulation of barriers at the electrodes. It is of course well known that double-injection of carriers in high resistance crystalline semiconductors can lead to an S shaped negative resistance [15]. Evidence for the existence of electronic effects comes from detailed study of the conditions in the switch during the switching pulses. Some experiments have shown that devices switch when a defined amount of charge accumulates [16], others that in the ON state a transient reduction of current to zero leaves an appreciable voltage across the switch [17]. Neither of these results can be explained on an electro-thermal theory.

A further result which is difficult to explain on a thermal theory is the temperature dependence of threshold voltage of thin film switches. V_T falls with temperature, and eventually the device fails to switch [10]. The temperature at which this occurs is close to T_G, the glass transition temperature. This suggests that thermal avalanching generates a filament with a temperature above T_G, and a highly conducting phase is then precipitated out of the matrix of glass. Dendrites give a connecting path in the conducting state, but the tips are thermodynamically unstable and redissolve when the temperature drops.

These theories for threshold switch operation lead naturally to explanations for the memory switch. This is made from a glass whose composition is near to the amorphous-crystalline boundary, so that the hot filament that forms after threshold switching is readily replaced by a line of crystals. The crystallization is field enhanced, and appears to start at the anode. When anode and

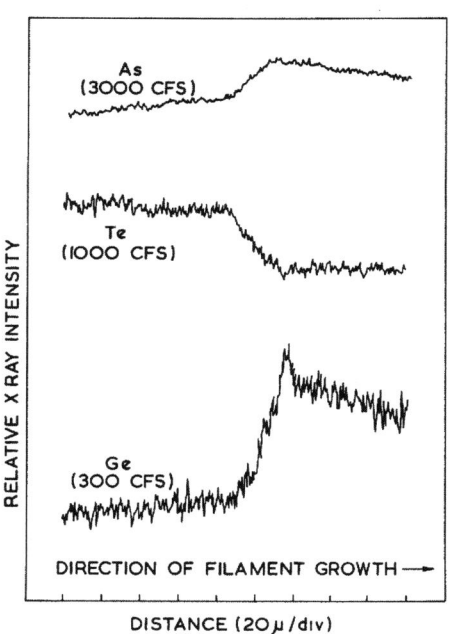

FIGURE 10. Arsenic, Tellurium and Germanium concentrations along a half-grown filament [18].

cathode are connected by crystals the device "locks on", with a slight reduction in resistance, and removal of the bias leaves the switch in the conducting state (Fig. 9). The existence of these crystalline filaments has been revealed by etching experiments, and concentration profiles have been measured along half-complete filaments [18]. Figure 10 shows that in a Ge-Te-As glass the crystalline precipitates are Tellurium rich. The resistive state is regained by a high current pulse which passes almost exclusively through the narrow crystalline filament, causing a steep temperature gradient. When the current pulse ends the filament cools very rapidly and the material freezes into the resistive state.

Development of glassy switches is mainly directed towards two applications, as the latching element in an electroluminescent display, and as an electrically addressed memory for a computer. The latching element may make possible a solid state display which can fulfil some of the functions of a cathode ray tube. Addressing an electroluminescent panel which has a large number of visual elements cannot give a high mean light output unless switching devices can be incorporated on the display, with one switch associated with each element. Since the display may be 20 cms or more square, and may contain 10,000 elements, it is not practicable to use conventional semiconductor switches. Glass switches can be deposited over the active display area on top of the polycrystalline ZnS electroluminescent layer, and they fulfil the function of a memory for each element. The simplest arrangement uses threshold switches, but for high resolution displays the holding current of present switches is too high. Various schemes using memory switches or combinations of the two types are under examination.

A read-mostly memory, in which the on-off states need to be altered infrequently, is an important part of many computers. A number of alternative approaches have been proposed to the existing ferrite cores or magnetic disk or tape stores. These include plated wires, magnetic bubbles, and MNOS arrays. The first two are not readily integrated with silicon chips, and the last is partially volatile, in that the stored information leaks over a few months. Glassy memory switches can be deposited on a silicon chip, so that the decoding circuitry can be incorporated. One specialized advantage that the switches have over conventional devices is that they are radiation-resistant. This gives a premium for military applications and for systems intended for operation in space. For normal commercial use glassy switches must prove competitive with MNOS memories, which appear to show the best balance of advantages of the different approaches. Two main problems face the designers of memory switches. The first is the life, which is about 10^6 reset operations. This is adequate for a read-mostly memory, but not for a display. The second problem is the high reset currents which are needed. Memory arrays need densities of 10^5 bits/cm^2, so that the devices must be small and the access lines to them narrow. This causes difficulties in heat dissipation and current transfer.

Though the two switches described above represent the most advanced glassy devices, there are other structures which may prove important. An adaptive memory has been described in which the device resistance can be varied electrically between that of the high resistance state and the conducting state [19]. Such a device would have many uses. A four-terminal glass switch is the equivalent of a relay. This exploits the fact that V_T decreases with temperature. A thin film resistor of low thermal capacity is incorporated in a threshold switch. The switch is biased near to V_T, but remains in the OFF-state. A pulse of current through the resistor heats the whole device, reduces V_T below the bias voltage, and induces switching. This device is readily developed from existing threshold switch technology.

Optical Imaging

Light can induce phase transitions in thin amorphous films [20]. There will be a number of ways in which the crystalline areas can be detected, and these define specific applications. The reflection or transmission change is adequate for microfiche duplication. A thermal development process either with or without additional light can enhance contrast. The techniques of electrostatic printing exploit the high resistivity of the untransformed amorphous material. Most of these transformations are reversible, in that a heating cycle converts the whole sensitive area back to the amorphous state. It has also been found that a laser beam, with suitable adjusted intensity, can first crystallize a glass and then erase the crystallized pattern. This effect can be exploited in a digital memory, in which the laser would be capable of creating a spot pattern, reading it and erasing it.

The optical sensitivity of amorphous films is rather less than that of Diazo materials, and is not high enough for holographic recording. The improvement in sensitivity which can be obtained by subsequent light exposure, as a separate development process, is not an unmixed blessing since this means that plates are light sensitive after their first exposure.

We can summarize this section by saying that there are definite applications for amorphous semiconductors in electro-photography and other techniques of document reproduction, lantern slide production, photolithography, and optically addressed computer memories. None of these except the first has yet reached the production stage, but considerable progress towards this has been made in the last two years.

CONCLUSION

Work on amorphous semiconductors has been expanding steadily recently, and progress on all fronts - fundamental physics, materials technology, device construction and applications - has been rapid. Whereas in 1968 it seemed unlikely that these uncontrolled and little understood materials could ever meet the stringent requirements of systems engineers accustomed to silicon integrated circuits, today we can clearly see areas where success is likely. Many scientists are now working in this field, and the literature grows too rapidly for one to follow all the developments. The physics of Amorphous Semiconductors is now a respectable subject, a study which can rank alongside the more ordered research on crystals.

ACKNOWLEDGMENTS

I am grateful to Dr. C. B. Thomas for valuable discussions and to Dr. J. Savage for the unpublished data included in Fig. 1.

RECOMMENDED READING

Journal of Non-Crystalline Solids, 4 (1970) and 8-10 (1972).

"Electronic Processes in Non-Crystalline Materials", N. F. Mott and E. A. Davis, Clarendon Press, Oxford (1971).

"Quantum Electron Theory of Amorphous Conductors", A. E. Gubanov, Consultants Bureau, New York (1965).

"Inorganic Glass-Forming Systems", H. Rawson, Academic Press, New York (1967).

REFERENCES

[1] J. A. Savage, J. Mat. Sci. 6, 964, 1971; 7, 64, 1972.
[2] J. T. Edmond, J. Non-Cryst. Solids 1, 39, 1968.
[3] M. H. Cohen, H. Fritzsche and S. R. Ovshinsky, Phys. Rev. Letts. 22, 1065, 1969.
[4] N. F. Mott, Advances in Physics, 16, 49, 1967; Phil. Mag. 22, 7, 1970.
[5] N. F. Mott and E. A. Davis, Electronic Processes in Non-Crystalline Materials, Clarendon Press, Oxford, 1971, Chapter 2.
[6] P. W. Anderson, Phys. Rev. 109, 1492, 1958.
[7] A. I. Gubanov, Quantum Theory of Amorphous Conductors, Consultants Bureau, New York, 1965.
[8] R. Shanks, J. Non-Cryst. Solids 2, 504, 1970.
[9] R. G. Neale, J. Non-Cryst. Solids 2, 558, 1970.
[10] J. R. Bosnell and C. B. Thomas, Solid-State Electronics 15, 1261, 1972.
[11] C. F. Drake and I. F. Scanlan, J. Non-Cryst. Solids 4, 234, 1970.
[12] J. M. Robertson and A. E. Owen, J. Non-Cryst. Solids 8-10, 439, 1972.
[13] D. M. Kroll and M. H. Cohen, J. Non-Cryst. Solids 8-10, 544, 1972.
[14] J. C. Male and A. C. Warren, Electronics Letters 6, 567, 1970.
[15] M. Lampert and R. B. Schilling, Semiconductors and Semi-Metals, Vol. 6, Academic Press, New York, 1970.
[16] D. R. Haberland and H. Stiegler, J. Non-Cryst. Solids 8-10, 408, 1972.

[17] H. K. Henisch, R. W. Pryer and G. J. Vendura, J. Non-Cryst. Solids, 8-10, 415, 1972.
[18] C. H. Sie, J. Non-Cryst. Solid 4, 548, 1970.
[19] E. J. Evans, J. H. Helbers and S. R. Ovshinsky, J. Non-Cryst. Solids 2, 334, 1970.
[20] S. R. Ovshinsky and P. H. Klose, J. Non-Cryst. Solids 8-10, 892, 1972.

CHAPTER 8

MAGNETIC SEMICONDUCTORS

C. Haas

Laboratory of Inorganic Chemistry
Materials Science Center of the University
Zernikelaan, Groningen, The Netherlands

INTRODUCTION

The applications of semiconductors are usually based on the fact that the charge carriers in a semiconductor form a very sensitive system, i.e. the charge carriers are easily disturbed by external influences. Examples of such disturbances are the injection of charge carriers by suitable electrical contacts, deviations of the charge carrier concentration from equilibrium as a result of the absorption of light, the thermoelectric effect resulting from the influence of a temperature gradient of the transport properties, the magnetoresistance, etc.

In magnetic semiconductors, the solid also contains magnetic moments. The application of an external magnetic field strongly influences the orientation of these magnetic moments. Thus, if there were an interaction between the charge carriers and the magnetic moments, one would expect a strong influence of a magnetic field on the transport properties. Effects of this type have been observed indeed in magnetic semiconductors. For example, the magnetoresistance of ferromagnetic semiconductors such as EuS or $CdCr_2Se_4$ is large, and shows a pronounced maximum at the Curie temperature [1-3]. The last point strongly suggests that the magnetoresistance is related to the ordering of the magnetic moments. The effects just mentioned have also been observed in ferrimagnetic semiconductors such as $FeCr_2S_4$ [1-3] and Cr_2S_3 [4].

The study of magnetic semiconductors began only a few years ago.* Although much data has been collected, the interpretation of the data is hampered by difficulties in obtaining good quality single crystals, or by complications due to impurities and deviations from stoichiometry. In this respect the area of magnetic semiconductors is still a primitive area if it is compared with the conventional semiconductors, such as Ge and InSb.

It is not only experimental difficulties which remain to be solved. Many theoretical problems have also not yet been answered in a satisfactory way. In the first place, little is known about the band structure of magnetic semiconductors. Secondly, an accurate theory of the transport properties of magnetic semiconductors has not so far been advanced.

In non-magnetic semiconductors, the transport properties are mainly governed by the interaction between charge carriers and phonons or impurities. These interactions are usually sufficiently weak to be treated in good approximation by perturbation-theoretical methods. This is not so for magnetic semiconductors; the interaction between the charge carriers and the localized magnetic moments is quite strong, and this makes the applicability of perturbation theory doubtful.

Nevertheless, in this chapter we will use perturbation theory to discuss the interaction between the electrical properties and the magnetic properties. This is not because one can expect that calculations of this type will reproduce accurately the observed phenomena, but merely because it is the simplest approach that describes in a qualitatively correct manner a number of interesting properties characteristic of magnetic semiconductors.

ELECTRONIC ENERGY LEVELS

Ferromagnetic Semiconductors

In this paragraph we discuss the influence of the magnetic properties on the electronic energy levels of a magnetic semiconductor. We shall treat only the case of electrons in a broad energy band or electrons in donor or acceptor levels, interacting with localized magnetic moments.

---

*A survey of theory and experimental data on magnetic semiconductors is given in refs. 1-3.

We assume that the interaction between a conduction electron and a magnetic atom of spin $\vec{S}$ can be written as a simple exchange interaction:

$$H' = -J(\vec{r} - \vec{R})\vec{s}\vec{S} \tag{1}$$

In this equation $\vec{s}$ represents the spin operator of the conduction electron, $\vec{S}$ that of the magnetic atom. The positions of electron and magnetic atom are $\vec{r}$ and $\vec{R}$, respectively. The function $J(\vec{r} - \vec{R})$ describes the dependence of the exchange interaction on the distance between electron and magnetic atom. The extension of this function is of the order of 1 Å, i.e. of the order of the extension of the orbitals of the electrons responsible for the spin $\vec{S}$. An interaction of the type, given in Eq. (1), has been used extensively in literature to discuss the interaction of conduction electrons with magnetic moments in transition and rare earth metals [5].

The magnitude of the interaction can be estimated from free ion data. Consider as an example the case of $CdCr_2Se_4$, and suppose that we wish to calculate the energy difference between a conduction electron with spin parallel and one with spin antiparallel to the magnetization in a completely magnetized crystal of the ferromagnetic semiconductor $CdCr_2Se_4$. The crystal contains as magnetic ions Cr^{3+}, with configuration d^3 and spin $S = 3/2$. For a free Cr^{2+} ion with configuration $3d^3 4s$ there are states of total spin $S = 2$ and $S = 1$, with the spin of the 4s electron parallel or antiparallel to the spin of the d electrons, respectively. Expressed in terms of Eq. (1), the energy difference between the $S = 2$ and the $S = 1$ states is

$$\Delta \varepsilon = 3/2 \int |\varphi_{4s}(\vec{r})|^2 J(\vec{r} - \vec{R}) dv(\vec{r}) \tag{2}$$

From free ion spectroscopic data one finds $\Delta \varepsilon = 0.9$ eV. Therefore, if the conduction band in a $CdCr_2Se_4$ crystal consisted mainly of 4s orbitals of Cr^{3+}, the expected energy difference between a conduction electron with spin parallel and spin antiparallel to the magnetization of a fully magnetized crystal would be of the order of 0.9 eV. This example shows that the influence of magnetic moments on the band structure is quite large.

We shall now discuss the interaction of conduction electrons with magnetic moments in a ferromagnetic semiconductor [6,7]. The magnetic atoms have spins $\vec{S}_n$, magnetic moments $-g\mu_B \vec{S}_n$, and are located at positions $\vec{R}_n$. The Hamiltonian for the interaction of a single conduction electron (position $\vec{r}$, spin $\vec{s}$) with all atomic spins is written as

$$H = H_o(\vec{r}) + H'(\vec{r},\vec{s},\vec{S}_n) + H_s(\vec{S}_n) \qquad (3)$$

In this equation $H_o(\vec{r})$ represents the Hamiltonian of the electron in the absence of the interaction H' with the magnetic moments. Since we consider electrons occupying states in a broad conduction band, the eigenfunctions of $H_o(\vec{r})$ will be Bloch functions $\varphi_{b\vec{k}}(\vec{r}) = u_{b\vec{k}}(\vec{r}) \exp(i\vec{k}\vec{r})$, where $\vec{k}$ is the wave vector and b an index labeling the energy bands. $u_{b\vec{k}}(\vec{r})$ is a periodic function, invariant for all translations which leave the crystal lattice unchanged. The functions are normalized in unit volume, i.e.

$$\int |\varphi_{b\vec{k}}(\vec{r})|^2 dv(\vec{r}) = 1 \qquad (4)$$

$$H_o \varphi_{b\vec{k}}(\vec{r}) = [(p^2/2m) + V_o(\vec{r})]\varphi_{b\vec{k}}(\vec{r}) = \varepsilon^o_{b\vec{k}} \varphi_{b\vec{k}}(\vec{r}) \qquad (5)$$

$V_o(\vec{r})$ is the periodic potential for the conduction electrons, p the momentum operator, and $\varepsilon^o_{b\vec{k}}$ the energy eigenvalue.

The interactions between the localized spins $\vec{S}_n$ are described by the Hamiltonian $H_s(\vec{S}_n)$, which contains exchange interactions, dipole-dipole interactions, magnetic anisotropy, etc. The eigenfunctions of H_s are called α, the corresponding energies ε_α, so that $H_s\alpha = \varepsilon_\alpha\alpha$. The number of states α is very large; it is $(2S+1)^N$ if there are N magnetic atoms of spin S in the crystal. Energy differences between these states are small; these states form a quasi continuum, containing the magnetic ground state and states with excited spin waves (magnons). At each temperature, there is a certain probability w_α for a state α to occur. Thus the statistical average of the value of the spin of the atom at R_n is given by

$$\langle \vec{S}_n \rangle = \sum_\alpha w_\alpha \langle \alpha | \vec{S}_n | \alpha \rangle \qquad (6)$$

$\langle \alpha | \vec{S}_n | \alpha \rangle$ represents the expectation value of $\vec{S}_n$ for the state α. The net magnetization of the crystal at a given temperature is given by

$$\vec{M} = -\sum_n g\mu_B \langle \vec{S}_n \rangle \qquad (7)$$

The term $H'(\vec{r}, \vec{s}, \vec{S}_n)$ is the interaction of the electron with the atomic spins, given by

$$H' = -\sum_n J(\vec{r} - \vec{R}_n) \vec{s} \vec{S}_n \qquad (8)$$

We now assume that the interaction term H' in the total Hamiltonian (Eq. (3)) is small, and that it can be treated as a perturbation. Consider an electron in an unperturbed band state $\varphi_{b\vec{k}}(\vec{r})$ of energy $\varepsilon^0_{b\vec{k}}$. The change of the electron energy as a result of the interaction H' is in first order given by the expression

$$\langle \varphi_{b\vec{k}}(\vec{r}) X\alpha | H' | \varphi_{b\vec{k}}(\vec{r}) X\alpha \rangle \qquad (9)$$

where X is the spin function of the electron. However we are not interested in the energy of the electron for a specific microscopic arrangement of the atomic spins, but rather in an average value corresponding to a certain temperature. That an averaging procedure of this type is allowed, is related to the fact that the dispersion of electron energies is much larger than the dispersion of the excitation energies of the spin system. This means that the electron moves rapidly compared to the slow motion of the magnetic moments, i.e. one can use the adiabatic or quasi-static approximation which considers the motion of an electron in a frozen arrangement of atomic spins.

The average first order change of the energy of an electron with wave vector $\vec{k}$ in energy band b is given by

$$\Delta\varepsilon^1_{b\vec{k}} = \sum_\alpha w_\alpha \langle \varphi_{b\vec{k}}(\vec{r}) X\alpha | H' | \varphi_{b\vec{k}}(\vec{r}) X\alpha \rangle \qquad (10)$$

Substituting H' and $\varphi_{b\vec{k}}(\vec{r}) = u_{b\vec{k}}(\vec{r}) \exp(i\vec{k}\vec{r})$ gives

$$\Delta\varepsilon^1_{b\vec{k}} = -\sum_n \int |u_{b\vec{k}}(\vec{r})|^2 J(\vec{r} - \vec{R}_n) dv(\vec{r}) \langle X|\vec{s}|X\rangle \sum_\alpha w_\alpha \langle \alpha|\vec{S}_n|\alpha\rangle \qquad (11)$$

Only collinear spin configurations will be considered; thus the average values of the spins $\vec{S}_n$ are all parallel to a common direction, which we call the z-axis. As spin functions for the conduction electron we choose X^+ and X^-, with spin parallel and

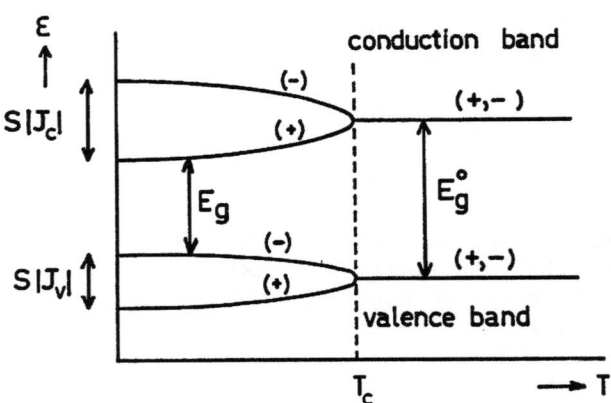

FIGURE 1. Spin splitting of the conduction band and the valence band of a ferromagnetic semiconductor. E_g is the energy gap.

---

antiparallel to the z-axis respectively: $s_z X^{\pm} = \pm \frac{1}{2} X^{\pm}$. If all magnetic atoms are equivalent, we can write

$$\int |u_{b\vec{k}}(\vec{r})|^2 J(\vec{r} - \vec{R}_n) dv(\vec{r}) = J_{b\vec{k}}/N \tag{12}$$

Substituting this result, and calculating the matrix elements of s, one obtains for the first-order energy change of electrons with spin parallel (+) or antiparallel (-) to the z-axis:

$$\Delta \varepsilon^{1,\pm}_{b\vec{k}} = \mp \tfrac{1}{2} S J_{b\vec{k}} (M/M_o) \tag{13}$$

where $M_o = -Ng\mu_B S$ is the saturation magnetization at $T = 0°K$, i.e. with all spins aligned parallel.

Because the magnetization is a function of the temperature, Eq. (13) introduces a temperature dependence of the energy of the band states. Moreover, because the energies of electrons with (+) and (-) spins are different for $M \neq 0$, there is a spin splitting of the band.

In Fig. 1 the results are sketched for a ferromagnetic compound. Below T_c, $M \neq 0$ and there is a spin splitting of the conduction and the valence bands. M decreases with increasing temperature, and so does the spin splitting of the bands. Above the Curie temperature T_c, $M = 0$ and the bands are degenerate for the two spin directions. However, in this region an applied magnetic field H induces a magnetization $M = \chi H$, where χ is the

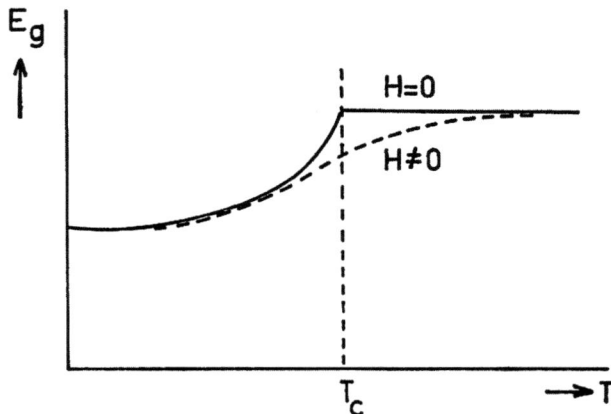

FIGURE 2. Temperature dependence of the energy gap E_g in a ferromagnetic semiconductor, with and without applied magnetic field.

---

magnetic susceptibility. Therefore a magnetic field induces a spin splitting of the bands proportional to χ and H at $T > T_c$:

$$\Delta \varepsilon^{1,\pm}_{b\vec{k}} = \mp \tfrac{1}{2} (SJ_{b\vec{k}}/M_o)\chi H \qquad (14)$$

This effect is largest at T_c, because at that temperature the susceptibility χ has a maximum.

From Fig. 1 it is clear that below T_c an anomalous temperature dependence of the energy gap E_g between conduction band and valence band is expected. The temperature dependence of E_g is sketched in Fig. 2. An applied magnetic field induces a change of the energy gap, given by

$$\frac{dE_g}{dH} = -S \{|J_c| + |J_v|\} (\chi/2M_o) \qquad (15)$$

So far we have discussed only first-order effects of H'. Because H' is not small, higher order terms will be important; these have been discussed by Rys, Helmann and Baltensperger [6]. The higher order terms depend on the short range order of the spins, and can be expressed in terms of spin correlation functions. These terms lead to energy shifts which extend well into the paramagnetic temperature region.

In the discussion we have also neglected interband terms of the type $\langle \varphi_{b\vec{k}} | H' | \varphi_{b'\vec{k}} \rangle$; this approximation is allowed if the spin splitting energies are small compared to interband energy differences $\varepsilon^o_{b\vec{k}} - \varepsilon^o_{b'\vec{k}}$.

If the conduction electron is localized in a finite region, it is able to produce a spin polarization of the magnetic moments in this region. For a strong interaction between electron and magnetic moments, such a localization may lead to a lowering of the total energy. Then the electron is self-trapped by the induced spin polarization, i.e. a magnetic polaron is formed [1].

An anomalous temperature dependence of E_g and a strong dependence of E_g on magnetic fields has been observed indeed for several magnetic semiconductors [1,2]. However, it is not certain that in all cases the observed effects are due to transitions between bands [8,9]; the effects in some of the compounds are attributed to magnetic exciton or impurity absorptions [10]. The spin splitting of the conduction band was also observed in experiments where electrons tunnel from metal electrodes into the conduction band of the europium chalcogenides [11].

Donors and Acceptors

For donor and acceptor energy levels one also expects a spin splitting, i.e. a different energy for electrons with spin parallel (+) and spin antiparallel (-) to the magnetization. Thus the energy level of a donor occupied by an electron with (+) or (-) spin can be written as

$$\varepsilon^{\pm}_d = \varepsilon^o_d \mp \tfrac{1}{2} J_d S(M/M_o) \qquad (16)$$

The value of J_d depends on the wave function of the electron in the donor state. In this section we give a qualitative discussion of the factors which determine J_d.

The Hamiltonian for an electron in a crystal with a donor atom is

$$H = H_o(\vec{r}) + H_s(\vec{S}_n) + H'(\vec{r},\vec{s},\vec{S}_n) + V_d(\vec{r}-\vec{R}_n) \qquad (17)$$

This Hamiltonian is that of the electron in a magnetic crystal (Eq. (3)) with, added to it, the interaction $V_d(\vec{r}-\vec{R}_o)$ of the electron with the donor atom located at $\vec{R}_o$. It is assumed that the presence of the donor atom does not change the interactions between the magnetic moments, contained in $H_s(\vec{S}_n)$, nor the interaction H'. This is a reasonable approximation if the donor atom substitutes for a non-magnetic atom, if the donor atom has the same spin and exchange interactions as the original lattice atom it replaces, or if the donor atom is a non-magnetic interstitial atom.

The ionized donor (i.e. donor atom minus electron) has an effective positive charge, and this leads to a Coulomb attraction between electron and ionized donor. At large distances between electron and donor this contribution to $V_d(\vec{r}-\vec{R}_o)$ can be written as $e^2/\varepsilon_o|\vec{r} - \vec{R}_o|$, where ε_o, the dielectric constant of the crystal, takes into account the electronic polarization. At short distances, $V_d(\vec{r} - \vec{R}_o)$ cannot be described by a simple Coulomb potential. In the first place the polarization effects at short distances are not adequately represented by a continuum description with a dielectric constant ε_o. The potential will also be modified by more specific interactions of the electron with other electrons of the donor atoms. Finally, local elastic deformations of the lattice around the donor will also change the energy of the electron, and contribute to $V_d(\vec{r} - \vec{R}_o)$.

In a non-magnetic semiconductor, the term H' is not present. Equation (17) then leads to a bound state, the donor level, at a certain energy ε_d^o [12]. This energy level is twofold degenerate because of the two possible spin orientations of the electrons. The ionization energy $E_D^o = -\varepsilon_d^o$ is the minimum energy required to raise the electron from the localized donor level into the conduction band. Although the donor level is twofold degenerate, it cannot be occupied by two electrons because of the repulsion between the two electrons.

The wave function of the electron in a donor state depends on $V_d(\vec{r} - \vec{R}_o)$. For the so-called shallow donors the ionization energy is small. This is particularly so for semiconductors with a large dielectric constant. The most important term in $V_d(\vec{r} - \vec{R}_o)$ is in that case $e^2/\varepsilon_o|\vec{r} - \vec{R}_o|$, and the Hamiltonian is similar to that of a hydrogen atom. The donor ionization energy is then given by $-\varepsilon_d^o = m^*e^4/2\hbar^2\varepsilon_o^2$. Examples are the shallow donors P, As, Sb in germanium, with donor ionization energies of about 0.01 eV. The electron in the bound state moves in a large orbit of radius $a = \varepsilon_o\hbar^2/m^*e^2$ around the donor atom. The wave function of the electron in the bound state consists of a combination of only slightly perturbed wave functions of the lower part of the conduction band [12].

For deep donors, with a large ionization energy, the radius of the orbit of the electron is small. The interaction with the donor atom $V_d(\vec{r} - \vec{R}_o)$ can no longer be approximated by a simple Coulomb potential. The electron is now strongly localized on the donor atom and a few neighboring atoms. It is even possible that the electron becomes localized almost completely on the donor atom, and that the donor states resemble the atomic states of the donor atom.

We now discuss the influence of H' on the donor states; this discussion differs on essential points from the discussion of the influence of H' on the band states. An electron in a band state is spread out over the entire crystal, and its influence on any individual spin will be very small. For the localized donor states, however, the electron is localized in a small volume around the donor. Therefore the electron is able to exert an appreciable influence on the atomic spins in the region around the donor atom.

Let us write for the wave function of an electron with spin ($\pm$) in the donor state $\varphi_d^{\pm}(\vec{r})$. The interaction energy of the spin of this electron with an atomic spin $\vec{S}_n$ at $\vec{R}_n$ is in first-order given by

$$\mp \tfrac{1}{2} \langle \varphi_d | J(\vec{r} - \vec{R}_n) | \varphi_d \rangle S_{nz} \qquad (18)$$

This is equivalent to the interaction of an effective magnetic field H_z^*, acting on the magnetic moment $-g\mu_B S_n$ at site $\vec{R}_n$:

$$H_z^*(\vec{R}_n) = \mp \tfrac{1}{2} (g\mu_B)^{-1} \langle \varphi_d | J(\vec{r} - \vec{R}_n) | \varphi_d \rangle \qquad (19)$$

This effective field H_z^* polarizes the spin $\vec{S}_n$. The total change of the energy of the electron consists of the interaction ε_1 of the electron with the atomic spins, and a term ε_2 which represents the energy required to produce the spin polarization. Therefore the donor energy is given by

$$\varepsilon_d^{\pm} = \varepsilon_d^o - \varepsilon_1^{\pm} + \varepsilon_2^{\pm} \qquad (20)$$

Here $\varepsilon_1^{\pm}$ is given by

$$\varepsilon_1^{\pm} = \mp \tfrac{1}{2} \sum_n \langle \varphi_d | J(\vec{r} - \vec{R}_n) | \varphi_d \rangle \langle S_{nz} \rangle \qquad (21)$$

where $\langle S_{nz} \rangle$ is the average value of the spin S_{nz} of an atom at $\vec{R}_n$ near the donor atom, taking into account the change in $\langle S_{nz} \rangle$ produced by the presence of the electron.

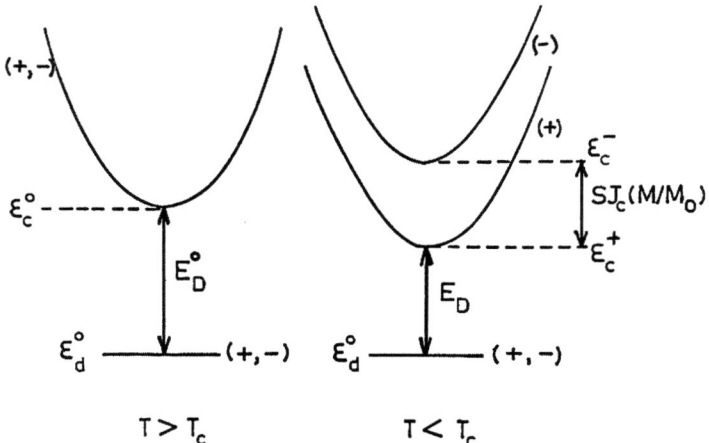

FIGURE 3. Change of donor ionization energy E_D for a donor level without spin splitting.

We now discuss qualitatively different types of donors. In the first place we consider the case that the donor electron is completely localized on a non-magnetic atom. Then the interaction with all magnetic moments vanishes, and $\varepsilon_d^\pm = \varepsilon_d^o$. In this case the donor level is degenerate for the two spin directions, and the energy of the donor level does not change with temperature. The energy of the bottom of the conduction band is given by

$$\varepsilon_c^\pm = \varepsilon_c^o \mp \tfrac{1}{2} SJ_c(M/M_o) \qquad (22)$$

and therefore the donor ionization energy E_D is given by (see Fig. 3)

$$E_D = \varepsilon_c^+ - \varepsilon_d^o = E_D^o - \tfrac{1}{2} SJ_c(M/M_o) \qquad (23)$$

If $E_D^o < \tfrac{1}{2} SJ_c(M/M_o)$ the bound donor state disappears in the conduction band. Because the magnetization changes as a function of temperature, this case can lead to a metal-semiconductor transition (Fig. 4). The transition temperature T_o is determined by the expression

$$E_D^o = \tfrac{1}{2} SJ_c M(T_o)/M_o \qquad (24)$$

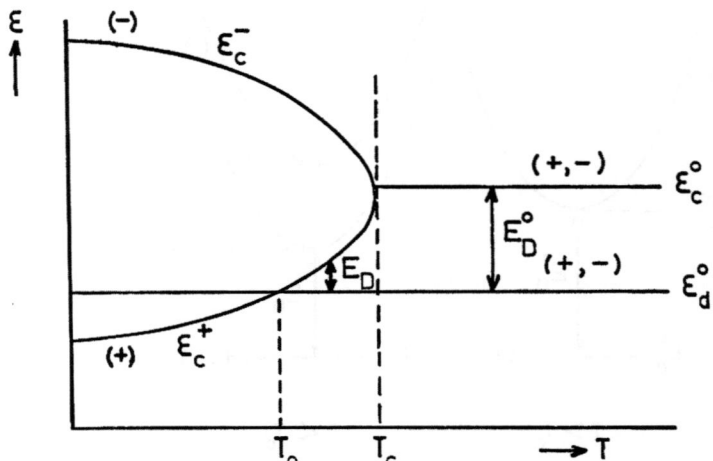

FIGURE 4. Energy of a conduction band with spin splitting, and a donor level without spin splitting. The donor ionization energy E_D vanishes below T_o.

The change of the number of charge carriers with temperature is shown schematically in Fig. 8.

Secondly the case of a shallow donor is discussed. The electron is now spread out over a region containing many atomic spins. Because the electron density on any of the magnetic atoms is small, the effective field H_z^* is small, and therefore the polarization of the spins by the electron will be small. Thus the average value $\langle S_{nz}\rangle$ of a spin S_n in the neighborhood of a donor will not deviate strongly from $S(M/M_o)$. Then $\varepsilon_2 \simeq 0$, and

$$\varepsilon_1^\pm = \mp \tfrac{1}{2} J_d S(M/M_o) \qquad (25)$$

with

$$J_d = \sum_n \langle \varphi_d | J(\vec{r} - \vec{R}_n) | \varphi_d \rangle \qquad (26)$$

The wave function of an electron is a shallow donor can be written as a linear combination of Bloch functions $\varphi_{\vec{K}} = u_{\vec{K}}(\vec{r}) \exp(i\vec{k}\vec{r})$ of the conduction band: $\varphi_d = \Sigma\, A_{\vec{K}} \varphi_{\vec{K}}$. Substituting this result, and using the effective mass approximation $u_{\vec{K}}(\vec{r}) \simeq u(\vec{r})$, one obtains $J_d = J_c$, so that

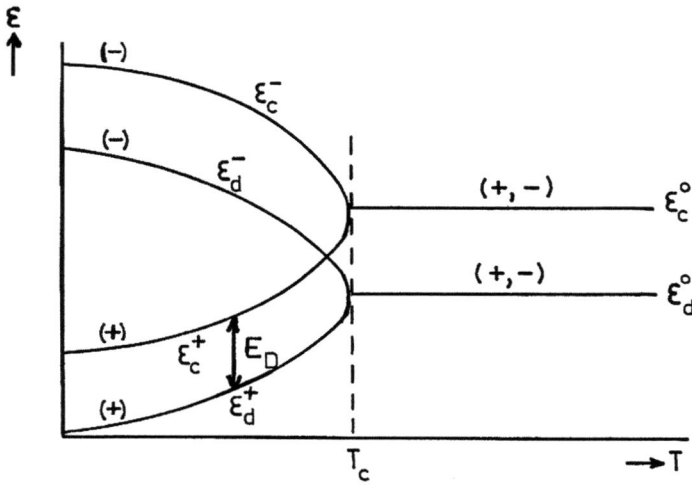

FIGURE 5. Temperature dependence of the energy of the conduction band and of the energy of a shallow donor level. The donor ionization energy E_D is independent of temperature.

$$\varepsilon_d^{\pm} = \varepsilon_d^o \mp \tfrac{1}{2} SJ_c(M/M_o) \qquad (27)$$

Therefore, in the limiting case of a very shallow donor, the spin splitting of the donor level is precisely the same as the spin splitting of the conduction band. The donor ionization energy remains unchanged in this case (Fig. 5).

In intermediate cases there will be a spin splitting of the donor level different from that of the conduction band. In this case it is also possible that a metal-semiconductor transition occurs (Fig. 6).

In the general case, the electron in the donor state moves over a restricted number of magnetic atoms. Thus the density of the electron on these magnetic atoms can be appreciable, leading to large values of H_z^*, and an appreciable spin polarization of these sites caused by the electron. At larger distances from the donor, φ_d, H_z^* and the polarization decrease. In this manner a kind of magnetic cluster is obtained.

The spin splitting of the donor level depends strongly on the spatial extension of the electron wave function. This spatial extension is in turn influenced by the spin polarization of the surrounding magnetic atoms, and should therefore be calculated in a self-consistent way. This problem is in many respects similar

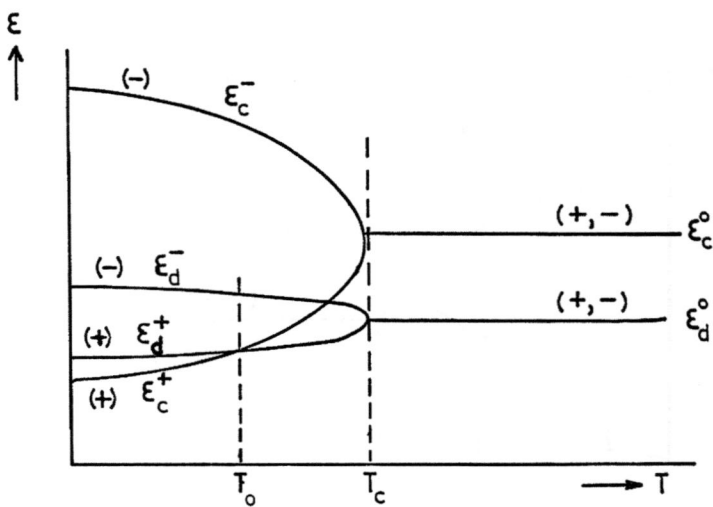

FIGURE 6. Temperature dependence of the conduction band and of the energy of a donor level. At T_c the donor ionization energy vanishes. There is metal-semiconductor transition at T_o.

to that of the magnetic polaron. A detailed discussion of the properties of such magnetic clusters, given by Yanase and Kasuya [13,14], was used for the interpretation of anomalous magnetic and electrical properties of europium chalcogenides.

Distribution of Charge Carriers

The spin splitting of energy bands and donor and acceptor levels influences the distribution of charge carriers [2,15]. As a result, the temperature dependence of the charge carrier concentration in magnetic semiconductors can differ appreciably from what is normally found in semiconductors. In this section we will discuss some examples.

Consider a n-type semiconductor, with N_d single donors per cm^3. The conduction band is split into two subbands with energies $\varepsilon_c^\pm = (\hbar^2 k^2/2m^*) \mp \frac{1}{2} S J_c (M/M_o)$, the spin splitting of the donor levels is given by $\varepsilon_d^\pm = \varepsilon_d^o \mp \frac{1}{2} S J_d (M/M_o)$. The value of J_d ranges from zero for very deep donors to $J_d = J_c$ for very shallow donors. For convenience we write $J_d = \gamma J_c$.

The distribution of the charge carriers is governed by the Fermi distribution function $f(\varepsilon) = [1 + \exp(\varepsilon-\zeta)/k_B T]^{-1}$, where $f(\varepsilon)$ is the probability for a state of energy ε to be occupied by an electron; ζ is the Fermi energy.

For a parabolic conduction band with effective mass m^*, the density of states for electrons with a given spin is $N(\varepsilon) = (1/4\pi^2)(2m^*/\hbar^2)^{3/2}(\varepsilon-\varepsilon_o)^{\frac{1}{2}}$, if ε_o is the energy of the bottom of the band.

We first consider cases with relatively small concentrations of charge carriers in the conduction band, so that for all states in the band $f(\varepsilon) \simeq \exp(\zeta-\varepsilon)/k_BT \ll 1$ (non-degenerate semiconductor). The concentration of charge carriers in a given subband for one spin direction is $n = \int_0^\infty N(\varepsilon)f(\varepsilon)d\varepsilon$, or $n = \frac{1}{2} N_c \exp(\zeta-\varepsilon_o)/k_BT$, with $N_c = 2(2\pi m^* k_B T/h^2)^{3/2}$. These results lead to the following expressions for the concentrations n^+ and n^- of electrons in the two subbands:

$$n^{\pm} = \tfrac{1}{2} N_c \exp\{\zeta \pm \tfrac{1}{2} SJ_c(M/M_o)\}/k_BT \tag{28}$$

and

$$n^+/n^- = \exp(SJ_c M/M_o k_B T) \tag{29}$$

For the calculation of the occupation of the donor states, it should be realized that each donor can bind only one electron. For the occupation of donors by electrons with (+) spin $(N_d - n_d^-)$ donor atoms are available, if n_d^- is the number of donor atoms occupied by an electron with (-) spin. Hence, the number n_d^+ of donors occupied by an electron with (+) spin is given by $n_d^+ = (N_d - n_d^-)f(\varepsilon_d^+)$. In the same way n_d^- is given by $n_d^- = (N_d - n_d^+)f(\varepsilon_d^-)$. The concentration of ionized donors N_i is $N_i = N_d - n_d^+ - n_d^-$. From these relations one finds:

$$n_d^{\pm} = N_i \exp\{\zeta - \varepsilon_d^o \pm \tfrac{1}{2}\gamma SJ_c(M/M_o)\}/k_BT \tag{30}$$

and

$$n_d^+/n_d^- = \exp(\gamma SJ_c M/M_o k_B T) \tag{31}$$

The number of ionized donors is also equal to the total concentration of electrons (at least if the crystal is uncompensated, i.e. if there are no acceptors present): $N_i = n^+ + n^-$.

From the equations given so far, it is possible to calculate all concentrations explicitely. For the case that only a small

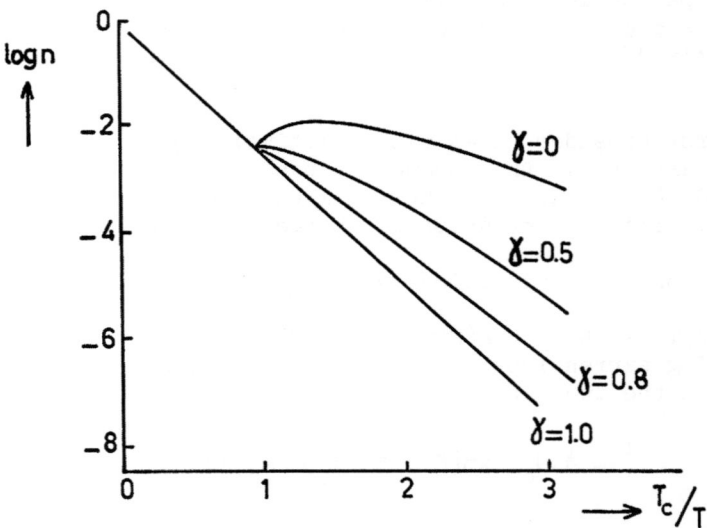

FIGURE 7. Charge carrier concentration (arbitrary units) in a ferromagnetic semiconductor, according to equation (32) [for $\varepsilon_d^o = -0.125$ eV, $SJ_c = 0.15$ eV, $T_c = 120°K$, and various values of γ.]

---

fraction of the donors is ionized (i.e. $n_d^+ + n_d^- \sim N_d$) the total concentration $n = n^+ + n^-$ of charge carriers in the conduction band is given by

$$n = (\tfrac{1}{2} N_c N_d)^{\tfrac{1}{2}} \exp(\varepsilon_d^o/2k_B T) \left\{ \frac{\cosh(\tfrac{1}{2} SJ_c M/M_o k_B T)}{\cosh(\tfrac{1}{2} \gamma SJ_c M/M_o k_B T)} \right\}^{\tfrac{1}{2}} \qquad (32)$$

The result for a given set of parameters is shown in Fig. 7. The spin splitting leads to an anomalous temperature dependence of the charge carrier concentration especially near T_c; this will of course directly influence the transport properties.

In the paramagnetic region $M = 0$, and n is given by

$$n = (\tfrac{1}{2} N_c N_d)^{\tfrac{1}{2}} \exp(\varepsilon_d^o/2 k_B T) \qquad (33)$$

If the spin splitting in the ferromagnetic region is large, so that $\gamma SJ_c M \gg M_o k_B T$, Eq. (32) reduces to

$$n = (\tfrac{1}{2} N_c N_d)^{\tfrac{1}{2}} \exp\{\varepsilon_d^o + \tfrac{1}{2}(1-\gamma)SJ_c(M/M_o)\}/2 k_B T \qquad (34)$$

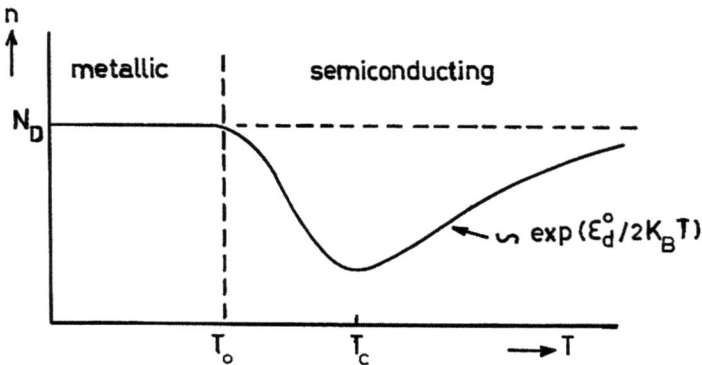

FIGURE 8. Charge carrier concentration as a function of temperature for a ferromagnetic semiconductor; the donor ionization energy vanishes below T_o.

It is possible to evaluate the temperature dependence of the charge carrier concentration for the case that the donor ionization energy vanishes below a given temperature T_o (i.e. energy levels corresponding to Fig. 6). In the paramagnetic region there is no spin splitting, and the charge carrier concentration is given by Eq. (33) for the temperature region where only a small fraction of the donors is ionized. At higher temperature the ionization of the donors increases, until at high temperature all donors are ionized: $n \simeq N_d$. Between T_c and T_o the donor ionization energy decreases and the charge carrier concentration increases. Below T_o all donors are ionized: $n = N_d$. Hence, in a crystal with a large concentration of not too shallow donors, there is metallic conductivity with $n = N_d$ below T_o, and semiconducting behavior at higher temperature (Fig. 8). A transition of this type was recently observed in slightly oxygen-deficient EuO, with oxygen vacancies acting as donors [16].

TRANSPORT PROPERTIES

Theory of Spin Disorder Scattering

In preceding sections we discussed the strong interaction between conduction electrons and magnetic moments. This interaction, which caused the spin splitting of energy levels, is also responsible for a strong scattering of the charge carriers [7,17-19]. Except at $T = 0°K$, the atomic spins are not completely ordered, and there will exist local fluctuations of the spins. These fluctuations are responsible for spin disorder scattering of the charge carriers.

Spin disorder scattering is usually quite strong because the interaction between electrons and magnetic moments is strong. In most magnetic materials, spin disorder scattering dominates over other types of scattering (phonons, impurity). In our discussion of the transport properties we will take into account only the spin disorder scattering.

The theory of spin disorder scattering of magnetic metals has been treated by several authors [17-19]. An important difference between metals and semiconductors is that in semiconductors only states with small values of the wave vector $\vec{k}$ are occupied (for a band with extremum at $\vec{k} = 0$), whereas in metals all states with $|\vec{k}| \leq \vec{k}_F$ are occupied ($\vec{k}_F$ is the Fermi wave vector, which is quite large in most metals). This difference has important consequences for the scattering. In a semiconductor the charge carriers can be regarded as particles with a long wavelength $\lambda = 2\pi/k$, and such particles are strongly scattered by long wavelength fluctuations of the magnetization. In a ferromagnet long wavelength fluctuations of the magnetization are particularly large near the Curie temperature T_c; they are the critical fluctuations associated with the second-order magnetic phase transition at T_c. Therefore, one expects that in a ferromagnetic semiconductor there will be a maximum in the spin disorder scattering and in the resistivity at T_c.

In this section we will calculate the spin disorder scattering using perturbation theory and the Boltzmann transport equation. We first calculate the transition probability P of a transition of an electron from a state $\vec{k}$ with spin (+) to a state $\vec{k}'$ with spin (+) or (-). As a result of this transition the state of the atomic spins will change from α to α'. We find

$$P(\vec{k}+\alpha, \vec{k}'+\alpha') = (2\pi/\hbar) |\langle \varphi_{\vec{k}}(\vec{r}) X^+ \alpha | H' | \varphi_{\vec{k}'}(\vec{r}) X^+ \alpha' \rangle|^2 \times$$

$$\times \delta(\varepsilon^+_{\vec{k}} + \varepsilon_\alpha - \varepsilon^+_{\vec{k}'} - \varepsilon_{\alpha'}) \qquad (35)$$

$$P(\vec{k}+\alpha, \vec{k}'-\alpha') = (2\pi/\hbar) |\langle \varphi_{\vec{k}}(\vec{r}) X^+ \alpha | H' | \varphi_{\vec{k}'}(\vec{r}) X^- \alpha' \rangle|^2 \times$$

$$\times \delta(\varepsilon^+_{\vec{k}} + \varepsilon_\alpha - \varepsilon^-_{\vec{k}'} - \varepsilon_{\alpha'}) \qquad (36)$$

In these equations $\varphi_{\vec{k}}(\vec{r})$ is the Bloch function of the electron, and $X^\pm$ its spin function. The processes in which the electron spin changes from (+) to (-) are called spin flip processes.

MAGNETIC SEMICONDUCTORS

For the interaction between electron and spins we use again the exchange interaction of Eq. (8). This leads to matrix elements of the type

$$\langle \varphi_{\vec{k}}(\vec{r}) X^+ \alpha | H' | \varphi_{\vec{k}'}(\vec{r}) X^+ \alpha' \rangle =$$

$$= -\tfrac{1}{2} \sum_n e^{i(\vec{k}'-\vec{k})\vec{R}_n} \langle \alpha | S_{nz} | \alpha' \rangle \int |u(\vec{r})|^2 e^{i(\vec{k}'-\vec{k})(\vec{r}-\vec{R}_n)} J(\vec{r}-\vec{R}_n) dv(\vec{r})$$

(37)

where in the Bloch functions $\varphi_{\vec{k}}(\vec{r})$ the effective mass approximation $\varphi_{\vec{k}}(\vec{r}) = u(\vec{r}) \exp(i\vec{k}\vec{r})$ is used. Writing the integral as

$$J(\vec{k}) \equiv N \int |u(\vec{r})|^2 J(\vec{r}-\vec{R}_n) e^{i\vec{k}(\vec{r}-\vec{R}_n)} dv(\vec{r}) \qquad (38)$$

we obtain

$$\langle \varphi_{\vec{k}}(\vec{r}) X^+ \alpha | H' | \varphi_{\vec{k}'}(\vec{r}) X^+ \alpha' \rangle = -(1/2N) J(\vec{k}-\vec{k}') \sum_n \langle \alpha | S_{nz} | \alpha' \rangle e^{i(\vec{k}'-\vec{k})\vec{R}_n}$$

(39)

In the same manner we obtain:

$$\langle \varphi_{\vec{k}}(\vec{r}) X^+ \alpha | H' | \varphi_{\vec{k}'}(\vec{r}) X^- \alpha' \rangle = -(1/2N) J(\vec{k}-\vec{k}') \sum_n \langle \alpha | \vec{S}_n^- | \alpha' \rangle e^{i(\vec{k}'-\vec{k})\vec{R}_n}$$

(40)

with $S_n^\pm = S_{nx} \pm iS_{ny}$.

The probability for a state $(\vec{k}\pm)$ to be occupied by an electron is called $f_{\vec{k}}^\pm$. The change of $f_{\vec{k}}^\pm$ due to the scattering processes is given by

$$\left(\frac{\partial f_{\vec{k}}^+}{\partial t}\right)_{scatt} = \sum_{\alpha}\sum_{\alpha'}\sum_{\vec{k}'} w_{\alpha'}[f_{\vec{k}'}^+(1-f_{\vec{k}}^+)P(\vec{k}'+\alpha', \vec{k}+\alpha) +$$

$$+ f_{\vec{k}'}^-(1-f_{\vec{k}}^+)P(\vec{k}'-\alpha', \vec{k}+\alpha)] -$$

$$- \sum_{\alpha}\sum_{\alpha'}\sum_{\vec{k}'} w_{\alpha}[f_{\vec{k}}^+(1-f_{\vec{k}'}^+)P(\vec{k}+\alpha, \vec{k}'+\alpha') +$$

$$+ f_{\vec{k}}^+(1-f_{\vec{k}'}^-)P(\vec{k}+\alpha, \vec{k}'-\alpha')] \qquad (41)$$

In an applied electric field $\vec{F}$ the electrons are accelerated; the equation of motion is $\hbar(d\vec{k}/dt) = e\vec{F}$. This leads to

$$\left(\frac{\partial f_{\vec{k}}^\pm}{\partial t}\right)_{field} = \left(\frac{\partial f_{\vec{k}}^\pm}{\partial \varepsilon_{\vec{k}}^\pm}\right)(\nabla_{\vec{k}}\varepsilon_{\vec{k}}^\pm)\left(\frac{d\vec{k}}{dt}\right) \qquad (42)$$

For an isotropic and parabolic energy band with $\varepsilon_{\vec{k}}^\pm = \varepsilon_o^\pm + (\hbar^2 k^2/2m^*)$, one obtains

$$\left(\frac{\partial f_{\vec{k}}^\pm}{\partial t}\right)_{field} = \left(\frac{e\hbar}{m^*}\right)\left(\frac{\partial f_{\vec{k}}^\pm}{\partial \varepsilon_{\vec{k}}^\pm}\right)(kF) \qquad (43)$$

In the stationary state the total change with time of the distribution function $f_{\vec{k}}^\pm$, due to the field and the scattering contributions, vanishes

$$\left(\frac{\partial f_{\vec{k}}^\pm}{\partial t}\right)_{field} + \left(\frac{\partial f_{\vec{k}}^\pm}{\partial t}\right)_{scatt} = 0 \qquad (44)$$

The equilibrium distribution (for $F = 0$) is the Fermi distribution function

$$f_{o\vec{k}}^\pm = [1 + \exp(\varepsilon_{\vec{k}}^\pm - \zeta)/k_B T]^{-1} \qquad (45)$$

where ζ is the Fermi energy.

For further calculations the following approximations will be used:

a) The field $\vec{F}$ is weak, so that $f_{\vec{k}}^{\pm}$ differs only slightly from $f_{o\vec{k}}^{\pm}$. If $f_{\vec{k}}^{\pm}$ is written as

$$f_{\vec{k}}^{\pm} = f_{o\vec{k}}^{\pm} + \left(\frac{\partial f_{o\vec{k}}^{\pm}}{\partial \varepsilon_{\vec{k}}^{\pm}}\right) g_{\vec{k}}^{\pm} \qquad (46)$$

only linear terms in $g_{\vec{k}}^{\pm}$ will be taken into account. This approximation corresponds to Ohm's law, i.e. an electrical current proportional to the applied electric field.

b) It is assumed that the electrical current has a negligible influence on the spin system, so that for the probabilities w_α the equilibrium distribution can be used. This assumption corresponds to neglecting magnon drag.

In equilibrium there is a detailed balance for all microscopic processes. This property can be used to eliminate $w_{\alpha'}$, leading to

$$\left(\frac{\partial f_{\vec{k}}^{+}}{\partial t}\right)_{scatt} = \sum_{\alpha} \sum_{\alpha'} \sum_{\vec{k}'} w_\alpha \left(\frac{\partial f_{o\vec{k}}^{+}}{\partial \varepsilon_{\vec{k}}^{+}}\right) \Bigg[P(\vec{k}+\alpha, \vec{k}'+\alpha') \left\{\frac{1-f_{o\vec{k}'}^{+}}{1-f_{o\vec{k}}^{+}}\right\} (g_{\vec{k}'}^{+} - g_{\vec{k}}^{+}) +$$

$$+ P(\vec{k}+\alpha, \vec{k}'-\alpha') \left\{\frac{1-f_{o\vec{k}'}^{-}}{1-f_{o\vec{k}}^{+}}\right\} (g_{\vec{k}'}^{-} - g_{\vec{k}}^{+}) \Bigg] \qquad (47)$$

For the calculations, the quasi-static approximation is used. This corresponds to neglecting the energy transfer from the charge carriers to the spin system. This approximation is valid because the excitation energies $h w_k$ of the spin system (the magnon energies) are much smaller than electrons energies of the same k.

We also use the closure relation to write

$$\sum_\alpha w_\alpha \sum_{\alpha'} \langle \alpha | V | \alpha' \rangle \langle \alpha' | W | \alpha \rangle = \sum_\alpha w_\alpha \langle \alpha | VW | \alpha \rangle = \langle VW \rangle \qquad (48)$$

for two operators V and W. Using this rule, and substituting the results obtained for the matrix elements, one obtains

$$\sum_{\alpha}\sum_{\alpha'} w_\alpha P(k+\alpha, k'+\alpha') = (2\pi/\hbar) \left|\frac{J(\vec{k}-\vec{k'})}{2N}\right|^2 \delta(\varepsilon^+_{\vec{k}} - \varepsilon^+_{\vec{k'}}) \times$$

$$\times \sum_n \sum_m \langle S_{nz} S_{mz} \rangle e^{i(\vec{k}-\vec{k'})(\vec{R}_n - \vec{R}_m)} \quad (49)$$

$$\sum_{\alpha}\sum_{\alpha'} w_\alpha P(k+\alpha, k'-\alpha') = (2\pi/\hbar) \left|\frac{J(\vec{k}-\vec{k'})}{2N}\right|^2 \delta(\varepsilon^+_{\vec{k}} - \varepsilon^-_{\vec{k'}}) \times$$

$$\times \sum_n \sum_m \langle S^+_n S^-_m \rangle e^{i(\vec{k}-\vec{k'})(\vec{R}_n - \vec{R}_m)} \quad (50)$$

The quantities $\sum_n \sum_m \langle S_{nz} S_{mz} \rangle e^{i(\vec{k}-\vec{k'})(\vec{R}_n - \vec{R}_m)}$ are Fourier components of spin correlation functions, describing magnitude and spatial extension of the spin fluctuations. Such correlation functions can be expressed in terms of a generalized magnetic susceptibility

$$\chi_{ij}(\vec{k}) = \left(\frac{g^2 \mu_B^2}{k_B T}\right) \sum_n \sum_m \{\langle S_{ni} S_{mj} \rangle - \langle S_{ni} \rangle \langle S_{mj} \rangle\} e^{i\vec{k}(\vec{R}_n - \vec{R}_m)} \quad (51)$$

where $i, j = x, y, z$.

It is possible to solve the transport equations by substituting $g^\pm_{\vec{k}} = e\hbar \tau^\pm_{\vec{k}} (kF/m^*)$; one obtains for the relaxation times $\tau^\pm_{\vec{k}}$:

$$\frac{1}{\tau_{\vec{k}}^{\pm}} = \left(\frac{2\pi}{\hbar}\right) \frac{k_B T}{(2Ng\mu_B)^2} \Big\{ \sum_{\vec{k}'} |J(\vec{k}-\vec{k}')|^2 \chi_z(\vec{k}-\vec{k}') \delta(\varepsilon_{\vec{k}}^{\pm} - \varepsilon_{\vec{k}'}^{\pm}) +$$

$$+ \sum_{\vec{k}'} |J(\vec{k}-\vec{k}')|^2 \chi_x(\vec{k}-\vec{k}') \delta(\varepsilon_{\vec{k}}^{\pm} - \varepsilon_{\vec{k}'}^{\mp}) +$$

$$+ \sum_{\vec{k}'} |J(\vec{k}-\vec{k}')|^2 \chi_y(\vec{k}-\vec{k}') \delta(\varepsilon_{\vec{k}}^{\pm} - \varepsilon_{\vec{k}'}^{\mp}) \Big\} \qquad (52)$$

(for a collinear spin system $\chi_{ij} = \chi_i \delta_{ij}$). The relation between the relaxation times $\tau_{\vec{k}}^{\pm}$ and the mobility μ for a cubic crystal is given by

$$\mu^{\pm} = \frac{e\hbar^2}{m^{*2}} \frac{\sum_{\vec{k}} (1/3)k^2 (\partial f_{o\vec{k}}^{\pm}/\partial \varepsilon_{\vec{k}}^{\pm}) \tau_{\vec{k}}^{\pm}}{\sum_{\vec{k}} f_{o\vec{k}}^{\pm}} \qquad (53)$$

As has been mentioned already, in semiconductors only states with small values of k are involved. For small k the generalized susceptibilities are given by

$$1/\chi_{ij}(\vec{k}) = 1/\chi_{ij} + A_{ij} k^2 \qquad (54)$$

where χ_{ij} is the ordinary susceptibility for homogeneous magnetic fields, defined as $\chi_{ij} = \partial M_i/\partial H_j$. It should be noted that the susceptibility is that of a one-domain crystal, i.e. not including contributions of domain wall motions, etc.

The constants A_{ij} in Eq. (54) can be calculated using a method given by de Gennes and Villain [20]. For a simple cubic lattice with only one type of exchange interaction between magnetic atoms separated by a distance b, one finds

$$A_{ij} = \delta_{ij} k_B T_c b^2 / 2N(g\mu_B)^2 S(S+1) \qquad (55)$$

In semiconductors with a small concentration of charge carriers one can easily show that usually it is permitted to neglect also the $A_{ij}k^2$ term in χ; this is found to be a very good approximation except for temperatures very close to T_c. For the same reason the k-dependence of J can be neglected: $J(k - k') \sim J$. Using these approximations leads to the following expression for the mobility of charge carriers in a non-degenerate ferromagnetic semiconductor:

$$\mu^{\pm} = \frac{8(2\pi)^{\frac{1}{2}}(Ng\mu_B)^2 e\hbar^4}{3(m^*)^{5/2}J^2(k_BT)^{3/2}} \int_0^{\infty} \frac{t e^{-t} dt}{\chi_z + 2f^{\pm}\chi_x[1 \mp (\delta/t)]^{\frac{1}{2}}} \quad (56)$$

The parameter δ in this equation is a measure of the spin splitting of the band

$$\delta = (\varepsilon_o^- - \varepsilon_o^+)/k_BT = (SJM/M_o k_BT) \quad (57)$$

The terms with χ_z and χ_x represent contributions of scattering within one subband $[(+) \to (+)$ or $(-) \to (-)]$, and between two subbands $[(+) \to (-)$ or $(-) \to (+)$, i.e. spin flip scattering]. The function $f^{\pm}$ is defined as $f^{\pm} = 1$ if $[1 \mp (\delta/t)]^{\frac{1}{2}}$ is real, and $f^{\pm} = 0$ if $[1 \mp (\delta/t)]^{\frac{1}{2}}$ is imaginary.

From the calculations it is found that the mobilities of (+) and (-) electrons are different. Therefore the resistivity ρ depends also on the distribution of charge carriers over the two subbands. With $1/\rho = n^+ e\mu^+ + n^- e\mu^-$, where n^+ and n^- are the concentrations of charge carriers with (+) and (-) spins, and with $n^+/n^- = \exp \delta$ and $n = n^+ + n^-$, one obtains

$$\frac{1}{\rho} = ne \left\{ \frac{\mu^+ e^{\delta/2} + \mu^- e^{-\delta/2}}{e^{\delta/2} + e^{-\delta/2}} \right\} \quad (58)$$

The mobility of charge carriers in a ferromagnetic semiconductor, calculated with Eq. (56), is shown in Fig. 9. The curve shows a sharp minimum of the mobility at T_c, due to critical scattering.

The spin splitting of the conduction band will usually be much larger than k_BT (except of course in a small temperature interval just below T_c). In that case, however, practically all charge carriers occupy states of the lower subband and spin flip scattering does not occur, so that $1/\rho = n^+ e\mu^+$ (or $n^- e\mu^-$, depending on the sign of J.)

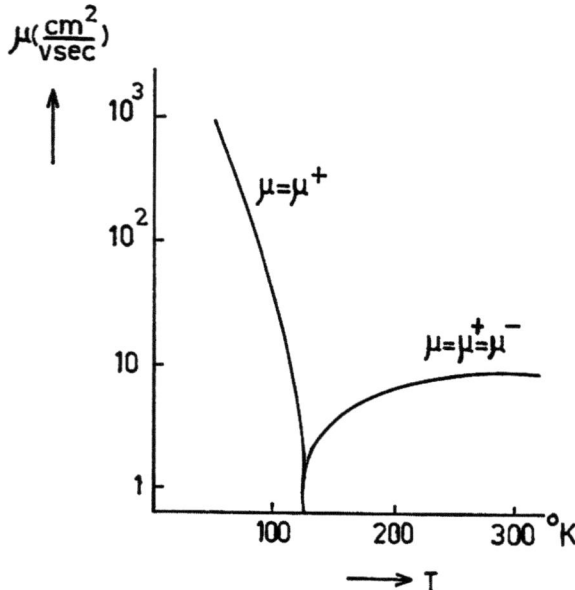

FIGURE 9. Mobility of charge carriers in a non-degenerate ferromagnetic semiconductor [for T_c = 120°K, J = 0.5 eV, S = 3/2, g = 2, m* = m_o . N = 1.3×10^{22} cm^{-3}].

For degenerate semiconductors the term $A_{ij}k^2$ cannot be neglected.

For metallic crystals the number of charge carriers is very large. In this case there will be contributions of all Fourier components of the fluctuations of the magnetization. The result of a perturbation calculation of the spin disorder resistivity of metals, but neglecting short range order, is given by [19].

$$\rho = \frac{m^{*2} k_F N}{\pi n e^2 \hbar^3} J^2 \left[S(S+1) - S^2(M/M_o)^2 - S(M/M_o) \tanh\left(\frac{3T_c M}{2TS(S+1)M_o}\right) \right]$$

(59)

The result is shown in Fig. 10. The resistivity of the metal rises gradually until at T_c the disorder of the spin system is complete (in this model without short range order). The resistivity remains constant at higher temperature.

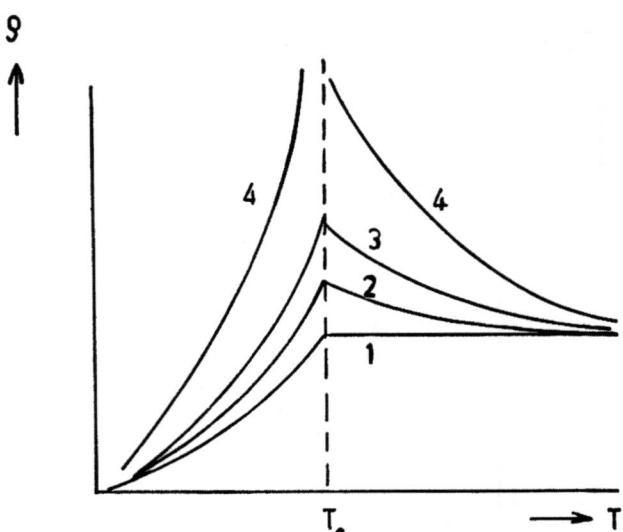

FIGURE 10. Resistivity ρ due to spin disorder scattering:
1) ferromagnetic metal, neglecting short range order; 2) and
3) ferromagnetic metal, taking into account short range order, for
two different values of k_F, with $k_F(2) > k_F(3)$; 4) ferromagnetic
semiconductor (with a temperature-independent charge carrier
concentration).

Short range order of the atomic spins leads to a maximum in
the resistivity of metals at T_c [18]; the peak is higher, the
smaller the carrier concentration, i.e. the smaller k_F (Fig. 10).
These results join smoothly with the results obtained above for
magnetic semiconductors: for these the k values are very small,
and the maximum of the resistivity at T_c is very large.

Magnetoresistance

An applied magnetic field influences the mobility and the
charge carrier concentration; both effects lead to a magnetoresistance.

Let us first consider the change of the charge carrier
concentration caused by an applied magnetic field. In a preceding
section it was shown that the ionization energy of donors (and of
acceptors) depends on the magnetization, and, therefore, also on
the magnetic field. For the case of donors which are ionized only
for a small fraction, Eq. (32) was derived. The corresponding

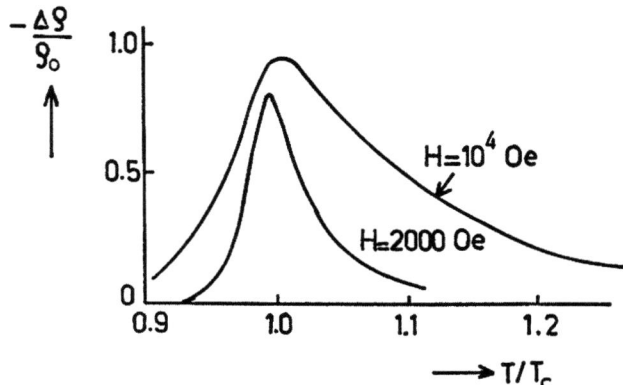

FIGURE 11. Magnetoresistance of a ferromagnetic semiconductor, assuming a field independent mobility [with $S = 3/2$, $J = 0.5$ eV, $\gamma = \frac{1}{2}$ and $\varepsilon_d^0 = 0.2$ eV].

magnetoresistance $\Delta\rho/\rho_0 = [\rho(H)-\rho_0]/\rho_0$, where $\rho(H)$ and ρ_0 are the resistivities in a field H and without field, respectively, is given by

$$\frac{\Delta\rho}{\rho} = \left\{\frac{\cosh\left(\frac{1}{2} SJ_c M(0)/M_o k_B T\right) \cosh\left(\frac{1}{2} \gamma SJ_c M(H)/M_o k_B T\right)}{\cosh\left(\frac{1}{2} \gamma SJ_c M(0)/M_o k_B T\right) \cosh\left(\frac{1}{2} SJ_c M(H)/M_o k_B T\right)}\right\}^{\frac{1}{2}} - 1 \tag{60}$$

In this equation M(H) is the magnetization in a field H, M(0) is the spontaneous magnetization. Curves of the magnetoresistance calculated with Eq. (60), are given in Fig. 11. The magnetoresistance shows a pronounced maximum at T_c.

The influence of a magnetic field on the mobility is calculated from the equations of the preceding section, taking into account the field dependence of the magnetic susceptibility. Figure 12 shows that the calculated magnetoresistance is negative, and has a maximum at T_c.

In the ferromagnetic region $T < T_c$, with all charge carriers in the lower subband, the equation for the magnetoresistance caused by the field dependence of the mobility is particularly simple:

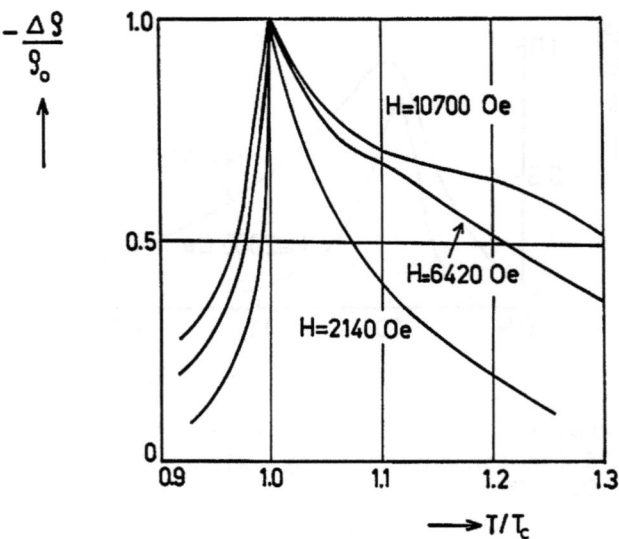

FIGURE 12. Magnetoresistance of a ferromagnetic semiconductor, assuming a field independent charge carrier concentration [with $S = 3/2$ and $J = 0.5$ eV].

$$\frac{\Delta\rho}{\rho} = \frac{\chi_z(H)}{\chi_z(0)} - 1 \qquad (61)$$

Thus the magnetoresistance below T_c can be calculated directly from the field dependence of the magnetic susceptibility.

REFERENCES

[1] Methfessel, S. and Mattis, D. C., Magnetic Semiconductors, in Encyclopedia of Physics, Vol. 18/1, Wijn, H. P. J., Ed., Springer Verlag, Berlin, 1968, 389.
[2] Haas, C., Critical Reviews in Solid State Sciences, 1, 47, 1970.
[3] Austin, I. G. and Elwell, D., Contemp. Phys., 11, 455, 1970.
[4] C. F. van Bruggen, M. B. Vellinga and C. Haas, J. Solid State Chem., 2, 303, 1970.
[5] Kasuya, T., Magnetism, Vol. 2B, Rado, G. T. and Suhl, H., Eds., Academic Press, New York, 1966, 1.
[6] Rys, F., Helman, J. S., and Baltensperger, W., Phys. Kondens. Materie, 6, 105, 1967.

[7]   Haas, C., Phys. Rev., 168, 531, 1968.
[8]   Harbeke, G. and Lehmann, H. W., Solid State Commun., 8, 1281, 1970.
[9]   Lehmann, H. W. and Harbeke, G., J. de Physique 32, Colloque no 1, 932, 1971.
[10] Busch, G., Streit, P. and Wachter, P., J. de Physique 32, Colloque no. 1, 926, 1971.
[11] Esaki, L., Stiles, P. J., and von Molnar, S., Phys. Rev. Letters, 19, 852, 1967.
[12] Kohn, W., Solid State Physics, Vol. 5, Seitz, F. and Turnbull, D., Eds., Academic Press, New York, 1957, 258.
[13] Yanase, A. and Kasuya, T., J. Phys. Soc. Japan, 25, 1025, 1968.
[14] Kasuya, T. and Yanase, A., J. Appl. Phys., 39, 430, 1968.
[15] Bongers, P. F., Haas, C., van Run, A. M. J. G., and Zanmarchi, G., J. Appl. Phys., 40, 958, 1969.
[16] Penney, T., Shafer, M. W. and Torrance, J. B., Phys. Rev. B5, 3669 (1972).
[17] Kasuya, T., Progr. Theor. Phys. (Kyoto), 16, 58, 1956.
[18] De Gennes, P. G., and Friedel, J., J. Phys. Chem. Solids, 4, 71, 1958.
[19] Van Peski-Tinbergen, T., and Dekker, A. J., Physica, 29, 917, 1963.
[20] De Gennes, P. G., and Villain, J., J. Phys. Chem. Solids, 13, 10, 1960.

CHAPTER 9

OPTICAL PROPERTIES OF SOLIDS

G. Chiarotti

Istituto di Fisica
dell'Università
Rome, Italy

THE OPTICAL CONSTANTS

The study of the optical properties (absorption, dispersion and emission) is a powerful tool for investigating the electronic and vibrational structures of solids [1].

The optical response of a solid material is characterized by the complex refractive index. A very useful constant, however, related to the imaginary part of the refractive index, is the absorption coefficient, which is susceptible of a simple experimental determination.

When a beam of light of intensity I goes normally through a slab of a medium of thickness dx:

$$dI = -\alpha I dx, \tag{1}$$

where α is the absorption coefficient which has the dimension of an inverse length and is generally measured in cm^{-1}. The beam of light attenuates, in passing through the medium, with the law (consequence of (1)):

$$I = I_o e^{-\alpha x}, \tag{2}$$

so that α can be obtained simply by measuring the ratio I_o/I of the intensities impinging on and emerging from the sample [2].

On the other hand the solution of Maxwell's equations appropriate to a sinusoidal wave propagating in an indefinite medium of (complex) refractive index

$$n = n' - in'',\qquad(3)$$

gives for the electric (or magnetic) field:

$$\vec{E} = \vec{E}_o e^{i(\vec{q}\cdot\vec{r}-\omega t)} = \vec{E}_o e^{i(n\vec{q}_o\cdot\vec{r}-\omega t)},\qquad(4)$$

where ω is the angular frequency of the electromagnetic wave, $\vec{q}$ its wave vector of modulus $\frac{2\pi}{\lambda}$, and $\vec{q}_o$ the wave vector in vacuum of modulus $\frac{2\pi}{\lambda_o}$. The electric field of the wave attenuates then as $e^{-n''\frac{2\pi x}{\lambda_o}}$ and its intensity as $e^{-n''\frac{4\pi x}{\lambda_o}}$. By comparing the last expression with (2), a relation between the absorption coefficient and the imaginary part of the refractive index is obtained, namely:

$$\alpha = \frac{4\pi}{\lambda_o} n''.\qquad(5)$$

In the same way, introducing the complex dielectric constant:

$$n^2 = \varepsilon = \varepsilon' - i\varepsilon'' = \varepsilon' - i\frac{4\pi\sigma}{\omega}\qquad(6)$$

and substituting into (5), we obtain:

$$\alpha = \frac{4\pi\sigma}{n'c} = \frac{\omega}{n'c}\varepsilon''\qquad(7)$$

c being the velocity of light, and σ the conductivity of the medium (or better, in the complex notation, its real part).

The imaginary part of the refractive index (or of the dielectric constant) is then directly related to the absorption of energy.

On the other hand, the real and imaginary parts of the refractive index are interdependent according to the well known Kramers-Kronig dispersion relations:

$$n''(\omega) = -\frac{2\omega}{\pi} \mathcal{P} \int_0^\infty \frac{n'(\Omega)-n'_\infty}{\Omega^2-\omega^2} d\Omega \qquad (8)$$

$$n'(\omega)-n'_\infty = \frac{2}{\pi} \mathcal{P} \int_0^\infty \frac{n''(\Omega)\Omega}{\Omega^2-\omega^2} d\Omega \qquad (8')$$

where $\mathcal{P}$ means the Cauchy principal value of the integral, and n'_∞ is the refractive index (real part) at infinite frequency (generally equal to 1) [3]. Equations (8) and (8') hold for real and imaginary parts of any physical quantity, provided the system is linear and obeys the principle of causality.

It is seen that a simple measurement of the ratio I_0/I, if carried on on a sufficiently extended range of frequencies, allows the determination of n' and n'' (through Eq. (2), (5) and (8')) that completely characterize the optical response at the system.

From the microscopic point of view, the absorption coefficient can easily be expressed in terms of the transition probability among the quantum states. If P_{ij} is the number of transitions per unit volume, per unit time induced by the electromagnetic field of frequency ω, from state $|i\rangle$ to state $|j\rangle$, then the removal of energy from the beam of intensity I is:

$$-dI = \hbar\omega \sum_{ij} P_{ij} \, dx, \qquad (9)$$

the sum being extended to all transitions $|i\rangle \rightarrow |j\rangle$ that occur at frequency ω. As is well known P_{ij} is easily obtained in quantum mechanics from first order time-dependent perturbation theory and has the following expression [4]:

$$P_{ij} = 4\pi^2 \hbar \frac{1}{n^2}\left(\frac{e}{m\omega}\right)^2 |\langle j|e^{i\vec{q}\cdot\vec{r}}\vec{\varepsilon}_q\cdot\nabla|i\rangle|^2 \rho(\omega)\delta(E_j-E_i\pm\hbar\omega), \qquad (10)$$

where n is the refractive index (real part) of the medium, m and e the mass and charge of the interacting particle (in general the electron), $\vec{\varepsilon}_q$ the polarization versor of the electromagnetic wave (which points in the direction of the electric field) and $\rho(\omega)$ the density of the electromagnetic energy in the medium. E_i and E_j are the energies of states $|i\rangle$ and $|j\rangle$, while the $\pm$ signs in the Dirac δ-function refer respectively to emission and absorption.

Recalling that the intensity of the light beam is simply:

$$I = \rho(\omega) \frac{c}{n} \tag{11}$$

and comparing (9) with (1), we obtain for the absorption coefficient the expression:

$$\alpha(\omega) = h^2 \frac{1}{nc\omega} \left(\frac{e}{m}\right)^2 \sum_{ij} |\langle j | e^{i\vec{q}\cdot\vec{r}} \vec{\varepsilon}_q \cdot \nabla | i \rangle|^2 \delta(E_j - E_i - \hbar\omega). \tag{12}$$

Expression (10) for the transition probability is derived within the usual dipole approximation, which neglects all terms but $\frac{e}{mc} \vec{A}\cdot\vec{p}$, in the interaction Hamiltonian, $\vec{A}$ being the vector potential of the electromagnetic wave. Two-photon transitions are then explicitly excluded and can be taken into account by carrying the perturbation to the second order and/or including the term $\frac{e^2}{2mc^2} \vec{A}^2$ in the Hamiltonian.

Since the atomic dimensions are in general much smaller than the wavelength of the electromagnetic radiation, the factor $e^{i\vec{q}\cdot\vec{r}}$ in (10) can be taken as unity. With this approximation, expression (10) is very often written in a simpler form, by introducing the dipole moment of the atomic system which, for electrons of coordinates $\vec{r}_i$ is:

$$\vec{M} = -e\Sigma_i \vec{r}_i. \tag{13}$$

The absorption coefficient then becomes:

$$\alpha(\omega) = \frac{4\pi^2 \omega}{nc} \sum_{ij} |\langle j | \vec{\varepsilon}_q \cdot \vec{M} | i \rangle|^2 \delta(E_j - E_i - \hbar\omega). \tag{14}$$

The theoretical study of the optical properties of solids is therefore mainly concerned with the dependence on frequency of expressions (12) or (14).

THE OPTICAL SPECTRUM

Both electrons and ions (in ionic solids) contribute to the optical absorption spectrum. However, since the ions are much heavier than the electrons, their absorption bands lie at low energy in the infrared. Between the lattice absorption and the onset of electronic transitions, there is a window in which the solid is transparent. In metals and highly doped semiconductors the free electrons, however, give rise to a characteristic absorption increasing quadratically with wavelength.

We shall divide the absorption spectrum of a solid according to the type of transitions that determine it.

a) Absorption due to interband transitions.
b) Absorption due to intraband transitions.
c) Exciton absorption.
d) Impurity absorption.
e) Free-carrier absorption.
f) Lattice absorption.

Absorption Due To Interband Transitions

The problem of applying Eq. (12) or (14) to the electronic transitions of a solid is in principle rather formidable since the wave functions of states $|i\rangle$ and $|j\rangle$ should depend on the coordinates of the nuclei as well as on those of the electrons. However, because of their large mass, the nuclei can be taken at rest during the electronic transitions (Franck-Condon principle). For each configuration of the nuclei, a (different) energy E_i and E_j can then be calculated. The lattice vibrations cause simply a temperature-dependent broadening of the absorption spectrum. When the spectrum is already broad as in most solids, the lattice broadening is unimportant and will be neglected in the following.

A further simplification is to assume that the states $|i\rangle$ and $|j\rangle$ are simply Bloch states $|\vec{k}\rangle$ and $|\vec{k}'\rangle$ belonging to different bands.

The matrix element:

$$\langle \vec{k}' | e^{i\vec{q}\cdot\vec{r}} \nabla | \vec{k} \rangle \tag{15}$$

is then zero unless:

$$\vec{k} - \vec{k}' + \vec{q} = \vec{g} \tag{16}$$

where $\vec{g}$ is a principal vector of the reciprocal lattice and can be taken as zero if the reduced zone scheme is adopted for the energy bands. Moreover, since $\lambda \gg a$, the lattice parameter, $q \ll k, k'$ so that Eq. (16) reduces to:

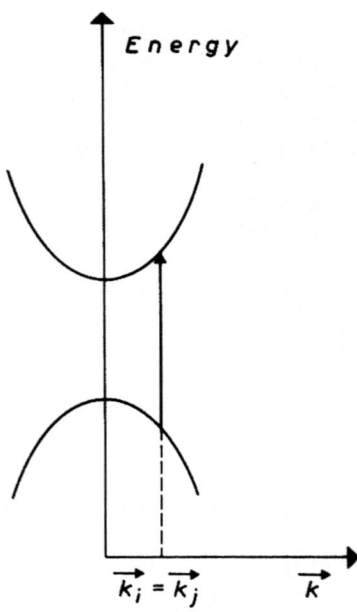

FIGURE 1. Interband vertical transition, showing the conservation of $\vec{k}$.

$$\vec{k} \simeq \vec{k}', \qquad (17)$$

that shows that the optical transitions are vertical in the reduced zone scheme, as sketched in Fig. 1. Because of (17) the $\sum_{ij}$ of Eq. (12) or (14), reduces to an integration over $\vec{k}$ extended to the first Brillouin zone, namely:

$$\alpha(\omega) = \frac{4\pi^2 \omega}{nc} \int \frac{d\vec{k}}{4\pi^3} |\langle j|\vec{\varepsilon}\cdot\vec{M}|i\rangle|^2 \delta(E_j - E_i - \hbar\omega) \qquad (18)$$

The integral in (18) can easily be calculated [5] if one assumes that $\langle j|\varepsilon_q\cdot M|i\rangle$ varies little with $\vec{k}$ and recalls the property of the δ-function:

$$\int_a^b g(x)\delta[f(x)]dx = \sum_{x_o} g(x_o) \left|\frac{\partial f}{\partial x}\right|^{-1}_{x=x_o} \qquad (19)$$

where the x_o's are the roots of the equation $f(x)=0$.

Equation (19) shows that the integral:

$$J(\omega) = \int \frac{d\vec{k}}{4\pi^3} \delta(E_j - E_i - \hbar\omega) \qquad (20)$$

called the "joint density of states", represents the number of states, per unit energy, whose energy difference E_j-E_i equals $\hbar\omega$. This is most easily seen in one dimension, where $\frac{d\vec{k}}{4\pi^3}$ becomes $\frac{dk}{\pi} \equiv dN$, the number of states between k and k+dk. The integral (19) is then simply $\left|\frac{dN}{dE}\right|_{E_j-E_i=\hbar\omega}$.

In three dimensions, the joint density of states is:

$$J(\omega) = \frac{1}{4\pi^3} \int \frac{dS}{|\nabla_k(E_j-E_i)|_{E_j-E_i=\hbar\omega}} , \qquad (21)$$

where dS is the element of the surface of constant energy defined by $E_j-E_i=\hbar\omega$.

The joint density of states becomes very large when $\nabla_k E_j = \nabla_k E_i$, i.e. in regions where the bands i and j are nearly parallel. Since the other factors appearing in (18) vary smoothly with frequency, the absorption coefficient is mainly determined by the behavior of the joint density of states. Optical absorption is then a direct probe for the band structure of solids.

As an example, let us consider the case of Ge. In Fig. 2 the absorption spectrum of Ge is shown, in the energy region corresponding to transitions from the valence to the conduction band. The dashed curve represents the joint density of states calculated by Brust, Phillips and Bassani [6] from the band structure of Fig. 3. The agreement is excellent and proves the correctness of the theory.

When the initial state is a core state its energy does not depend upon $\vec{k}$ and the optical absorption depends simply upon the density of the final states. The interpretation of the optical data is then straightforward. Transitions from core states, however, occur in the very far vacuum-ultraviolet region of the spectrum and their experimental study is much more difficult.

The dipole matrix element $\langle j|\vec{\varepsilon}\cdot\vec{M}|i\rangle$ is generally taken as a constant. However, it is identically zero when $|i\rangle$ and $|j\rangle$ have

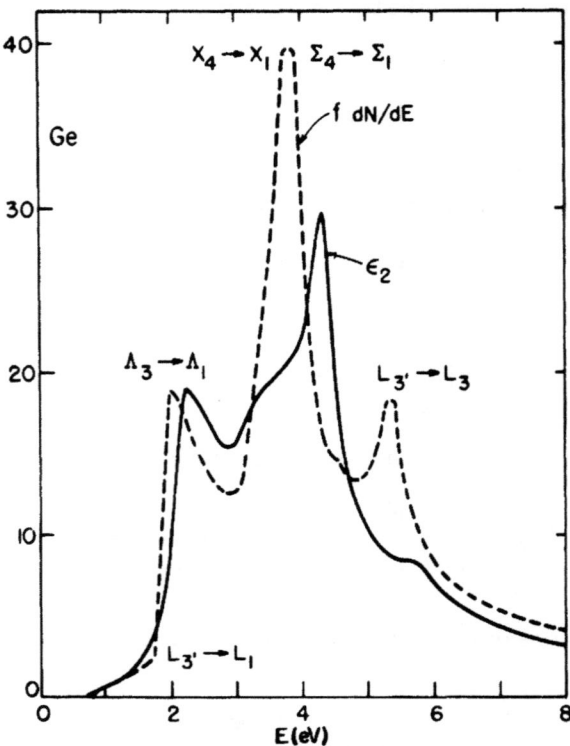

FIGURE 2. The absorption spectrum of Ge. The imaginary part of the dielectric constant (obtained from the experimental value of the absorption coefficient through Eq. (7)) is compared with the joint density of states for transitions from the valence to the conduction band, dashed curve [6].

---

the same parity, thus forbidding transitions between certain bands [7]. Symmetry plays then a fundamental role in the interpretation of the optical spectra [5].

The absorption coefficient due to allowed band-to-band direct (vertical) transitions is generally very large reaching values of $10^5 cm^{-1}$. It can be easily measured only for thin films. More often, the reflectivity of the solid is measured and the absorption coefficient is then calculated by Kramers-Kronig analysis.

In many solids (e.g. Si, Ge, AgBr, etc.) the maximum of the valence band and the minimum of the conduction band occur at different points of the Brillouin zone. The minimum energy of excitation across the gap corresponds then to an indirect

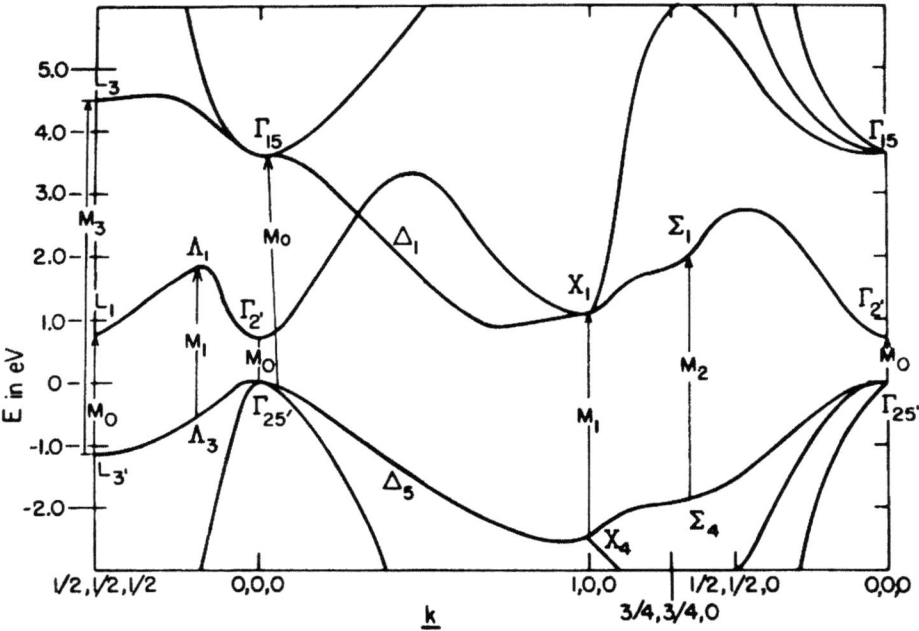

FIGURE 3. The band structure of Ge [6].

transition with $\vec{k}_j \neq \vec{k}_i$, forbidden by the selection rule [17]. However, in such a case, the conservation of the wavevector of the electron can occur through the interaction with the vibrational quanta of the crystals.

The transition probability for indirect transitions is different from zero only at second order. It is then much smaller than for direct transitions. This effect is in part compensated by the relaxation of selection rule [17] in the evaluation of the joint density of states. From a given $\vec{k}_i$ a direct transition takes the electron to a state of the upper band with the same $\vec{k}$, while for the indirect transition many $\vec{k}_j$ are allowed, provided energy and wave-vectors are conserved.

The interaction Hamiltonian for indirect transitions includes a term due to electron-phonon interaction, besides the usual electromagnetic term $H^1 = \frac{e}{cm} \vec{A} \cdot \vec{p}$, and can be written as [8]:

$$H = H^1 + H^2, \qquad (22)$$

where:

$$H^2 = \tfrac{1}{2}(\vec{U}_s \cdot \nabla V(\vec{r})) e^{i(\vec{q}_s \cdot \vec{r} - \omega_s t)}, \tag{23}$$

$\vec{U}_s$ being the amplitude of the vibrational wave of frequency ω_s and wavevector $\vec{q}_s$; $V(\vec{r})$ is the lattice potential. The transition probability, for transitions from the valence to the conduction band, is given to second order by [8]:

$$\frac{2\pi |H^1_{vc}(\vec{k}_i,\vec{k}_i)|^2 \cdot |H^2_{cc}(\vec{k}_i,\vec{k}_j)|^2}{\hbar^2 [E_c(\vec{k}_i)-E_v(\vec{k}_i)-\hbar\omega]^2} \delta(E_c(\vec{k}_j)-E_v(\vec{k}_i)\pm\hbar\omega_s-\hbar\omega)$$

$$+ \frac{2\pi |H^1_{vc}(\vec{k}_j,\vec{k}_j)|^2 \cdot |H^2_{vv}(\vec{k}_j,\vec{k}_i)|^2}{\hbar^2 [E_c(\vec{k}_j)-E_v(\vec{k}_j)-\hbar\omega]^2} \delta(E_c(\vec{k}_j)-E_v(\vec{k}_i)\pm\hbar\omega_s-\hbar\omega) \tag{24}$$

The ± signs in the δ-function represent creation and annihilation of phonon, respectively.

The first term of (24) refers to the situation sketched in Fig. 4 that can be described by the following processes:

a) the interaction with the electromagnetic field induces a transition from the state $\vec{k}_i$ of the valence band to the virtual state $\vec{k}_i$ of the conduction band;

b) the electron-phonon interaction induces a transition from the virtual state $\vec{k}_i$ to the final state $\vec{k}_j$ both in the conduction band.

For the second transition:

$$\vec{k}_i - \vec{k}_j \pm \vec{q}_s + \vec{q} = 0 \tag{25}$$

where the ± signs represent here processes with annihilation and creation of a phonon, respectively.

In transition a) energy is not conserved; the intermediate virtual state should then be sufficiently short-lived so that its energy indetermination is larger than ΔE (defined in the figure). If $\Delta E \sim 0.2 eV$, $\tau \sim 10^{-15} s$.

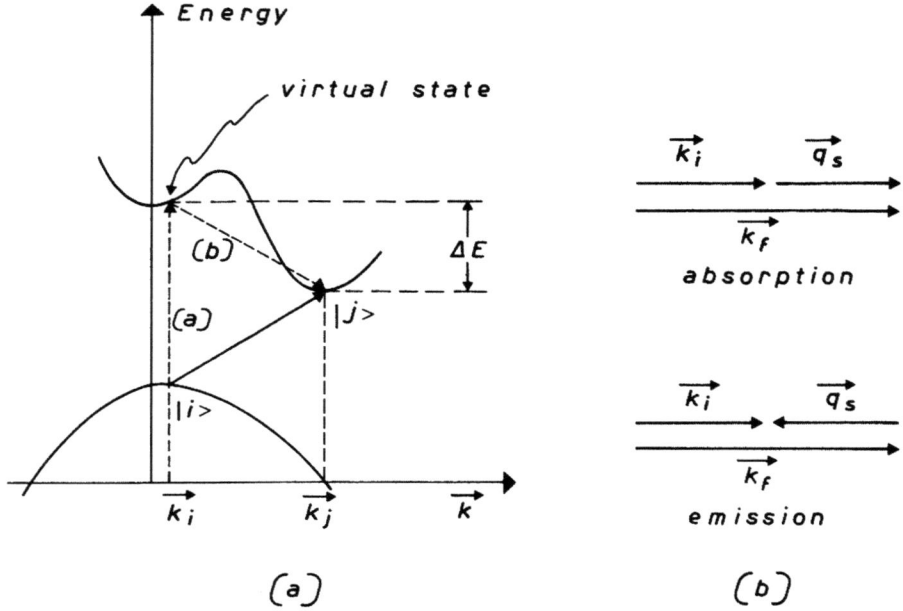

FIGURE 4. Schematic representation of an indirect transition given by the first term of Eq. (24); part (a) of the figure. The transition can occur either with absorption or with emission of a phonon. The conservation of wave-vectors in the two cases is shown in part (b) of the figure.

The processes due to the second term in (24) are similar and are sketched in Fig. 5.

Processes occurring with annihilation of a phonon depend on the number n_s of phonons present in the crystal. They vary with temperature according to

$$n_s = \frac{1}{\exp\left(\frac{\hbar\omega_s}{kT}\right) - 1} \qquad (26)$$

which vanishes at $T \to 0$.

Processes involving the creation of a phonon depend on

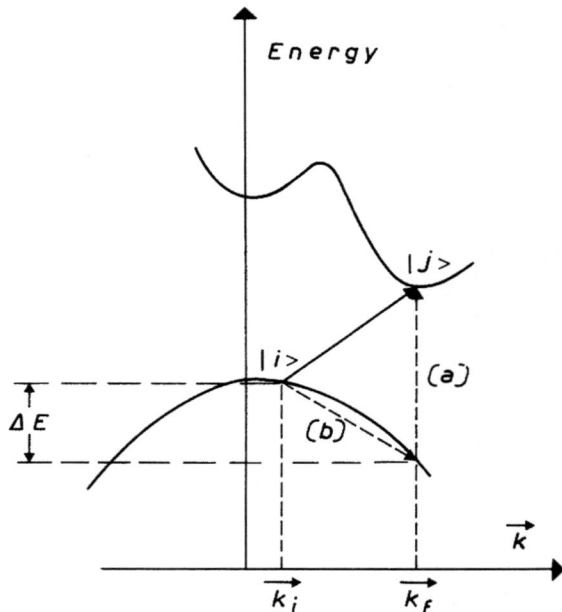

FIGURE 5. Schematic representation of an indirect transition given by the second term of Eq. (24).

$$n_s + 1 = \frac{1}{1-\exp\left(-\frac{\hbar\omega_s}{kT}\right)} \qquad (27)$$

and are nearly independent on temperature.

Indirect transitions, being much less probable than direct ones, are significant only at the energy of the edge of the absorption spectrum.

In Fig. 6 the indirect absorption edge of Ge is plotted for two different temperatures, and shows processes occurring with the annihilation of a phonon which depend on temperature and process with creation of a phonon, independent of temperature [9].

With the modern techniques of modulation spectroscopy [10], the indirect absorption edge of most solids shows various structures which can be attributed to the phonons of the various branches [11].

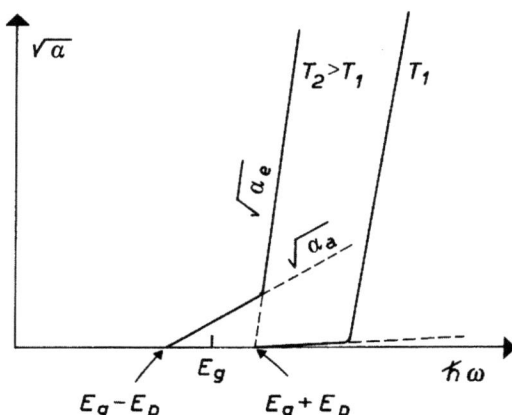

FIGURE 6. The square root of the absorption coefficient as a function of energy for the indirect edge of Ge at two different temperatures. The steepest part of the curve corresponds to emission of a phonon. It does not depend on temperature and is displaced to higher energy at lower temperatures because of the increase of energy gap.

Near the top of the valence band or the bottom of the conduction band, the energy versus $\vec{k}$ curve can be approximated by its parabolic expansion. Under such conditions the joint density of states can be easily evaluated, thus obtaining, for the dependence on energy of the absorption coefficient, the expressions:

$$\alpha_{dir}(\omega) = C(\hbar\omega - E_g)^{\frac{1}{2}} \qquad (28)$$

($\hbar\omega > E_g$, the energy gap) for direct, allowed transitions and

$$\alpha_{indir}(\omega) = C'(T)(\hbar\omega - E_g - \hbar\omega_s)^2 + C''(T)(\hbar\omega - E_g + \hbar\omega_s)^2 \qquad (29)$$

for indirect allowed transitions [12]. In (29) it has been assumed that a simple phonon of energy $\hbar\omega_s$ is involved. The first term in Eq. (29) corresponds to the emission of a phonon and holds for $\hbar\omega > E_g + \hbar\omega_s$, while the second term corresponds to annihilation of a phonon and holds for $\hbar\omega > E_g - \hbar\omega_s$.

Figure 6 shows that the square root of the absorption coefficient varies linearly with ω according to Eq. (29), thus confirming that we are dealing with indirect transitions.

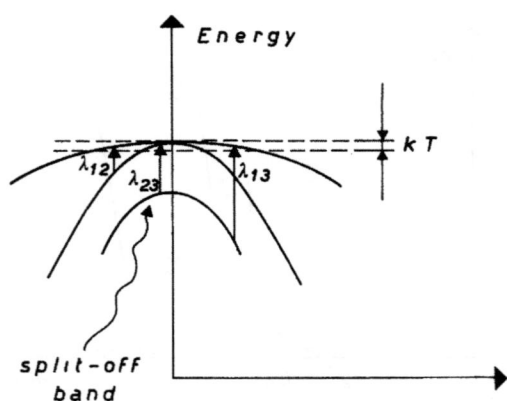

FIGURE 7. The E vs. $\vec{k}$ curve for the valence band of Ge. The various transitions responsible for intraband absorption are shown.

In many solids, however, the absorption edge does not follow the simple dependence (28) and (29), or the analogous ones for forbidden transitions, but show a temperature dependent exponential tail extending into the gap. In a number of materials, the exponential tail obeys the so called Urbach's rule, namely: $\frac{d(\log\alpha)}{d(\hbar\omega)} = \frac{1}{kT}$. Such a tail is now believed to be due to the effect of the electric fields due to ionic vibrations and/or to vibrating impurities [13].

Absorption Due To Intraband Transitions

This absorption is due to electronic transitions among the levels of the same band.

Up to now, we have neglected statistical occupation of the states thus tacitly implying that the initial state $|i\rangle$ is occupied and the final state $|j\rangle$ empty.

This is no longer correct in dealing with transitions among the levels of the same band. Moreover, very often the levels of a single band have the same parity (at least at $\vec{k}=0$), so that we have to do with parity forbidden transitions.

As an example, we shall treat the hole absorption in Ge [14]. At $\vec{k}=0$ the valence band of Ge consists of three sub-bands, two (degenerate at $\vec{k}=0$) having the symmetry of $p_{3/2}$ orbitals and the third, shifted by the spin-orbit interaction, having the symmetry of $p_{1/2}$ orbitals. The three bands are shown as a function of $\vec{k}$ in Fig. 7. Transitions among the three sub-bands are forbidden at

OPTICAL PROPERTIES OF SOLIDS 213

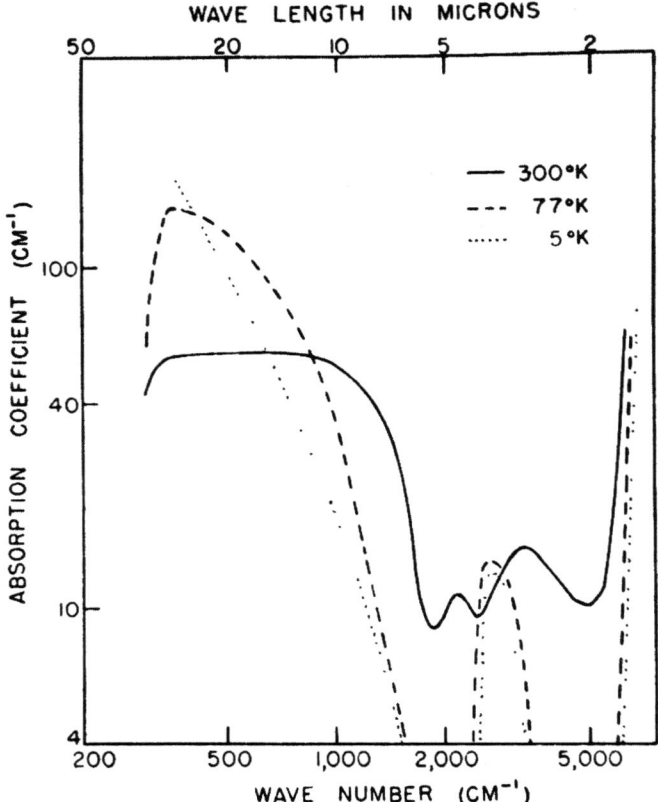

FIGURE 8. The absorption spectrum of p-type Ge, showing the bands arising from the various transitions of Fig. 7 [15].

$\vec{k}=0$ by the selection rule $\Delta \ell = \pm 1$. However, for $\vec{k} \neq 0$ the symmetry is lower, so that the classification according to the azimutal quantum number is no longer correct and the transitions among the three sub-bands become partially allowed.

The absorption spectrum of p-type Ge, caused by intraband transitions, is shown for two different temperatures in Fig. 8 where the contributions of the various transitions are also shown [15].

The peaked structure of the absorption is caused by the competition between the parity selection rule (which is more "relaxed" as $\vec{k}$ departs from zero) and statistics which increases the occupation of the bands the larger is $\vec{k}$.

The transition λ_{23} tends to disappear at low temperatures because the light-hole band becomes fully occupied.

Intraband transitions between the levels of the conduction band of n-type GaP at point $\langle 100 \rangle$ have also been observed [16].

Exciton Absorption

In the last sections, the evaluation of the matrix element in Eq. (10) or (12) has been done under the hypothesis that $|i\rangle$ and $|j\rangle$ were the Bloch's states of band theory. This is a one-electron theory in which the states of the conduction band are those appropriate to an additional electron introduced into the solid. Moreover, the electron is assumed to experience the average field of the nuclei and the other electrons.

When the electron is created by a photon, a hole in the valence band is left behind. The interaction between the electron and the hole is not taken into account in the theory previously outlined which is appropriate only when the electron and the hole generated by the optical transition are free and move away in opposite directions with the same momentum [17]. The band-to-band transitions correspond then to atomic transitions that ionize the atom. We expect, however, that transitions analogous to non-ionizing optical excitations of atoms occur also in solids. The excitation thus produced can move freely and is called an exciton [18]. The optical spectrum of excitons consists of a series of lines at energies lower than those of the corresponding band-to-band transitions. When the region where the excitation has occurred is large with respect to the atomic dimensions, the exciton can be visualized as a hydrogen-like complex consisting of an electron and a hole revolving around their center of mass.

Let us assume, in a first approximation, that the atoms of the solid do not interact. The wave function of the ground state can then be written as a product of atomic wavefunctions centered at various lattice positions.

$$\psi_o = \varphi_o(\vec{R}_1)\varphi_o(\vec{R}_2) \cdots \varphi_o(\vec{R}_N). \tag{30}$$

The lowest excited state is then:

$$\psi_n^L = \varphi_o(\vec{R}_1)\varphi_o(\vec{R}_2) \cdots \varphi_n(\vec{R}_L) \cdots \varphi_o(\vec{R}_N), \tag{31}$$

OPTICAL PROPERTIES OF SOLIDS

in which L-th atom is excited into the n-th level [19]. Any one of the atoms of the solid can be excited so that the state (31) is highly degenerate.

If we let the atoms interact, the degeneracy is lifted and the perturbed wave function of the excited state is a linear combination of the (31), namely:

$$\psi_n(\vec{k}_{ex}) = \frac{1}{\sqrt{N}} \sum_L e^{i\vec{k}_{ex} \cdot \vec{R}_L} \psi_n^L , \qquad (32)$$

Eq. (32) indicates that the excitation propagates with the wave vector $\vec{k}_{ex}$.

When the interaction between the electron and the hole is small, like in solids with very high dielectric constants, a theory of the exciton can be worked out in the frame of the effective mass approximation [20]. The energy of the exciton, with respect to the bottom of the conduction band, is then:

$$E_n(k_{ex}) = \frac{1}{2} \frac{\hbar^2}{m_e^* + m_h^*} k_{ex}^2 - \mu \frac{e^4}{2\hbar^2 \varepsilon^2} \frac{1}{n^2} , \qquad (33)$$

where m_e^* and m_h^* are the effective masses of electrons and holes, $\mu = \frac{m_e^* m_h^*}{m_e^* + m_h^*}$ their reduced mass and n an integer. For direct transitions $\vec{k}_{ex} = \vec{q} \sim 0$, so that the exciton spectrum consists of a series of relatively sharp lines given by the second term in Eq. (33). The selection rule $\vec{k}_{ex} \sim 0$ has a very simple interpretation: for direct transitions the electron and the hole left behind should move in opposite direction. However, if they form an exciton, they should move together, so that the only possibility is $\vec{k}_e = \vec{k}_h = \vec{k}_{ex} = 0$.

On the other hand for solids with indirect gaps, optical transitions involve phonons which can provide the momentum for the kinetic energy given by the first term of Eq. (33). The exciton spectrum consists then of broad bands with a definite lower energy. In indirect excitons the transition probability is simply proportional to the density of states of the valence band [21], i.e.:

$$\alpha_{ex} = C(\hbar\omega - E_g + E_{ex} \pm E_s)^{\frac{1}{2}} \qquad (34)$$

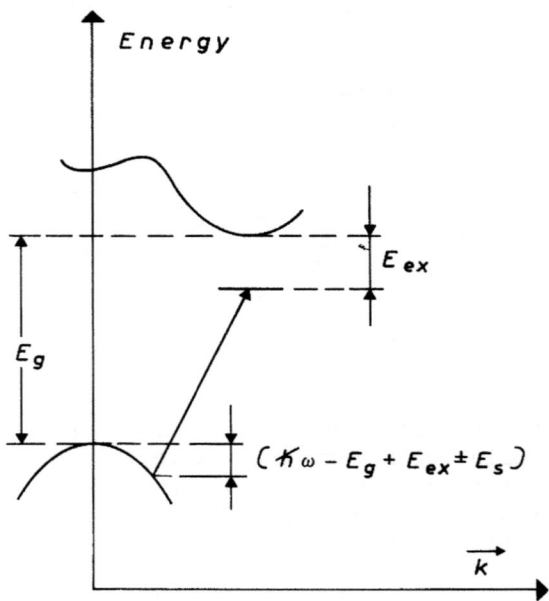

FIGURE 9. Schematic representation of the creation of an indirect exciton. In the energy balance giving the position of the initial state, the energy of the phonon has been taken into account.

---

where E_s is the energy of the phonon that conserves momentum, and E_{ex} is the binding energy of the exciton (Fig. 9). The absorption coefficient for an indirect edge is proportional to the square of $(\hbar\omega - E_g \pm E_s)$ according to Eq. (29) so that the occurrence of an indirect exciton is evidenced by a step in the absorption coefficient versus energy curve.

Obviously the exciton does not carry a current so that excitonic transitions are often detected by comparing the optical absorption with the photoconductivity curves.

Figure 10 shows the excitonic absorption of RbI which is an example of a direct exciton rather localized on the halide ion [22]. Transitions with n=1, n=2 and n=3 are clearly seen. The second part of the figure shows the onset of photoconductivity due to band-to-band transitions [23].

Figure 11 gives the spectrum of the so-called yellow series of Cu_2O, an example of an extended exciton with sharper lines [24]. In Cu_2O the exciton forms in a P state so that the term n=1 is missing. Figure 12 shows the spectrum of direct excitons in GaAs [25] and Fig. 13 that of indirect excitons in GaP [26].

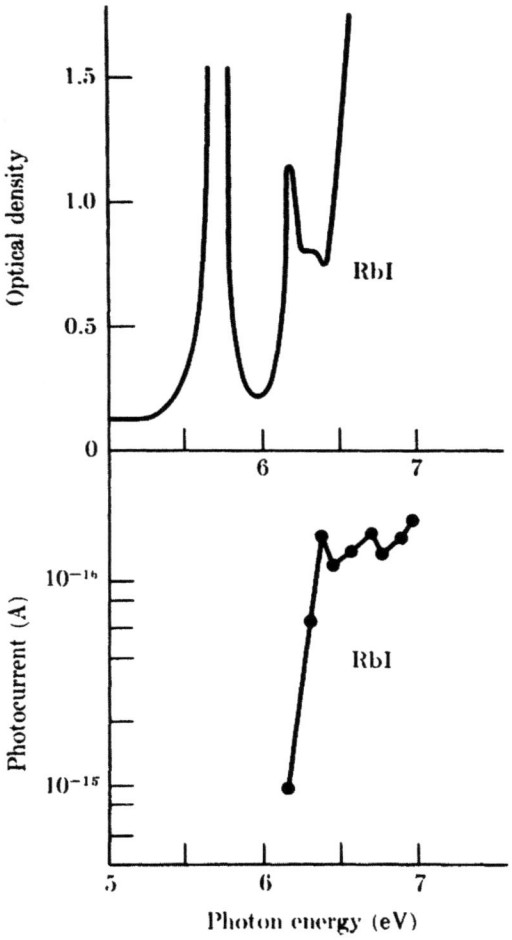

FIGURE 10. Exciton spectrum and intrinsic photoconductivity of RbI at 10°K [22,23].

Impurity Absorption

In insulators and semiconductors, impurities and lattice defects give rise to localized electronic states with energies in the gap, between the valence and the conduction band. In fact, the levels of the conduction band correspond to the ionized electron, so that the bound states have energies below the bottom of the conduction band. Optical transitions among the levels of the impurities (when they are occupied be electrons) give rise to absorption bands in the spectral region of the transparent window [27].

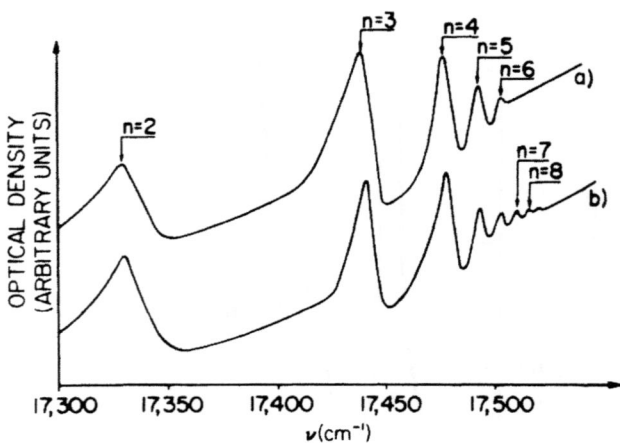

FIGURE 11. The so-called yellow series of excitons in Cu_2O at 4.2°K [24].

Typical examples of impurity absorption are donors and acceptors in group-IV semiconductors, and the F-center in alkali halides. Impurities in semiconductors are important as recombination centers, especially in indirect gap semiconductors where the direct electron-hole recombination is forbidden by the k-conservation rule.

In order to discuss the electronic problem of an impurity or a defect, we cannot longer neglect the interaction with the lattice vibrations. For any configuration of the lattice around the defect the electronic energy has a different value. A great simplification, however, is brought about the so called adiabatic approximation which allows the separation of electronic and nuclear coordinates [28]. The separation is possible because the motion of the nuclei is much slower than that of the electrons.

As a consequence of the adiabatic theorem the following procedure can be used to solve the problem of an electron trapped at a lattice defect or at an impurity [27]:

i) Keeping the nuclei fixed in a given position, say $\vec{R}_\alpha$, we find the potential in which the electron moves and solve the Schrödinger equation for the electron to obtain the eigenvalues $\varepsilon(\vec{R}_\alpha)$ for all possible $\vec{R}_\alpha$.

ii) The dynamics of the nuclear motion is then obtained by assuming $\varepsilon(\vec{R}_\alpha)$ as the potential energy in the Schrödinger equation for the nuclei.

OPTICAL PROPERTIES OF SOLIDS

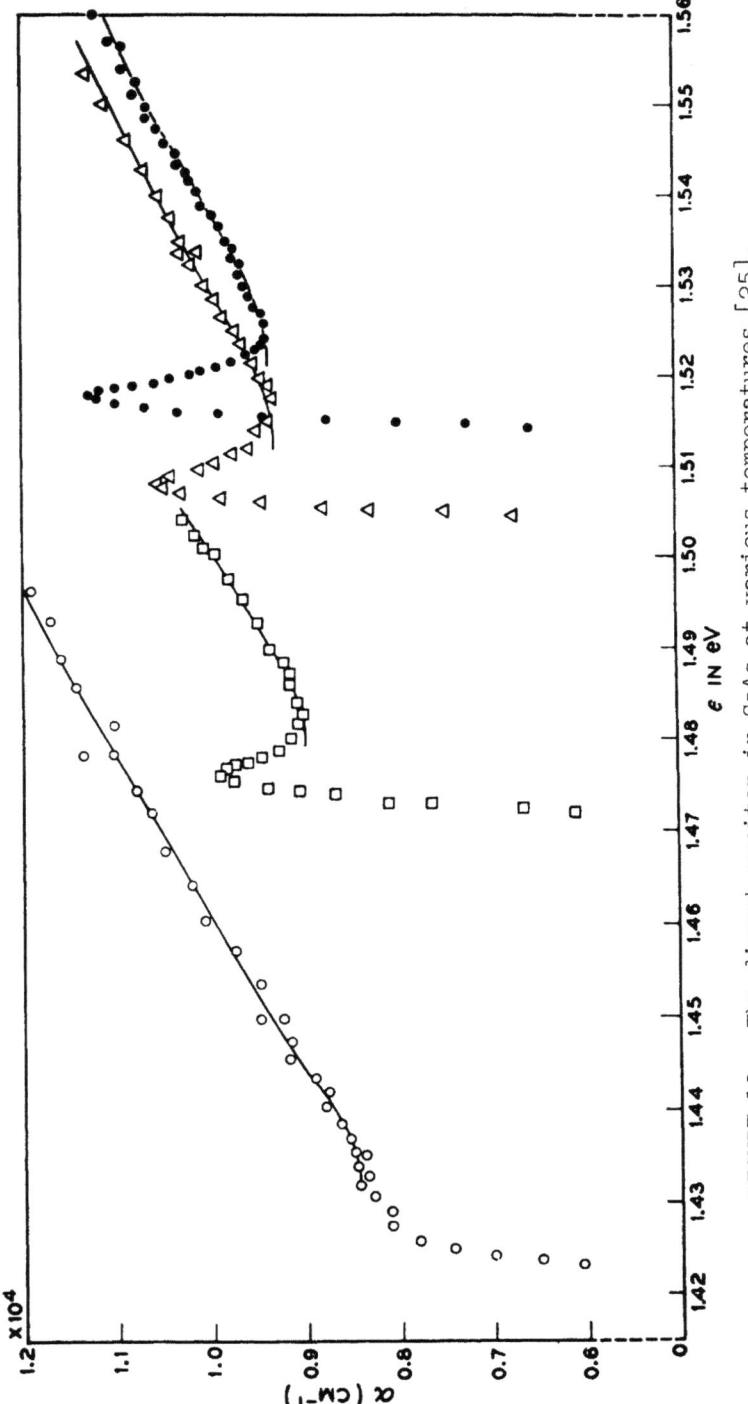

FIGURE 12. The direct exciton in GaAs at various temperatures [25].

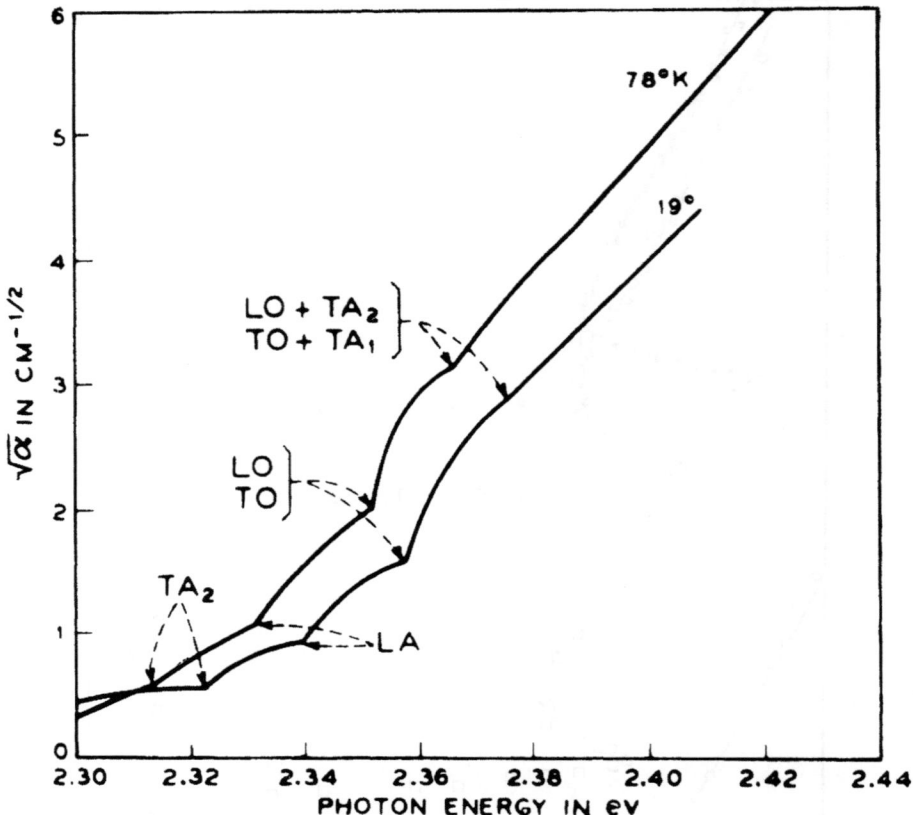

FIGURE 13. Threshold energies for the indirect excitons in GaP at two temperatures [26].

It can be demonstrated that the eigenvalues of the nuclear problem represent the total energy of the system (nuclei+electron); the optical transitions occur then among the vibrational levels of the nuclei. This is illustrated in Fig. 14 where a single nuclear coordinate $\vec{R}$ has been assumed for simplicity. It is seen that the minimum of the $\varepsilon_i(\vec{R})$ curve is at a different position than that of $\varepsilon_j(\vec{R})$. This is because the charge distribution of the excited state differs in general from that of the ground state.

Optical transitions for absorption and emission are shown as vertical arrows according to the Franck-Condon principle; one notices that the energy of the emitted photon is much smaller than that of the absorbed photon (large Stokes shift) and that several phonons are created during the optical process. The Stokes shift increases with increasing the difference between $\vec{R}^o_{(i)}$ and $\vec{R}^o_{(j)}$.

OPTICAL PROPERTIES OF SOLIDS

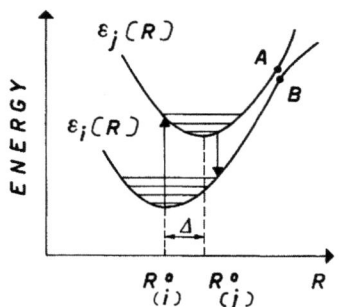

FIGURE 14. Configurational coordinate diagram for a system with two electronic states, showing the various vibrational levels and the optical transitions in absorption and emission.

When this difference is very large (strong electron-phonon interaction) it may be that the excited electron reaches a point higher than A of curve $\varepsilon_j(\vec{R})$. The electron makes then a (non-adiabatic) transition to point B of curve $\varepsilon_i(\vec{R})$ and falls to the ground state without emitting a photon. Such transitions are called radiationless transitions.

For strong electron-phonon interaction, the absorption spectrum consists of broad bands, with the emission of many phonons and a large Stokes shift as sketched in Fig. 15, at T=o; W and W' are the band-widths in absorption and emission; in the example shown 12 phonons are created in the absorption process.

The opposite case is that of weak electron-phonon interaction: the $\varepsilon_i(R)$ and $\varepsilon_j(R)$ curves are but slightly displaced, there is a very small Stokes shift, only few phonons are emitted and the spectrum consists of narrow lines (Fig. 16). The last result is a consequence of the fact that, according to the adiabatic theorem, the wavefunction of states $|i\rangle$ and $|j\rangle$ can be written as a product of a function of the electronic coordinates times a function of the nuclear coordinates. Moreover, the last can be expressed as a product of harmonic oscillator wavefunctions. These do not depend on electronic coordinates, can be taken out of the matrix element in Eq. (14) and being orthonormal make it vanish unless $\Delta n=0$. The situation sketched in Fig. 16 then ensues.

As an example of strong electron-phonon interaction we consider the F-center in alkali halides. The F-center consists of an electron trapped at a negative-ion vacancy. Its absorption spectrum consists of a broad band caused by a transition of the type 1s→2p, followed by a shoulder due to transitions to higher excited states. Approximately 30 phonons are emitted in KCl while the Stokes shift is larger than 1eV.

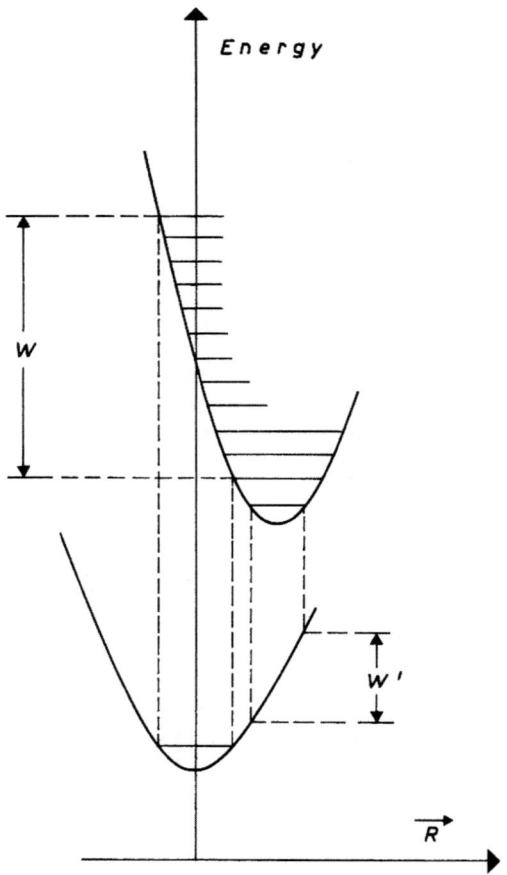

FIGURE 15. Configurational coordinate diagram for the case of strong electron-phonon interaction, showing the width of the absorption and emission bands W and W'.

An example of weak electron-phonon interaction is the R_2-center in LiF, whose spectrum is shown [29] in Fig. 17. The R_2-center (or F_3-center) consists of three F-centers forming an equilateral triangle in (111) planes. Its spectrum shows zero-phonon lines superimposed to a broad continuum.

The case of donors (or acceptors) in group-IV semiconductors is simpler. Ge and Si have a large dielectric constant that strongly reduces the Coulomb force; orbits are then large and a hydrogenic model is well obeyed. The spectrum of B in Si is shown as an example in Fig. 18 [30].

OPTICAL PROPERTIES OF SOLIDS

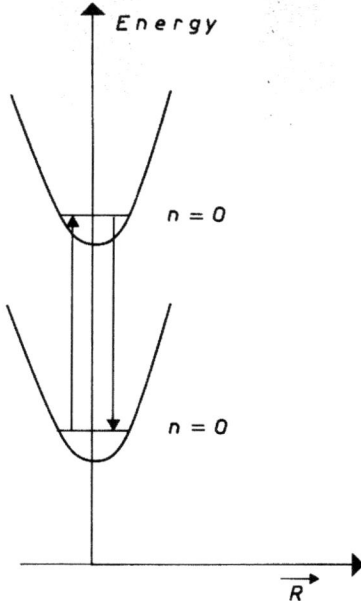

FIGURE 16. Configurational coordinate diagram for the case of weak electron-phonon interaction. Since now $\Delta n=0$ the optical transitions are narrow lines, with no phonon emission.

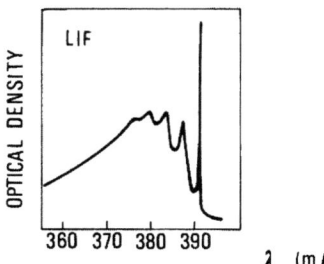

FIGURE 17. The absorption spectrum of the R_2-band in LiF at 4.2°K, showing the zero-phonon lines [29].

Transitions from localized levels to the conduction band (or from the valence band to localized levels) are also observed and are important in photoconductive devices.

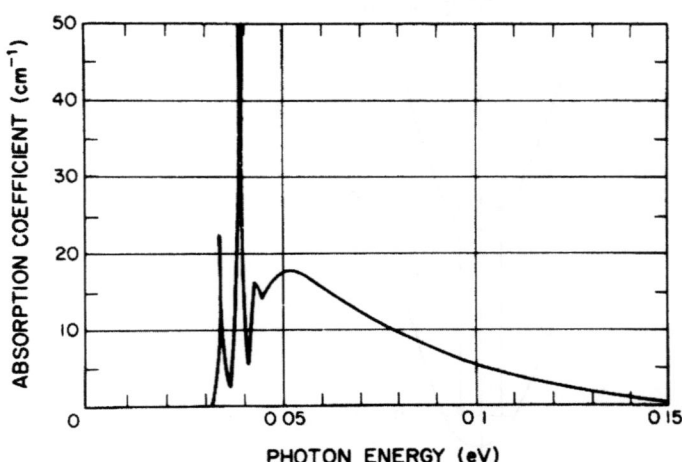

FIGURE 18. Absorption spectrum of B-doped Si [30].

Free Carrier Absorption

Free-electron absorption is caused by indirect transitions among the levels of the conduction band (Fig. 19). It is then a forbidden transition which becomes allowed through the interaction with the phonons. Free-carrier absorption is important in metals and heavily doped semiconductors.

The quantum theory has been developed first by J. Bardeen [31]. However, a semiclassical theory based on simplified transport equations gives essentially correct results [21].

The classical equation of motion of a free electron damped by the interaction with the lattice, in an electric field of angular frequency ω is:

$$\dot{\vec{v}} + \frac{\vec{v}}{\tau} = - \frac{e\vec{E}_o}{m^*} e^{i\omega t}, \qquad (35)$$

where τ is the relaxation time and m^* the effective mass. When τ is a constant, Eq. (35) has the solution:

$$\vec{v} = - \frac{e}{m^*} \frac{\tau}{1+i\omega\tau} \vec{E}_o e^{i\omega t}. \qquad (36)$$

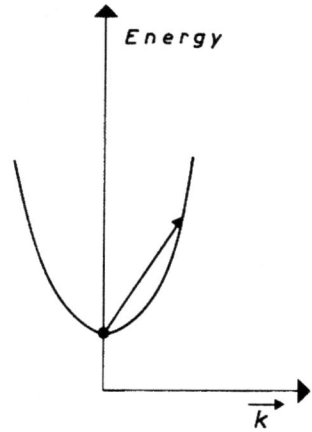

FIGURE 19. Indirect transition responsible for free-carrier absorption.

From (36) one obtains easily the current density $\vec{J}$ and the conductivity, whose real part is:

$$\sigma = \frac{Ne^2}{m^*} \frac{\tau}{1+\omega^2\tau^2}, \qquad (37)$$

N being the density of the free carriers.

Equation (7) allows then the determination of the absorption coefficient for free-carrier absorption. In the infrared and visible regions, where $\omega\tau \gg 1$, one obtains the simplified expression:

$$\alpha = \frac{Ne^2}{\pi c^3 n' m^* \tau} \lambda_o^2, \qquad (38)$$

which gives a characteristic spectrum with a λ_o^2 dependence.

In Ge such a dependence is well obeyed for $\lambda_o > 5\mu$. However, more refined theories, taking into account intervalley transitions, the dependence of τ on energy and the non-parabolicity of the bands, are required in order to explain the details of the spectrum in most semiconductors.

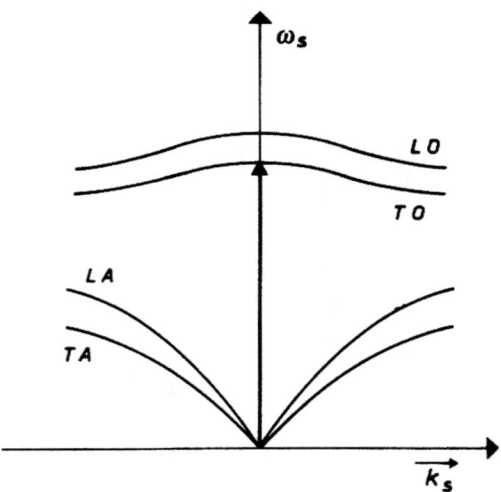

FIGURE 20. Excitation of a phonon of the transverse optical branch by an electromagnetic wave. The $\vec{k}$-conservation law requires that $\vec{k}_s = 0$.

Lattice Absorption

We have already seen many effects of the lattice vibrations on the optical absorption.

In addition the vibrating lattice can directly interact with electromagnetic waves [32], provided the lattice displays an electric dipole moment, so that the matrix element in Eq. (14) does not vanish. We then expect strong lattice absorption only for ionic or partially ionic solids. Alkali halides and compound semiconductors are classical examples. Covalent semiconductors like Si and Ge show a much weaker lattice absorption as a result of second order processes.

Again we have the $\vec{k}$-conservation rule

$$\vec{q}_s = \vec{q} \simeq 0, \qquad (39)$$

which shows that the created phonon has a nearly zero wave vector. The optical transition shown in Fig. 20 excites directly a phonon of the transverse optical mode which corresponds to ions of the two signs moving in opposite directions perpendicularly to $\vec{k}_s$. The transition of Fig. 20 gives rise to a strong absorption line (absorption coefficient of the order of 10^3-$10^4 cm^{-1}$), which

OPTICAL PROPERTIES OF SOLIDS

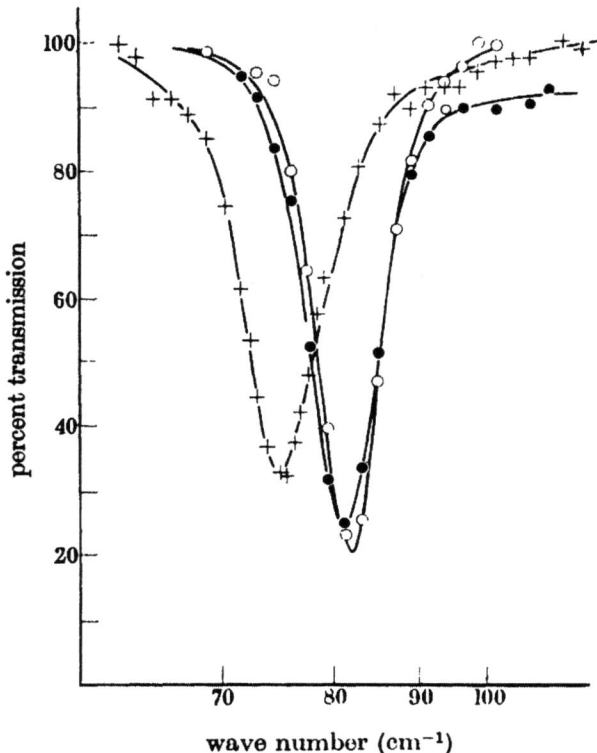

FIGURE 21. The transmission I/I_0 of a thin film of RbI at various temperatures showing the Reststrahalen band [33].

sometimes is called Reststrahlen absorption. The Reststrahlen band of RbI is shown in Fig. 21 for a thin film at different temperatures [33].

The line is broadened by multiphonon processes which also cause the appearance of a complicated satellite structure strongly dependent on temperature. A two-phonon process is illustrated in Fig. 22 where an optical and an acoustic phonon of opposite wavevectors are created by the photon whose frequency should then be the sum of those of the created phonons. Difference processes are also possible with the creation of an optical phonon and the annihilation of an acoustic phonon with the same wavevector. Since there are several branches in the phonon dispersion curves, the number of combinations is large and the absorption spectrum rather complex. The multi-phonon part of the spectrum of GaAs is shown in Fig. 23 at various temperatures [34].

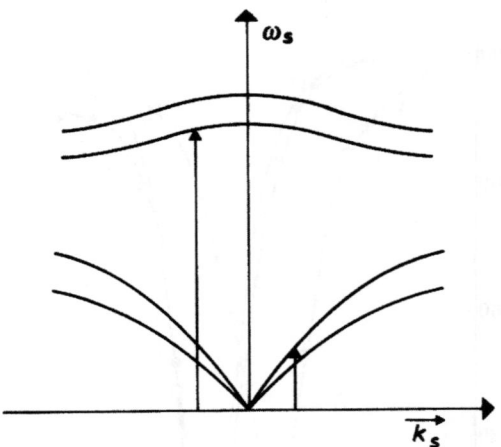

FIGURE 22. Schematic representation of a two phonon process. An optical and acoustic phonon with opposite wave-vectors are created by the electromagnetic wave.

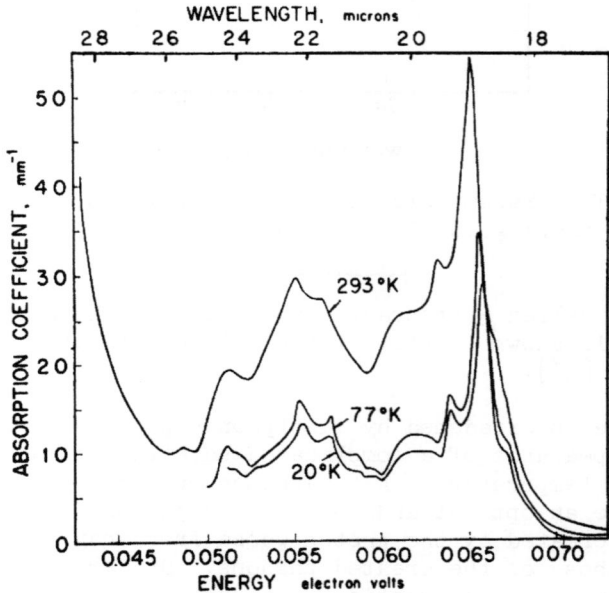

FIGURE 23. Lattice absorption spectrum of GaAs due to multiphonon processes [34].

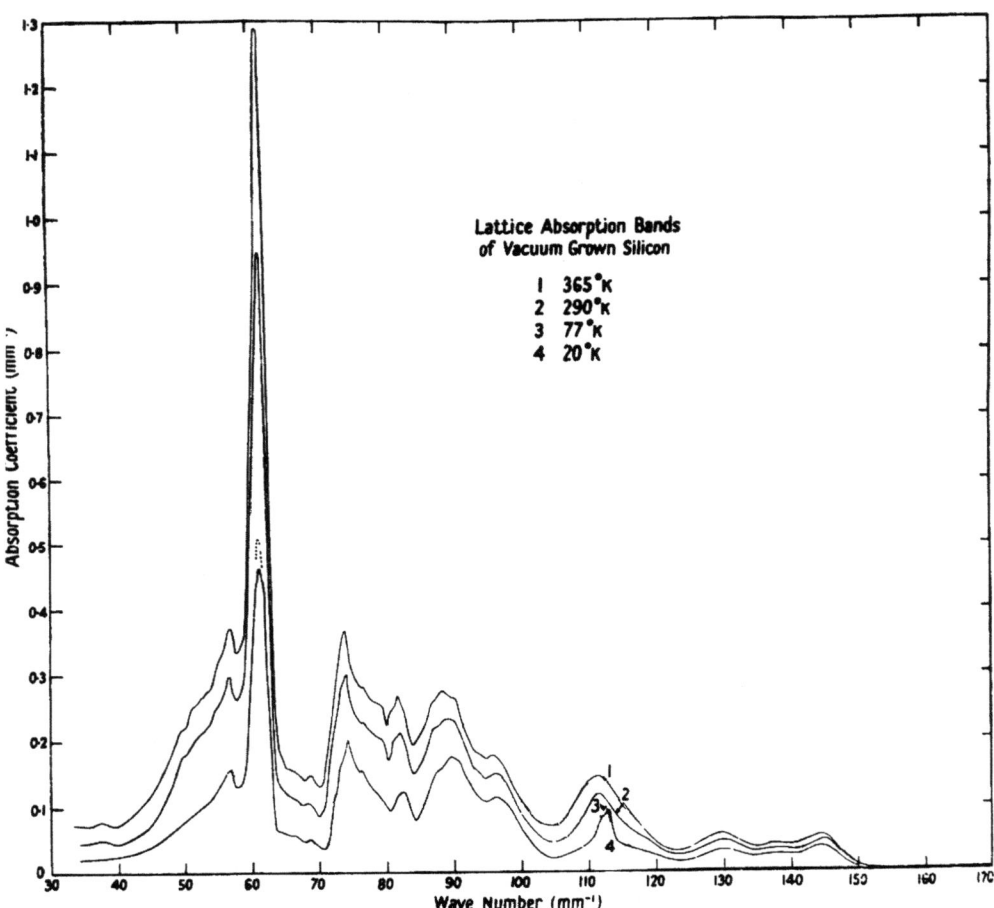

FIGURE 24. Lattice absorption spectrum of Si [35].

The vibrational modes of the diamond-type semiconductors do not develop an electric dipole moment and are not active in infrared. However, a small lattice absorption due to second order processes is present also in diamond-type semiconductors. Apparently, the radiation induces a dipole that in turn couples with the electromagnetic field. The absorption spectrum of Si is shown in Fig. 24 [35].

PHOTOCONDUCTIVITY

Electrons raised by optical transitions to the conduction band (or removed from the valence band) alter the population

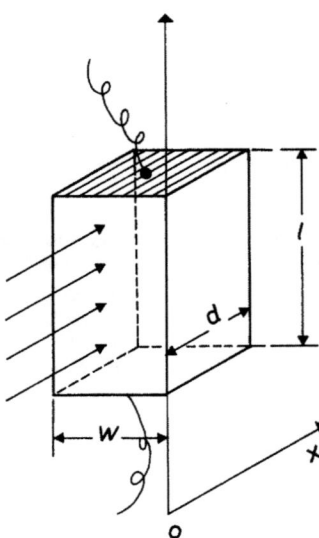

FIGURE 25. The geometry of the sample for the calculation of ΔG in Eq. (42).

of carriers and give rise to photoconductivity. Semiconductors (and insulators) can then be used as detectors of electromagnetic radiation. Band-to-band, impurity-band, and intra-band transitions can be used for this purpose.

The electrons excited into the conduction band recombine with holes with a characteristic time τ, so that the stationary extra-population of electrons is simply $N\tau$, N being the number of electrons raised by the electromagnetic radiation per unit time, per unit volume. The extra conductivity is then:

$$\Delta\sigma = eN\tau(\mu_e + \mu_h), \qquad (40)$$

μ_e and μ_h being the mobilities of electrons and holes. On the other hand, if we assume that any absorbed photon generates a hole-electron pair as in band-to-band transitions, Eq. (1) and (2) give immediately:

$$N = \alpha I = \alpha I_o(I-R)e^{-\alpha x}, \qquad (41)$$

where, now, I is measured in photons per sec per cm^2.

The variation of conductance G for the sample shown in Fig. 25 can then be calculated:

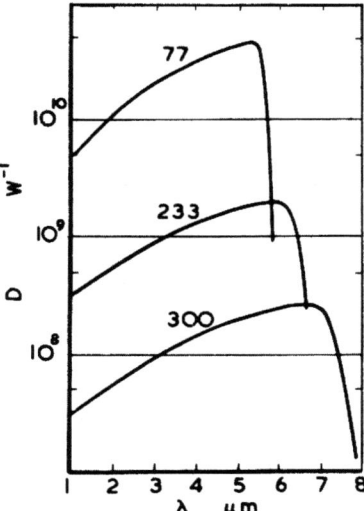

FIGURE 26. Detectivity of an InSb photodetector at various temperatures [37].

$$\Delta G = \left(\frac{w}{l}\right) e(\mu_e + \mu_h) \tau \alpha I_o (1-R) \int_0^d e^{-\alpha x} dx = \left(\frac{w}{l}\right) e(\mu_e + \mu_h) \tau I_o (1-R) [1 - e^{-\alpha d}].$$

(42)

Very often $\alpha d \gg 1$, so that ΔG depends upon the frequency of the photon only through the factor $(1-R)$ which is a smooth function of ω. Thus the photoconductivity curves do not display the richness of structures seen in absorption spectra but vary smoothly with frequency as is shown in Fig. 26 for InSb at various temperatures [36].

A variation in the conductance of the sample can also be obtained by changing the mobility of the carriers. In free-carrier absorption the excited hot electrons have in general a larger mobility. At very low temperatures, where the coupling with the lattice is small, it is possible to maintain, under illumination, a stationary population of electrons with energies well above those appropriate to thermal equilibrium [37].

PHOTOEMISSION

When the energy of the final state in an optical transition is above the vacuum level, the excited electron can escape from the solid giving rise to photoemission [38].

The process of photoemission consists of the following sequence i) photoexcitation of the electron in the bulk: ii) diffusion of the electron to the surface; iii) escape through the surface barrier. In process ii) the electron can loose energy through inelastic scattering.

In photoemission spectroscopy, the distribution in energy of the emitted electrons is measured for a fixed value of the frequency of the exciting photon. The energy of the electrons is generally measured by means of a retarding potential V that allows only the faster electrons to reach the anode. A small a.c. voltage in series allows the determination of $\frac{dI}{dV}$ as a function of V, I being the external current. If we introduce the energy distribution curve (EDC) for the emitted electrons, $N(E,\omega)$, which, of course, depends on the frequency ω of the exciting photon, we can write:

$$I = e \int_{-eV^*} N(E,\omega)dE, \qquad (43)$$

V* being the potential difference between the surface of the sample and the collector (not necessarily equal to the applied voltage V). A measure of

$$\frac{dI}{dV} = \frac{dI}{dV^*} = e^2 N(-eV^*,\omega) \qquad (44)$$

gives immediately the energy distribution curve [39].

From the point of view of the theory $N(E,\omega)$ can be written as:

$$N(E,\omega) = T(E)[P_o(E,\omega) + P_s(E,\omega)] \qquad (45)$$

where $P_o(E,\omega)$ is the energy distribution of internal electrons, $P_s(E,\omega)$ the secondary internal distribution which originates from $P_o(E,\omega)$ through inelastic scattering and $T(E)$ is the escaping function. The three functions correspond to the processes i), ii) and iii).

$P_o(E,\omega)$ can be directly correlated with the band structure of the solid through a joint density of states analogous to that introduced in the section entitled Absorption Due to Interband Transitions. However, the integral in (21) is now calculated along a line in $\vec{k}$-space (instead of on a surface of constant energy) since we are interested only in final states with energy

OPTICAL PROPERTIES OF SOLIDS   233

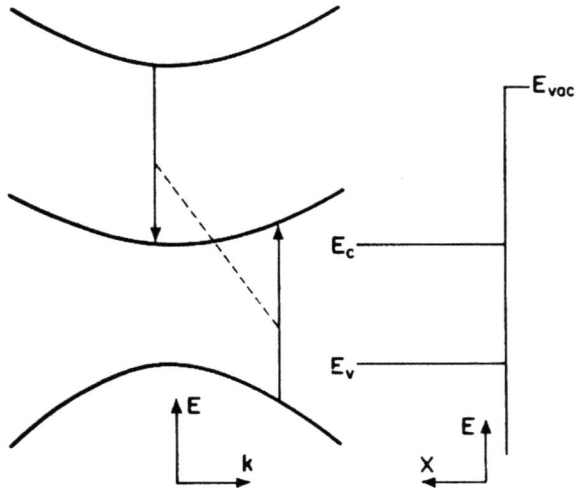

FIGURE 27. Electron-electron interaction through an Auger mechanism. The excited electron loses its energy to an electron of the valence band which is raised to the conduction band.

$E_j = E$ (and not in a constant energy difference $E_j - E_i = \hbar\omega$). The line of integration is the intersection of the energy surface:

$$E_j(\vec{k}) - E_i(\vec{k}) = \hbar\omega, \qquad (46)$$

with the electron energy surface:

$$E_j(\vec{k}) = E. \qquad (47)$$

T(E) can be calculated with the methods of quantum mechanics: it is zero for $E < E_F + \varphi$, φ being the work function of the solid and is nearly constant for $E > E_F + \varphi$.

$P_s(E,\omega)$ is the energy distribution of the electrons inelastically scattered before reaching the surface. In metals the scattering is mainly due to electron-electron interaction, while in semiconductors the electron-phonon interaction and the Auger mechanism of Fig. 27 predominate. In the last process the excited electron loses its energy to an electron of the valence band which is then raised into a higher band.

The analysis of the experimental data allows in general the separation of P_o and P_s giving a joint density of states that can

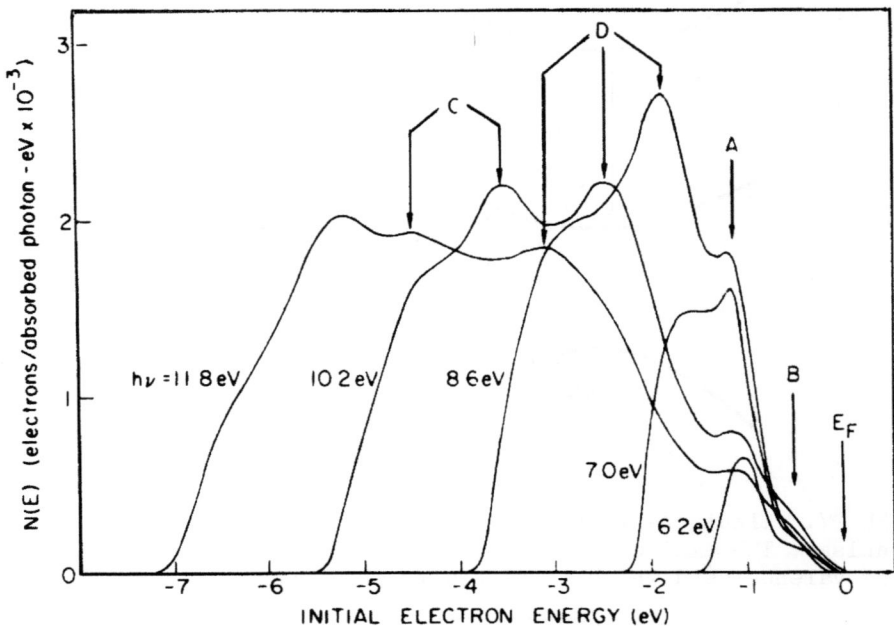

FIGURE 28. Energy distribution curves for electrons photoemitted from Si, for various frequencies of the exciting photon. Energies are measured with respect to the electrons emitted from levels at the Fermi energy [41].

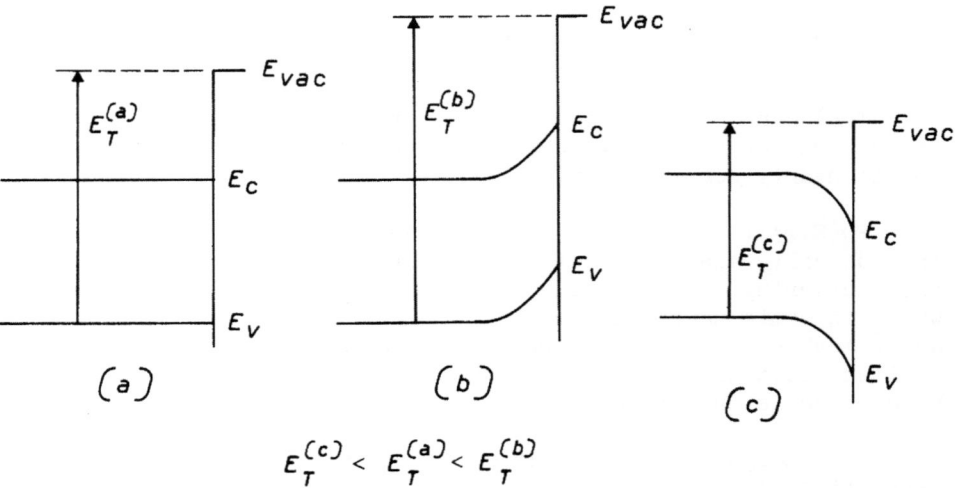

$$E_T^{(c)} < E_T^{(a)} < E_T^{(b)}$$

FIGURE 29. Threshold energy E_T for a semiconductor with flat bands (a), p-type surface (b) and n-type surface (c).

be referred to the density of states of the valence band. EDC curves for Si are shown in Fig. 28 for different frequencies of the exciting photon [41].

A parameter that is of great technical interest is the threshold $\hbar\omega_0$ for photoemission. The threshold depends strongly upon the conditions of the surface.

In semiconductors the existence of surface states determines a localization of charge at the surface and therefore a bending of the bands [42]. Acceptor-like states make the bands bend up while donor-like states make them bend down. The threshold energy is increased in the former and decreased in the latter case, as is shown schematically in Fig. 29.

Though intrinsically present at the surface of semiconductors, the surface states are strongly affected by the treatment of the surface and by the impurities present in it. Surfaces of GaAs covered by a monolayer of Cs show a remarkable decrease of threshold energy that becomes approximately equal to the band gap ($\sim$1.5eV) [43].

REFERENCES

[1] Many books and articles on the optical properties of solids are available. Among them: J. Tauc, Editor, The Optical Properties of Solids, Proc. Int. School E. Fermi, Academic Press, New York (1966); D. L. Greenaway and G. Harbeke, Optical Properties and Band Structure of Solids, Pergamon Press, Oxford, 1968; S. Nudelman and S. S. Mitra, Editors, Optical Properties of Solids, Plenum Press, New York 1969; J. I. Pankove, Optical Processes in Semiconductors, Prentice Hall, Englewood Cliffs, New Jersey (1971); J. C. Phillips, Solid State Physics, Vol. 18, Academic Press, New York (1966).

[2] When the reflectivity of the sample is not negligible, part of the light will be reflected from the two surfaces. It can be easily shown that, for normal incidence,

$$I/I_o = \frac{(1-R)^2 e^{-\alpha d}}{1-R^2 e^{-2\alpha d}}$$

where d is the thickness of the sample and R its reflectivity. The above formula reduces to (2) when R=0.

[3] For a demonstration of the Kramers-Kronig dispersion relations see for example the article by M. Cardona, Optical Constants of Insulators: Dispersion Relations, in the book by S. Nudelman and S. S. Mitra, Editors, quoted in Ref. (1).

[4] See for example, F. Seitz, Modern Theory of Solids, McGraw-Hill, New York and London (1940), p. 215 ff.
[5] See for example the article by G. F. Bassani, Band Structure and Interband Transitions in the book by J. Tauc, Editor, quoted in Ref. (1).
[6] D. Brust, J. C. Phillips, F. Bassani, Phys. Rev. Letters, $\underline{9}$, 94 (1962).
[7] This selection rule is most easily stated in the language of group theory: if the direct-product representation of $\vec{r}$ and $|i\rangle$ does not contain $|j\rangle$, the transition is forbidden.
[8] See for example: R. A. Smith, Wave Mechanics of Crystalline Solids, Chapman Hall, 1961, p. 444 ff.
[9] G. G. Macfarlane and V. Roberts, Phys. Rev. $\underline{97}$, 1714 (1955).
[10] M. Cardona, Modulation Spectroscopy, Sol. State Physics, Supplement 11 (1969).
[11] A. Frova and P. Handler, Phys. Rev. Letters $\underline{14}$, 178 (1965).
[12] See for example: J. I. Pankove, quoted in Ref. (1), p. 34 ff.
[13] J. D. Dow and D. Redfield, Phys. Rev. Lett. $\underline{26}$, 762 (1971).
[14] The interpretation that follows is due to H. B. Briggs and R. C. Fletcher, Phys. Rev. $\underline{87}$, 1130 (1952) and A. Kahn, Phys. Rev. $\underline{97}$, 1647 (1955).
[15] W. Kaiser, R. J. Collins and H. Y. Fan, Phys. Rev. $\underline{91}$, 1380 (1953).
[16] W. G. Spitzer, M. Gershenzon, C. J. Frosh and D. F. Gibbs, J. Phys. Chem. Solids, $\underline{11}$, 339 (1959).
[17] In fact the group velocity $\vec{v} = \frac{1}{h} \nabla_k E(\vec{k})$ has opposite directions, for the electron and the hole generated in the transition of Fig. 1.
[18] The theory of the exciton has been developed by J. Frenkel, Phys. Rev. $\underline{37}$, 17 (1931). Among the various review articles, see for example: S. Nitkitine, Excitons, in the book by S. Nudelman and S. S. Mitra, Editors, quoted in Ref. (1) and R. Knox, Theory of Excitons, in Sol. State Physics, Suppl. 5 Academic Press, New York (1963).
[19] To be true, expressions (30) and (31) should be antisymmetrized and substituted by a Slater determinant.
[20] G. Dresselhaus, J. Phys. Chem. Solids $\underline{1}$, 14 (1956); R. J. Elliot, Phys. Rev. $\underline{108}$, 1384 (1957).
[21] See, for example, R. A. Smith, Semiconductors, Cambridge University Press (1959), Chapter 7.
[22] G. Baldini, data published in Ref. (23).
[23] G. R. Huggett and K. Teegarden, Phys. Rev. $\underline{141}$, 797 (1966).
[24] S. Nikitine, Ref. (18).
[25] M. D. Sturge, Phys. Rev. $\underline{127}$, 768 (1962).
[26] M. Gershenzon, D. G. Thomas and R. E. Dietz, Proc. Int. Conference on Semiconductor Physics, Exeter (1962).
[27] For a review see: G. Chiarotti, Spectroscopy of Localized States in "Theory of Imperfect Crystalline Solids", IAEA, Vienna (1971).

[28] M. Born and K. Huang, Dynamical Theory of Crystal Lattices, Oxford University Press, London and New York (1954).
[29] D. B. Fitchen, R. H. Silsbee, T. A. Fulton, E. L. Wolf, Phys. Rev. Letters 11, 275 (1963).
[30] E. Burstein, G. S. Picus and N. Sclar, Proc. Photoconductivity Conference, Atlantic City, J. Wiley, New York (1956).
[31] J. Bardeen, Phys. Rev. 79, 216 (1950).
[32] For a review see the article by S. S. Mitra, Infrared and Raman Spectra due to Lattice Vibrations in, Optical Properties of Solids, S. Nudelman and S. S. Mitra, Editors quoted in Ref. (1).
[33] G. O. Jones, D. H. Martin, P. H. Mawer and C. H. Perry, Proc. Roy. Soc. (London) A 261, 10 (1961).
[34] W. Cochran, S. J. Fray, F. A. Johnson, J. E. Quarrington and N. Williams, J. Appl. Phys. 32, 2102 (1961).
[35] F. A. Johnson, Proc. Phys. Soc. (London) 73, 265 (1959).
[36] F. D. Morten and R. E. J. King, Appl. Opt. 4, 659 (1965).
[37] E. H. Putley, Infrared Photoconductivity, in Optical Properties of Solids, S. Nudelman and S. S. Mitra, Editors, quoted in Ref. (1).
[38] For a review, see for example D. E. Eastman, Photoemission Spectroscopy of Metals, IBM Yorktown Heights, Report RC 2987.
[39] W. E. Spicer and C. N. Berglund, Rev. Sci. Instrum. 35, 1665 (1964).
[40] P. O. Nilsson, The Study of the Electronic Band Structure of Solids by means of Optical Absorption and Photoemission, Lectures of the Graduate School, University of Rome (1972).
[41] L. F. Wagner and W. E. Spicer, Phys. Rev. Letters 28, 1381 (1972).
[42] For the dependence of photoconductivity on surface parameters see Pankove J. I., op. cit. Chapters 13 and 16.
[43] J. J. Scheer and J. Van Laar, Solid State Comm. 3, 189 (1965).

CHAPTER 10

NONLINEAR OPTICS

P. A. Wolff

Massachusetts Institute of Technology

Cambridge, Massachusetts

INTRODUCTION

Nonlinear properties of Maxwell's constitutive relations,

$$\vec{D} = \overleftrightarrow{\epsilon}(\vec{E}) \cdot \vec{E} \quad ; \quad \vec{B} = \overleftrightarrow{\mu}(\vec{H}) \cdot \vec{H} \tag{1}$$

have been known for many years. A variety of low frequency phenomena and devices have their origins in the nonlinearity of these equations. However, it is only in the past decade that nonlinear <u>optical</u> phenomena have been demonstrated and exploited. The first measurement of this kind was the (now classic) experiment of Franken, Hill, Peters, and Weinreich [1] in which ultraviolet light (at $\lambda = 3470 A°$) was produced by a ruby laser beam ($\lambda = 6940 A°$) traversing a quartz crystal. This process is now termed, for obvious reasons, second harmonic generation. It is a truly nonlinear process in which two photons of the pump (ruby) frequency are converted into a single photon at the doubled (UV) frequency.

Nonlinear optics revolves around the nonlinear constitutive equations, particularly that relating the polarization to the electric field:

$$\vec{P} = \overleftrightarrow{\chi}(\vec{E}) \cdot \vec{E}. \tag{2}$$

Here $\overleftrightarrow{\chi}(\vec{E})$ is the (nonlinear) susceptibility tensor. For moderate fields, in the dipole approximation, this expression can be expanded in the form

$$\vec{P} = \overleftrightarrow{\chi}^{(1)} \cdot \vec{E} + \overleftrightarrow{\chi}^{(2)} : \vec{E}\vec{E} + \ldots \quad . \tag{3}$$

$\overleftrightarrow{\chi}^{(1)}$ is the ordinary linear susceptibility of the medium, $\overleftrightarrow{\chi}^{(2)}$ is the lowest order <u>nonlinear</u> susceptibility, etc. The $\overleftrightarrow{\chi}^{(2)}$ term in this expansion is responsible for the second harmonics observed in the Franken experiment.

There are two sorts of questions one may ask concerning this nonlinear relation between $\vec{P}$ and $\vec{E}$. First of all, one may inquire as to the origin of $\overleftrightarrow{\chi}^{(1)}$, $\overleftrightarrow{\chi}^{(2)}$, ... - their magnitude, their frequency dependence, their symmetry properties, their variation from material to material, etc. Here one is, in effect, asking about the physics of the linear and nonlinear susceptibilities. A second set of questions concerns new phenomena, such as second harmonic generation, which arise from this nonlinear $\vec{P}$-$\vec{E}$ relation. To study this question, one assumes a nonlinear relation between $\vec{P}$ and $\vec{E}$, consistent with crystal symmetry, and investigates how the solutions of Maxwell's equations are changed by it. Notice that one uses here the standard Maxwell equations. The only new feature is the nonlinear relation between $\vec{P}$ and $\vec{E}$. We will see, however, that this nonlinearity gives rise to a host of new phenomena - and devices - which would not occur in its absence.

It should also be realized that nonlinear optics is a subject which could hardly exist without the laser. Most nonlinear optical phenomena are relatively weak and can only be observed with strong sources, such as a laser. Even more important, is the coherence of the laser beam. We will see presently that the most pronounced nonlinear optical effects are produced when all the nonlinear dipoles of a medium oscillate with prescribed phases relative to one another. Such precise phasing can only be achieved if the driving field (of the laser) has a well-defined phase; i.e., is coherent. Thus, nonlinear optics depends upon two of the most important features of the laser source-power and coherence.

THE NONLINEAR SUSCEPTIBILITY

As a start, let us focus our attention on the first nonlinear term in the expansion of $\overleftrightarrow{\chi}(\vec{E})$ in powers of $\vec{E}$. It can be written in the form [2]

$$P_i^{\omega_3} = d_{ijk} E_j^{\omega_1} E_k^{\omega_2}$$
$$\equiv d_{ijk}(\omega_3 = \omega_1 + \omega_2) E_j^{\omega_1} E_k^{\omega_2} , \tag{4}$$

where $P_i^{\omega_3}$, $E_j^{\omega_1}$, $E_k^{\omega_2}$ are Fourier components at frequencies ω_3, ω_1 and ω_2, respectively. d_{ijk} (a function of ω_1, ω_2 and ω_3) is a third rank tensor which must have the symmetry of the crystal to which it refers. Most importantly, d_{ijk} <u>vanishes for any material which has a center of inversion</u>. This fact immediately tells us that the most interesting nonlinear optical materials are to be found in crystal classes which lack a center of inversion. For example, Si and Ge (which are centro-symmetric) are rather uninteresting from the nonlinear optical viewpoint, whereas the III-V compounds (which are acentric) have strong and important nonlinear optical properties. The crystal classes which permit second order optical nonlinearity are the same as those which permit piezoelectricity, and are also the classes which contain ferroelectrics [3]. It is not surprising, therefore, that many important nonlinear optical materials are ferroelectrics (KDP, ADP, $LiNbO_3$, $Ba_2NaNb_5O_{15}$, etc.)

We next consider the symmetry of the tensor d_{ijk} under index interchanges. Here one is concerned with such questions as, does $d_{ijk} = d_{ikj}$? In the most general case, there is no symmetry of this kind. However, in many practical situations there is approximate symmetry, which greatly aids in simplifying the form of the d_{ijk} tensor. First of all, it should be observed that in the special case $\omega_1 = \omega_2$, $\omega_3 = \omega_1 + \omega_2 = 2\omega_1$, (second harmonic generation) the formula for $P(\omega_3)$ takes the form

$$P_i^{\omega_3} = d_{ijk} E_j^{\omega_1} E_k^{\omega_1}$$
$$= \tfrac{1}{2} (d_{ijk} + d_{ikj}) E_j^{\omega_1} E_k^{\omega_1} \qquad (5)$$

(here the summation convention on indices j and k is being used). We see, therefore, that if $\omega_1 = \omega_2$, the tensor $\overleftrightarrow{d}$ has the symmetry $d_{ijk} = d_{ikj}$. Its symmetry is then the same as that of the piezoelectric tensor. Even if $\omega_1 \neq \omega_2$, this symmetry usually holds if the frequencies ω_1 and ω_2 fall below any band gaps in the crystal, but are well above the reststrahl frequencies. This is the usual situation in optical mixing experiments. One then expects $\overleftrightarrow{d}$ to be nearly independent of ω_1 or ω_2 - just as the linear dielectric constant is independent of frequency, for frequencies below the band gap. Hence, again, we have

$$P_i^{\omega_3} = d_{ijk}(\omega_3;\omega_1;\omega_2)E_j^{\omega_1}E_k^{\omega_2}$$

$$= \tfrac{1}{2}[d_{ijk}(\omega_3;\omega_1;\omega_2) + d_{ikj}(\omega_3;\omega_2;\omega_1)]E_j^{\omega_1}E_k^{\omega_2}$$

$$\simeq \tfrac{1}{2}(d_{ijk} + d_{ikj})E_j^{\omega_1}E_k^{\omega_2} \qquad (6)$$

i.e., that $d_{ijk} \simeq d_{ikj}$. The symmetry $d_{ijk} = d_{ikj}$, which holds in almost all cases of practical interest, reduces the number of independent elements in the tensor from 27 to 18. It is then convenient to write $\overleftrightarrow{d}$ in the Voigt form, as follows:

$$\begin{pmatrix} P_x \\ P_y \\ P_z \end{pmatrix} = \begin{pmatrix} d_{11} & d_{12} & d_{13} & d_{14} & d_{15} & d_{16} \\ d_{21} & d_{22} & d_{23} & d_{24} & d_{25} & d_{26} \\ d_{31} & d_{32} & d_{33} & d_{34} & d_{35} & d_{36} \end{pmatrix} \begin{pmatrix} E_x^2 \\ E_y^2 \\ E_z^2 \\ 2E_y E_z \\ 2E_z E_x \\ 2E_x E_y \end{pmatrix} \qquad (7)$$

In a cubic crystal, such as GaAs, $\overleftrightarrow{d}$ has only a single non-vanishing component:

$$\overleftrightarrow{d} = \begin{pmatrix} 0 & 0 & 0 & d_{14} & 0 & 0 \\ 0 & 0 & 0 & 0 & d_{14} & 0 \\ 0 & 0 & 0 & 0 & 0 & d_{14} \end{pmatrix}. \qquad (8)$$

Another important case is that of the KH_2PO_4 (KDP) crystal, for which

$$\overleftrightarrow{d} = \begin{pmatrix} 0 & 0 & 0 & d_{14} & 0 & 0 \\ 0 & 0 & 0 & 0 & d_{14} & 0 \\ 0 & 0 & 0 & 0 & 0 & d_{36} \end{pmatrix}. \qquad (9)$$

The Voigt form is convenient, but it should be clearly understood that the elements of this 3×6 matrix are not the components of a tensor.

We have indicated above that the symmetry condition $d_{ijk} = d_{ikj}$ follows, if $\omega_1 = \omega_2$ or if $\overleftrightarrow{d}$ is independent of ω_1 and ω_2. If $\overleftrightarrow{d}$ is completely independent of frequency ($\omega_3 = \omega_1 + \omega_2$, as well as ω_1 and ω_2) further symmetries apply, and d_{ijk} is invariant to any interchange of i, j, and k. This is the Kleinman [2] symmetry condition which seems to apply, at least in the visible range, to most nonlinear materials. It reduces the number of independent components in d_{ijk} to 10. We will not derive the Kleinman condition – interested readers are referred to his paper.

A SIMPLE MODEL FOR NONLINEAR PROPERTIES

To get some feel for the size of nonlinear optical coefficient we consider a simple, but often studied, model of a nonlinear system – the anharmonic oscillator [4]. Its equation of motion, in one dimension, is

$$\ddot{x} + \Gamma\dot{x} + \omega_o^2 x + vx^2 = \frac{e}{m}\sum_i (E_i e^{-i\omega_i t}), \qquad (10)$$

where Γ is a damping constant, v the strength of the nonlinearity, and the E_i's are optical fields. The linear (ignoring vx^2) solution of this equation is

$$x = \sum_i \left[\frac{eE_i}{m(\omega_o^2 - \omega_i^2 - i\Gamma\omega_i)} e^{-i\omega_i t}\right]. \qquad (11)$$

For brevity, we will often use the notation $\mathcal{D}(\omega) \equiv (\omega_o^2 - \omega^2 - i\Gamma\omega)$ for the denominator of this expression.

A straightforward iteration gives the first nonlinear term in the solution of the equation of motion:

$$x(\omega_i + \omega_j) = -\frac{e^2 v}{m^2} \sum_{ij} \left[\frac{E_i E_j e^{-i(\omega_i+\omega_j)t}}{\mathcal{D}(\omega_i)\mathcal{D}(\omega_j)\mathcal{D}(\omega_i+\omega_j)}\right]. \qquad (12)$$

The second harmonic polarization is,

$$P(2\omega) = -\frac{e^3 v n_o}{m^2} \frac{E^2(\omega)}{\mathcal{D}^2(\omega)\mathcal{D}(2\omega)} \equiv \chi^{NL}(2\omega)E^2(\omega), \quad (13)$$

where n_o is the density of anharmonic oscillators, per unit volume. It is interesting to rewrite this nonlinear (second harmonic) susceptibility in terms of the linear susceptibility defined by

$$\chi(\omega) = P(\omega)/E(\omega) = \frac{e^2 n_o}{m\mathcal{D}(\omega)}. \quad (14)$$

One then finds

$$\chi^{NL}(2\omega) = -\delta \chi^2(\omega)\chi(2\omega), \quad (15)$$

where $\delta = \left(\frac{mv}{e^3 n_o^2}\right)$. The constant δ is made up of quantities which should not vary much from material to material. Thus, assuming that the harmonic oscillator model has some relevance, one expects nonlinear coefficients to scale as $\chi^2(\omega)\chi(2\omega)$. This is the Miller [5] rule which is remarkably well obeyed throughout a wide variety of compounds. For materials whose nonlinear coefficients vary over four orders of magnitude, the coefficient δ is constant to within about a factor of two. Miller's rule is a useful guide in searching for new nonlinear materials. Basically, it tells one that any acentric crystal with a large electronic dielectric constant will also have a large nonlinear coefficient.

The quantity δ in Miller's rule has the dimensions of reciprocal electric field. The experimental value is about 10^{-8} cm/volt. This fact tells us that the dimensionless expansion parameter in the formula

$$\vec{P} = \overleftrightarrow{\chi}^{(1)} \cdot \vec{E} + \overleftrightarrow{\chi}^{(2)} : \vec{E}\vec{E} + \ldots \quad (16)$$

is (E/E_a), where $E_a \cong 10^8$ volt/cm is a typical atomic field. In most laser experiments the electric field intensities are small compared to 10^8 volts/cm, so in these cases the power series expansion of $\overleftrightarrow{\chi}(\vec{E})$ is justified. A field of 10^8 volts/cm implies a power density greater than 10^{13} watts/cm^2. Such power densities can be achieved with tightly focused, high power lasers - for example, a 100 MW ruby beam focused to an area $(10\mu)^2$. However, the intensity is then so great that it destroys all solid materials. The threshold for material damage is always well below the condition $(E/E_a) = 1$.

NONLINEAR OPTICS

Finally, it should be emphasized that in the preceding discussion it was tacitly assumed that all frequencies (ω_1, ω_2, $\omega_3 = \omega_1 + \omega_2$) lay in the optical range. This need not be the case. In fact, an exceedingly important application of nonlinear optics arises when one of the three waves (say ω_1) has a low frequency- in the microwave range or below. One is then concerned with the <u>linear electrooptic</u> (or Pockels) <u>effect</u>, in which the low frequency field modulates the refractive index of the medium. This effect plays a central role in several types of light modulators and, for this reason, has been extensively studied [6]. Readers interested in pursuing this subject further are referred to the bibliography.

MAXWELL'S EQUATIONS IN NONLINEAR MEDIA

In the preceding section, we have discussed some properties of the nonlinear susceptibility tensor, $d_{ijk}(\omega_3; \omega_1, \omega_2)$. We now wish to study the effects of this nonlinear interaction on the solutions of Maxwell's equations. The basic equations are

$$\vec{\nabla} \times \vec{E} = -\frac{\dot{\vec{B}}}{c}$$

$$\vec{\nabla} \times \vec{H} = \frac{4\pi \vec{j}}{c} + \frac{1}{c}\left(\frac{\partial \vec{E}}{\partial t} + 4\pi \frac{\partial \vec{P}}{\partial t}\right). \quad (17)$$

These are the familiar Maxwell equations - the only new feature is the nonlinear relation between $\vec{P}$ and $\vec{E}$ (we assume $\vec{B} = \vec{H}$; i.e., $\mu = 1$). It is convenient to write

$$\vec{P} = \overleftrightarrow{\chi_L} \cdot \vec{E} + \vec{P}_{NL}, \quad (18)$$

where $\overleftrightarrow{\chi_L}$ is the linear susceptibility tensor. We also assume that $\vec{j} = \overleftrightarrow{\sigma} \cdot \vec{E}$. After eliminating $\vec{H}$, the wave equation takes the form

$$\nabla^2 \vec{E} - \vec{\nabla}(\vec{\nabla} \cdot \vec{E}) = \frac{4\pi \overleftrightarrow{\sigma}}{c^2} \cdot \frac{\partial \vec{E}}{\partial t}$$

$$+ \frac{(1 + 4\pi \overleftrightarrow{\chi_L})}{c^2} \cdot \frac{\partial^2 \vec{E}}{\partial t^2} + \frac{4\pi}{c^2} \frac{\partial^2 \vec{P}_{NL}}{\partial t^2}. \quad (19)$$

In general, this is a complicated, nonlinear differential equation. To simplify matters, we will specialize to a one-dimensional situation $\left(\frac{\partial}{\partial x} = \frac{\partial}{\partial z} = 0\right)$, and consider the mixing of three transverse, travelling waves of frequencies ω_1, ω_2, and $\omega_3 = (\omega_1 + \omega_2)$:

$$\vec{E}^{\omega_1}(y,t) = \vec{E}_1(y)e^{i(k_1 y - \omega_1 t)} + \text{c.c.}$$

$$\vec{E}^{\omega_2}(y,t) = \vec{E}_2(y)e^{i(k_2 y - \omega_2 t)} + \text{c.c.}$$

$$\vec{E}^{\omega_3}(y,t) = \vec{E}_3(y)e^{i(k_3 y - \omega_3 t)} + \text{c.c.} \qquad (20)$$

Now, for example, the wave at frequency ω_3 satisfies the equation

$$\frac{\partial^2}{\partial y^2}\vec{E}^{\omega_3} - \frac{4\pi\overleftrightarrow{\sigma}}{c^2} \cdot \frac{\partial \vec{E}^{\omega_3}}{\partial t} - \left(\frac{1+4\pi\overleftrightarrow{\chi}_L}{c^2}\right) \cdot \frac{\partial^2 \vec{E}^{\omega_3}}{\partial t^2}$$

$$= \frac{4\pi}{c^2}\frac{\partial^2}{\partial t^2}[\overleftrightarrow{d}:\vec{E}^{\omega_1}(y,t)\vec{E}^{\omega_2}(y,t)] \qquad (21)$$

where $\overleftrightarrow{d}$ is the nonlinear susceptibility. We next assume that $\vec{E}_3(z)$ varies <u>slowly</u>, in the sense that

$$\left[\left|\frac{dE_3(y)}{dy}\right| \frac{1}{|E_3(y)|}\right] \ll k_3. \qquad (22)$$

This condition states that the amplitude of the wave $\vec{E}_3$ does not appreciably change within a wave length. Experimentally, this criterion is always well satisfied. With the assumption of a slowly varying field, the wave equation for $\vec{E}_3$ becomes

$$\left[-k_3^2 + \frac{\omega_3^2}{c^2}(1+4\pi\overleftrightarrow{\chi}_L)\right]\vec{E}_3$$

$$+ 2ik_3\frac{\partial \vec{E}_3}{\partial y} + \frac{4\pi i \overleftrightarrow{\sigma}\omega_3}{c^2} \cdot \vec{E}_3$$

$$= -\frac{4\pi\omega_3^2}{c^2}\overleftrightarrow{d}:\vec{E}_1(y)\vec{E}_2(y)e^{i(k_1+k_2-k_3)y}. \qquad (23)$$

Choosing $k_3^2 = \frac{\omega_3^2}{c^2}(1+4\pi\overleftrightarrow{\chi}_L) \equiv \frac{\varepsilon_3 \omega_3^2}{c^2}$, one finds

$$\frac{\partial \vec{E}_3}{\partial y} + \frac{2\pi \overleftrightarrow{\sigma}_3 \cdot \vec{E}_3}{c\sqrt{\varepsilon_3}}$$

$$= -\frac{2\pi\omega_3}{ic\sqrt{\varepsilon_3}} \overleftrightarrow{d} : \vec{E}_1(y)\vec{E}_2(y) e^{i(k_1+k_2-k_3)y}. \quad (24)$$

There are similar equations relating $\vec{E}_1$ to $\vec{E}_2$ and $\vec{E}_3$, and $\vec{E}_2$ to $\vec{E}_1$ and $\vec{E}_3$. For example

$$\frac{\partial \vec{E}_1}{\partial y} + \frac{2\pi \overleftrightarrow{\sigma}_1 \cdot \vec{E}_1}{c\sqrt{\varepsilon_1}}$$

$$= -\frac{2\pi\omega_1}{ic\sqrt{\varepsilon_1}} \overleftrightarrow{d} : \vec{E}_2^*(y)\vec{E}_3(y) e^{i(k_3-k_1-k_2)y}. \quad (25)$$

Thus, even in this much simplified three wave approximation, Maxwell's equations remain a set of nonlinear, coupled differential equations.

SECOND HARMONIC GENERATION

Let us now consider the simplest application of Eqs. (21-23) — to the problem of optical second harmonic generation. Here we assume an intense pump beam (at frequency ω_1) of the form

$$\vec{E}^{\omega_1} = \vec{E}_1 e^{i(k_1 y - \omega_1 t)}. \quad (26)$$

We ignore any spatial variation of $\vec{E}_1$, assuming that the power lost by second harmonic conversion is small. Then we need only consider the equation for $\vec{E}_3$, where $\omega_3 = 2\omega_1$. One finds (neglecting $\overleftrightarrow{\sigma}_3$ since the medium is assumed transparent at ω_3)

$$\frac{\partial \vec{E}_3}{\partial y} = -\frac{2\pi\omega_3}{ic\sqrt{\varepsilon_3}} \overset{\leftrightarrow}{d}:\vec{E}_1\vec{E}_1 \, e^{i(2k_1-k_3)y}$$

$$\equiv -\frac{2\pi\omega_3}{ic\sqrt{\varepsilon_3}} \overset{\leftrightarrow}{d}:\vec{E}_1\vec{E}_1 \, e^{i\Delta ky}. \tag{27}$$

Here $(\Delta k)y = (2k_1-k_3)y$ is the difference in phase between the nonlinear polarization wave $(2k_1 y)$ and the naturally propagating wave $(k_3 y)$ at the frequency ω_3. This phase mismatch occurs because of dispersion in the index of refraction. That is

$$2k_1 - k_3 = \frac{2\omega_1\sqrt{\varepsilon_1}}{c} - \frac{\omega_3\sqrt{\varepsilon_3}}{c}$$

$$= k_3\left(\frac{n_1-n_3}{n_3}\right). \tag{28}$$

Here n_1 and n_3 are the indices of refraction at frequencies ω_1 and ω_3.

The solution of the preceding wave equation, with the boundary condition $\vec{E}_3 = 0$ at $y = 0$ is

$$\vec{E}_3 = \frac{2\pi\omega_3}{c\Delta k\sqrt{\varepsilon_3}} \overset{\leftrightarrow}{d}:\vec{E}_1\vec{E}_1 [e^{i\Delta ky}-1]$$

and

$$|\vec{E}_3(y)|^2 = \left\{ \frac{(2\pi\omega_3)^2}{\varepsilon_3 c^2} |\overset{\leftrightarrow}{d}:\vec{E}_1\vec{E}_1|^2 y^2 \right.$$

$$\left. \times \frac{\sin^2(\Delta ky/2)}{(\Delta ky/2)^2} \right\}. \tag{29}$$

Note that if $\Delta k \neq 0$, the second harmonic power oscillates in space between zero and a rather small finite value (Fig. 1). This oscillation is a result of the phase mismatch between the driving second harmonic polarization (which varies as $e^{2ik_1 y}$) and the free second harmonic wave (whose spatial variation is $e^{ik_3 y}$). When $y = 0$, the phase factor $e^{i\Delta ky}$ is unity, and energy flows from the pump beam into the second harmonic wave. However, at the point $\Delta ky = \pi$, $e^{i\Delta ky} = -1$ and the energy flow reverses energy is pumped from the second harmonic wave back into the fundamental. The length defined by the condition

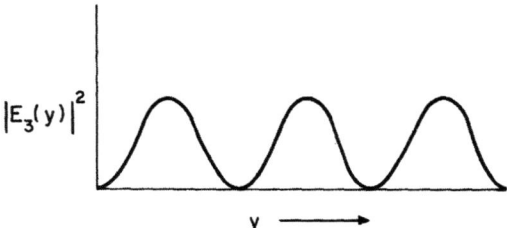

FIGURE 1. Intensity of second harmonic radiation as a function of distance in a non-phase-matched situation.

$$\Delta k \, \ell_{coh} = \pi$$

or

$$\ell_{coh} = \pi/\Delta k \tag{30}$$

is termed the <u>coherence length</u>. It is the distance the waves can travel before the energy flow from $\omega_1 \rightarrow \omega_3$ reverses.

It is clear from the preceding equations that efficient second harmonic generation can only be achieved if $\Delta k = 0$. This condition is termed <u>phase matching</u>. Note that in the phase matched case ($\Delta k = 0$) the intensity of the second harmonic beam grows as the square of the interaction length. Generally speaking, it is not possible to achieve phase matching in optically isotropic (cubic) crystals. In almost all cases, the refractive index of a transparent medium is an increasing function of frequency so that $n_3 > n_1$ and $\Delta k \neq 0$. When this is the case, phase matching is not possible. The difficulty can be avoided in birefringent crystals [7]. To understand how birefringence can be used to solve the phase matching problem, we consider the relatively simple case of a uniaxial crystal. Biaxial crystals can also be used for phase matching, but are more complicated, and possess certain inherent disadvantages.

We begin by recalling the optical properties of a uniaxial crystal. They are described by a dielectric tensor of the form:

$$\varepsilon = \begin{pmatrix} \varepsilon_1(\omega) & 0 & 0 \\ 0 & \varepsilon_1(\omega) & 0 \\ 0 & 0 & \varepsilon_2(\omega) \end{pmatrix}. \tag{31}$$

The electromagnetic waves which can propagate in a uniaxial medium are determined by Maxwell's equations. Assuming

$$\vec{E}(\vec{r},t) = \vec{E}\, e^{i(\vec{k}\cdot\vec{r}-\omega t)}, \tag{32}$$

one then finds

$$\left[k^2 \vec{E} - \vec{k}(\vec{k}\cdot\vec{E}) - \frac{\omega^2}{c^2}\overleftrightarrow{\varepsilon}\cdot\vec{E} \right] = 0. \tag{33}$$

This is a set of three simultaneous, linear, homogeneous equations for the components (E_i) of the electric field vector, which have a solution if, and only if,

$$\det\left[k^2 \delta_{ij} - k_i k_j - \frac{\omega^2}{c^2}\varepsilon_{ij} \right] = 0. \tag{34}$$

Equation (34) is the dispersion relation for electromagnetic waves in a medium with dielectric tensor ε_{ij}. To solve these equations, it is convenient to define an effective refractive index $n = \frac{ck}{\omega}$ and the unit vector $\alpha_i = \frac{k_i}{|k|}$. The dispersion relation then becomes

$$\det[n^2(\delta_{ij} - \alpha_i \alpha_j) - \varepsilon_{ij}] = 0. \tag{35}$$

For the uniaxial case, this formula simplifies to

$$\begin{vmatrix} [n^2(1-\sin^2\theta)-\varepsilon_1] & 0 & n^2\sin\theta\cos\theta \\ \\ 0 & n^2-\varepsilon_1 & 0 \\ \\ n^2\sin\theta\cos\theta & 0 & [n^2(1-\cos^2\theta)-\varepsilon_2)] \end{vmatrix} \tag{36}$$

where θ is the angle between the propagation vector and the c-axis $\left(\cos\theta = \frac{k_z}{|k|}\right)$. Expanding the determinant yields the result:

$$[n^2-\varepsilon_1][n^2(\varepsilon_2\cos^2\theta + \varepsilon_1\sin^2\theta) - \varepsilon_1\varepsilon_2] = 0 \tag{37}$$

or

$$n^2 = \varepsilon_1, \left(\frac{\varepsilon_1\varepsilon_2}{\varepsilon_2\cos^2\theta + \varepsilon_1\sin^2\theta}\right). \tag{38}$$

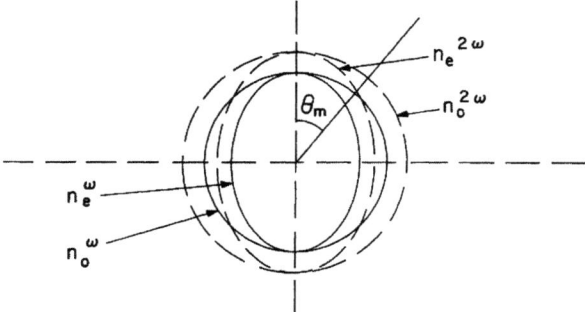

FIGURE 2. Index ellipsoids at the fundamental and second harmonic frequencies in a uniaxial crystal.

Thus, there are two index surfaces — one spherical, the other ellipsoidal in shape. They are illustrated in Fig. 2 for a negatively birefringent crystal such as KDP. At the second harmonic frequency, these surfaces have about the same shape, but are larger in size than those at the fundamental because the indices increase, assuming normal dispersion with frequency. The second harmonic index surfaces are the dotted lines of the figure (the birefringence is considerably exaggerated in this drawing). We note the important fact that the ordinary index surface at the fundamental frequency (n_o^ω) intersects the extraordinary surface at the second harmonic frequency ($n_e^{2\omega}$). At this intersection point, the wave vectors of the forced, second harmonic, ordinary polarization wave and the free, extraordinary wave at 2ω are equal to one another; i.e., $\Delta k = 0$ and one has achieved phase matching. The angle (θ_m) at which matching occurs is determined by the condition

$$\left[\frac{1}{n_e^{2\omega}(\theta_m)}\right]^2 = \left[\frac{1}{n_o^\omega}\right]^2 , \qquad (39)$$

which can be rewritten in the form

$$\sin^2\theta_m = \frac{\left[\frac{1}{\varepsilon_1(\omega)} - \frac{1}{\varepsilon_1(2\omega)}\right]}{\left[\frac{1}{\varepsilon_2(2\omega)} - \frac{1}{\varepsilon_1(2\omega)}\right]}$$

$$= \left(\frac{\text{dispersion}}{\text{birefringence}}\right) . \qquad (40)$$

Phase matching is possible provided

$$\left[\frac{1}{\varepsilon_1(2\omega)} - \frac{1}{\varepsilon_1(\omega)}\right] < \left[\frac{1}{\varepsilon_1(2\omega)} - \frac{1}{\varepsilon_2(2\omega)}\right]. \tag{41}$$

(Equation (41) is valid for the case of negative birefringence). Notice that phase matching is only possible if the birefringence exceeds the dispersion.

It is important to realize that the intensity of second harmonic generation is enormously enhanced when phase matching is achieved. This fact is confirmed by many experiments [8], and is also implied by Eq. (29). If $\Delta k \neq 0$, $|\vec{E}_3(y)|^2$ has a maximum value

$$|E_3|^2 \leq \frac{(2\pi\omega_3)^2}{\varepsilon_3 c^2} |\overleftrightarrow{d}:\vec{E}_1\vec{E}_1|^2 \left[\frac{4}{(\Delta k)^2}\right]. \tag{42}$$

In crystals where phase matching is not possible (for example, GaAs) the quantity Δk typically has a value of order 10^3-10^4cm^{-1}, depending upon the frequencies involved. As a result, only dipoles within a region of thickness 10^{-3}-10^{-4}cm radiate in phase to generate the second harmonic wave. On the other hand, if $\Delta k = 0$ one finds

$$|\vec{E}_3(y)|^2 = \frac{(2\pi\omega_3)^2}{\varepsilon_3 c^2} |\overleftrightarrow{d}:\vec{E}_1\vec{E}_1|^2 y^2 \tag{43}$$

indicating that all the dipoles in the crystal are contributing in phase. The ratio, $\frac{y^2(\Delta k)^2}{4}$, of phase-matched to non-phase-matched second harmonic powers can easily be 10^6. It is this large numerical factor which makes phase matching so crucial in non-linear optical experiments.

The fact that $|\vec{E}_3(y)|^2$ varies quadratically with y (in the phase matched case, $\Delta k = 0$) is an indication that second harmonic generation is a coherent process. In coherent processes, the intensity of radiation varies as the square of the number of radiating dipoles or, for a given number of dipoles per unit volume, as the square of the volume. This is precisely what the formula for $|\vec{E}_3(y)|^2$ implies. Note also that $|\vec{E}_3(y)|^2 \sim |\vec{E}_1|^4 \sim P_1^2$, where P_1 is the power density at frequency ω_1. This quadratic variation ($P_3 \sim P_1^2$) is typical of a two photon nonlinear process.

OPTICAL PARAMETRIC OSCILLATION

We next consider a slightly more complicated application of the three wave, nonlinear equations. This is the <u>optical parametric oscillator</u>. Such a device consists of a nonlinear crystal pumped with an intense, coherent optical beam at a frequency ω_3. In the presence of the nonlinear medium, fields at frequencies ω_1 (signal) and ω_2 (idler), which satisfy the condition $\omega_1 + \omega_2 = \omega_3$, are coupled to one another via the pump field. For example, the fields $\vec{E}_1^*$ and $\vec{E}_3$ combine to generate a nonlinear polarization at frequency ω_2:

$$\vec{P}_{NL}^{\omega_2} = \overset{\leftrightarrow}{d}:\vec{E}_3\vec{E}_1^*. \tag{44}$$

Similarly, $\vec{E}_2^*$ and $\vec{E}_3$ generate a nonlinear polarization at ω_1. Thus, if fields $\vec{E}_1$ and $\vec{E}_2$ are present, each will drive the other, through mixing with the pump field at ω_3. This bootstrap operation has a net gain for pump powers above a certain threshold, which is determined by the losses of the system. When this critical threshold is passed, the device breaks into oscillation at frequencies ω_1 and ω_2.

There are, of course, many ways in which a pump photon (at ω_3) can split into signal (ω_1) and idler (ω_2) photons with energy conservation ($\omega_3 = \omega_1 + \omega_2$). Thus, we need an additional condition to determine the frequencies (ω_1 and ω_2) at which a parametric oscillator operates. The second condition is the phase matching condition;

$$k_3 = k_1 + k_2, \tag{45}$$

or

$$n(\omega_3)\omega_3 = n(\omega_1)\omega_1 + n(\omega_2)\omega_2, \tag{46}$$

where $n(\omega)$ is the index of refraction of the nonlinear medium. As in second harmonic generation, this momentum condition cannot be satisfied, because of color dispersion, in optically isotropic crystals. Parametric oscillation does not occur in such materials. Again one must use birefringence to achieve phase matching. For this reason, the materials which give efficient second harmonic generation are often also good candidates for parametric generation (e.g., KDP, $LiNbO_3$, Te, etc.). Note, however, that once phase matching is achieved, one may adjust the momentum balance equation, Eq. (46), by varying external parameters such as crystal orientation, crystal temperature or external fields. These modifications change the frequencies (ω_1 and ω_2) at which phase matching occurs - and thus tune the frequency of the parametric oscillator. The parametric oscillator is, therefore, a <u>tunable laser</u>. It is this feature which makes it an important device, and has stimulated much of the work on optical paramps.

To determine the threshold for parametric oscillation, we may make the simplifying assumption that the pump field is constant in space. This approximation is valid below threshold, and for low level oscillation ($|\vec{E}_1|, |\vec{E}_2| \ll |\vec{E}_3|$). However, once the signal and idler fields have built up to intensities comparable to the pump, one must take account of the fact that energy has been transferred from the latter to the former. This phenomenon (called <u>pump depletion</u>) is often observed in parametric oscillators. A more sophisticated theory than that we present here is required to discuss pump depletion.

If one ignores pump depletion, two equations (for $\vec{E}_1$ and $\vec{E}_2^*$) suffice to determine the behavior of a parametric oscillator. They are:

$$\frac{\partial \vec{E}_1}{\partial y} + \frac{2\pi\sigma_1 \vec{E}_1}{c\sqrt{\varepsilon_1}} = -\frac{2\pi i \omega_1}{c\sqrt{\varepsilon_1}} \overleftrightarrow{d} : \vec{E}_3 \vec{E}_2^*(y) e^{-i\Delta k y} \qquad (47)$$

$$\frac{\partial \vec{E}_2^*}{\partial y} + \frac{2\pi\sigma_2 \vec{E}_2^*}{c\sqrt{\varepsilon_2}} = \frac{2\pi i \omega_2}{c\sqrt{\varepsilon_2}} \overleftrightarrow{d} : \vec{E}_3^* \vec{E}_1(y) e^{i\Delta k y}, \qquad (48)$$

where $\Delta k = (k_1 + k_2 - k_3)$. These equations have solutions of the form

$$\vec{E}_1(y) = \vec{E}_1 e^{\gamma y} e^{-\frac{i\Delta k y}{2}},$$

$$\vec{E}_2^*(y) = \vec{E}_2^* e^{\gamma y} e^{\frac{i\Delta k y}{2}}. \qquad (49)$$

Substitution yields the following linear, homogeneous equations for the coefficients $\vec{E}_1$ and $\vec{E}_2^*$:

$$\left[\gamma - \frac{i\Delta k}{2} + \frac{2\pi\sigma_1}{c\sqrt{\varepsilon_1}}\right] \vec{E}_1 + \frac{2\pi i \omega_1}{c\sqrt{\varepsilon_1}} \overleftrightarrow{d} : \vec{E}_3 \vec{E}_2^* = 0$$

$$-\frac{2\pi i \omega_2}{c\sqrt{\varepsilon_2}} \overleftrightarrow{d} : \vec{E}_3^* \vec{E}_1 + \left[\gamma + \frac{i\Delta k}{2} + \frac{2\pi\sigma_2}{c\sqrt{\varepsilon_2}}\right] \vec{E}_2^* = 0, \qquad (50)$$

NONLINEAR OPTICS

which have a solution if, and only if, the determinant of coefficients is equal to zero. This secular equation determines the growth rate, γ.

As an example, we consider the $LiNbO_3$ case, where ordinary and extraordinary waves propagating perpendicular to the optical axis are coupled by the d_{15} component of the nonlinear tensor. Phase matching of extraordinary pump waves ($E_3 = E_{3z}$) and ordinary signal and idler waves ($E_1 = E_{1x}$, $E_2 = E_{2x}$) can also be achieved under such circumstances. This feature makes $LiNbO_3$ an especially valuable nonlinear material. The wave equations take the form

$$\left[\gamma - \frac{i\Delta k}{2} + \frac{2\pi\sigma_1}{c\sqrt{\varepsilon_1}}\right] E_{1x} + \left(\frac{2\pi i\omega_1}{c\sqrt{\varepsilon_1}}\right) d_{15} E_{3z} E_{2x}^* = 0$$

$$-\left(\frac{2\pi i\omega_2}{c\sqrt{\varepsilon_2}}\right) d_{15} E_{3z}^* E_{1x} + \left[\gamma + \frac{i\Delta k}{2} + \frac{2\pi\sigma_2}{c\sqrt{\varepsilon_2}}\right] E_{2x}^* = 0, \quad (51)$$

and the secular equation is

$$\left(\gamma - \frac{i\Delta k}{2} + \alpha_1\right)\left(\gamma + \frac{i\Delta k}{2} + \alpha_2\right)$$

$$- \frac{(2\pi)^2 \omega_1 \omega_2}{\sqrt{\varepsilon_1 \varepsilon_2} c^2} d_{15}^2 |E_{3z}|^2 = 0, \quad (52)$$

where $\alpha_1 = \frac{2\pi\sigma_1}{c\sqrt{\varepsilon_1}}$ and $\alpha_2 = \frac{2\pi\sigma_2}{c\sqrt{\varepsilon_2}}$. Usually, ω_1 and ω_2 are fairly close to one another so that one may assume $\alpha_1 \simeq \alpha_2 \simeq \alpha$. One then finds

$$\gamma = \left[\frac{(2\pi)^2 \omega_1 \omega_2}{\sqrt{\varepsilon_1 \varepsilon_2} c^2} d_{15}^2 |E_{3z}|^2 - \frac{(\Delta k)^2}{4}\right]^{1/2} - \alpha,$$

$$\equiv \left[g^2 - \frac{(\Delta k)^2}{4}\right]^{1/2} - \alpha. \quad (53)$$

where

$$g^2 = \frac{(2\pi)^2 \omega_1 \omega_2}{c^2 \sqrt{\varepsilon_1 \varepsilon_2}} d_{15}^2 |E_{3z}|^2 \qquad (54)$$

The quantity g has the units cm^{-1}, and may be thought of as a <u>gain</u> for the parametric process. In $LiNbO_3$, g has the value $g \sim 4\times 10^{-4} \ P_3 cm^{-1}$, where P_3 is the pump power in watts/cm^2. This estimate of g assumes λ_1, $\lambda_2 \sim 1.0\mu$ as in the experiments of Giordmaine and Miller. It is clear from Eq. (53) that parametric oscillation can only be achieved in phase matched situations ($\Delta k = 0$). Otherwise the threshold condition ($\gamma = 0$) requires fantastic pumping powers. Even when phase matching does occur, as in $LiNbO_3$, strong pump fields are needed to overcome absorption and cavity losses (represented by α in the above equation). In the $LiNbO_3$ experiments, the effective loss was about 1 cm^{-1}. Calculated threshold pump powers (assuming a doubled Nd laser pump at 0.53μ and oscillation near 1μ) are about 10^6 watts/cm^2, in reasonable agreement with the observed value of $P = 4\times 10^5$ watts/cm^2. Such powers are only attainable in pulsed operation, and most parametric oscillators operate in this way. Recently, however, CW optical parametric oscillation has been achieved in both $LiNbO_3$ and $Ba_5NaNb_5O_{15}$ by carefully controlling losses and placing the nonlinear crystal within the pump laser cavity to make use of the large circulating power which occurs there.

It is important to consider the meaning of the solution to the coupled wave equations [Eqs. (51)]. These equations have another solution, $\vec{E}_1 = \vec{E}_2^* \equiv 0$, in addition to that derived above. So long as the oscillator is below threshold ($\gamma < 0$), the solution $\vec{E}_1 = \vec{E}_2^* = 0$ is a stable one, in the sense that small thermally or quantum mechanically induced fluctuations about $\vec{E}_1 = \vec{E}_2^* = 0$ will decay away in space (or time). On the other hand, above threshold the solution $\vec{E}_1 = \vec{E}_2^* = 0$ is unstable. Fluctuations grow exponentially, leading to a situation in which large amplitude signal ($\vec{E}_1$) and idler ($\vec{E}_2^*$) fields are present. This growth is ultimately stabilized by higher order effects, such as pump depletion. Thus, the preceding calculation should be thought of as a stability analysis leading to a threshold condition for oscillation, but is not a valid calculation of the fields in the oscillating regime.

THE STIMULATED RAMAN EFFECT

As a final topic in this survey of nonlinear optics, we briefly consider the Raman effect and the problem of stimulated Raman emission. Raman scattering is an <u>inelastic</u> light scattering process in which the frequency of the scattered radiation is shifted up in frequency (anti-Stokes scattering) or down in frequency (Stokes radiation) compared to that of the incident radiation. The frequency shift is determined by a characteristic energy (phonon, magnon, electronic excitation, etc.) of the medium. The ordinary Raman effect was discovered many years ago, and has been extensively exploited as a tool for probing the nature of excitations in condensed matter. This area of research (inelastic light scattering) has been revitalized in the last decade by the development of the laser, and is now an active branch of physics.

Under the usual experimental conditions, Raman scattering is a feeble phenomenon. Typically, the intensity of the scattered radiation is about 10^{-8} times that of the primary beam. However, if the power in the primary beam is increased (to values of the order of 10^8 watts/cm^2), the character of the phenomenon changes, and efficient (Stokes) Raman scattering occurs. This effect was first discovered experimentally by Eckhardt et al [10], who found that intense Raman scattering took place when focussed ruby laser beams traversed liquids such as nitrobenzene. The phenomenon is termed **stimulated Raman scattering**. Stimulated scattering differs from ordinary Raman scattering in several respects: 1) it only occurs above a certain threshold intensity of the exciting light, 2) the scattered light is highly directional (usually parallel to the primary beam), 3) the output is monochromatic, 4) the conversion efficiency is large - often as big as 10-20%. These features are typical of laser action, and the devices which embody them are often termed **Raman lasers**.

To understand Raman lasers, we must consider a higher order optical nonlinearity, which is known as the Raman nonlinearity. If an intense pump field ($\vec{E}_p e^{-i\omega_p t}$) and a Stokes field ($\vec{E}_s e^{-i\omega_s t}$) are simultaneously present in a nonlinear medium, there is an induced, nonlinear polarization at frequency ω_3 of the form [11]

$$\vec{P}^{\omega_3}_{Raman} = \chi_s |\vec{E}_p|^2 \vec{E}_s. \tag{55}$$

Note that χ_s describes a **third order nonlinearity**, in contrast to the second order nonlinearity discussed previously. The symmetry properties of the third order susceptibility tensor differ from those of the second order susceptibility. In particular, χ_s **does not vanish** in a medium with a center of inversion. Stimulated Raman scattering is therefore observed in media (such as diamond or CCl_4) which are centrosymmetric.

The Raman susceptibility (χ_s) has a sharp resonance [11] when the frequency shift, $\omega_p - \omega_s$, is equal to the characteristic frequency (ω_v) of the condensed medium. Most often ω_v refers to a vibrational (phonon) mode of the system, but stimulated Raman scattering involving purely electronic excitations has also been observed [12]. At resonance ($\omega_s = \omega_p - \omega_v$) the Raman susceptibility has its largest value, and is purely imaginary with negative sign. Thus, when substituted into Maxwell's equation for the Stokes wave;

$$\nabla^2 \vec{E}_s + \frac{\omega_s^2 \epsilon(\omega_s) \vec{E}_s}{c^2} = \frac{4\pi\omega_s^2}{c^2} \vec{P}_{NL}$$

$$= \frac{4\pi\omega_s^2}{c^2} \chi_s |\vec{E}_p|^2 \vec{E}_s, \qquad (56)$$

the Raman susceptibility provides gain at the Stokes frequency. This gain is due to inelastic photon scattering from the primary beam (E_p) into the Stokes wave (E_s). Note that the gain term [right hand side of Eq. (56)] is proportional to $\vec{E}_s$, indicating that stimulated scattering is involved. The threshold for Raman laser action is reached when the Raman gain exceeds the losses (due to absorption, mirror losses, etc.) at the Stokes frequency. Since the Raman gain is proportional to pump laser [Eq. (56)], this condition sets a definite pump power threshold for Raman laser action. Above this threshold, the Raman laser has all the desirable characteristics of coherence, directionality, monochromaticity, etc., that are found in a conventional laser. Estimates of the Raman threshold (based on known Raman cross sections) yield values of about 10^8 watts/cm^2 and are in reasonable agreement with experiment. The thresholds are large - though easily attainable with present laser powers - because, in most cases, the Raman cross sections are quite small.

Stimulated Raman scattering has been observed in many systems, and the effect is often used to provide coherent light sources at a variety of frequencies. In most cases, the threshold power is sufficiently high that only pulsed operation is possible. However, even with this limitation the effect is a useful one and has been exploited in a variety of ways.

REFERENCES

[1] P. A. Franken, A. E. Hill, C. W. Peters and G. Weinreich, Phys. Rev. Letters 7, 118 (1961).
[2] D. A. Kleinman, Phys. Rev. 128, 1761 (1962).
[3] The properties of the nonlinear optic tensor are discussed by Amnon Yariv, Quantum Electronics, John Wiley and Sons, New York, 1967, Chap. 21. See also J. F. Nye, Physical Properties of Crystals, Oxford University Press, New York, 1957.

[4] N. Bloembergen, *Nonlinear Optics*, W. A. Benjamin, Inc., New York, 1965, p. 5.
[5] R. C. Miller, Appl. Phys. Letters $\underline{5}$, 17 (1964).
[6] I. P. Kaminow and E. H. Turner, Proc. I.E.E.E. $\underline{54}$, 1374 (1966); S. H. Wemple, *Laser Handbook* (F. T. Arrecchi and E. O. Schulz-DuBois, Eds.). To be published.
[7] J. A. Giordmaine, Phys. Rev. Letters $\underline{8}$, 19 (1962). P. D. Maker, R. W. Terhune, M. Nisenoff, and C. M. Savage, Phys. Rev. Letters $\underline{8}$, 21 (1962).
[8] Ref. 7. See also A. Ashkin, G. D. Boyd, and J. M. Dziedzic, Phys. Rev. Letters $\underline{11}$, 501 (1963).
[9] J. A. Giordmaine and R. C. Miller, Phys. Rev. Letters $\underline{14}$, 973 (1965).
[10] G. Eckhardt, R. W. Hellworth, F. J. McClung, S. E. Schwarz, D. Weiner, and E. J. Woodbury, Phys. Rev. Letters $\underline{9}$, 455 (1962).
[11] N. Bloembergen and Y. R. Shen, Phys. Rev. $\underline{133}$, 210 (1964). See also reference 4.
[12] C. K. N. Patel and E. D. Shaw, Phys. Rev. Letters $\underline{24}$, 512 (1970). C. K. N. Patel, *Proceedings of the Tenth International Conference on the Physics of Semiconductors* (National Technical Information Service, National Bureau of Standards, U. S. Dept. of Commerce, Springfield, Virginia, 22151 (1970)), p. 746.

CHAPTER 11

OPTICAL MATERIALS

N. B. Hannay

Bell Laboratories, Murray Hill, New Jersey 07974

The last decade has seen an enormous increase in interest in optical materials, beginning with the invention of the laser. The possibility of optoelectronic applications, combined with new research interests in optics and optical materials, has brought this about. In other chapters the basic physics (Chapters 9 and 10), semiconductor and dye lasers (Chapters 2 and 12), and the application of optical materials (Chapters 2, 12, 13, 20 and 22) are discussed. Here we will be concerned with optically pumped laser materials and nonlinear optical materials.

OPTICALLY PUMPED LASER MATERIALS [1]

The first laser utilized ruby, Al_2O_3:Cr, a three-level system, Fig. 1(a). It was soon found that much lower thresholds in optically pumped systems could be achieved in four-level systems, Fig. 1(b), as population inversion does not have to be reached with respect to the ground state, but only with respect to some relatively unpopulated higher lying level. The Nd^{3+} ion, in various host lattices, is a four-level system of this kind. The pump energy may be absorbed by the host lattice and transferred to the active ion. Alternatively, it may be absorbed by another ion or sensitizer, which then transfers energy to the active ion by radiative transfer or by nonradiative resonance transfer, and this may provide for more efficient utilization of the pumping radiation. An example of this is yttrium aluminum garnet (YAG) containing Nd^{3+} and Cr^{3+}; the Cr^{3+} acts as a sensitizer, transferring its absorbed energy to the Nd^{3+}.

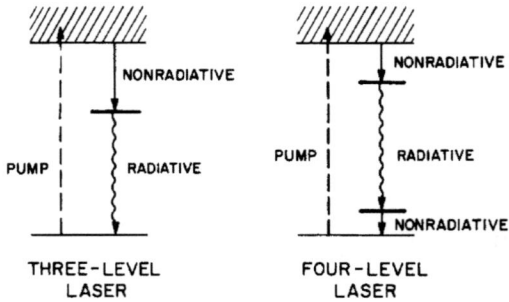

FIGURE 1. Electronic energy levels for optically pumped systems: (a) three-level system, (b) four-level system.

A number of active ions have been used for optically pumped lasers. These include both divalent and trivalent ions, from several groups of the Periodic Table. Similarly a number of host lattices have been used which provide both divalent and trivalent sites. It is of interest to note that they are in every case either oxides or fluorides. Rather high quality crystals are usually required and parallel energy loss paths which detract from the laser efficiency are to be avoided.

In the simplest systems the active ion substitutes for host ions of the same valence, and an example is ruby, in which Cr^{3+} substitutes for Al^{3+}. At low Cr^{3+} concentrations isolated ions only are important and it is a three-level system, operating at 0.69 μ. At higher chromium concentrations (0.5%, as against 0.05%) second and fourth nearest neighbor interactions produce new electronic energy levels and these lead to four-level systems, which can lase because of the high efficiency of energy transfer to these interacting pairs of Cr^{3+} ions. Ruby shows a broad absorption spectrum throughout the visible region of the spectrum. Ruby crystals can be grown by Verneuil and Czochralski crystal growth methods, the latter being preferred because it leads to fewer imperfections and a higher efficiency.

Another host lattice that is particularly useful as an optically pumped laser material is garnet, $Y_3Al_5O_{12}$ (YAG), doped with Nd^{3+}. This has the lowest threshold for an optically pumped laser system, and the highest output power. It can be operated CW (continuous) at 300°K with a 75-watt tungsten lamp, and pumping with a semiconductor diode can be achieved with as little as 5 watts input power. The output power can be over 1000 watts at 300°K. The emission is at 1.06 μ.

There are three crystal sites for the cations in YAG. The Y^{3+} is on a dodecahedral site and the Al^{3+} on octahedral and tetrahedral sites. The Nd^{3+} substitutes for Y^{3+}. A typical composition is $Y_{2.95}Nd_{0.05}Al_5O_{12}$. Sensitization with a small amount of Cr^{3+} improves the efficiency of utilization of the pump light but it degrades the optical quality of crystal, so little or no net gain results; the Cr^{3+} substitutes for Al^{3+} on octahedral sites. The best crystals of YAG:Nd are grown by the Czochralski method, although flux methods have also been used, with lead oxide and lead fluoride fluxes.

A number of other garnet lasers using other rare earth combinations have also been investigated. Other host lattices have also been used where the active ion has the same valence as the cation of the host lattice for which it substitutes; an example is $LaF_3:Er^{3+}$.

There are also optically pumped laser systems in which the active ion has a different valence from that of the cation of the host lattice. An example of this is scheelite, $CaWO_4$:Nd. This was the first Nd^{3+} laser and also the first solid state 300°K CW laser. It does not have as low a threshold as YAG:Nd, but the material is considerably cheaper so it may have advantages for some purposes. The best crystals are grown by the Czochralski technique but many other crystal-growing methods have also been used. There has been extensive study of imperfections and composition control in this material.

The $CaWO_4$:Nd system is an instructive example of control over the solid state chemistry, as a trivalent Nd^{3+} ion is substituted for a divalent Ca^{2+} ion. Charge compensation, of course, is required. In the absence of other mechanisms this will occur through introduction of cation vacancies, and one cation vacancy will appear for each two Nd^{3+} ions introduced into the lattice ($Ca_{1-3x}Nd_{2x}\square_x WO_4$ - see Fig. 2). Another, more efficient way to achieve charge compensation is to incorporate in the lattice an amount of a monovalent ion, such as Na^+, exactly equal to the concentration of Nd^{3+} ($Ca_{1-2x}Nd_xNa_xWO_4$). This one-for-one charge compensation allows Nd^{3+} to be introduced with ease into the lattice. An equivalent result can be achieved by substituting along with Nd^{3+} an equal amount of Nb^{5+}, which substitutes for the W^{6+} in the lattice ($CaNd_xW_{1-x}Nb_xO_4$). Figure 3 shows the distribution coefficient of Nd^{3+} vs the concentration of Na^+ in the melt. Na^+ is the most effective impurity for this purpose because its ionic size is close to that of calcium. Control of the Nd^{3+} concentration can be calculated by the Law of Mass Action. Typical concentrations used are about 2% neodymium and sodium in the crystal, with 3-4% neodymium and 12-15% sodium in the melt.

	Ca^{2+}	WO$_4^{2-}$	☐	WO$_4^{2-}$
(a)	WO$_4^{2-}$	Nd^{3+}	WO$_4^{2-}$	Nd^{3+}
	Ca^{2+}	WO$_4^{2-}$	Ca^{2+}	WO$_4^{2-}$

	Ca^{2+}	WO$_4^{2-}$	Nd^{3+}	WO$_4^{2-}$
(b)	WO$_4^{2-}$	Na$^+$	WO$_4^{2-}$	Ca^{2+}
	Ca^{2+}	WO$_4^{2-}$	Ca^{2+}	WO$_4^{2-}$

	Ca^{2+}	WO$_4^{2-}$	Ca^{2+}	WO$_4^{2-}$
(c)	WO$_4^{2-}$	Nd^{3+}	NbO$_4^{3-}$	Ca^{2+}
	Ca^{2+}	WO$_4^{2-}$	Ca^{2+}	WO$_4^{2-}$

FIGURE 2. Charge compensation in CaWO$_4$:Nd (from Nassau [1]).

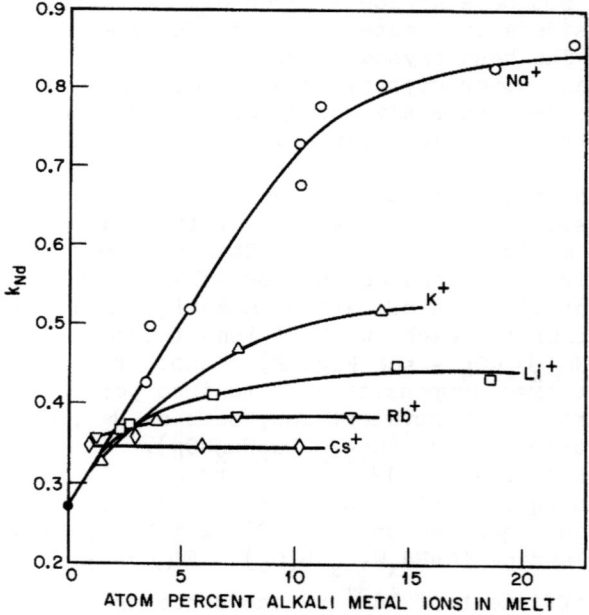

FIGURE 3. Distribution coefficient of neodymium in CaWO$_4$ with varying concentration of sodium in the melt from which the crystal is grown (from Nassau [1]).

Another useful optically pumped laser material with complex substitution is CaF$_2$:Er^{3+}. Only half of the available calcium lattice sites are filled, and the empty calcium sites are available for substitutions. To put in RE^{3+}, charge compensation is required; this occurs through the insertion of extra F^- ions at the empty Ca^{2+} sites in the lattice. Calcium fluoride crystals can be grown by both the Czochralski and the Bridgman techniques with good results. More complex hosts have also been used, such as $CaF:YF_3$; and other active ions, such as U^{3+}, have been utilized.

Another optically pumped laser that has been of interest is Nd^{3+} in glass as the host lattice. The important transitions are the same as in crystals using Nd^{3+}. The Nd^{3+} linewidth in glasses is generally greater, although there is enough local symmetry in glass to make lasing possible. The advantage of the glass host is that it is easy to make large rods and they can be of high optical quality. A great deal of work has been done on glass systems and on laser design using glass as the host matrix. High output energies on a pulsed basis have been obtained.

In Table 1 the parameters of a number of the most interesting optically pumped solid state lasers are compared [2], for pulsed and CW operation. The merits of the YAG:Nd system are apparent.

NONLINEAR OPTICAL MATERIALS [3]

Nonlinear optical materials are of importance for frequency mixing, modulation, and other optoelectronic circuit functions. The physics of nonlinear optical materials has been discussed in Chapter 10. In this chapter, we are concerned with the solid-state chemistry of the materials and with the properties of solids which make them useful as nonlinear optical materials. The search for useful materials has been active in the last several years and a number of crystals have been discovered. The search criteria were at first mainly chemical, in the absence of adequate physical theory. More recently the physical basis has been developed.

In a nonlinear optical material there is a nonlinear response of the electrons to the E-field of the light beam and the objective is to maximize this effect. The polarization is a function of the electric field,

$$P = \chi E + dE^2 + RE^3 + \dots \qquad (1)$$

χ is the linear optical susceptibility, and the higher order terms describe the nonlinearity. The nonlinear terms give rise to effects such as second harmonic generation and the electrooptic effect. For second harmonic generation it is the quadratic term, i.e. d, which is important, and for high output intensity a large value of d is needed. A criterion for second harmonic generation

TABLE 1

	λ	Pulsed Operation			Continuous Operation		
		Temp (°K)	Threshold (joules)	Output Energy (joules)	Temp (°K)	Threshold (watts)	Output Power (watts)
Al_2O_3:Cr	0.69	300	>100	50 joules	300	1400	2
$CaWO_4$:Nd	1.06	300	4	~2	300	1500	0.5
$CaWO_4$:Nd,Na	1.06	300	3	~2	300	1000	2
YAG:Nd	1.06	300	1.5	~2	300	75	>1000
Glass:Nd	1.06	300	50	>10^3	300	1300-1500	0.01-2
YAG,Er,Tm:Ho	2.1μ	300	100	~2	77	30	50

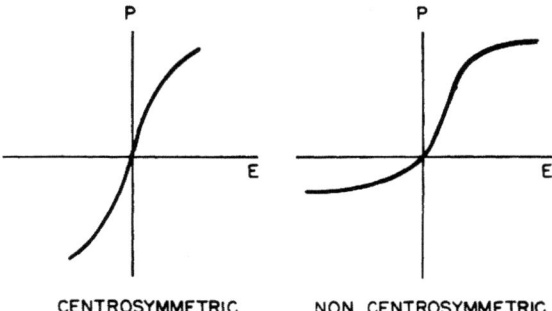

FIGURE 4. Electric field dependence of the polarization in centrosymmetric and noncentrosymmetric crystals (from Bergman and Kurtz [3]).

follows (Fig. 4); if the crystal is centrosymmetric a reversal of the E-field changes the sign of the polarization but not the magnitude, and therefore all the even terms in Eq. (1) must vanish and d is identically zero. Hence, second harmonic generation will occur only in a noncentrosymmetric crystal, in which both odd and even terms are present. Twenty-one out of thirty-two crystal classes are noncentrosymmetric (about one in five inorganic compounds) and meet this requirement. It should be noted that this is also a criterion for piezoelectricity.

Birefringence is also required for second harmonic generation. To avoid destructive interference the phase velocity at frequency ω must be equal to that of the second harmonic at frequency 2ω, and therefore the refractive indices at these two frequencies must be equal. In general this will not be true, as n varies with ω. However, if the crystal is birefringent the index of refraction for the extraordinary ray and the ordinary ray will not be equal. Comparison of the dispersion curves for the ordinary and extraordinary rays in a birefringent crystal (Fig. 5) shows that it is possible to find a condition where the index of refraction from the ordinary ray at frequency ω is equal to that for the extraordinary ray at frequency 2ω; this condition is found for a particular direction in the crystal. Thus high birefringence is also a criterion for second harmonic generation, and the birefringence must exceed the dispersion.

Finally, a high index of refraction is desirable. An empirical rule due to Miller [4]

$$d = \Delta \chi^3 \qquad (2)$$

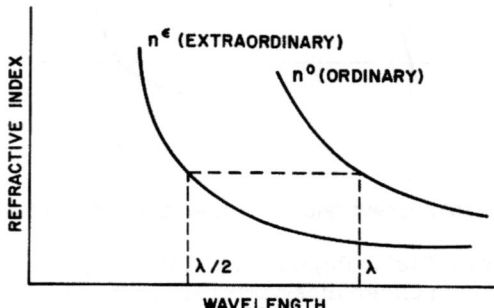

FIGURE 5. Dispersion curves for a birefringent crystal, showing the phase-matching condition for second harmonic generation (from Bergman and Kurtz [3]).

relates the nonlinear susceptibility to the linear susceptibility through a constant Δ that has approximately the same value for a number of materials. The linear susceptibility is related to the index of refraction by

$$4\pi\chi = n^2 - 1 \qquad (3)$$

Therefore

$$d \sim (n^2 - 1)^3 \qquad (4)$$

This shows that maximum d is obtained for a high n.

Second harmonic generation allows the use of parametric methods for the generation of frequencies other than those fundamental to the laser (Chapter 10).

Another nonlinear effect that is of great interest is the electrooptic effect, as this can be used to modulate an optical beam through the application of an electric field to a crystal. Here again it is the quadratic term in the polarization that is important. This can be seen as follows: if the polarization is written in terms of both the E-field of the light beam (frequency ω) and an applied electric field (frequency Ω),

$$P = \chi E + dE^2 + \ldots$$

Let $E(t) = E_1 \cos \Omega t + E_2 \cos \omega t$

$$P = \chi[E_1 \cos \Omega t + E_2 \cos \omega t] +$$
$$d[E_1^2 \cos^2 \Omega t + E_2^2 \cos^2 \omega t + 2 E_1 E_2 \cos \Omega t \cos \omega t]$$
$$= \chi[E_1 \cos \Omega t + E_2 \cos \omega t] +$$
$$d\left[E_1^2 \left(\frac{1 + \cos 2\Omega t}{2} \right) + E_2^2 \left(\frac{1 + \cos 2\omega t}{2} \right) \right.$$
$$\left. + E_1 E_2 \cos(\omega + \Omega)t + E_1 E_2 \cos(\omega - \Omega)t \right] \quad (5)$$

The linear susceptibility term shows an oscillating polarization at each of the two frequencies. However, the quadratic term gives both second harmonics and, in the cross term, a modulation of ω, that is, the oscillating charges radiate at frequencies $(\omega+\Omega)$ and at $(\omega-\Omega)$ and modulation is achieved. Thus, a large d is also required for the electrooptic effect, but in this case birefringence is not needed since there is no phase-matching requirement.

The nonlinear optical coefficient for material cannot be calculated from first principles, although Miller's Rule provides an empirical guide for the search for useful nonlinear materials. Most nonlinear optical materials therefore had to be found using essentially chemical criteria to guide the search [5]. In order that the electric field should have a big effect on the index of refraction the material should be highly polarizable. The polarizability may be either electronic or ionic. Materials showing large electronic polarizabilities will be those which have atoms with a large number of electrons so that the valence electrons are far from the nucleus, and these would include halogens and chalcogens in the lower right-hand corner of the Periodic Table. Materials with large ionic polarizability will usually show large permanent dipoles, thus piezoelectrics and ferroelectrics will be of particular interest. The crystals in all cases should be noncentrosymmetric and should have a large index of refraction (i.e. dielectric constant). As a variation one might look also for stereochemical arrangements that combine several of the factors, and these could include acentric molecules that include polarizable ions, as building blocks. This last guideline led to the discovery of iodates, in which the iodine resides in an acentric structure arising from the distorted iodate group arising from the lone pair of electrons. These chemical criteria are not sufficiently restrictive to make the search for nonlinear optical materials straightforward, but they allow some selection of materials as a simplification of the search process. A practical consideration in the screening of potential nonlinear optical materials is the ease with which reasonable quality crystals can be grown.

A number of materials have been discovered [6] using these criteria, as shown in Table 2. Some of these meet the high ionic polarizability requirement, and others have high electronic polarizabilities.

TABLE 2. Properties Found in Materials with Large Polarizability.

Material	Harmonic Generator	Electro-optic	Pyro-electric Detector	Acousto-optic	Piezo-electric
Lithium niobate	X				
Lithium tantalate		X			X
Barium sodium niobate	X				
Barium strontium niobate		X	X		
Lead molybdate				X	
Bismuth germanate				X	
Tellurium dioxide				X	
Iodic acid				X	
Lithium iodate	X				
Silver gallium sulfide	X				
Silver gallium selenide	X				

OPTICAL MATERIALS

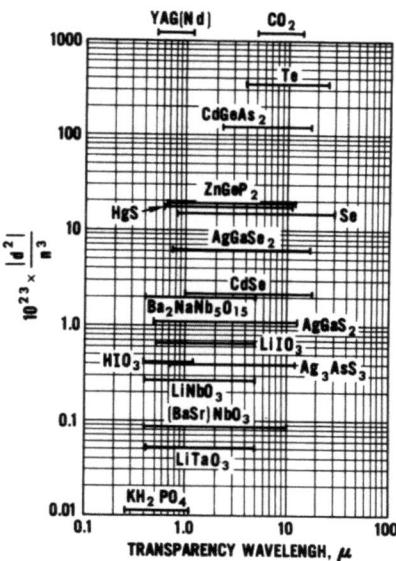

FIGURE 6. Figures of merit and useful frequency range for a number of second harmonic generation materials; operating range (fundamental and second harmonic) for YAG:Nd and CO_2 laser sources is indicated at top (from Wernick [7]).

A figure of merit for second harmonic generation is $\frac{d^2}{n^3}$. This is the critical factor controlling gain in the parametric amplification of plane waves. Figure 6 shows [7] comparisons of a number of nonlinear optical materials using this figure of merit. Also shown are the wavelength ranges for the CO_2 and Nd^{3+} lasers (fundamental and second harmonic frequencies). Some of the materials shown are easily prepared as good crystals, in other cases they are difficult to prepare and may exhibit chemical instability. The niobates are particularly useful as they have relatively good figures of merit, they are transparent in both the visible and the infrared, and good crystals can be grown by the Czochralski technique. $Ba_2Na_2Nb_5O_{15}$ is especially valuable, and can be used to convert the 1.06 μ line from YAG:Nd to 0.53 μ with 100% efficiency. $LiNbO_3$ will also do this. The iodates are also good second harmonic generators and large crystals can easily be grown from water solution; however, the crystals are susceptible to damage from humidity.

A number of materials are of interest in the infrared. Among the first discovered were proustite, Ag_3AsS_3, and CdSe. The ternary semiconductor compounds have been of recent interest for this purpose, because they are uniaxial and hence phase matchable (Chapter 10); crystal growth and composition control in these materials are still a problem and are currently under investigation. Tellurium shows high free-carrier absorption, which presents a problem. Selenium forms glasses and this is a serious disadvantage from the materials standpoint. In Table 3 several nonlinear optical materials are compared [6] as electrooptic modulator materials, in terms of the half-wave voltage.

TABLE 3. Electrooptical Materials.

Material	Modulator Voltage
Quartz	30,000
KH_2PO_4	8,000
HIO_3	2,900
$LiNbO_3$	2,900
ZnTe	2,700
$LiTaO_3$	2,500
HCOOLi	2,000
$Ba_2NaNb_5O_{15}$	1,500
$(Ba,Sr)Nb_2O_6$	100

Another nonlinear optical effect that is of great interest is the acousto-optic effect. A figure of merit for this is [8]

$$M = \frac{n^6 p^2}{\rho v^3} \qquad (6)$$

where p is the photoelastic component, ρ the density, and v the acoustic velocity in the crystal. Also required is a low acoustic loss. In order to maximize the figure of merit a low v is desirable, which suggests that soft materials, with a low Debye temperature, should be examined. The density and the chemical composition allow theoretical estimation of v, p, and n. Alternatively, n is usually known and so experimental values can be used. This figure of merit provides the elements of a search strategy for acousto-optic materials and its use has led to discovery of one of the most useful materials, lead molybdate (Table 4). TeO_2 (paratellurite) shows an even higher figure of merit and appears to be the best material for device applications below 1 Ghz [9].

OPTICAL MATERIALS

TABLE 4. Acousto-Optic Materials.

Material	Figure of Merit
Fused silica	1
$LiNbO_3$	5
GaP	30
HIO_3	55
$PbMoO_4$	25
TeO_2	85
Ge-As-Se glass	160

A very different approach [10] led to the chalcogenide glasses as acousto-optic materials. This began with the observation that the lowest loss material known was SiO_2. Chemical substitutions were investigated with the aim of retaining the low loss while increasing the acousto-optic coefficient. Group IV elements were substituted for the oxygen, Group IV elements for the silicon, and, in order to convert the material to the glassy state, a small amount of a Group V element was added. The Ge-As-S(Se) glasses have high acousto-optic efficiency, with a coefficient three times that for iodic acid and seven times that for lead molybdate.

Pyroelectricity is closely related to nonlinear optical phenomena, and therefore it is not surprising that useful pyroelectric materials have also been discovered in the search for nonlinear optical materials. A useful pyroelectric material is one in which there is a high pyroelectric coefficient, that is $\frac{dP}{dT}$. In order to achieve this, a large polarizability is desirable, with a Curie temperature somewhere in the vicinity of 300°K so that a maximum coefficient can be achieved in a convenient working temperature range (Fig. 7). $(Ba,Sr)Nb_2O_6$ has proved to be a particularly interesting pyroelectric material [11] since a change in the composition leads to changes in the Curie temperature and the pyroelectric coefficient. Table 5 compares several compositions in this system with a number of pyroelectric materials.

Finally, it is expected also that new piezoelectrics will be discovered using the search criteria outlined above, and in fact this has been the case [6]. Table 6 lists the piezoelectric coefficients of several materials. $LiTaO_3$ emerges as a particularly promising new piezoelectric; its coefficient is quite insensitive to temperature and this makes it superior to

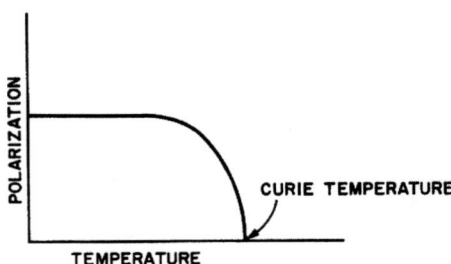

FIGURE 7. Schematic of temperature-dependence of polarization.

TABLE 5. Pyroelectric Detector Materials.

		$T_c(°K)$	$\frac{dP_s}{dT}$ ($\mu C/cm^2/°C$)
$Sr_{1-x}Ba_xNb_2O_6$	x = 0.27	317	0.28
	x = 0.33	335	0.11
	x = 0.40	351	0.085
	x = 0.52	388	0.065
	x = 0.75	469	0.030
Triglycine sulphate		322	0.035
$Li_2SO_4 \cdot H_2O$		--	0.01
$LiTaO_3$		891	0.019
Polyvinylidene Fluoride		--	0.0024
$P_{0.65}Z_{0.35}T$ Ceramics		648	0.035
$P_{0.65}Z_{0.35}T$ + 8% La		370	0.17

OPTICAL MATERIALS   275

TABLE 6. Piezoelectrics.

Material	Coupling Efficiency
Quartz	12%
ZnO	36%
$LiTaO_3$	50%
$Ba_2NaNb_5O_{15}$	60%

$Ba_2NaNb_5O_{15}$, which has a larger coefficient but one with greater temperature sensitivity.

We now turn to the theoretical basis for the occurrence of large nonlinear optical effects in solids. As noted earlier Miller's Rule provided an empirical correlation which has been useful for predicting the nonlinear susceptibility from the known linear optical susceptibility. As indicated in Chapter 10, this relationship can be derived from a simple harmonic oscillator model if a nonlinear term is included and it is assumed that the strength of this term is approximately the same for all materials; no physical basis for this assumption can be readily identified, however. A fuller theoretical explanation for Miller's Rule has more recently been given by Levine [12].

A starting point for this theory is the Phillips-Van Vechten theory of the chemical bond (Chapter 19). Levine considers the effect of the application of an electric field. This displaces the bond charge q by some distance Δr, and the effect of this is to induce a net polarization in the bond.

$$P = Nq\Delta r \tag{7}$$

where N is the number of bonds per cm^3. This modulates C, the ionic energy gap, so that the total energy gap in the presence of the E-field becomes

$$E_g^2 = E_h^2 + (C+\Delta C)^2 \tag{8}$$

ΔC is derived from Eq. (7) of Chapter 19,

$$\Delta C = -\frac{16}{3\sqrt{3}}\, b\, e^{-kR}\left(\frac{Z_\alpha + Z_\beta}{q}\right) e^2\, a\chi E \tag{9}$$

$$= \text{constant} \times (E\chi)$$

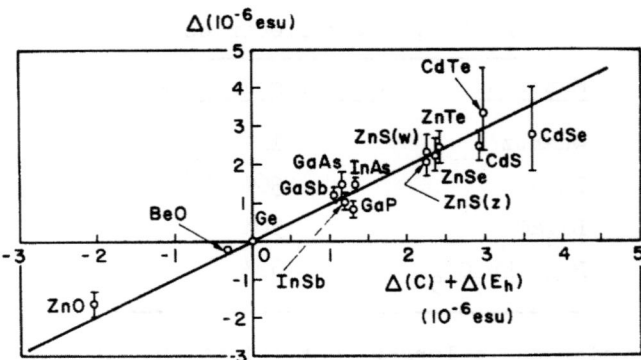

FIGURE 8. Miller's Δ compared with theory (from Levine [12]).

In addition, there is a change in χ, from the zero-field value; the change in χ is derived from the relation $\chi = (\hbar\Omega_p)^2/E_g^2$, where Ω_p is the plasma frequency, so that

$$\chi(E) = \chi\left[1 - \frac{2C\Delta C}{E_g^2} - \frac{(\Delta C)^2}{E_g^2}\left(1 - \frac{4C^2}{E_g^2}\right)\right] \quad (10)$$

By making the appropriate substitution of Eq. (9) in Eq. (10), and comparing with Eq. (1), i.e., $P = \chi(E)E = \chi E + dE^2 + RE^3$, the coefficients of the quadratic term in the E-field dependence of P can be identified, with a result that is similar in form to Miller's Rule and that shows that the Miller Δ is proportional to the Phillips-Van Vechten ionic energy gap C. In Fig. 8 the observed Miller Δ's for a number of compounds are plotted against theory, and it is seen that the agreement is good. In a few cases, notably including zinc oxide, it was found that the results were anomalous and that the value of the quadratic nonlinear coefficient d even changed sign. Levine has explained these anomalies also and finds that they occur when there is a large size difference between the ions, as is the case with zinc and oxygen. In such cases it is necessary to take into account not only the change in the ionic energy gap C, from the application of an electric field, but also the change in the homopolar part of the energy gap, E_h.

SOLID STATE CHEMISTRY OF NONLINEAR OPTICAL MATERIALS

A number of problems were identified at an early stage in nonlinear optical materials. These included a range of phenomena such as optical damage in the laser beam, and the inability to grow uniformly good crystals, especially with respect to

OPTICAL MATERIALS 277

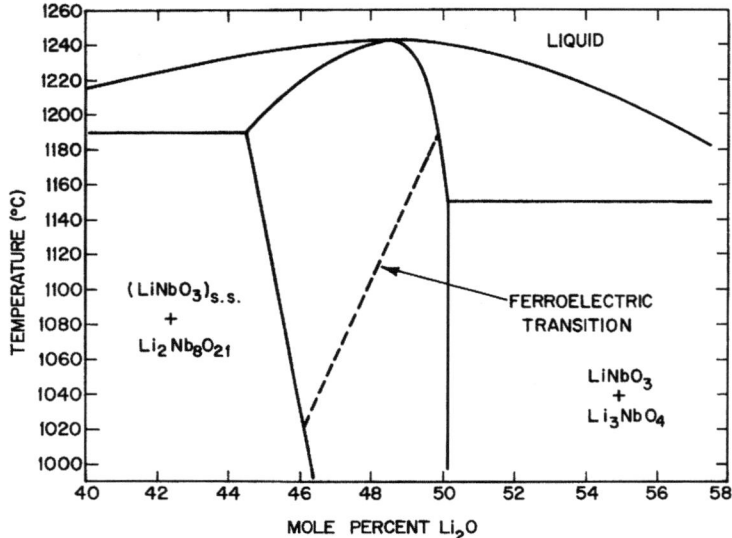

FIGURE 9. Phase diagram for lithium niobate (from Carruthers et al [13]).

nonuniformities of phase-matching temperature, Curie temperature, χ, n, birefringence, etc. This led to an intense study of the materials and of the solid state chemistry. Lithium niobate serves as a good example because it has been investigated more thoroughly than other materials [13], although the phenomena observed in $LiNbO_3$ apply rather generally to a number of other related materials.

The basis for many of the optical problems in lithium niobate, as well as their solution, can be understood by reference to the phase diagram for lithium niobate shown in Fig. 9. It is seen that the congruent melting point occurs off-stoichiometry, at 48.6% Li_2O. Thus the phase diagram is like that discussed in Chapter 19 (Fig. 11). There is a region of single phase material, with two solid phases encountered outside this range. The ferroelectric transition temperature is shown as a function of composition.

The phase diagram was originally suggested [14] by X-ray observations, which showed that there was a range of compositions for single-phase lithium niobate. Two kinds of experiments were used to fill out the details of the phase diagram. One of these was a measurement of the Curie temperature. Ceramics of various compositions were prepared, using long firing times to insure complete reaction and uniformity of the material, and these ceramics were compared with single crystals grown from melts of

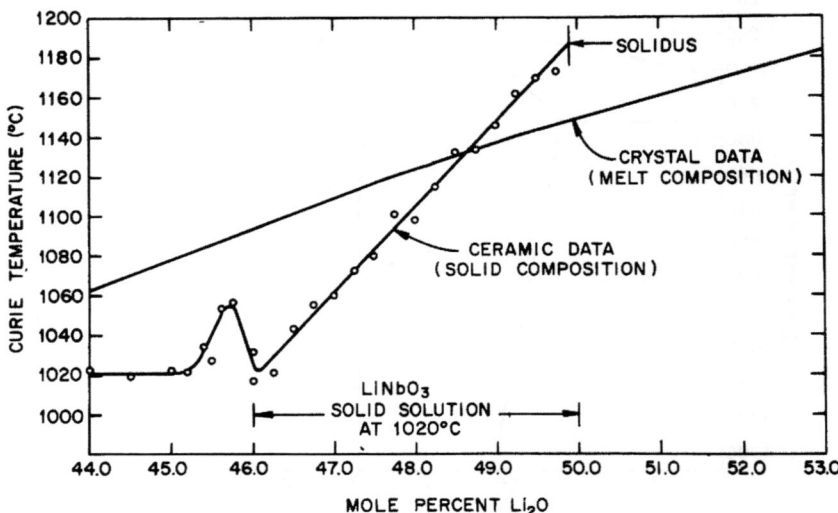

FIGURE 10. Variation in Curie temperature with $LiNbO_3$ composition (from Carruthers et al [13]).

different composition (Fig. 10). It was assumed that a given T_c corresponded to a particular composition in both the ceramic and single crystal. Thus it was possible to deduce the composition of the single crystal resulting from a particular melt composition and this established phase diagram tie-lines. Similar experiments were performed using nuclear magnetic resonance, using the Nb^{93} linewidth (Fig. 11). The reason for the variation in linewidth is the variation in the local environment for the Nb^{93} resonance; as the defect concentration increases, with increasing departure of the crystal from stoichiometry, the linewidth increases. Again, the comparison of ceramic data with the single crystal data provided information about phase diagram tie-lines. From these experiments the phase diagram in Fig. 9 was determined.

It is seen that the growth from the melt of a uniform but nonstoichiometric crystal will occur with a melt composition of 48.6% Li_2O. A stoichiometric crystal can be grown at melt compositions of 58% lithium oxide, although if the melt is allowed to change composition as the crystal grows (which will be the case, in conservative crystal growth procedures), the crystal will soon depart from stoichiometry because the melt is enriched in Li_2O. Ordinary growth at a 50% melt composition will therefore result in a nonuniform crystal with an approximately linear composition variation down the length of the crystal. This accounts for the variation in optical properties that has been

OPTICAL MATERIALS

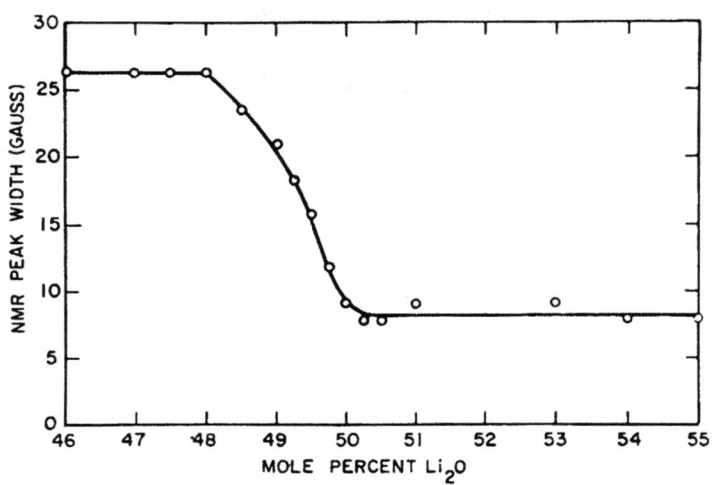

FIGURE 11. Variation in Nb^{93} NMR linewidth with $LiNbO_3$ composition (from Carruthers et al [13]).

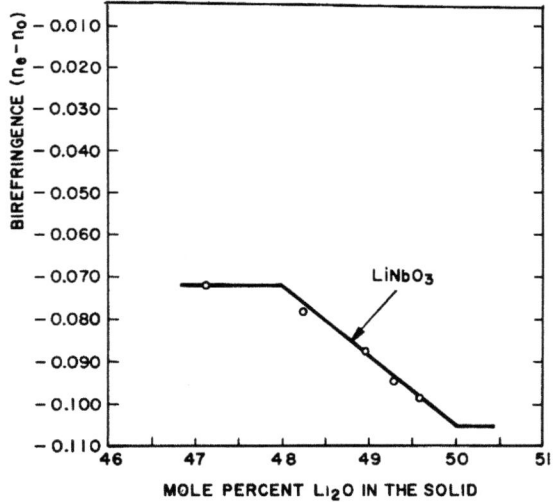

FIGURE 12. Variation in birefringence of $LiNbO_3$ with composition (from Carruthers et al [13]).

observed. Figures 12 and 13 show the variation in birefringence, refractive index, phase-matching temperature and Curie temperatures as a function of composition. The growth conditions for lithium niobate have been studied extensively, with attention not only to stoichiometry but also to the avoidance of difficulties due to

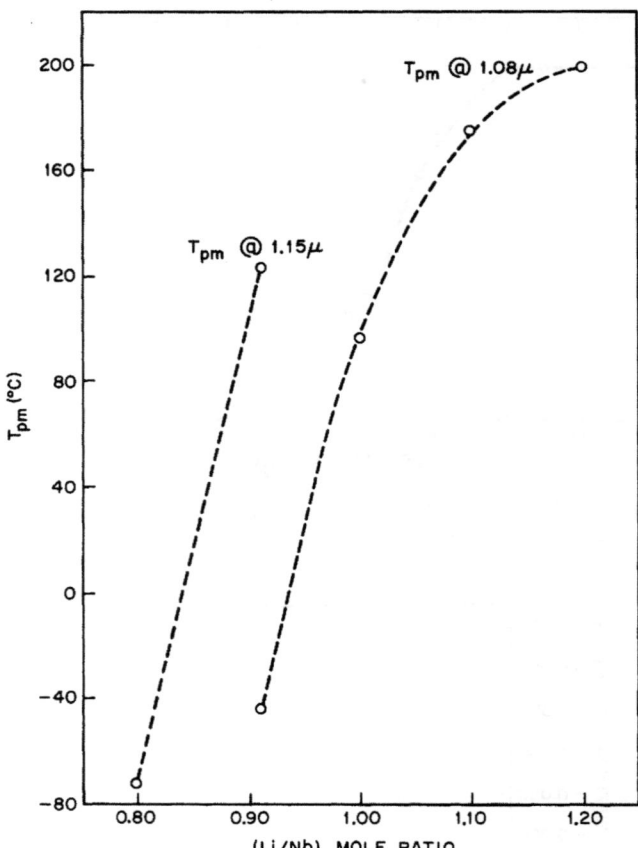

FIGURE 13. Variation in phase-matching temperature with $LiNbO_3$ composition (from Laudise [3]).

constitutional supercooling and to growth striations resulting from compositional variations.

The phase-matching temperature for second harmonic generation is high in stoichiometric crystals and low in nonstoichiometric crystals, as seen in Fig. 13. However, optical damage from laser beams (see below) occurs at low temperatures, although it can be removed by annealing at high temperatures. Hence the use of stoichiometric crystals with a high phase-matching temperature is of interest, as the optical damage is constantly removed in operation at these temperatures. Unfortunately, however, uniformly stoichiometric crystals are particularly difficult to grow, and it is often observed that they contain striations which detract from their optical perfections. Another way to achieve a similar result is to add MgO to the crystal [15]. This raises the phase-matching temperature. Since the distribution coefficient of magnesium oxide is greater than one, its concentration decreases along the crystal, while the Li_2O concentration solid increases down the length of the crystal because its distribution coefficient is less than one. As the phase-matching temperature is raised both by MgO and Li_2O, the decrease in concentration of one compensates for the increase of the other and this leads to a uniform and higher phase-matching temperature.

Optical damage has been studied extensively in lithium niobate [16]. Reversible changes in refractive index occur under illumination with a laser beam. It was assumed that this was due to a defect, or trap, and that the mechanism involved the photo-excitation and trapping of carriers. However, it turned out the optical damage was independent of stoichiometry and this ruled out the obvious possibility that the defect is associated with non-stoichiometry. EPR experiments were used to look at impurities in $LiNbO_3$ and it was found that the ordinary material contains a considerable concentration of Fe^{3+} (Fig. 14). When the iron is removed the crystal is no longer damaged by a laser beam. The damage results from the photo-ionization of electrons from Fe^{2+} under the influence of light beam, $Fe^{2+} + h\nu \rightarrow Fe^{3+} + e^-$. The released electrons drift off to another region of the crystal where they are trapped. As the refractive index depends upon the concentration of Fe^{2+} vs Fe^{3+}, there is a change in the refractive index as a result of this photo-ionization process. The avoidance of damage, of course, is achieved through removal of iron from the crystal. Also, since the Fe^{2+} is the species that leads to damage, the susceptibility of the crystal to damage can be influenced by oxidation or reduction, as this converts $Fe^{2+} \rightleftharpoons Fe^{3+}$. It is of interest to note that for some purposes, such as the holographic storage of information, $LiNbO_3$ might be made useful by increasing the susceptibility to refractive index change under the influence of a laser beam, through deliberate additions of iron to the crystal. Lithium tantalate, barium sodium niobate, and perhaps other materials in this class show similar stoichiometry effects.

A different problem encountered in the growth of barium sodium niobate that had to be overcome before this material could be grown into the large single crystals required for second harmonic generation was a persistent cracking as the crystal was removed from the Czochralski apparatus. The explanation for this [17] lay in the stresses that were developed when the crystal was cooled in a normal way. The expansion coefficients are shown in Fig. 15 as a function of temperature, and it is seen that there is a large anisotropic change as the crystal passes through the Curie temperature. Also, when the melt composition is

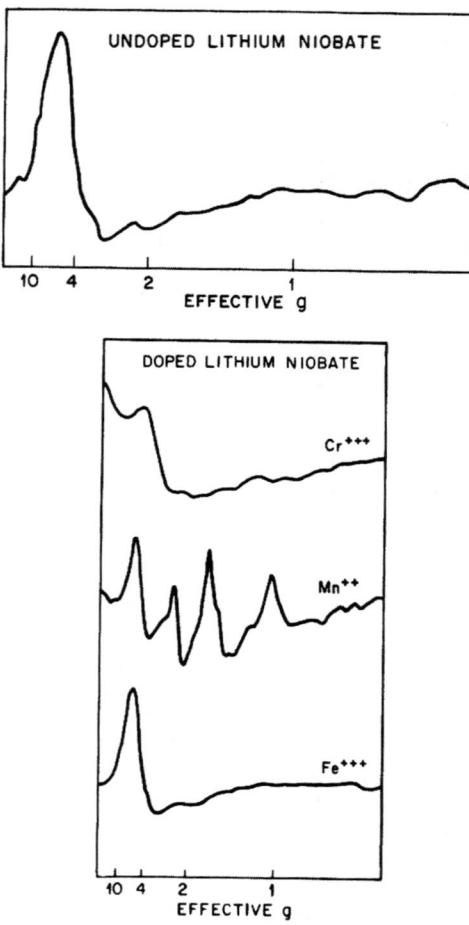

FIGURE 14. EPR spectrum of ordinary $LiNbO_3$ compared with samples deliberately doped with impurities (from Glass, Peterson and Negran [16]).

stoichiometric the crystals often exhibit growth striations, and therefore variations in T_c; in such crystals the transformation occurs at different temperatures in different regions of the crystal. The avoidance of cracking is achieved by choosing a melt composition closer to the congruent melting composition (≈69%), and by maintaining the crystal above the Curie temperature after growth, so that no part of it is allowed to cool below T_c during growth. After the crystal is completely grown it is maintained at a uniform temperature which is lowered very slowly through the Curie temperature. This minimizes the effect of the internal stresses that lead to cracking.

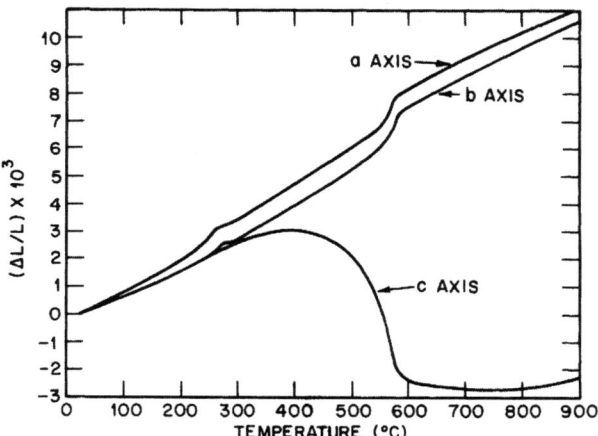

FIGURE 15. Expansion coefficients of barium sodium niobate (from Ballman, Carruthers and O'Bryan [17]).

REFERENCES

[1] K. Nassau, "Applied Solid State Science", Vol. 2 (p. 173), Academic Press, New York (1971); A. Yariv, "Quantum Electronics", John Wiley & Sons, New York (1967).
[2] J. E. Geusic, private communication.
[3] R. A. Laudise, Proc. of Chania Internat'l. Conf. on Growth and Characterization of Electronic Materials, Crete, Greece (6/69); R. A. Laudise, Bell Labs Record (Jan. 1968); J. G. Bergman and S. K. Kurtz, Materials Science & Engineering 5, 235 (1969/70); E. G. Spencer, P. V. Lenzo and A. A. Ballman, Proc. IEEE 55, 2074 (1967); J. R. Carruthers, Ency. of Chem. Tech., Suppl. Vol., 2nd ed., John Wiley & Sons, New York (1971).
[4] R. C. Miller, Appl. Phys. Letters 5, 17 (1964).
[5] R. A. Laudise, ref. 3.
[6] R. A. Laudise, private communication.
[7] J. H. Wernick, in "Treatise on Solid State Chemistry", Vol. 1; N. B. Hannay, Ed., Plenum Publishing Corp., New York (to be published); see also J. R. Carruthers, ref. 3.
[8] R. W. Dixon, J. Appl. Phys. 38, 5149 (1967); D. A. Pinnow, IEEE J. Quantum Electron. QE-6, 223 (1970).
[9] W. A. Bonner, S. Singh, L. G. Van Uitert, and A. W. Warner, J. Electr. Mat. 1, 155 (1972).
[10] J. T. Krause, C. R. Kurkjian, D. A. Pinnow and E. A. Sigety, Appl. Phys. Letters 17, 367 (1970); R. W. Dixon, ref. 8.
[11] A. M. Glass, J. Appl. Phys. 40, 4699 (1969); A. M. Glass and R. L. Abrams, Proc. of Symposium on Submillimeter Waves, Poly. Inst. of Brooklyn (1970).

[12] B. F. Levine, Phys. Rev. Letters 22, 787 (1969) and 25, 440 (1970).
[13] J. R. Carruthers, G. E. Peterson, M. Grasso and P. M. Bridenbaugh, J. Appl. Phys. 42, 1846 (1971).
[14] P. Lerner, C. Legras and J. P. Dumas, J. Cryst. Growth 3-4, 231 (1968).
[15] P. M. Bridenbaugh, J. R. Carruthers, J. M. Dziedzic and F. R. Nash, Appl. Phys. Letters 17, 104 (1970).
[16] A. M. Glass, G. E. Peterson and T. J. Negran, NBS Special Publication: Damage in Laser Materials 1972 (to be published).
[17] A. A. Ballman, J. R. Carruthers and H. M. O'Bryan, J. Cryst. Growth 6, 184 (1970); W. A. Bonner, J. R. Carruthers and H. M. O'Bryan, Mat. Res. Bull. 5, 243 (1970).

CHAPTER 12

ORGANIC DYES IN LASER TECHNOLOGY

D. J. Bradley

The Queen's University of Belfast

Belfast BT7 1NN, Northern Ireland

INTRODUCTION

While capable of producing high powers ruby and neodymium lasers suffer from the serious defect of fixed frequencies, apart from the high intrinsic cost of the active materials. The emission spectra of fluorescent dyes are broad, and there is a large number of dyes with fluorescence covering the optical spectrum from 340 to 1170 nm (see Fig. 1). While the fluorescence efficiency of these organic compounds is very high, it was not until 1966 that Sorokin and Lankard [1] and Schafer, Schmidt and Volze [2] achieved laser action in dyes pumped by giant pulse lasers. The dye molecular energy levels are shown schematically in Fig. 2. Because of very fast radiationless internal conversion within the manifold of excited singlet states, the lasing transitions occur from the bottom vibrational level of the first excited electronic singlet state to the ground state vibrational levels, and the resulting broad band emission is homogeneously broadened. The short fluorescence lifetimes, of a few nanoseconds, severely restrict the storage of energy in the active media, and to this extent, pulsed dye lasers may be regarded as operating quasi-continuously. The design of highly efficient amplifiers for pulsed systems is correspondingly more difficult, requiring very fast optical pumping or multipass arrangements.

Intersystem crossing leads to the accumulation of molecules in the triplet state with the long lifetimes typical of phosphorescence. The resulting triplet state absorption spectrum (Fig. 3) overlaps the fluorescence emission, eventually leading to quenching of laser action. Because of these absorption losses

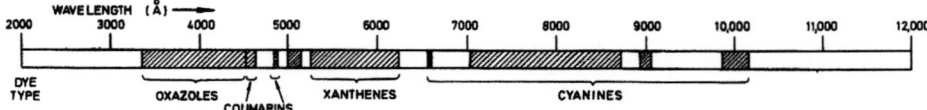

FIGURE 1. Frequency range of dye lasers giving the main operating spectral regions. Most of the spectral range from 340 to 1200 nm can be covered by flashlamp or laser pumped systems.

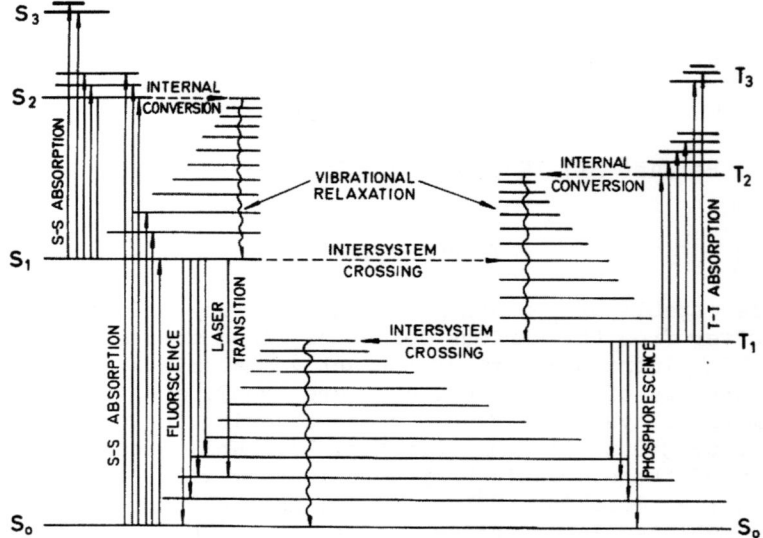

FIGURE 2. Electronic levels of dye molecules.

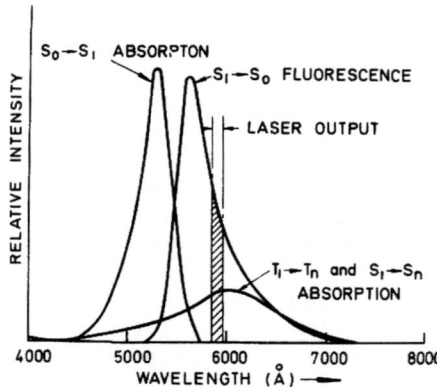

FIGURE 3. Absorption and fluorescence spectra of Rhodamine 6G.

ORGANIC DYES IN LASER TECHNOLOGY

associated with the accumulation of molecules in the triplet state, pulsed laser action would be expected to be achieved only by pumping with pulsed lasers or fast rise-time flashlamps [3].

By quenching the triplet state with molecular oxygen [4] or cyclooctatetraene [5], long pulse operation has been obtained and Peterson, Tuccio and Snavely [6] first achieved CW action in a water solution of Rhodamine 6G excited by an argon ion laser. With the Rhodamine dyes, which lase in the green to red spectral regions, thermal heating has proved to be more troublesome than triplet state losses, but this problem has been overcome by careful design of the dye cell combined with rapid flowing of the dye solution. Since 1966 dye lasers have been developed and applied in many laboratories and the scientific literature now contains hundreds of publications dealing with this type of laser. (See for example the recent survey by C. De Michelis [7]). In this paper it will not be possible to cover all the recent advances and for convenience, the present state of the art will be illustrated, where possible, by results obtained in our laboratories at Belfast. Since one of the most useful and attractive properties of dye lasers is the present availability of a large number of lasing dyes covering the optical spectrum from 330 nm to 1170 nm, combined with the frequency tunability of each dye over its broad fluorescence spectrum, most emphasis will be placed on frequency control, both narrowing and tuning. Alternatively, the broad output spectra can be mode-locked to produce ultra-short (1-2 picoseconds) tunable-frequency high-power pulses, with reliability and reproducibility exceeding that obtained to date from solid state and gas lasers. While previously it was customary to discuss and treat dye lasers according to the method of excitation (laser pumped, flashlamp pumped, C.W. etc.) with our present greater understanding of the photochemical and physical processes involved, it would seem more appropriate to deal rather with the basic techniques employed to produce the laser characteristics required for a particular application or experiment.

FREQUENCY NARROWING AND TUNING

Despite the use of liquid active media, by employing rapid flow techniques in various arrangements single-transverse mode, single-longitudinal mode operation, first achieved with a ruby laser pumped dye laser system [8], has been extended to both flashlamp pumped and C.W. dye lasers. In both cases the effects of thermal distortion have been sufficiently reduced by employing water-based solutions of Rhodamine dyes. In the pulsed laser a 50 cm long dye cell was pumped, in a cylindrical elliptical pumping reflector, with a fast rise-time linear quartz flashtube.

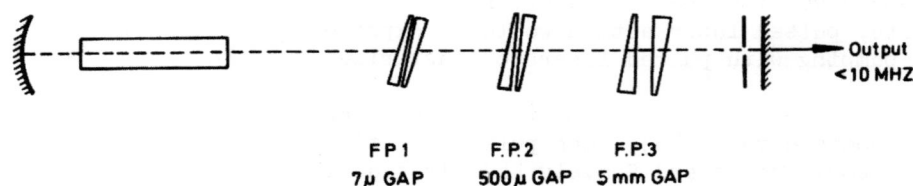

FP1 F.P.2 F.P.3
7μ GAP 500μ GAP 5 mm GAP

FIGURE 4. TEM$_{oo}$ single longitudinal-mode flashlamp pumped dye laser.

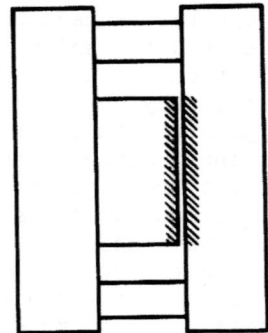

FIGURE 5. Optically contacted, thermally compensated quartz Fabry-Perot interferometer filter.

In a generalized confocal laser cavity, with a 1.5 mm circular aperture close to the plane output mirror (Fig. 4), TEM$_{oo}$ operation is easily achieved in a mode of 1 mm diameter, with a half-angle beam divergence of 0.4 milliradian [9]. An electrical input energy of 20 J, produces a 400 mJ, laser output pulse of peak power 1 MW cm^{-2}. The 8 nm spectral bandwidth obtained with broadband reflecting laser mirrors can be frequency narrowed and tuned, with low loss, by the insertion into the laser cavity of the three tilted, optically contacted permanently adjusted [10], Fabry-Perot interferometer filters of plate separations 10 μm, 500 μm and 5 mm respectively. Details of the construction of the narrower gap interferometers are shown in Fig. 5. We have found that when diffraction gratings are employed for tuning high energy, flashlamp pumped dye lasers, the thermal refraction effects in the dyes can lead to the appearance of satellite lines in the output spectra. With high energy lasers (output > 0.5 J) bandwidths of < 0.3 nm have been obtained [11] with single narrow gap interferometers (gaps of 4-10 μm). Inserting an interferometer filter, tilted at an angle of ~5° in the vertical plane to prevent optical feedback, the spectrally narrowed laser output is easily tuned over the free spectral range of the interferometer.

ORGANIC DYES IN LASER TECHNOLOGY

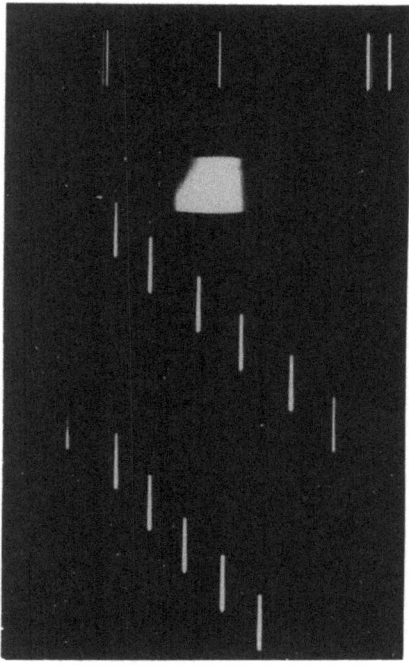

FIGURE 6. Successive exposures of spectra (1 m Ebert spectrograph) of Rhodamine 6G dye laser tuned with a single 7 μm gap Fabry-Perot interferometer. By rotation of the interferometer through ≈30° the free spectral range of 270 nm is scanned through twice. Top two spectra are a mercury lamp and the untuned laser respectively. Untuned laser bandwidth = 6.5 nm at 595 nm. Tuned bandwidth = 0.3 nm.

Figure 6 shows the tuned spectra obtained when the interferometer was rotated through a total angle of ∼30°, corresponding to nearly two free spectral ranges. Employing 3 layer dielectric coatings on the interferometer plates, the tuning efficiency curves are remarkably flat over a range of >25 nm for each dye and with a single interferometer it is possible to cover a range of >2000 Å of the spectrum employing different dyes. Thus both the coumarin family of dyes in the blue and the Rhodamines in the yellow, orange and red can be tuned with the same interferometer. Using three etalons as in Fig. 4, the laser energy is compressed into a single longitudinal mode of the 50 cm long cavity. Each etalon frequency narrows the laser by approximately a factor of 40 with little reduction of output energy and the resulting 14 MHz spectral bandwidth obtained can be seen in the defocussed spherical Fabry-Perot interferogram [12] of Fig. 7. Again rotation of the three etalons permits tuning over a substantial portion of the dye fluorescence bandwidth, from 587 nm to 605 nm for Rhodamine 6G. By enclosing the complete laser in a pressure

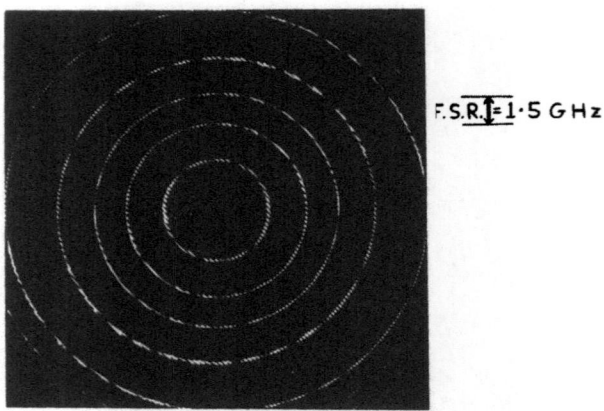

FIGURE 7. Defocussed spherical Fabry-Perot (plate separation 5 cm) interferogram of single-transverse, single-longitudinal mode laser of Fig. 4. Linewidth ~14 MHz.

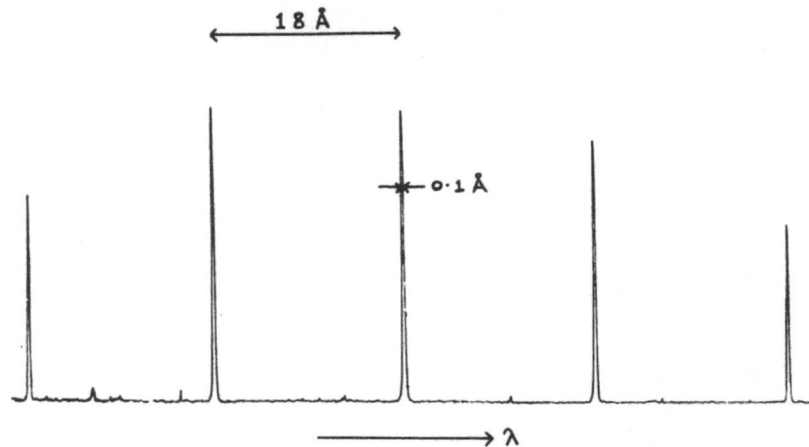

FIGURE 8. Microdensitometer trace of multiple line spectrum produced by the single-transverse mode dye laser with a single 100 μm gap intracavity interferometer filter. Spectral width of each component is 0.1 Å. Corresponding coherence length is ~2 cm.

---

and temperature stabilized environment accurate control of the output wavelength is obtained with fine tuning (in steps of 10^{-3} cm^{-1}) over several wavenumbers, by variation of the chamber pressure without adjustment of the interferometers. If a single etalon, with a plate separation of 100 μm, only is used the spectrum then consists of 5 main lines, spaced by the interferometer free spectral range of 1.8 nm, (Fig. 8) and each of the individual components has a linewidth of ~0.01 nm and a corresponding coherence length of ~2 cm. This multiple line narrow-band laser source should have immediate application for improved contour holography [13].

Single-transverse, single-longitudinal mode operation of a C.W. dye laser has also been achieved by Hercher and Pike [14], with a bandwidth of 35 MHz, while Hansch [15] has succeeded in narrowing to 300 MHz a Rhodamine dye laser, repetitively pumped by a nitrogen laser, by employing the combination of a beam expanding telescope, tilted etalon and a Littrow mounted diffraction grating as cavity elements.

By frequency doubling and mixing, the tuning range achievable with dyes can be extended to the ultra-violet [16] and the infrared [17]. Megawatt second harmonic and sum frequency generation with 10% efficiency, and frequency tunable from 280 nm to longer wavelengths, has been obtained [18] in our laboratories and the laser output spectrum has been narrowed to 0.04 nm.

While the energies obtained from flashlamp-pumped dye lasers have been limited to a few joules [8], with peak powers of ~1 MW, employing a battery of flashtubes, energies in excess of 100 J (>100 MW peak power) should be available for selective photochemical processing, atmospheric scattering and plasma studies. The use of multiple flashlamps leads to better beam quality with greatly reduced beam divergence [19] and these improvements should facilitate frequency tuning and narrowing.

MODE-LOCKING AND THE GENERATION OF FREQUENCY TUNABLE PICOSECOND PULSES

While flashlamp-pumped Rhodamine dye lasers had been passively mode-locked to produce pulses of durations <5 psec [20,21], as is the case with mode-locked ruby and neodymium lasers [22,23], the laser spectral bandwidths of 30-100 Å were up to 100 times greater than the Fourier-transform limit set by the pulse durations. By frequency narrowing a mode-locked Rhodamine 6G dye laser with a single, 8 μm gap, Fabry-Perot interferometer filter, the pulse bandwidth-duration discrepancy was removed and picosecond pulses of transform-limited durations ($\Delta t \sim 3$ psec, $\Delta t \Delta \nu \sim 0.5$), frequency tunable over a range of 23 nm, were obtained [24]. The tuned output spectra are shown in Fig. 9 and the microdensitometer traces of, simultaneously recorded, two-photon fluorescence tracks and spectra are given in Fig. 10. More recently [25], by employing different polymethine dyes as saturable absorbers and a range of intracavity Fabry-Perot interferometers, picosecond pulses frequency tunable from 580 nm to 700 nm have been generated with the three lasing dyes Rhodamine 6G, Rhodamine B and Cresyl-Violet. The spectral coverage obtained

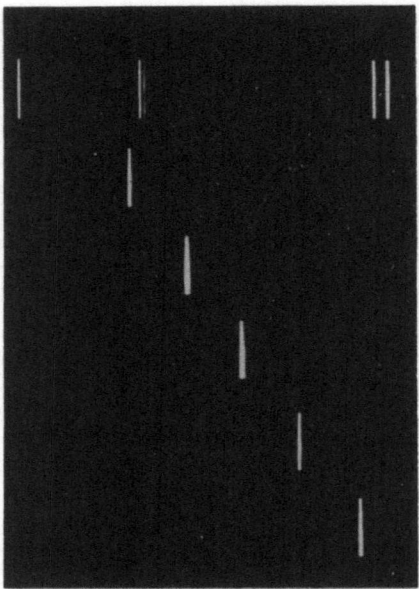

FIGURE 9. Spectra showing tuning of transform-limited picosecond pulses. Mercury calibration spectrum at top. Plate of 1-m spectrograph was moved vertically as laser was tuned from 6030 to 6250 Å.

with different dye combinations is given in Table I and the relevant normalized fluorescence spectra and absorption spectra,

TABLE 1. Spectral Ranges Covered by Various Combinations of Lasing Dyes and Saturable Absorbers.

Laser Dye	Saturable Absorber	Mode-Locked Tuning Range (nm)
Rhodamine 6G	DODCI	584 - 625
Rhodamine B	DQTCI	605 - 639
	DODCI	615 - 645
Cresyl-Violet	DTDCI	652 - 704
	DDCI	644 - 680
+ Rhodamine 6G	DOTCI	644 - 680

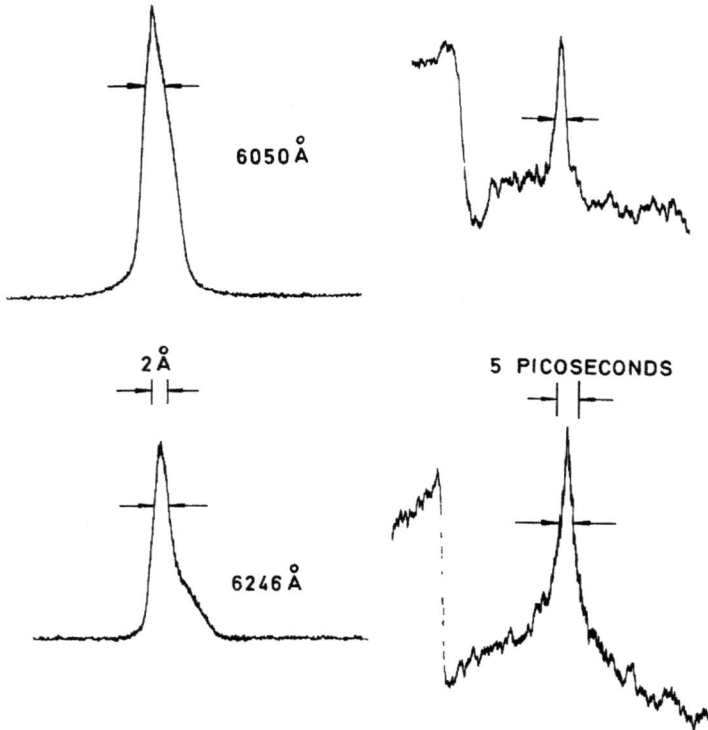

FIGURE 10. Microdensitometer traces of two pairs of simultaneously recorded TPF tracks and spectra of transform-limited Rhodamine 6G picosecond pulses. Top: spectral width 2.4 Å at 6050 Å. Pulse duration 3.3 psec for Gaussian-shaped pulse; 2.3 psec if Lorentzian-shaped pulse. Bottom: spectral width 1.8 Å at 6246 Å. Pulse duration 4.7 psec (Gaussian); 3.4 psec (Lorentzian). All ordinates are linear density scales. Time-calibration mark of TPF tracks gives directly duration for a Lorentzian pulse. For a Gaussian-shaped pulse the measured half-width has to be multiplied by $\sqrt{2}$ to give correct duration. ×2.4 intensity steps on left-hand sides of TPF traces. (Background variation is due to non-uniform spatial response of image-tube intensifier employed to record the fluorescence tracks.)

respectively, are shown in Fig. 11. The trains of mode-locked picosecond pulses can be generated with great reliability and reproducibility and employing these pulses permitted [26] the first direct unambiguous demonstration of the linear measurement of laser pulse durations with an electro-optical streak camera [27,28] with a camera instrumental width <5 psec. The camera and the pulse generating laser system have since been further developed to measure a time-resolution limit approaching one

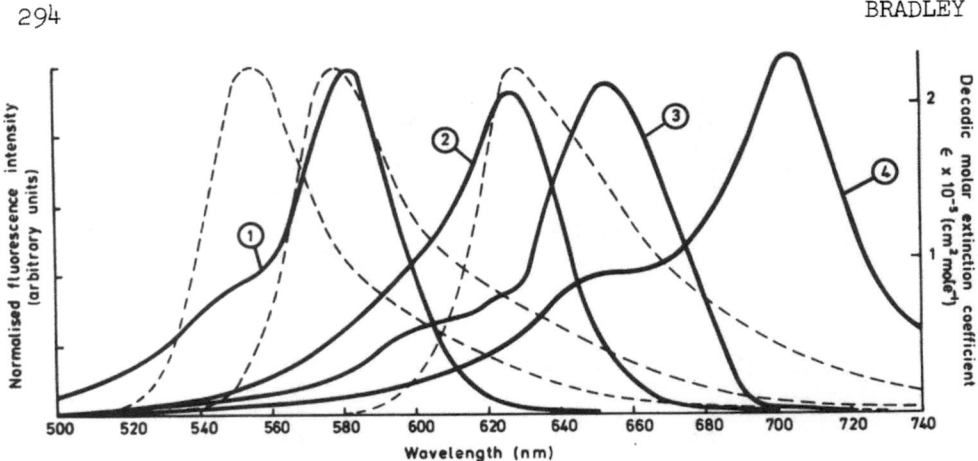

FIGURE 11. Normalized fluorescence spectra of laser dye solutions.
(a) Rhodamine 6G; (b) Rhodamine B; (c) Cresyl-violet and
absorption spectra of polymethine dyes employed as saturable
absorbers. (1) DODCI; (2) DQTCI; (3) DTDCI; (4) DDCI.

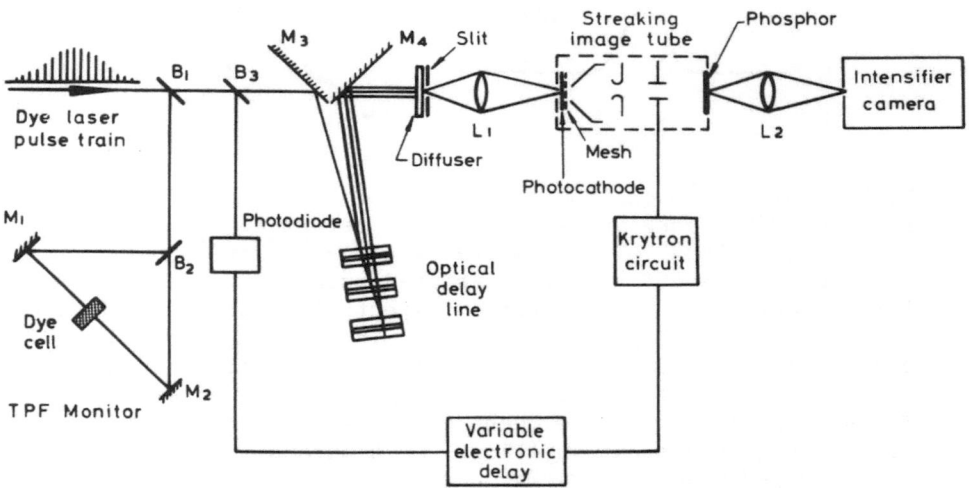

FIGURE 12. Experimental arrangement for testing picosecond streak camera.

picosecond, with increased reliability and improved synchronization. With the very high light gain available (the over-all gain of the camera system is sufficient to detect every photo-electron from the photocathode of the streaking image tube) we have been able to measure the pulses from a passively mode-locked C.W. dye laser of peak power of only 1 watt and to study the development of picosecond pulses from the initial background photon noise of high-power ($\sim$50 MW) mode-locked pulsed dye lasers.

FIGURE 13. Photograph of picosecond streak camera and mode-locked twin-flashlamp pumped dye laser.

DIRECT MEASUREMENT OF PICOSECOND LASER PULSES WITH AN ELECTRO-OPTICAL STREAK CAMERA

The experimental arrangement for testing the streak camera performance can be seen in Fig. 12 and 13, which show details of the image-tubes employed, the streak deflection voltage and synchronizing circuits, the optical coupling lenses and the optical delay line for the generation of sub-pulses, of accurately known delays, for calibration of the linearity of the camera streak and for accurately determining the system time-resolution achieved under differing operating conditions. To test the camera to its limit of time-resolution, the mode-locked dye laser, pumped with a 2000 J single ablative quartz lamp previously employed [24,25], was replaced with a more efficient passively mode-locked laser [19] pumped with twin, 200 J, sealed-off Xenon flashlamps [8] and employing a fast capacitor discharge triggered with a pressurized spark gap. Pulses of 1-2 psec duration are reliably produced by this laser. With a streak writing speed (at the phosphor of the first image-tube) of 10^{10} cm sec^{-1}, a dynamic spatial resolution of 5 line pairs/mm was achieved for an exposure (on Ilford HP4 Film) sufficient to produce a photographic record intense enough for accurate microphotometry. The streak record of Fig. 14(a) shows a group of 4 pulses obtained when the delay line was arranged to produce, from a single laser pulse, pairs of

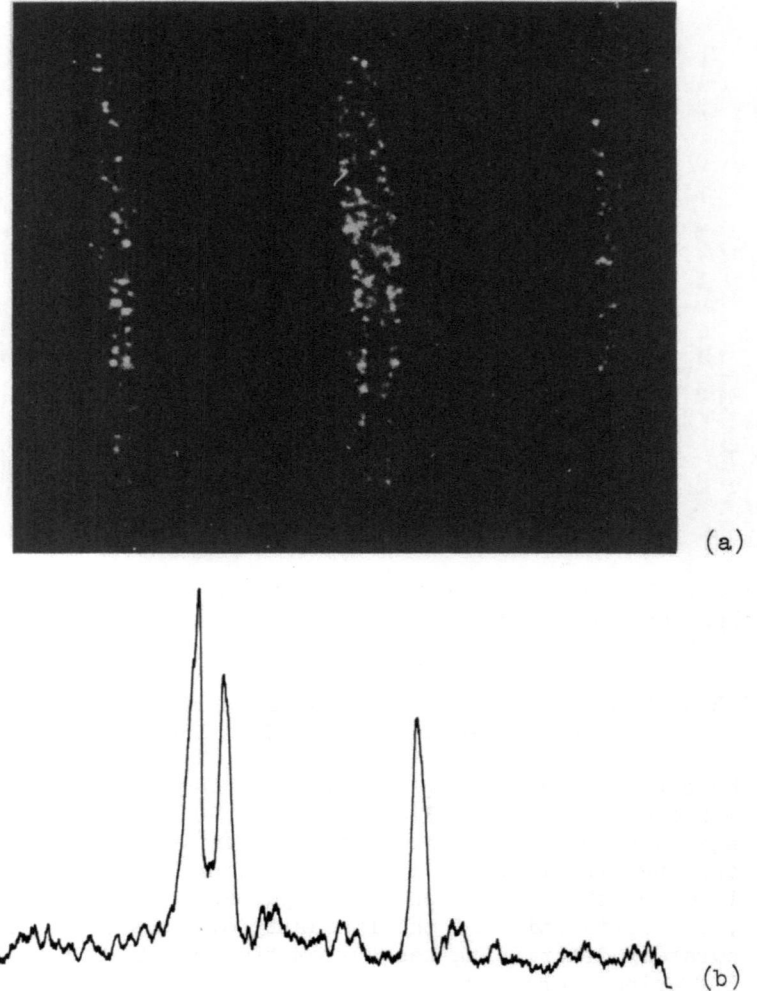

FIGURE 14. (a) Photograph of streak showing subpulses generated in the optical delay line (see Fig. (12)) from a single dye laser pulse, with separations 6 psec and 60 psec respectively.
(b) Microdensitometer trace of streak photograph of Fig. (14a). Note complete resolution of pulses separated by 6 psecs. Estimated streak-camera time resolution is ∼1 psec. (Recorded width of the laser "pulse" is ∼3 psec.)

---

sub-pulses separated by 6.6 psec, with single sub-pulses spaced 60 psec from each close pair. The central pair are clearly resolved and from the microdensitometer trace of Fig. 14(b) the

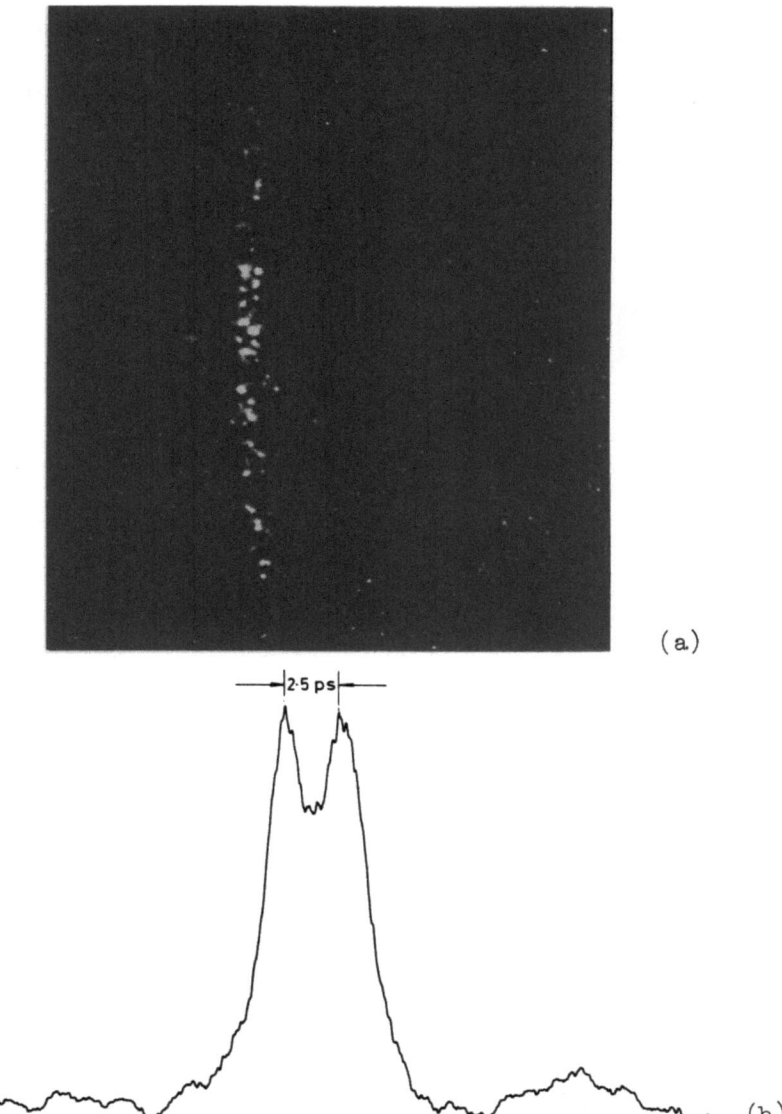

FIGURE 15. (a) Enlarged photograph of streak of Fig. 14(a) showing resolution of two pulses separated by 2.5 psecs. (b) Microdensitometer trace of Fig. 15(a).

recorded pulse half-width was measured to be 2.5 psecs. Closer examination of the central "pair" of pulses showed that, in fact, each "single pulse" from the dye laser consisted of two pulses

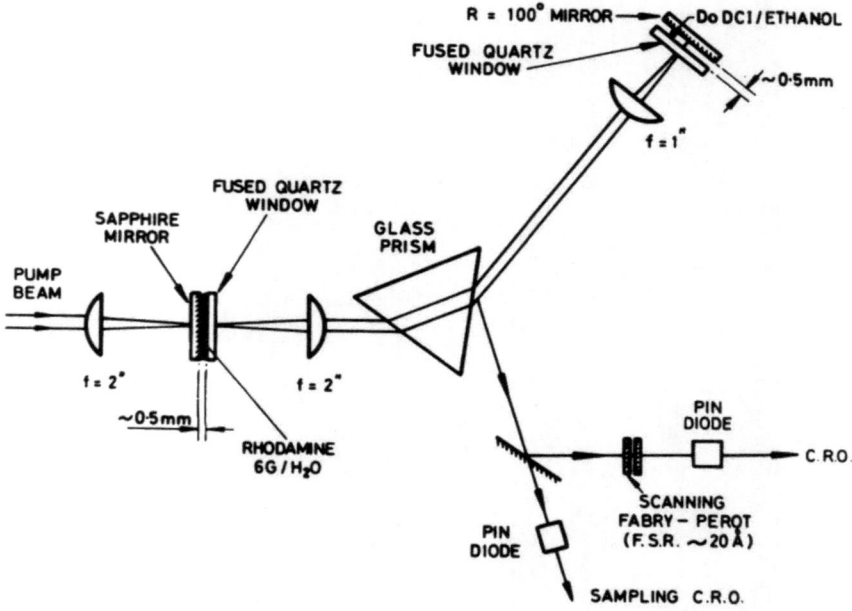

FIGURE 16. Passively mode-locked C.W. dye laser experimental arrangement.

separated by 2.5 psec. These pulses were in turn clearly resolved near the end of the streak, where the writing speed was ~10% greater and the exposure was optimum. The enlarged print and the corresponding microdensitometer trace are shown in Fig. 15(a) and 15(b). From these results, deconvolving [28] a dynamic spatial resolution limit of 1.8 psec (writing speed 1.1×10^{10} cm sec^{-1}, spatial resolution 5 line pairs/mm) and assuming a minimum laser pulse duration of 1 psec gives a time-dispersion limit of 1.4 psec. The laser had been tuned [24,25] to operate at 615 nm and for an S11 photocathode and light of this wavelength, for an extraction field of 8000 V cm^{-1} the calculated time-dispersion spread [28] is in good agreement with this value. Thus the determination of the durations of ultra-short light pulses as short as 1 picosecond is now possible. Time can then be measured to a fraction of a picosecond.

C.W. MODE-LOCKED DYE LASER

C.W. dye laser [6] have been mode-locked by employing intracavity modulators [29,30] to produce pulses of durations ~55 picoseconds [29]. Passive mode-locking of Rhodamine 6G C.W. lasers has been recently obtained [31] with the same mode-locking dye (DODCI) as employed with the pulsed lasers [24,25]. Employing the arrangement shown in Fig. 16, a passively mode-locked C.W. train

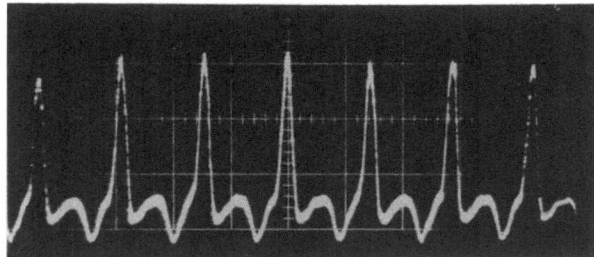

FIGURE 17. C.W. mode-locked dye laser pulse train recorded on sampling oscilloscope. Horizontal time scale 5 nsec per major division. (Sub-pulses are due to electrical ringing in detector circuit.)

of ultra-short pulses has also been achieved [32]. The rise-time of the oscillogram of Fig. 17 is instrument limited and to measure the time durations of the C.W. laser pulses, the sampling oscilloscope was replaced by a picosecond camera with a streak image tube (manufactured by Instrument Technology Ltd.) with S20 spectral response. The laser beam was focussed with a 13 cm focal length lens to a spot on the 50µ slit and each pulse of the train was sub-divided into two pulses separated by 200 psec and displaced in the direction of the slit length. Despite the low peak power (∼1 watt) of the pulses reflected from the tuning prism face, care had to be taken to avoid over-exposure. The laser was most stable and produced shortest pulses when tuned to the red (610 nm) and under these conditions isolated single pulses with high signal-to-background ratios [33], separated by the round trip time (7 nsecs) of the laser cavity were obtained. The streaked photograph and microdensitometer trace of Fig. 18 show a recorded pulse width of 10 picoseconds for a writing speed of 3×10^9 cm sec^{-1} and deconvolving the corresponding camera instrumental width gives a laser pulse duration of 6 psecs. Thus this direct linear detection of C.W. laser pulse durations gives an unambiguous measurement, unlike nonlinear correlation techniques based upon second-harmonic generation [29,31]. Moreover with such a nonlinear pulse duration measurement it is not possible to detect low intensity but long lasting sections of the laser output and the presence of a substantial proportion of the laser energy outside the ultra-short pulses themselves would not be revealed. With the aid of the streak camera it is relatively easy to optimize the C.W. laser mode-locking and it should be possible to generate continuous trains of pulses of durations <2 psecs, frequency tunable over the same spectral regions (580-700 nm) as for pulsed lasers. Transform limited pulse durations would also be expected with this C.W. laser which from the point-of-view of mode-locking operation is essentially the same as the mode-locked flashlamp pumped dye lasers. These tunable frequency, picosecond pulses should prove useful for

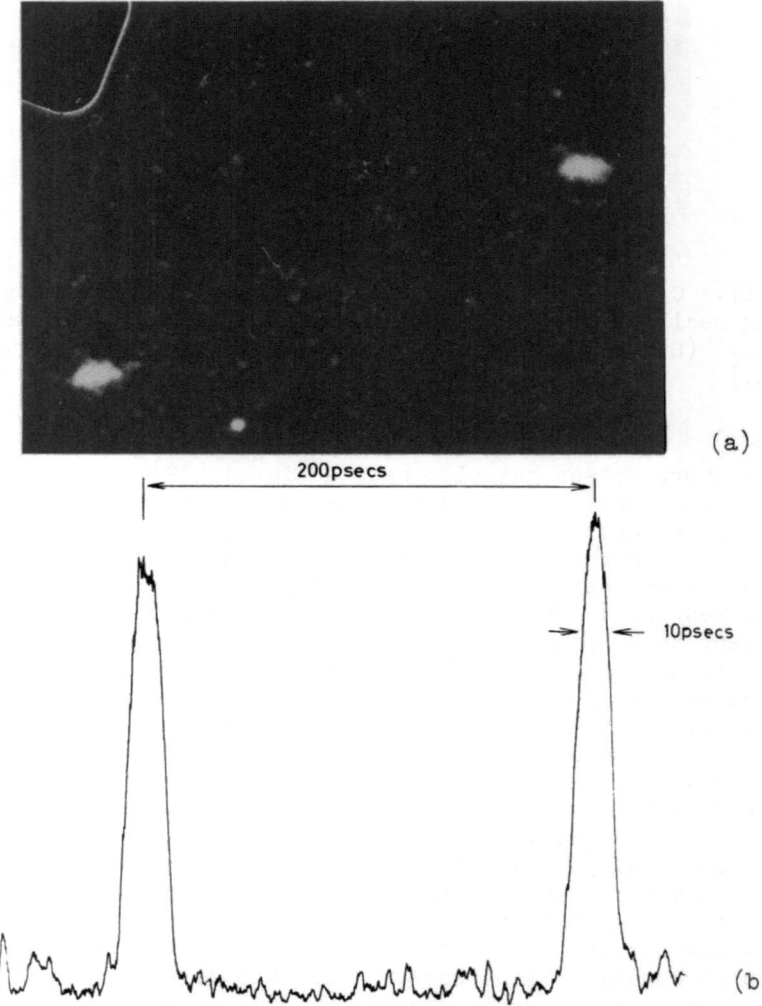

FIGURE 18. (a) Streak of two sub-pulses separated by 200 psecs, generated from a single pulse of the C.W. mode-locked dye laser. Pulses were imaged on two different positions of the streak-camera slit. (b) Microdensitometer trace of 18(a) showing recorded pulse width of <10 psecs. Deconvolving the camera instrumental width gives a laser pulse duration of ∼6 psecs.

---

quantitative investigations of molecular excited states, particularly if employed for time-resolved spectroscopy with the electron-optical streak camera as well as for studies of excited states in solids. These advances in the development of picosecond

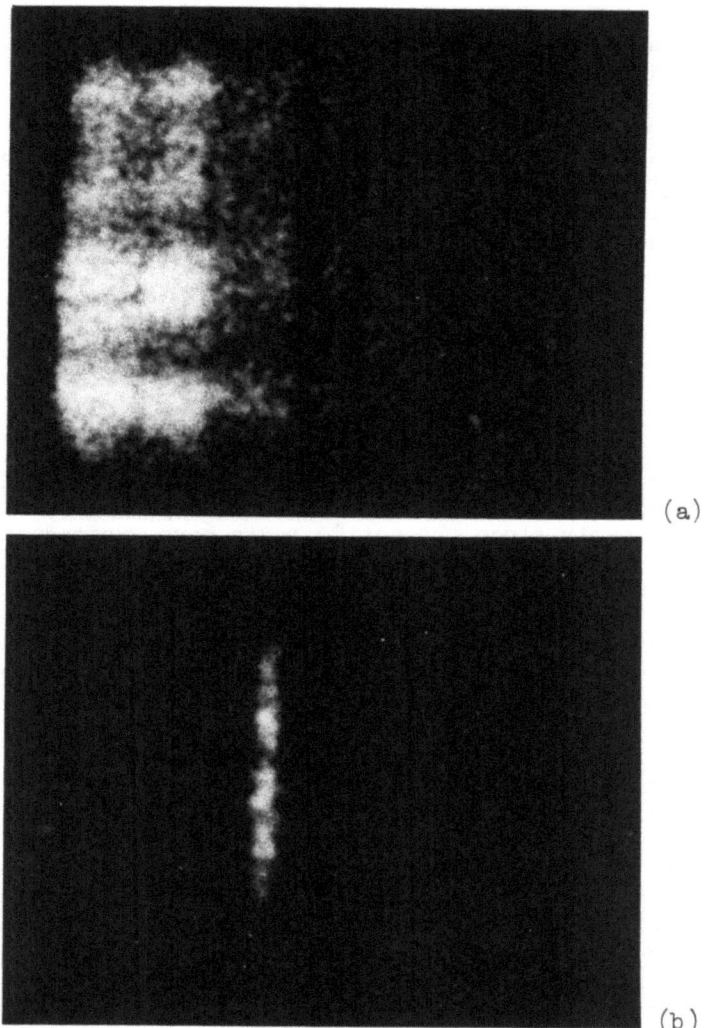

FIGURE 19. Streaks of output pulses of mode-locked flashlamp pumped Rhodamine 6G dye laser (a) 100 nsec and (b) 250 nsec from the start of laser action.

light sources and diagnostic apparatus should generally extend the experimental study of intermolecular and intramolecular excited state and relaxation phenomena [34] and in turn perhaps lead to other applications of organic dyes to laser technology. With the high powers available from the pulsed lasers directly (∼50 MW), or after amplification [8] (∼1 GW), generation of U.V. harmonics with high efficiency should extend the frequency range into the U.V. The high quality mode structure obtainable with the C.W. system should also be advantageous for efficient frequency doubling and tripling.

With the aid of the streak-camera it has also been possible to investigate [35] the effect in a pulsed mode-locked laser of bandwidth limitation by an intracavity interferometer. As expected, limitation of the lasing bandwidth is clearly seen to result in the generation of random noise spikes near the beginning of a pulse train, when the mode-locking build-up is slowed down by reducing the laser cavity Q-value.

STUDIES OF LASER MODE-LOCKING

The dye laser differs in its mode-locking behavior from ruby and neodymium solid-state lasers. Even pulsed dye lasers operate quasi-continuously since the active media have excited-state lifetimes of a few nanoseconds and, as a consequence, saturable amplification as well as saturable absorption plays a role in the mode-locking process. We have found the picosecond streak-camera a valuable diagnostic tool for the study of the build-up of the picosecond pulses from the initial laser background noise. A 1.5×10^{-4} M solution of Rhodamine 6G circulating in a 5.5 mm diameter, 21 cm long quartz laser dye cell was pumped by a linear quartz ablative flashlamp, capable of withstanding up to 2000 J input electrical energy, in an elliptical pumping reflector. The mode-locking saturable absorber dye solution (10^{-4} M DODCI in ethanol) was contained in a 2 mm path-length cell in optical contact with the 100% reflectivity laser mirror. By rotating an intracavity optically contacted 7 μm gap Fabry-Perot interferometer filter the laser was tuned to operate at 611 nm with a bandwidth of 0.4 nm. By reducing the cavity Q-value, the mode-locking was slowed down so that the peak of the pulse train envelope was reached ∿250 nsec from the start, compared with the usual build-up time of ∿50-100 nsec. The two streak photographs of Fig. 19(a) and (b) show the development of the laser pulse after 100 nsec and 250 nsec, corresponding to 42 and 70 round-trips, respectively, of the laser cavity. The series of micro-densitometer traces of streaks taken at times of 50 ns, 80 ns, 100 ns, 200 ns and 250 ns (Fig. 20) clearly show (for the first time) the development of a single picosecond pulse. Near the beginning of a pulse train the streak-camera photograph shows mainly "uniform" noise with components of durations <10 picoseconds. A fairly smooth pulse of duration ∿200 psecs appears after 100 nsecs and then an isolated single pulse of duration <5 psec evolves. When the laser is optimized for mode-locking this build-up of a single picosecond pulse inside the cavity occurs within ∿100 nsec from the beginning of laser action [24,25].

ORGANIC DYES IN LASER TECHNOLOGY

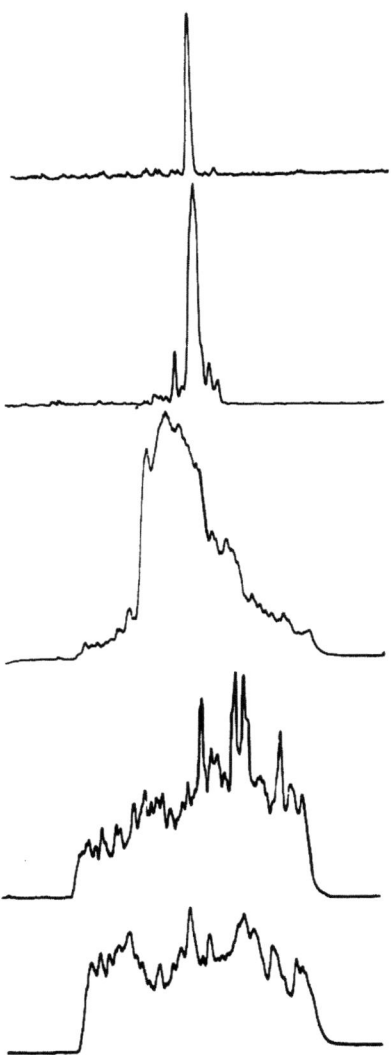

FIGURE 20. Successive stages of build-up (from bottom to top) of mode-locking in flashlamp pumped Rhodamine 6G dye laser. Total streak in each case is 500 ps. Laser was bandwidth limited with an intracavity 7 μm gap Fabry-Perot interferometer filter and the mode-locking process was slowed down by reducing the cavity Q-value.

When the interferometer is removed the laser bandwidth increases to 3 nm and, as expected, the laser noise structure was no longer resolved by the streak camera. (A bandwidth of 3 nm corresponds to a pulse structure of ∽0.3 psec.)

ACKNOWLEDGMENTS

Several members of the Queen's University laser and optics groups have been involved in obtaining the results (some which have still to be published fully) used in this paper. The principal scientific investigators were E. G. Arthurs, W. G. I. Caughey, P. Ewart, G. M. Gale, J. V. Nicholas, F. O'Neill, A. G. Roddie, J. R. D. Shaw, W. Sibbett, W. E. Sleat and J. I. Vukusic. Contributions to laser design and construction were made by W. Boyd, T. McClements, W. McDade, W. A. Montgomery, R. S. Morrison, C. Shaw and R. Stringer; chemical preparations and measurements were carried out by R. Compton; microdensitometry and photographic reproduction by R. Frame, and laser mirrors were manufactured by D. Blanc.

REFERENCES

[1] Sorokin, P. P., Lankard, J. R., Hammond, E. C. and Morruzi, V. L., "Laser-pumped stimulated emission from organic dyes: experimental studies and analytical comparisons", IBM J. Research and Develop., Vol. 11, pp. 130-147, 1967.

[2] Schafer, F. P. Schmidt, W. and Volze, J., "Organic dye solution laser", Appl. Phys. Letters, Vol. 9, pp. 306-309, 1966.

[3] Sorokin, P. P. and Lankard, J. R., "Flashlamp excitation of organic dye lasers", IBM J. Research and Develop., Vol. 11, p. 148, 1967.

[4] Snavely, B. B. and Schafer, F. P., "Feasibility of C.W. Operation of Dye-Lasers", Phys. Letters, Vol. 28A, pp. 728-729, 1969.

[5] Pappalardo, R., Samelson, H. and Lempicki, A., "Long pulse laser emission from Rhodamine 6G using Cyclooctatetraene", Appl. Phys. Letters, Vol. 16, pp. 267-269, 1970.

[6] Peterson, O. G., Tuccio, S. A. and Snavely, B. B., "C.W. operation of an organic dye solution laser", Appl. Phys. Letters, Vol. 17, pp. 245-247, September 1970.

[7] De Michelis, C., "A survey of organic dye laser research", Euratom-C.E.A. Report EUR-CEA-FC-617, November 1971.

[8] Bradley, D. J., "Recent developments in dye lasers and their applications", Proceedings of the Technical Programme, Electro-Optics '71 International Conference, pp. 1-8, 1971.

[9] Gale, G. M. (private communication).

[10] Bates, B., Bradley, D. J., Kohno, T. and Yates, H. W., "An optically contacted permanently adjusted high finesse Fabry-Perot interferometer", J. of Sci. Instruments, Vol. 43, pp. 476-477, July 1966.

[11] Bradley, D. J., Caughey, W. G. I. and Vukusic, J. I., "High efficiency interferometric tuning of flashlamp-pumped dye lasers". Optics Communications, Vol. 4, pp. 150-153, October 1971.

[12] Bradley, D. J., "Quasi-linear dispersion spherical Fabry-Perot interferograms for giant pulse laser spectroscopy", Nature, Vol. 215, pp. 499-501, July 1967.

[13] Zelenka, J. S. and Varner, J. R., "A new method for generating depth contours holographically", Appl. Optics, Vol. 7, pp. 2107-2110, October 1968.

[14] Hercher, M. and Pike, H. A., "Single mode operation of a continuous tunable dye laser", Optics Communications, Vol. 3, pp. 65-67, March 1971.

[15] Hansch, T. W., "Repetitively pulsed tunable dye laser for high resolution spectroscopy", Appl. Optics, Vol. 4, pp. 895-898, April 1972.

[16] Huth, B. G., Farmer, G. I., Taylor, L. M. and Kagan, M. R., "Tunable second harmonic generation from an organic dye laser", Spectroscopy Letters, Vol. 1, pp. 425-432, January 1969.

[17] Dewey, C. F. and Hocker, L. O., "Infrared difference-frequency generation using a tunable dye laser", Appl. Phys. Letters, Vol. 18, pp. 58-60, January 1971.

[18] Bradley, D. J., Nicholas, J. V. and Shaw, J. R. D., "Megawatt tunable second harmonic and sum frequency generation at 280 nm from a dye laser", Appl. Phys. Letters, Vol. 19, pp. 172-173, September 1971.

[19] Bradley, D. J., Sibbett, W. and Sleat, W. E., Unpublished.

[20] Bradley, D. J. and O'Neill, F., "Passive mode-locking of flashlamp pumped Rhodamine dye lasers", J. Opto-Electronics, Vol. 1, pp. 69-74, February 1969.

[21] Bradley, D. J., Durrant, A. J. F., O'Neill, F. and Sutherland, B., "Picosecond pulses from mode-locked dye lasers", Phys. Letters, Vol. 30A, pp. 535-536, December 1969.

[22] Shapiro, S. L. and Duguay, M. A., "Observation of subpicosecond components in the mode-locked Nd:glass laser", Phys. Letters, Vol. 28A, pp. 698-699, February 1969.

[23] Bradley, D. J., New, G. H. C. and Caughey, S. J., "Subpicosecond structure in mode-locked Nd:glass lasers", Phys. Letters, Vol. 30A, pp. 78-79, September 1969.

[24] Arthurs, E. G., Bradley, D. J. and Roddie, A. G., "Frequency-tunable transform-limited picosecond dye laser pulses", Appl. Phys. Letters, Vol. 19, pp. 480-481, December 1971.

[25] Arthurs, E. G., Bradley, D. J., Roddie, A. G., "Passive mode-locking of flashlamp-pumped dye lasers tunable between 580 and 700 nm", Appl. Phys. Letters, Vol. 20, pp. 125-127, February 1972.

[26] Bradley, D. J., Liddy, B., Sibbett, W., and Sleat, W. E., "Picosecond electron-optical chronography", Appl. Phys. Letters, Vol. 20, pp. 219-221, March 1972.

[27] Bradley, D. J., U.K. Provisional Patents Specification 31167/70, 1970.

[28] Bradley, D. J., Liddy, B. and Sleat, W. E., "Direct linear measurement of ultra-short light pulses with a picosecond streak camera", Optics Communications, Vol. 2, pp. 391-395, January 1971.

[29] Dienes, A., Ippen, E. P. and Shank, C. V., "A mode-locked C.W. dye laser", Appl. Phys. Letters, Vol. 19, pp. 258-260, October 1971.

[30] Kuizenga, D. J., "Mode-locking of the C.W. dye laser", Appl. Phys. Letters, Vol. 19, pp. 260-262, October 1971.

[31] Shank, C. V., Ippen, E. P. and Dienes, A., "Passive mode-locking of the C.W. dye laser", VII International Quantum Electronics Conference, May 1972. Digest of Technical Papers p. 7 and results reported at the Conference.

[32] O'Neill, F., "Picosecond pulses from a passively mode-locked C.W. dye laser", Optics Communications, Vol. 6, pp. 360-363, 1972.

[33] Bradley, D. J., Liddy, B., Roddie, A. G., Sibbett, W. and Sleat, W. E., "Direct measurement of duration and background energy content of dye laser picosecond pulses", Optics Communications, Vol. 3, pp. 426-428, August 1971.

[34] Bradley, D. J., Hutchinson, M. H. R., Koetser, H., Morrow, T., New, G. H. C. and Petty, M. S., "Interactions of picosecond laser pulses with organic molecules. I: Two-photon fluorescence quenching and singlet states excitation in Rhodamine dyes". Proc. R. Soc. Lond. A., Vol. 328, pp. 97-121, May 1972. I: Two-photon absorption cross-sections", Proc. R. Soc. Lond. A., Vol. A329, pp. 105-119, July 1972.

[35] Arthurs, E. G., Bradley, D. J. and Roddie, A. G., Unpublished.

CHAPTER 13

OPTICAL FIBER WAVEGUIDES

N. B. Hannay

Bell Laboratories, Murray Hill, New Jersey 07974

The development of optical communications systems is of great current technological interest. A system operating at the frequency of visible or near infrared light would have a capacity that is exceedingly large, and this is of course a principal reason for interest in such systems. Others include freedom from electromagnetic interference and the potentiality of achieving small size in the transmission medium. There would be many elements in such a system, including a source of light, a means for modulating the light beam, a transmission system, and a detector, as well as other components, perhaps; in this chapter we will be concerned only with the transmission aspects.

Optical beams could be transmitted in a number of ways [1]. In free space there is a serious attentuation problem due to water (fog, rain, snow, etc.) that would prevent long distance communication. Hollow metallic or dielectric guides are a possibility but the loss is sensitive to bends in the guide. Gas lenses have been investigated extensively and they might have a place in optical communication systems, although they are rather cumbersome to use. Optical lenses are a possibility for long distance systems, although there are several technical problems: there is loss at each surface and each lens of course has two surfaces, and the system has to be carefully aligned and maintained, even in the event that temperature changes and gradients occur. The most attractive scheme is a continuous dielectric, that is, a fiber [1,2].

PHYSICAL PRINCIPLES

A simple construction for a fiber waveguide is achieved in principle through a concentric structure employing two media of different refractive indices. If the outer refractive index (n_2) is less than that of the inner core (n_1) then a light beam at less than a critical angle will propagate with total internal reflection at the interface, as shown in Fig. 1. There will be an evanescent field in the cladding but no energy will be lost if the interface reflects properly. The critical angle will be given by Snell's Law, $\cos\theta_c = n_2/n_1$. There can be a number of modes, depending upon the ratio of the radius of the fiber to the wavelength λ/n_1 (in the core medium). Thus the rays for the different modes would follow different zigzag paths, like those shown in Fig. 2.

The single mode condition has been derived [2]

$$ka\left(n_1^2 - n_2^2\right)^{1/2} < 2.4 \qquad (1)$$

where a is the radius and k is $2\pi/\lambda$. When this is satisfied, only the lowest order mode can propagate and the fiber is a single mode structure. In order to achieve a single mode, therefore, the difference in refractive index should not be great; this is one reason for applying a cladding to the structure, rather than just simply using air as the outer dielectric. (Another very practical reason, of course, is to provide physical support for the inner core.) The condition for a single mode fiber can be illustrated numerically. If it is assumed that $n_1 = 1.500$ and $n_2 = 1.485$, then the critical angle is $8.1°$; if the radius of the fiber is no more than a few times the wavelength then the inequality Eq. (1) is fulfilled and only a single mode will propagate. Assuming a laser source in the visible or near infrared, this means that the core diameter would be in the range of a few microns, at most. The cladding would be chosen to be several hundred wavelengths thick, and as the evanescent field falls off exponentially with radius it will then be close to zero at the outer boundary of the cladding. Both single mode and multimode structures are of practical interest.

A single mode fiber is suitable for laser transmission. This combination avoids the delay distortion that is inherent in the different paths for different rays (modes) which are present when a nonlaser source is used with a multimode structure, shown schematically in Fig. 2; the distortion results from the different path lengths and different propagation times for the various modes. Multimode structures and electroluminescent diode sources which can radiate into many modes make an attractive combination,

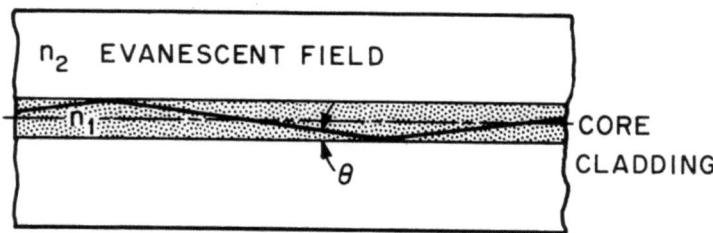

FIGURE 1. Optical waveguide utilizing total internal reflection (from Gloge [2]).

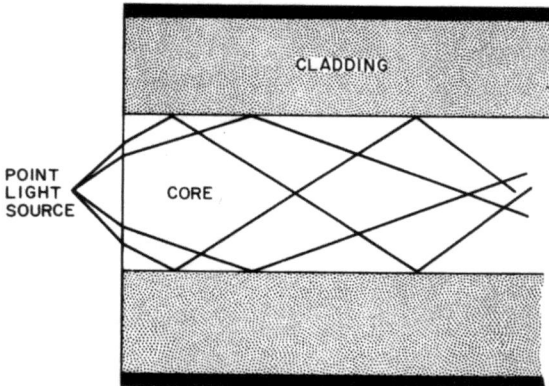

FIGURE 2. Modes in optical waveguides; internally reflected rays for two propagation modes are shown (from Li and Marcatili [2]).

however, from a practical point of view, and if the dispersion in such structures turns out to be small they may be exceedingly important in optical communications systems. For pulsed operation a single mode is essential, if the pulses are on a nanosecond scale and are to be kept separated in time, over long distances. The spreading of a pulse, in a single mode structure, due to optical dispersion in the medium, that is due to the material itself because of the dependence of refractive index on frequency, will cause each component of a message to go at a different speed. However this spreading will be in the picosecond range for laser sources, for kilometer lengths. For electroluminescent diode sources it will be greater, in the nanosecond range.

Another possible clad structure makes use of a graded index of refraction within the fiber [1-4]. Of special interest is a parabolic gradation of refractive index with radius, because this acts like a lens.

$$n = n_o(1-\Delta\, r^2/a^2) \tag{2}$$

where n_o is the refractive index on the fiber axis and $n_o(1-\Delta)$ is the index at the surface. A beam injected off axis oscillates about the axis. This structure has the interesting property that both low and high order modes have essentially the same group velocity so there is little spreading in time. Also, the ideal structure allows spatial separation of beams with the same frequency. A difficulty, however, is that this structure has a high sensitivity to systematic variations in the profile of the refractive index. Both simple clad structures and the parabolic graded-index structure are under investigation at the present time. It appears that the former is somewhat simpler to control and at least initially may be favored.

The physical principles governing propagation of optical beams in fibers apply also to optical paths in thin-film optical circuitry, or "integrated optics" (see Chapter 20).

LOSSES [5,6]

In order for fibers to be useful for long distance communication, losses must be substantially below 20 db per kilometer. The best commercial optical fibers and glasses characteristically show losses in excess of 1000 db per kilometer. This indicates that a major improvement in the quality of the optical glass is necessary for use in fibers for long-distance systems. It should be noted that loss requirements in thin-film optical circuitry are much less difficult to meet, because of the short path lengths.

Losses arise from two sources, absorption and scattering. Absorption intrinsic to the glass itself is quite small in glasses that are transparent in the visible. Absorption losses are due mainly to trace quantities of transition metal ions in the glass, although water may also be a problem (OH^- ion absorption) at certain frequencies. Relative absorptions for various impurities in glass at the wavelengths of interest for optical communications are shown in Fig. 3. The absorption depends upon the valence state of the ion; the worst cases are shown. It has been concluded that impurities must be substantially below the one part per million region in order for the absorption loss to be within the desired range. This is seen to be an achievable level of purity, in comparison with standards commonly encountered in electronic materials.

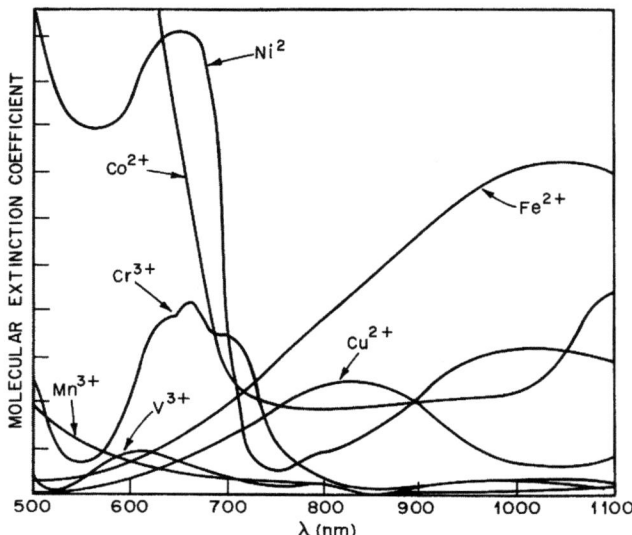

FIGURE 3. Relative absorption coefficients for several impurities.

Scattering losses may come from several sources. Rayleigh scattering is due to microscopic inhomogeneities in the dielectric. The dimensions of these inhomogeneities are small compared with the wavelength and the scattering occurs in all directions. This scattering would be absent in single crystals and it occurs in gases, liquids and glasses as a result of the lack of structural order. Rayleigh scattering would contribute a maximum of several db per kilometer loss at 0.63 µ, and about 1 db per kilometer at 0.9 µ. This then sets a lower limit on the loss to be expected in a glass system.

Scattering may also be the result of imperfections; these can be crystallites, phase separations, bubbles, or foreign particles. Thus scattering is controlled by the method of the glass preparation so as to avoid bubbles or inclusions, and by the choice of the glass system, to avoid phase separation and crystallization. These latter topics are considered in the next section. Finally, scattering may occur at the core-cladding interface if this departs from a perfect surface. The scattering can be into the surrounding medium or it can lead to a mixing of modes and a variation in mode propagation velocity. More generally, this scattering mechanism arises from any variations in the refractive index profile as a function of radius, with imperfections at the interface in the discontinuous refractive index structure a troublesome example. Control over losses due to interfacial scattering must be achieved in the fiber-pulling

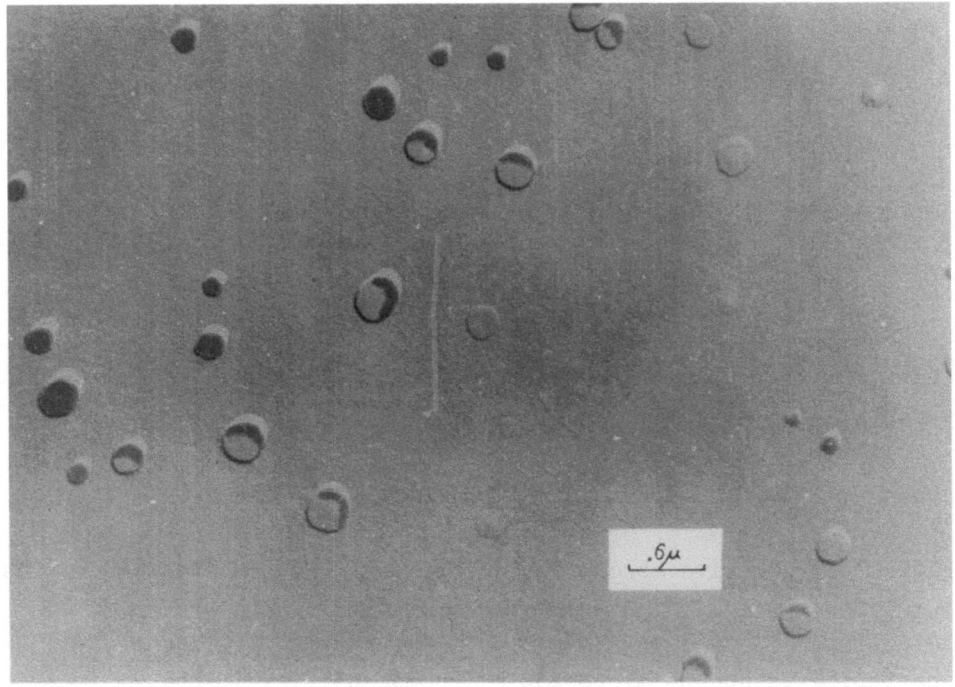

FIGURE 4. Phase separation in soda-lime-silicate glass (photo by Bagley).

process and in the cladding operation. Close attention must be paid to diameter control and to perfection and cleanliness of the surfaces.

GLASSES

Glass compositions of interest for optical fibers include pure and doped fused silica, and compound glasses. The advantage of the fused silica is that it is simple chemically and therefore purification presents less of a problem. On the other hand, it has a high melting temperature and is not as easily worked. Compound glasses, on the other hand, are more difficult to purify, but offer easy workability as a possible advantage. They also entail the hazards of phase separation (Fig. 4). The best results to date have been achieved in silica, with losses of only 5 to 10 db per kilometer in high quality bulk silica and silica fibers. The best results reported to date in compound glasses are of the order of 50 db per kilometer, although further progress may be expected as purities improve.

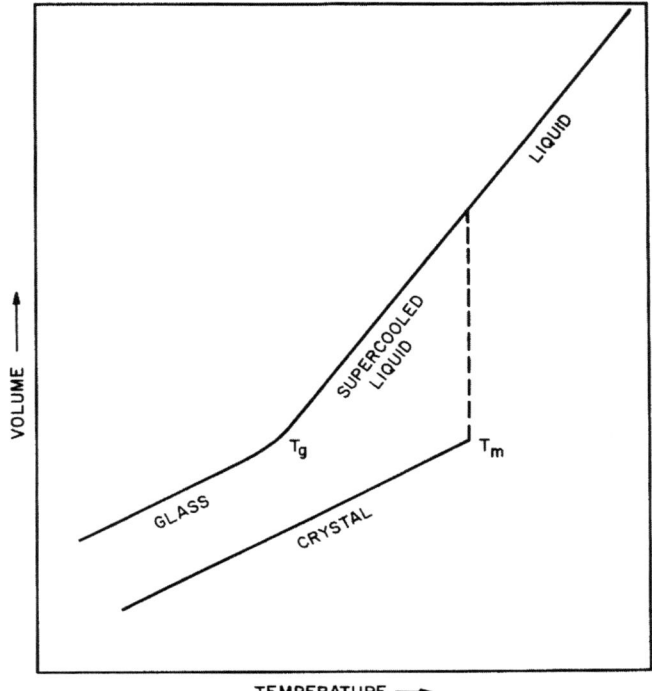

FIGURE 5. Schematic showing conditions relating to glass formation.

Pure silica obviously presents no phase separation problem, although there are limitations to the amount of doping possible, in the single-phase system. Compound glasses are chosen from a region of the phase diagram in which there is no miscibility gap, so as to avoid the possibility of phase separation. Most of the work has been in the soda-lime-silicate system, although it is possible that good results could be obtained in other systems.

The avoidance of partial crystallization is also a requirement in glasses for optical fibers [7]. Figure 5 shows schematically the significant temperatures relating to glass formation. Heating of the crystal leads to melting, at the melting point, T_m. As the melt is cooled, however, it can remain liquid and supercool below the melting point; eventually, at the glass transition temperature T_g it solidifies into a glass. In the supercooled liquid the kinetics are relatively fast, so that crystallites easily nucleate and grow. On further cooling these are frozen into the glassy solid matrix, and this results in a state of partial crystallization. Thus an optical inhomogeneity can be produced. Once the material is in the glassy state, however, the kinetics are very slow. To minimize crystallization, therefore, it is desirable to have a short T_m-T_g range. This is somewhat inconsistent with the features desired for workability, as glass is most easily handled when there is a long range of liquid with a high viscosity. The values for these quantities are shown in Table 1 for SiO_2 and a silicate glass. The particular

TABLE 1

	T_m	T_g	η_m
Silica	1710°C	1200°C	10^7 poise
73.5% SiO_2 21.3% Na_2O 5.2% CaO	725°C	550°C	10^7 poise

composition of soda-lime-silicate shown is a eutectic composition, chosen to minimize T_m-T_g. The viscosities for both systems are sufficiently high to be attractive.

FIBERS

Glass fibers are easily made by heating the end of a solid glass rod and pulling out the fiber [6]. Clad fibers can be made in any of several ways. A preform technique, Fig. 6(a), can be used, with a solid rod of glass of the higher refractive index enclosed in a hollow cylinder of glass of lower refractive index, and the two pulled simultaneously into a fiber. A second technique makes use of a double crucible, Fig. 6(b). The crucibles have nozzles at the bottom and contain glasses of different refractive indices; the fiber emerging from the inner crucible is coated by the glass contained in the outer crucible, as it is drawn. Finally [8], a clad fiber can be made by coating the inside of a glass tube with a doping impurity and then collapsing the tube into a fiber, Fig. 6(c). As it is drawn, the interior region is doped with the impurity, which is chosen so as to raise the refractive index. Each of the methods has apparent advantages and disadvantages. The preform technique is subject to difficulties resulting from imperfections at the interface which can lead to scattering losses, and the possibility of recrystallization in slow passage of the preform through the furnace. However, it is easy to do and it gives good cross-section control. The double crucible technique can result in core cross-section variations and dimensional control is not simple. However, interface problems are minimized and this

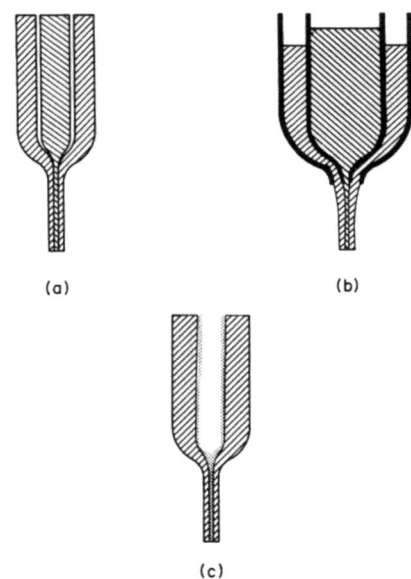

FIGURE 6. Three methods for pulling clad fibers: (a) preform method, (b) double-crucible method, and (c) collapsed tube method (after Pearson and French [6]).

is a considerable advantage. The use of crucibles may increase the possibility of contamination. Finally, the collapsed-tube technique could increase the loss in the core, where it is most important to avoid it, unless suitable dopants are used. There is a danger that bubbles will be entrapped. It may also be difficult to control the geometry. However, it is a simple technique and has been shown to give excellent results. In order to produce strain-free fibers, there must be a good match in thermal expansion and a suitable relation between the softening point and viscosity, for the two kinds of glass. Single-mode fibers are more difficult to prepare than are multimode fibers, because of the smaller core diameter, with the preform technique offering the most difficulty.

Another kind of low-loss fiber has been achieved by filling a hollow fiber with a liquid [9], and for this purpose several organic liquids, such as tetrachlorethylene, have been used. Because of the difficulty of filling small diameter fibers only multimode fibers with relatively large diameters have been made. Low-loss structures have been achieved by this means.

Graded-index structures can also be made. An attractive way of achieving this is by ion exchange methods. A system [4] in which this has been done uses lithium aluminum silicate glass. The Li^+ in the glass exchanges with Na^+ in a fused $NaNO_3$-$LiNO_3$ salt bath and this results in a parabolic distribution, with respect to radial position, of the sodium and therefore of the refractive index, also. The so-called "Selfoc" fiber is a graded-index structure [3]. Fibers with discontinuous refractive index structure may also show a grading of n as a result of diffusion during the fiber-drawing process.

Clad fibers have been made with losses low enough to make them attractive for optical communication purposes. Solid core multimode SiO_2 fibers with losses of the order of 4 db per kilometer have been made [10]. Liquid core fibers with losses of 14 db per kilometer have been reported [9]. Low loss single-mode structures have been constructed also, but no recent loss data have been published.

REFERENCES

[1] R. Kompfner, Applied Optics (Special Issue), to be published; S. E. Miller, Science 170, 685 (1970).
[2] D. Gloge, Proc. IEEE 58, 1513 (1970); T. Li and E. A. J. Marcatili, Bell Labs Record (Dec. 1971).
[3] T. Uchida, M. Furukawa, I. Kitano, H. Koizumi, and H. Matsumura, IEEE J. Quantum Electron. 6, 606 (1970); H. Kita, I. Kitano, T. Uchida and M. Furukawa, J. Amer. Cer. Soc. 54, 321 (1971).
[4] A. D. Pearson, W. G. French and E. G. Rawson, Appl. Phys. Letters 15, 76 (1969); W. G. French and A. D. Pearson, Amer. Cer. Soc. Bull. 49, 974 (1970).
[5] A. R. Tynes, A. D. Pearson and D. L. Bisbee, J. Opt. Soc. Amer. 61, 143 (1971).
[6] A. D. Pearson and W. G. French, Bell Labs Record (April 1972).
[7] B. G. Bagley, in "Amorphous and Liquid Semiconductors", J. Tauc, Ed., Plenum Publishing Corp., New York (to be published).
[8] R. D. Maurer and P. C. Schultz, U.S. Patent 3,659,915 (filed 5/11/70); D. B. Keck and P. C. Schultz, U.S. Patent 3,711,262 (issued 1/16/73).
[9] J. Stone, Appl. Phys. Letters 20, 239 (1972); IEEE J. Quantum Electron. QE-8, 386 (1972).
[10] Corning Glass Press Release (8/14/72); R. D. Maurer, European Electro-optics Markets and Technology Conf., Geneva (9/72), and NEREM Meeting of IEEE, Boston (11/72).

CHAPTER 14

LARGE AREA DISPLAY MATERIALS AND DEVICES

G. Marie

Laboratoires d'Electronique et de Physique Appliquée

94, Limeil-Brévannes, France

INTRODUCTION

This chapter attempts a survey of the state of the art in display materials and devices and an assessment of what could be the future in this field. The range both of display applications and of physical phenomena that may be employed is vast. Therefore this chapter will deal only with dynamic electronic displays in which electrical signals control a visual output which can be changed without delay. Furthermore it will also be restricted to large area displays and the accent will be put on new promising technics rather than on well known technics. Before examining in detail various large area displays, let us consider what are the main features of a display and what are the general technics involved.

Main Features of a Display

The first features are the display size and the resolution. What is the limit of large area displays? Of course, TV or data projections on screen areas ranging between 1 and 40 square meters are large area displays. They are needed in application such as military and aviation control centers, flight simulators, teaching, and for large audience displays in general. These represent, however, just a few of the potential displays of the future. Likewise flat direct view displays, which are expected for the future, with a picture area ranging between 0.2 and 1 square meter and with a resolution reaching the order of 500×500 elements, will also be large area displays. Their application field will be

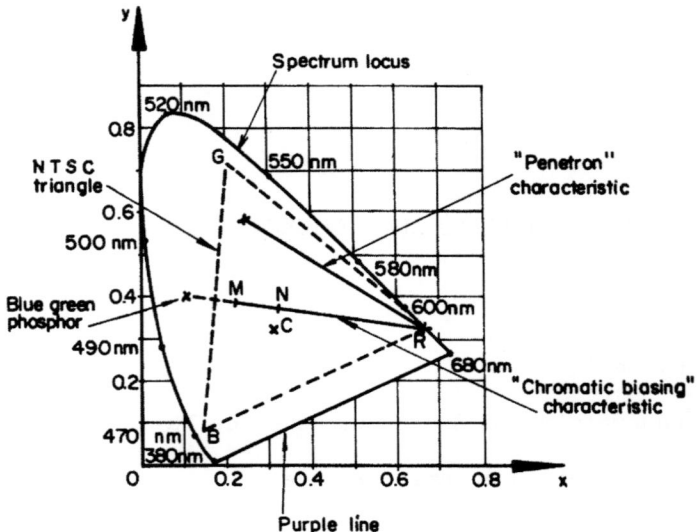

FIGURE 1. Color diagram showing the RGB triangle defined by the NTSC, a typical "Penetron" characteristic and a "chromatic biasing" characteristic obtained using a blue-green phosphor and a red external illumination.

---

very large, from home TV to all the kinds of graphic displays associated with computers. Lastly direct view cathode ray tube displays, with a picture area ranging between 0.1 and 0.2 square meter and a resolution of more than 500×500 elements can be considered as large area displays, particularly when one considers their information capacity. In fact they have been, over the past decades, the almost only displays used worldwide in applications for which one now is thinking of flat displays. However, because CRT are well known, displays involving CRT will not be examined in detail in this lecture except when new display aspects will be associated with them.

Brightness is an important feature. In this field the requirements depend on the application: in a dark room, for instance, a white area brightness ranging between 25 and 30 cd/m^2 (7 to 9 FL) is comparable to that obtained in cinema pictures and is sufficient for TV picture projections; with a Lambert law emission, this corresponds to a luminous emittance in the order of 100 $lumens/m^2$. For home TV pictures, seen in a moderately illuminated room, a brightness roughly ten times higher is necessary and, for displays in a high ambient illumination, a still higher brightness is needed.

Contrast is another main feature and can be characterized by the ratio of high level to low level output; this ratio has to be in the order of 30 to 50, for TV pictures, but can be in the order of 10 or less, for data displays without half tones. In fact the contrast needed depends on whether the data are seen on a dark or on a lit background and particularly if different colors are displayed.

The possibility of displaying colors is a very important feature. It is often thought that color displays are generally about 3 times as complex and as expensive as black and white and monochrome devices. In fact it depends on the color gamut wanted for the display and it is only true in the case of color TV pictures in which it is necessary to reproduce most of the hues included in a color triangle RGB defined by the NTSC, as shown on Fig. 1. Three independently variable parameters are then needed to reproduce those hues with different amplitudes; these parameters can be, for instance, the amplitudes of three primaries, red, green and blue, whose color coordinates are close to that of the vertices of the NTSC triangle.

The situation is completely different in the case of data displays where colors are desired only to introduce a distinction between different kinds of data, such as planes flying at different altitudes in a radar image. Then, due to the low visual acuity in the blue, it is not necessary to reproduce this color and it is generally sufficient to reproduce only hues whose representative points in the color diagram are located on a line; this requires two different variable parameters, when half tones are desired, and even only one parameter when half tones are not necessary. This can result in a device whose complexity and cost are comparable to that of a monochrome one.

A well-known example is given by the beam penetration tube, called Penetron, which is a CRT including two superimposed phosphor layers, emitting different colors, and in which the penetration depth of the electron beam can be controlled by the acceleration voltage [1]. The representative point of the emitted light is situated on a line, as shown on Fig. 1, for recent tubes, and varies for instance from red, for 6 kV acceleration voltage, to green, for 12 kV acceleration voltage, with the possibility of one or two other distinguishable hues: yellow and orange.

The Penetron itself is almost as simple as a monochrome CRT, but its use requires fast commutation of high voltages and of the scanning current almplitude. Another possibility, allowing easier use, is the employment of current sensitive polychromatic phosphors, obtained by mixing different phosphors exhibiting different luminance versus current density characteristics [2]. In such CRT, color display is obtained using only one independent variable parameter, the electron beam density, but the color range is less extended than that of the "Penetron".

It must be noted that, when a monochrome display emits a sufficiently saturated color, it is always possible to obtain a polychrome display using a method, sometimes called "chromatic biasing", which consists of an addition of a uniform light component, on the screen, coming from an external source and exhibiting a hue as complementary as possible of the original display hue [3]. As an example, Fig. 1 shows the case of a "chromating biasing" using a blue-green phosphor and a red external source. In the absence of displayed data, a mean excitation is applied to the phosphor resulting, with the addition of the external source, in a neutral hue N, near white (point C). One kind of data is characterized by the suppression of the phosphor excitation, resulting in the display of a well saturated red hue. The second kind of data is characterized by a more powerful excitation of the phosphor, resulting in a medium saturated blue-green display (point M on the Figure). It must be emphasized that data appear on a lit background and that, in order to obtain sufficient contrast, the saturation of the hues displayed must be greater than that of dark background displays (which is the case with the two phorphor screens). On the other hand, the mean consumption of the system is greatly increased; as brightness and lifetime of displays depend more on the mean than on the peak dissipation, useful brightness is in general greatly reduced, in this system; thus, it is not sure that the perception is really better than if a classical monochrome display, with a dark background, were used.

General Techniques Involved

Let us consider now what can be the general techniques involved in a large area display. Table I is an attempt to classify the different kinds of display by choosing six principal parameters: the type of display, the addressing means, the light source, the kind of modulating device used, the physical phenomena involved in the modulation and the modulating medium.

This table, which is certainly not complete, shows that a large variety of combinations is possible. Let us give some examples:

TABLE 1

Type of Display	Active			Passive	
	Direct view	Projection		Direct view	
Addressing means	Electron beam	Cross-bar	Light + photoconductor		Light deflection
Light source	Electroluminescent	Photoluminescent (UV, IR)	Gas discharge	External source ↔ Image light valve	Laser ↔ Electro-optic modulator
Modulating Device		Power supply			
Physical phenomenon	Light absorption			Light deflection diffraction or scattering	Change of polarization state
Modulating medium	Solids Powders	Particles in liquid or plastic	Liquids or plastics	Solid Membranes / Liquid crystals	Ferroelectric ceramics / Electro-optic crystals

a) the characteristics of the classical CRT display lie at the left of the table: it is an active display, observed in direct view, addressed by an electron beam, using an electroluminescent phosphor as a light source and where the signal modulation is applied to the electron beam intensity, that is to the power supply of the display;

b) on the other hand, a laser display is an active projection display where the light source is a laser whose output is modulated, for instance, by an electro-optic modulator and where the address is obtained by deflection of the light beam.

Starting from a common characteristic, a great number of different displays is sometimes possible. For instance, let us consider displays using liquid crystals as the modulating medium. The physical phenomenon involved in the modulation process can be either a scattering of the light, which appears when an ionic current passes through the medium, or a change of the polarization state of the light due to a birefringence induced by an alternating electric field. Then the medium acts as an image light-valve modulating, in space and time, light coming either from the ambient light, resulting in a passive display, or from a separate source, resulting in an active display which can work in direct view or in projection. Finally the address can be done either by an electron beam, using a CRT with a wire-mosaic face plate, by a cross-bar system, or by projecting a picture on a photoconductive layer sandwiched with the liquid crystal between two transparent plates. In this case, it can be seen that about 15 different displays can be built using the same liquid crystal medium.

In view of the great number of combinations possible we will only be able to consider here typical cases. We will, in the following sections, consider three main display classes distinguished by their type and their use: passive displays, active flat displays and projection displays, the first one being, however, of lesser importance compared to the two others.

PASSIVE DISPLAYS

Passive displays act as image light valves modulating the ambient light and their application field lies essentially in displays used in very high ambient illumination. We have just mentioned passive displays using liquid crystals and we will discuss it further on in more detail. First, however, let us consider other possibilities offered in Table 1. For passive displays, the change of the polarization state of the light, which involves the use of crossed polarizers, would result in a picture with too low brightness. On the other hand, as passive displays are seen in direct view, the area of the modulating medium has to be large; when the modulation is obtained by light deflection, diffraction or scattering, this large area is practically only possible with liquid crystals. When the modulation is obtained by light absorption, the absorption can be induced in a solid layer or can occur in a toner powder or in particles in suspension in a liquid or a plastic medium. Let us consider successively these various possibilities.

Cathodochromic Display

Cathodochromic display makes use of light absorption in color centers induced by a high energy electron beam in a polycrystalline layer. An example is the well known "Skiatron" developed for radar display 25 years ago and using an alkali halide evaporated layer of KCl where electrons, with about a 20 kV acceleration potential, induce F centers [4]. These centers consist of electrons trapped in negative ion (Cl^-) vacancies, with a band absorption around 560 nm wavelength resulting in a strong absorption of green and yellow light, the maximum contrast being in the order of 4. This tube was not widely used, due essentially to the long time, 10 seconds, needed for the erasure, performed by heating the layer. New versions of cathodochromic tubes are under study, at present, using new materials, such as bromide sodalite ($Na_4Al_3Si_3O_{12}Br$) which exhibits the same kind of F center when an electron replaces one Br^- ion [5]. With this material, the contrast ratio can exceed 10 and the thermal erasing time can be reduced to 2 seconds when the sodalite is deposited on a mica foil, placed inside the tube and heated by passing an electric current in a transparent conducting layer deposited on it.

Photochromic Display

The photochromic display makes use of a light absorption which appears in certain molecules when submitted to visible or UV radiations. The more common photochromic materials are films of organic dyes in which UV radiations produce a change in the molecular structure and which can be put against a CRT incorporating a UV phosphor and a fiber optics window [6]. Due to the cost of the window, the area of such a display cannot be large; besides the lifetime of organic dyes is limited. Use has been made also of inorganic photochromic materials incorporated in a glass inside the CRT and resulting in a longer lifetime. Erasure can be obtained in several seconds by heating or by exposure to red or infrared radiation.

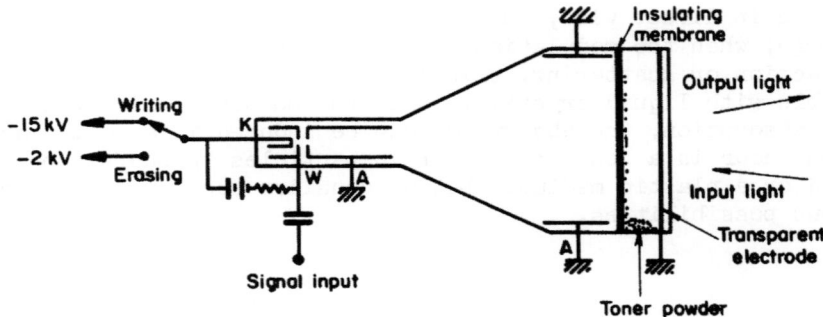

FIGURE 2. Reflection mode Xerographic display tube; negative charges are deposited on the insulating membrane during writing, using a fast electron beam; erasing is obtained by flooding the membrane with 2 kV accelerated electrons.

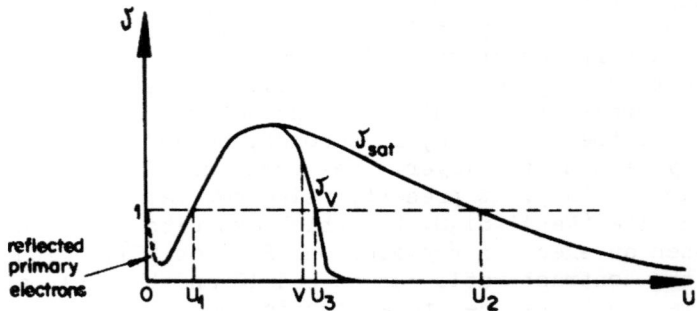

FIGURE 3. Secondary emission ratio as a function of primary electron acceleration voltage U; the δ_{sat} curve is obtained when the anode potential is always higher than the target potential; the δ_V curve is obtained when the anode potential V is comprised between the potentials U_1 and U_2 for which δ_{sat} is unity.

Xerographic Display

Passive displays where light absorption occurs in a powder make use of the Xerographic process in which powder grains are retained by electrostatic charges [7]. Such a display tube is described in Fig. 2. The target consists of a thin insulating membrane separating the tube in two parts with a vacuum maintained in each of them. Assuming that the starting potential of the membrane is that of the grounded anode, writing is obtained by scanning the surface with a highly accelerated electron beam, say by a 15 kV potential, so that the ensuing secondary emission ratio is lower than unity. Local areas of the membrane thus acquire a negative charge proportional to the beam current. Development in order to produce a visible image is obtained by "cascading" the toner powder on the second face of the membrane with the help of a mechanical rotation or of a "magnetic brush" if the powder incorporates magnetic particles. Erasing is obtained by flooding the membrane with electrons with an acceleration voltage in the order of 1 or 2 kV.

The process of writing a charge pattern on a dielectric surface and of erasing it is very commonly used, for instance in storage tubes, and can be explained with the help of Fig. 3 which shows the variations, versus the acceleration voltage U, of the secondary emission ratio at saturation δ_{sat} of a target; δ_{sat} is the ratio between secondary electrons leaving the target and primary electrons hitting it, the secondary electrons being collected by an anode at a potential always higher than that of the target. At very low and very high voltages, δ_{sat} is lower than unity. For most dielectrics, δ_{sat} reaches a maximum value, higher than unity, for a voltage in the order of 1 kV. For two voltages U_1 and U_2, called first and second cross-over and which are in the order of 50 V and 10 kV respectively, δ_{sat} is equal to unity.

In the writing mode, the acceleration voltage is chosen higher than U_2 and the points of the target hit by the electron beam get negatively charged with respect to ground, with a charge roughly proportional to the electron current. In the erasing mode, the potential difference between the cathode and the grounded anode is chosen equal to V, comprised between U_1 and U_2; the potential differences between the various points of the target and the cathode are comprised between U_1 and V. Then δ_V is higher than 1, the target loses electrons and its potential increases. When this potential difference reaches and then exceeds V, that is that of the anode, the secondary emission ratio δ_V decreases quickly, reaches 1 for a potential difference U_3, which is 2 to 5 volts higher than V, and becomes less than unity when the difference between target and anode potentials becomes higher than the mean energy of the secondary electrons. For an isolated cathode, U_3 is then an equilibrium potential which means that the target, when flooded by slow electrons, gets uniformly charged to a potential approximately equal to the grounded potential of the anode. Then the toner powder can be removed by cascading it again.

Particle Orientation Display

Another passive display working similarly to the previous one uses light absorption in a suspension of plate like opaque

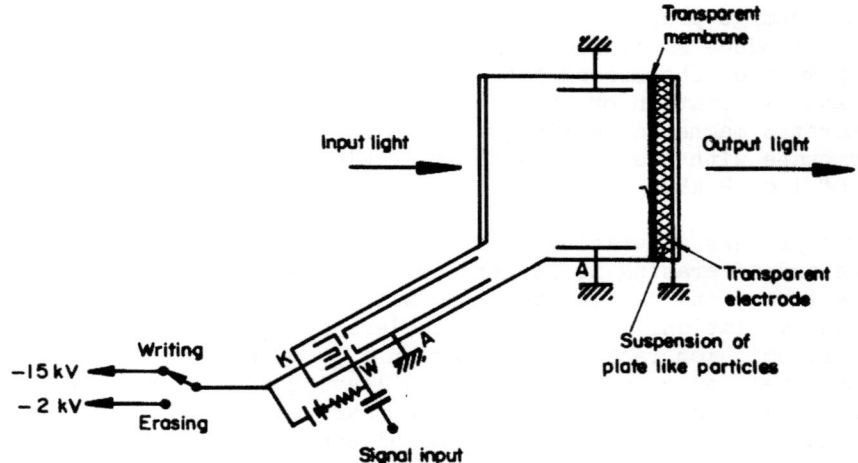

FIGURE 4. Transmission mode particle orientation display tube; writing and erasing operations are similar to that of Fig. 2.

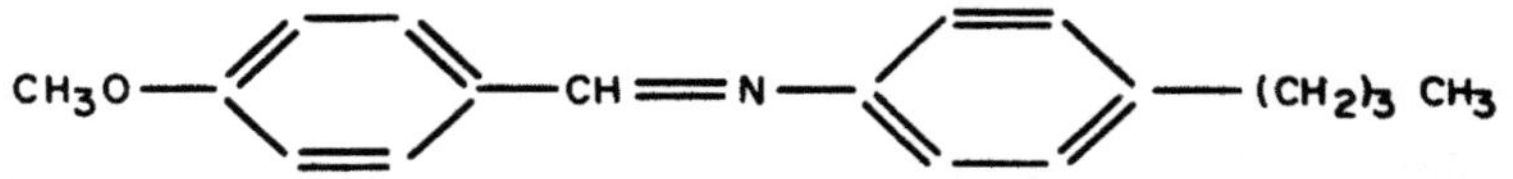

FIGURE 5. Formula of the methoxybenzilidenebutylaniline (MBBA) nematic liquid crystal.

---

particles [8,9]. Such a display operating in transmission mode is described on Fig. 4. Disc-shaped particles, in suspension in an oil, are placed between the face plate and a mica sheet which could be replaced by a wire mosaic plate. When an electric field is applied, the particles, initially randomly oriented, tend to rotate so that their long dimension is parallel to the field, resulting in a strong decrease of the absorption. The electric field is locally generated by charges deposited by an electron beam following the mechanism described in the case of the Xerographic display. As previously, erasing is obtained by flooding the target with low energy electrons.

In practice, the various types of light absorption passive displays just described did not have a large development due particularly to their slowness and were generally built only in small dimensions. They are better adapted to storage applications; in this field it is interesting to mention a version of the last device described where particles are in suspension in a thermoplastic material heated during writing and then cooled down in order to record pictures [10].

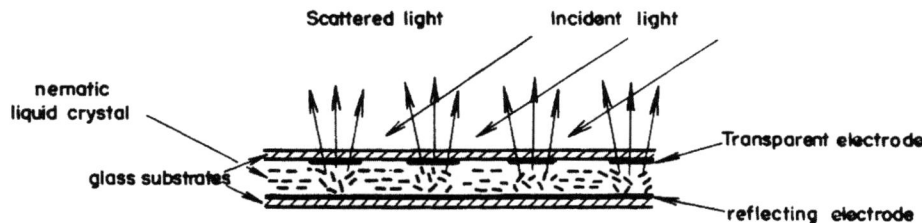

FIGURE 6. Dynamic scattering operation in a nematic liquid crystal cell.

Nematic Liquid Crystal Passive Display

For passive large area display, the possibilities offered by the scattering of light in liquid crystals are much more promising. Use is made, in general, of liquid crystals in the nematic phase where long rod-like molecules are arranged with their long axes parallel [11]. The liquid crystal is sandwiched between two conductive coated plates, one of them being transparent, spaced between 5 and 50 μm apart. Use is made usually of mixtures comprising in general one or several nematic liquid crystals whose molecules have two aligned phenyl rings; Fig. 5 shows, for example, the formula of the methoxybenzylidenebutylaniline, called simply MBBA.

Such a liquid crystal in a nematic phase is optically equivalent to a birefringent uniaxial crystal with refraction indices n_o and n_e, the optical axis being parallel to the long axis of the molecule; electrically it exhibits a dielectric anisotropy, that is a difference between the dielectric constants $\mathcal{E}_{||}$ and $\mathcal{E}_{\perp}$, for electric fields respectively parallel and perpendicular to the molecule axis. In the absence of an electric field, the molecules of the liquid crystal are oriented by wall effects and the liquid crystal is transparent. When a voltage is applied between the electrodes, a negative ion current appears which induces a molecular flow from cathode towards anode [12]; when the voltage is increased, the flow becomes turbulent resulting in the scattering of light as seen on Fig. 6. This phenomenon, called "dynamic scattering", occurs for electric fields in the order of 10^4 V/cm, corresponding to a voltage of 10 V for a 10 μm thickness, and for frequencies lower than the relaxation frequency of the molecules which lies in the 0.1 to 5 kHz range. When the electric field is suppressed, the scattering effect disappears with a time constant ranging between 0.1 and several seconds.

During the last four years a lot of results on liquid crystal displays have been published. Lifetimes up to 5000 hours were

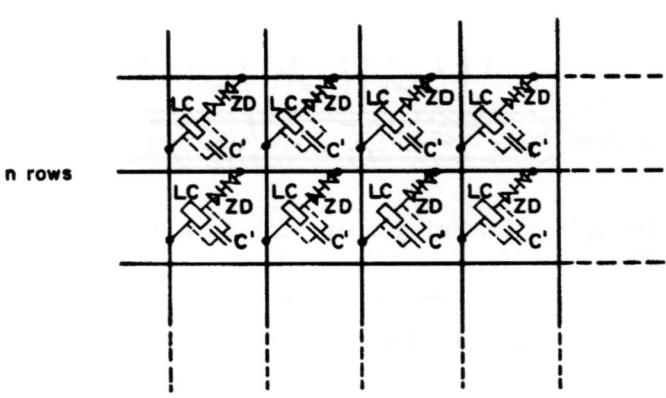

FIGURE 7. Cross-bar matrix addressing of liquid crystal cells LC using two zener diodes ZD in series with each cell and, if memory is desired, an additional capacitance C' in parallel.

---

obtained [13] using a.c. driving, at 60 Hz, which avoids the electrolysis effect of d.c. driving. In the case of passive displays operated in a reflection mode (Fig. 6) the contrast ratio depends strongly on the angular distribution of the ambient light and on the angular aperture in which the display is observed; in practice contrast ratios ranging between 5 and 40 have been obtained.

Liquid crystal displays seem well adapted to cross-bar XY addressing. For low resolution displays, the existing threshold in the contrast versus voltage characteristic curve [13] allows direct XY control of the cross-bar. Nevertheless this direct control is not possible anymore for high resolution displays, due to a lack of threshold in the admittance of the cells. Thus, for a n rows and m columns matrix, the admittance between a row and a column is about $\frac{nm}{n + m - 1}$ times the admittance of the addressed element. To overcome this difficulty several possibilities have been described in the literature; a first classical possibility is to introduce a diode in series with each element; in a better solution, shown on Fig. 7, the diode is replaced by two zener-diodes ZD in opposition exhibiting a threshold U equal or superior to the maximum useful signal [14]; the addressing signal is then equal to the sum of U and of the useful signal V. The disposition of Fig. 7 allows the a.c. operation by alternating the sign of the addressing signal at each frame; moreover, by addition, in parallel, of a capacitance C' to each liquid crystal cell LC, the discharge time constant of each element could be made higher than the frame period, and the display could work with memory between two successive addressing instants. Of course, the zener-diodes could be replaced by field effect transistors.

For the moment the decay time of liquid crystal limits operation to 10 frame/s. In the future, it is expected that TV rates of 25 to 30 frame/s could be achieved. But the decay time is not the only limitation and it is necessary to take into account the rise time and, also, a delay time between the excitation and the response, which are in the order of milliseconds. In fact, if the time constant RC of each cell, with or without additional capacitance C', exceeds the sum of the rise and delay times, an addressing pulse of 60 μs duration is sufficient. It is then possible to address the whole TV picture in 33 to 40 ms by using a line-at-a-time addressing, that is by applying, simultaneously to the m columns, the m different signals corresponding to one line, during 60 μs.

The nematic liquid crystal large area passive display panel seems therefore very promising but needs a large effort in the development of the control circuitry. Due to the large dimensions, diodes or transistors would probably have to be made by thin-film semiconductor evaporation techniques, using the backplate of the liquid crystal as a substrate for the circuitry. At present inexpensive solutions have not been found for all the problems encountered.

Cholesteric Liquid Crystal Passive Display

Another kind of liquid crystal offers interesting properties for displays: the cholesteric liquid crystals in which the molecules have their long axes in the plane xy of the layers of molecules and are rotated from layer to layer as indicated on Fig. 8 with a pitch equal to p along the z direction. The cholesteric phase reflects white light to give brilliant iridescent colors, which is attributed to light scattering by the ordered molecules. When the pitch p of the structure is equal to the wavelength λ of a visible radiation propagating along z, this radiation is strongly reflected.

The pitch of the structure depends on the temperature. It can also be modified by applying an electric field in the z direction; this always results in the reduction of the pitch as shown in Fig. 9. Thus, if the liquid crystal reflects the red, in the absence of an electric field, it can reflect the green and then the blue, when an electric field is applied. Experiments have been conducted using electron beam addressing [15] and

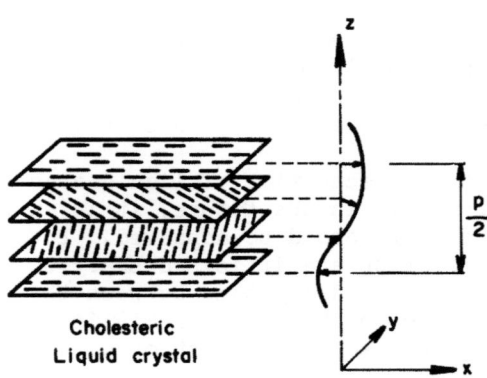

FIGURE 8. Cholesteric liquid crystal; molecules are rotated from layer to layer with a pitch p along the z direction.

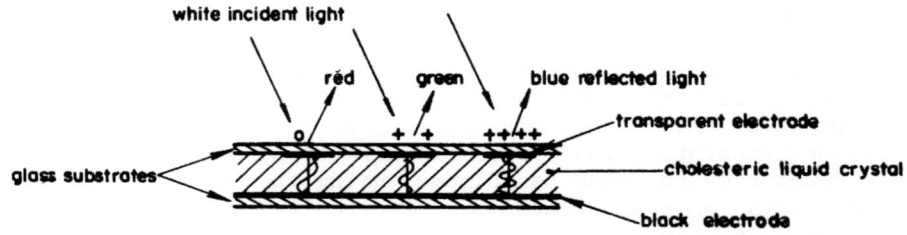

FIGURE 9. Cholesteric liquid crystal cell; the pitch p and, consequently, the hue of the reflected light depend on the electric field.

---

cholesteric mixtures such as cholesteryl chloride (30%) - cholesteryl oleyl carbonate (70%); red, yellow, green and blue reflections were obtained by applying a voltage ranging from 0 to 150 V for a cell thickness comprised between 10 and 20 μm. One can think that, in the future, flat passive color display could be made using cholesteric liquid crystals. It must be noted, however, that, due to the higher voltages needed, the difficulties encountered for the realization of the addressing matrix would be even greater than in the case of nematic liquid crystal displays.

FLAT ACTIVE DISPLAYS

Let us look again at the Table 1 to examine the possibilities offered for flat active displays. Such displays are observed in direct view and addressed by means of a cross-bar matrix. In the case where the light source energy comes from the power supply of the display, the light can be generated by:

- electroluminescent semiconductor diodes;
- electroluminescent phosphors;
- photoluminescent phosphors excited by infrared emitting diodes;
- photoluminescent phosphors excited by UV emitting gas discharges;
- gas discharges emitting visible light.

In the case where light comes from an external source the only practical possibility for large dimensions is offered by liquid crystals. However, in contrast with passive displays, two physical phenomena can be used for modulation in a liquid crystal active display:

- dynamic scattering as previously described;
- electrically induced birefringence resulting in a change of the polarization state of a linearly polarized incident light.

Liquid Crystal Display

Let us consider first the active liquid crystal display. When use is made of dynamic scattering, the operation is similar to that described in the case of passive displays, but the panel is illuminated by a source which can be powerful enough to give a high brightness display, without difficulty. Color can be obtained either by different monochromatic sources, if the display operates in transmission, or by colored band filters with operation in transmission or in reflection; color possibility results, however, in a complication of the display and of the control electronics. Besides, operation in transmission raises a very difficult problem of transparency when the cross-bar addressing system involves the use of diodes or transistors on each display element. Perhaps, in the future, a solution could be found in the use of transparent nonlinear elements, such as ferroelectric capacitances, in a way similar to the present research in the field of electroluminescent phosphor flat displays, as we will see in a moment.

The second way for utilizing nematic liquid crystals was proposed more recently [16,17]. We stated above that such a liquid crystal was optically equivalent to a birefringent uniaxial crystal with an optical axis parallel to the long axis of the molecules. Usually, in a liquid crystal cell and in the absence of an electric field, the molecules orient themselves parallel to the electrode planes. However, by choosing proper compounds or mixtures of liquid crystals and proper surface states of the electrodes, bondings can occur which result in an orientation of

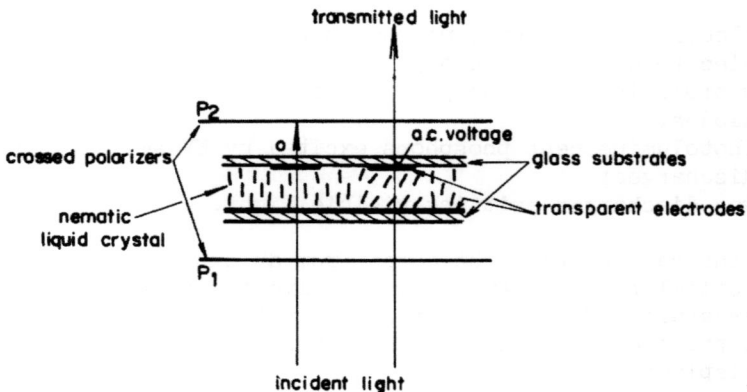

FIGURE 10. Nematic liquid crystal cell operating in polarized light and making use of the birefringence induced by the tilt of the molecules submitted to an a.c. electric field.

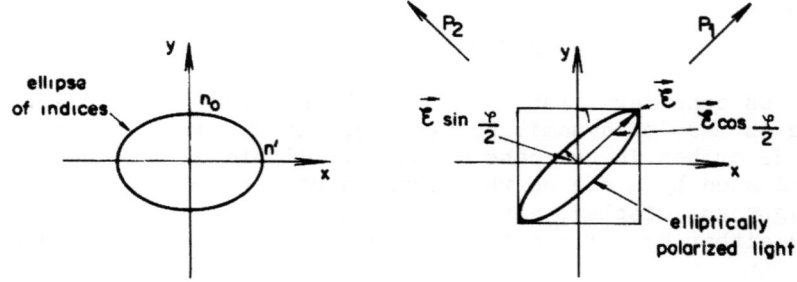

FIGURE 11. Left: optical indicatrix in the xy plane of the liquid crystal cell. Right: elliptically polarized light emerging from the cell, for an input light polarized parallel to the first bisectrix of x and y.

---

the molecule axes perpendicular to the electrode planes, as shown on Fig. 10. The cell is then equivalent to an uniaxial crystal plate perpendicular to its optical axis and no light is transmitted when this plate is examined between two crossed polarizers, the light propagating roughly parallel to the optical axis. Indeed, for an electric vector of the incident light perpendicular to the optical axis, the index of refraction is equal to the ordinary index n_O and the light is not perturbed.

Now let us look at what appears when a voltage is applied across the cell. If the molecules present a negative dielectric anisotropy, as is the case with MBBA, that is if $\varepsilon_{\|}$ is lower than $\varepsilon_{\perp}$, then the molecules tend to orient their axes perpendicular to the electric field. If the voltage applied is an a.c. voltage at a frequency higher than the relaxation frequency, for instance 10 kHz, the ionic current and the associated dynamic scattering would not occur and the only effect will be a tilting of the molecule axes, as shown on Fig. 10. Then the direction of the light is no longer parallel to the optical axis and, in the plane xy of the cell, the variations of the index can be described by an ellipse, the index for an electric vector parallel to y remaining equal to n_o while the index for an electric vector parallel to x becoming equal to n', comprised between the ordinary and the extraordinary indices n_o and n_e, as shown on Fig. 11. There appears a birefringence $\Delta n = n' - n_o$, which increases when the tilt of the molecules increases. If the input polarizer P_1 selects the light whose electric vector $\vec{\mathcal{E}}$ is parallel to the first bisectrix of x and y axes, this light is then divided into two components with vectors parallel to x and y (Fig. 11) propagating with different velocities. At the output of the cell, these components exhibit a phase difference equal to:

$$\varphi = 2\pi \frac{l \Delta n}{\lambda} ; \qquad (1)$$

then, they combine themselves to give, in general, an elliptically polarized light with the axis of the ellipse proportional respectively to:

$$\cos \frac{\varphi}{2} \quad \text{and} \quad \sin \frac{\varphi}{2} ; \qquad (2)$$

if the output polarizer P_2 is crossed with the input one, as shown on Fig. 11, the only transmitted component is that whose amplitude is proportional to $\sin \frac{\varphi}{2}$. The transmittance is then equal to:

$$T = \frac{r^2}{2} \sin^2 \frac{\varphi}{2} = \frac{r^2}{2} \sin^2 \frac{\pi l \Delta n}{\lambda} \qquad (3)$$

where the factor $\frac{1}{2}$ comes from the polarization of the light and the factor r from the yield of each polarizer for the transmitted component. When "Polaroids" are used, r ranges between 0.75 and 0.8 and the value of $r^2/2$ is close to 0.3.

For a given electric field the birefringence Δn is not the same for all the wavelengths of visible light. Its variation depends on the dispersion of the medium but, in practice, the variation of Δn versus wavelength can be neglected in comparison with the variation of the wavelength λ itself, in the above

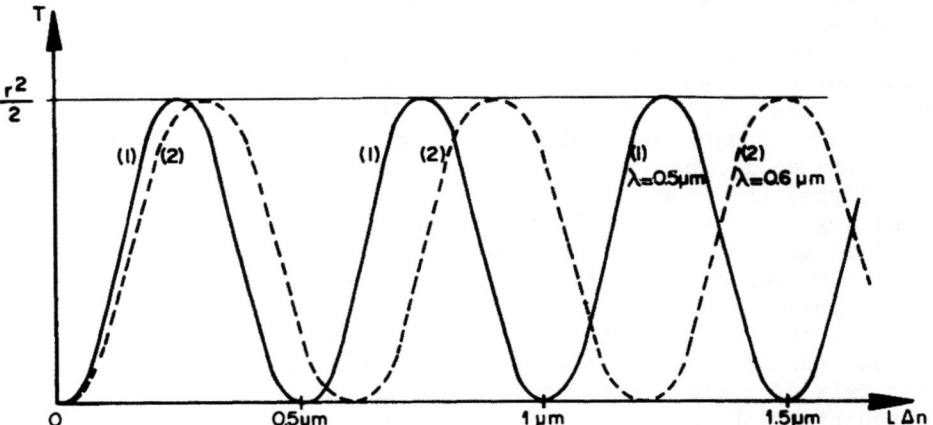

FIGURE 12. Transmission of the nematic liquid crystal cell versus the retardation $l\Delta n$ for 0.5 µm (1) and 0.6 µm (2) wavelengths.

formula. In a first approximation, φ varies inversely with λ and the transmission characteristics, versus the retardation $l\Delta n$, are close to that shown on Fig. 12 for 0.5 and 0.6 µm wavelengths respectively. It must be noted that $l\Delta n$ is not proportional to the voltage applied to the cell; there is a threshold corresponding to a voltage in the order of 4 to 5 volts and the product $l\Delta n$ can reach 1.5 µm for a voltage in the order of 10 to 15 V, for instance. If the range of variation of $l\Delta n$ is limited to 0.2 µm, then the characteristics are practically quadratic which means that they are roughly proportional. For a white light source, the transmitted light is also white, with a slight dominance of the blue components, and no observable change in the hue appears during modulation.

In contrast, if the variation of $l\Delta n$ reaches 1.5 µm, for instance, the transmission of the green component, at 0.5 µm, is zero and that of the red component, at 0.6 µm, is maximum. Then, by choosing a variation range of the signal corresponding to a variation range of $l\Delta n$ between 1.2 and 1.5 µm, it is possible to display data in green, yellow and red hues, as in the case of "Penetron tubes". Of course, in this type of use, the amplitude modulation of the light is replaced by a hue modulation and half tones cannot be reproduced; consequently this type of display could be very useful for data displays but will not be convenient for full color displays. Besides, it must be recalled that this type of liquid crystal display, which has to be used in transmission, raises a lot of problems concerning the manufacturing of the addressing matrix.

FIGURE 13. Electroluminescent phosphor display using light amplification.

Electroluminescent Phosphor Display

In the category of self-light-emitting flat displays, the first which comes to mind is that using phosphors incorporated in an XY matrix and excited by direct or by alternating voltages. An interesting compound is, for example, zinc sulfide doped with copper, halogen and manganese [18]. Zinc sulfoselenide has been also proposed. TV picture display experiments have been carried out with 60 μs pulse excitation, well adapted to the line-at-a-time addressing. The electroluminescent phosphor was incorporated between two glass substrates supporting respectively the X and Y electrodes of the matrix. Cross-talk was avoided either by the use of a non-linear impedance layer, in barium titanate, sandwiched with the phosphor layer [19], or by the use of a very non-linear electroluminescent material [20]. The experimental devices exhibited a yellow-orange luminescence with a contrast ratio of about 20 and a highlight brightness of 30 cd/m^2 (10 FL), allowing comfortable observation in a dark room only. The consumption was in the order of 100 W for a 23×30 cm^2 screen, which corresponds to an overall efficiency close to $7.5 \cdot 10^{-2}$ lm/W that is about 100 times less than cathode ray tube efficiency. Other experiments have been carried out using ferroelectric capacitances as power transfer elements with an approximately square hysteresis loop [21]. Due to the dissipation inside these capacitances, the overall efficiency was only in the order of 3.10^{-2} lm/W, with a highlight brightness of 55 cd/m^2 (16 FL) and a good contrast.

These performances are far from those required for home television which needs highlight brightness in the order of 300 cd/cm^2 for comfortable observation in a well-lit room. As it is difficult to increase voltages and currents carried through the addressing circuits, solutions incorporating a light amplifier have been proposed [22]: above a phosphor incorporating a cross-bar similar to the one previously described, are placed a photoconductive material and another electroluminescent phosphor sandwiched between "power" electrodes, the upper one being a transparent conductive layer, as shown on Fig. 13. An a.c. high voltage is continuously

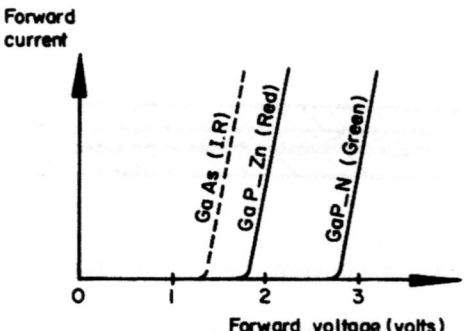

FIGURE 14. Light emitting diode forward characteristics.

applied between the "power" electrodes. A light blocking layer, conductive only in a direction perpendicular to its surface, is placed between the photoconductive and the electroluminescent materials so that the photoconductor is only sensitive to light coming from the cross-bar. By a proper choice, the photoconductor could exhibit a decay time of the order of 20 ms, allowing a high brightness of the "power" phosphor, with an excitation time of a fraction of μs only, which would allow the use of a simple element-at-a-time addressing. On a small sample a brightness of 170 cd/m^2 has been obtained which is not far from the desired brightness; however, the overall efficiency was still very low. Probably, in the future, electroluminescent phosphor display panels will be usable for home television only if the phosphor efficiency exceeds 1 lm/W; that value would correspond to a mean consumption of the order of 300 W/m^2 for a picture with a 300 cd/m^2 highlight brightness and with a mean brightness of about a third of this value.

Light Emitting Diode Display

At a first glance, electroluminescent semiconductor diodes seem well adapted to matrix addressing due to two main reasons:

- they are conducting in only one direction and their forward characteristics exhibit a threshold, as shown on Fig. 14, which allows direct XY addressing without the need of supplementary diodes or transistors on each element;

- their yield increases, in general, with the driving current, which means that they are well adapted to a sequential addressing.

However, three main difficulties appear for the realization of large area displays:

- the cost;
- the low efficiency of the diodes;
- the low impedance of the driving circuits.

Light emitting diodes can be relatively cheap in the case of alphanumeric displays where a number of diodes, 35 for instance, corresponding to 1 character, are simultaneously made on the same semiconductor substrate. In the case of large displays, sufficiently large semiconductor substrate does not exist and techniques, such as beam lead, where individual diodes are arranged in a matrix, have to be used. At present, however, it is not sure that economic solutions are available for displays including 500×500 elements, for instance.

The efficiency of the diodes is also a problem. As above, the same lower limit of the luminous efficiency, 1 lm/W, can be given, which corresponds to a yield of about 0.2% in the green. At present, the yield of green emitting nitrogen doped gallium phosphide diodes reaches 0.6%, at 560 nm, and that of the red emitting zinc doped gallium phosphide diodes reaches 6%, at 690 nm, corresponding respectively to luminous efficiencies of about 4 lm/W, in the green, and 1 lm/W, in the red. These characteristics are therefore sufficient for monochrome or dichrome displays. In the case of TV trichrome displays, the luminous efficiency of each primary can be in the ratio of its contribution to white, that is about 30%, 60% and 10% for red, green and blue respectively; this means that 0.1 lm/W, corresponding approximately to a yield of about 0.2%, would be sufficient for a blue emitting diode. Unfortunately adequate blue electroluminescent diodes do not exist at present [23].

However a promising solution is expected from combinations of an infrared emitting diode and photoluminescent phosphors, allowing the up-conversion from infrared to visible light [24,25]. Such phosphors are generally based on mixed rare-earth fluorides or tungstates. In an Ytterbium-Erbium compound, for instance, Yb^{3+} ion has a strong absorption at 0.97 μm and, when an Er^{3+} ion is at a short distance from it, 10 Å for instance, there is a large probability of transfer of energy from an excited Yb^{3+} ion to Er^{3+}. Moreover, as the Er^{3+} first excited level has a relatively long lifetime, in the order of 1 ms, and as Er^{3+} possesses a second excited level distant from the first one by an energy corresponding to about the 0.97 μm wavelength, there is a probability of transfer to Er^{3+} of the energy of a second photon absorbed by Yb^{3+}, as shown on Fig. 15. The Er^{3+} emits a green radiation, at 0.54 μm, when it returns to its fundamental level.

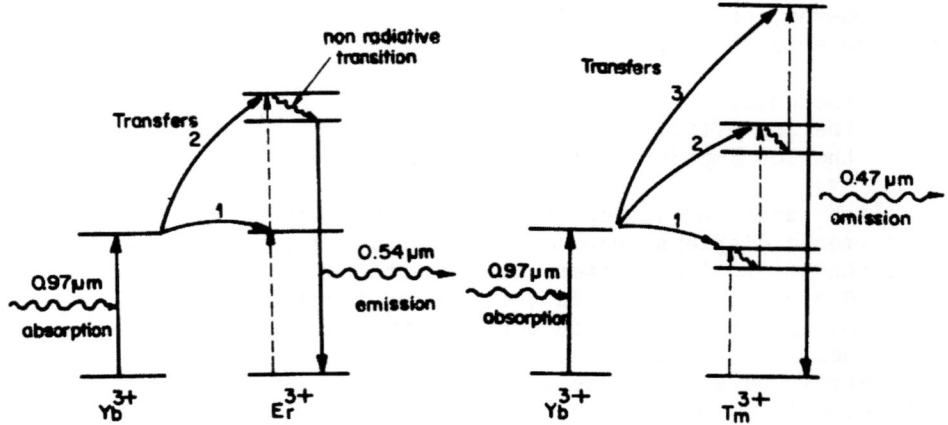

FIGURE 15. Up-conversion from infrared to visible light involving a two-step (Yb-Er compound) or a three-step (Yb-Tm compound) transfer of energy in photoluminescent phosphors.

When excited by a GaAs diode, the overall efficiency reaches 0.2%, at present. By a better coupling and a better adaptation of absorption and emission spectra, 1% efficiency for green light is expected in the future.

A similar transfer energy mechanism with one more step occurs also, as shown on Fig. 15, in Ytterbium-Thulium compounds, allowing the emission of a 0.47 μm blue radiation with an overall efficiency of 0.01% at present; in the future, an overall efficiency of 0.1 to 0.2% is expected, thus solving the problem of the blue source.

The low impedance of the sources raises another problem in the realization of a large area diode matrix. Indeed, as shown on Fig. 14, the excitation voltage ranges between 1 and 3 volts. In the case of a matrix of 500×500 green emitting diodes, with an area of 0.1 m^2, for example, a highlight brightness of 300 cd/m^2 corresponds to a luminous emission of about 90 lm. With an efficiency of 3 lm/W, which is the best value at present, it corresponds to a peak consumption of 30 W that is 10 amperes for 3 volts. In the case of an element-at-a-time addressing, each diode and each X or Y bar must be able to withstand this 10 A current; in the case of a line-at-a-time addressing, the maximum diode and X bar current is only 0.02 A but the Y bars must be able to withstand 10 A again. As, for the same brightness, the current is proportional to the area, the low impedance of the light emitting diodes limits the area of the displays using them.

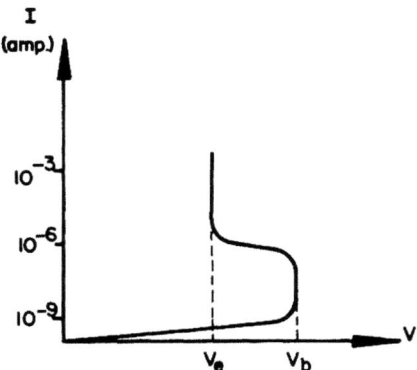

FIGURE 16. Typical current/voltage characteristic of a gas discharge cell.

Gas Discharge Display

Due to their higher impedance and their expected low cost, the gas discharge panels do not seem limited, in area, as are the semiconductor diode matrices. A gas discharge cell consists of two electrodes in an envelope containing gas at a pressure ranging between 0.1 and 0.5 atmosphere. A typical current/voltage characteristic is shown on Fig. 16; it exhibits two useful thresholds: the breakdown and the extinction potentials, ranging typically between 200 and 300 V for the first one V_b and between 100 and 200 V for the second one V_e. The existence of these two thresholds allows not only an easy cross-bar addressing but also storage capabilities.

Light is emitted from the discharge as a result of excitation and ionization of the gas and radiative deexcitation and recombination of ions and electrons. The emitted light spectrum depends on the gas or the gas mixture used. In the case of neon, the light has a characteristic red-orange hue and the efficiency ranges between 0.2 and 1 lm/W which is just sufficient for large area displays seen in a well lit room, as was pointed out above.

There are two main types of gas discharge panels based on d.c. and a.c. driving respectively. In the first case, when an addressed pulse is applied between a row and a column of the XY matrix, a delay of the order of 10 μs occurs before breakdown takes place which limits the practical use of the continuous refresh mode to displays with a maximum of 200×200 elements. For larger displays it is preferable to operate in a storage mode which can be performed by continuously applying a voltage comprised between V_b and V_e, between the rows and the columns. Due to the

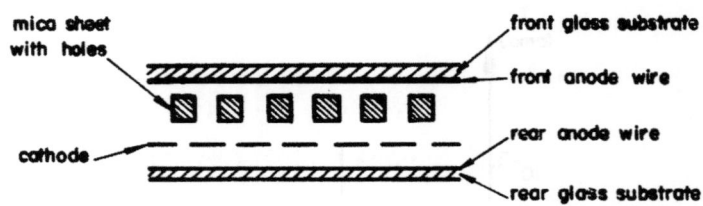
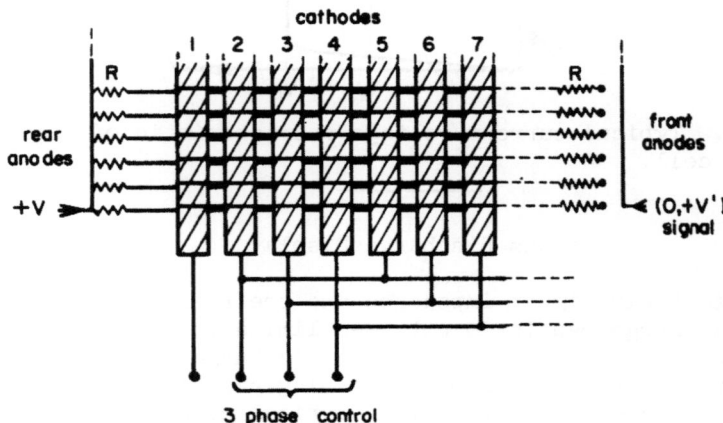

FIGURE 17. Principle of the "Self-Scan" panel; it makes use of a rear cross-bar continuously scanned by means of a three-phase control.

negative slope of the characteristic curve (Fig. 16), it is then necessary to limit the current by adding a resistor in series with each individual cell, while only one resistor by row was necessary without storage. These resistors can be manufactured as an integral part of the panel; nevertheless the manufacturing process is complicated and large d.c. panels with storage have not yet been achieved.

A promising version of the d.c. gas discharge is the "Self-Scan" panel [26] whose structure is shown on Fig. 17. It incorporates a double cross-bar with one X cathode array between two Y anode arrays situated on the front glass and the rear glass respectively. There are only four cathode leads, connected respectively to:

- the 1st cathode;
- the cathode with $(3n + 2)$ indices;
- the cathode with $(3n + 3)$ indices;
- the cathode with $(3n + 4)$ indices;

where n is an integer or zero.

The rear cross-bar is continuously scanned; at the beginning a negative pulse with an amplitude higher than $(V_b - V)$ is applied to the first cathode, called reset cathode. Then glow discharges appear between the first cathode and all the rear anodes which are connected to the voltage V through the resistances R. At the end of the first pulse a second pulse is applied to the cathodes with $(3n + 2)$ indices; then, due to the presence of excited atoms coming from K_1 in the neighborhood of K_2, a glow discharge occurs at K_2 before K_5, K_8, etc; then, due to limiting resistances R, the applied voltage drops to V_e preventing the ignition of cathodes K_5, K_8, etc. The complete scanning of the panel is thus obtained by applying a pulse successively to lead numbers 1, 2, 3, 4, 2, 3, 4, 2, 3... The glow discharge occurring in the rear cross-bar is not visible from the front face of the display; thus a visible dot appears only when a discharge occurs in the front cross-bar through the matrix of holes in the mica sheet. Due to the diffusion of excited atoms around the lit cathode, the front discharge occurs when a positive voltage V' is applied to the desired front anodes and firstly in front of the lit cathode; then, due to the presence of resistances R, the applied voltage drops to V_e' preventing the ignition between the front anodes and the other cathodes. The front anodes can be addressed either sequentially or simultaneously. Alphanumeric displays using the "Self-Scan" principle are already marketed and it may be expected that, in the future, larger area displays will be made.

The a.c. driving technique [27] offers storage capabilities without needing the use of individual resistors in series with each cell. They are replaced by a capacitive impedance provided by a glass insulating layer deposited on each array, as shown on Fig. 18. An a.c. sustaining voltage, at a frequency in the order of 50 kHz, is applied, via capacitances, between the rows and the columns. The peak voltage between rows and columns is such that, taking into account the capacitive dividing effect of the insulating layers, the voltage $V(t)$ between the cell walls is below the breakdown voltage V_b. Assuming that no discharge has previously appeared, nothing occurs.

To initiate a discharge between a row and a column, a pulse is applied symmetrically, using the X and Y switching circuits, at an instant t_1 so that this pulse and the a.c. component add up to give a resultant voltage higher than V_b between the cell walls (Fig. 18); then, the breakdown occurs, for a short period, providing an electric charge to the insulating layer capacitances. At the following half cycle, the voltage corresponding to this wall charge is in the same direction as $V(t)$, initiating a new breakdown of the cell and a new wall charge in the opposite direction. Thus, without the need of an additional starting pulse, this process is repeated at each half cycle giving an

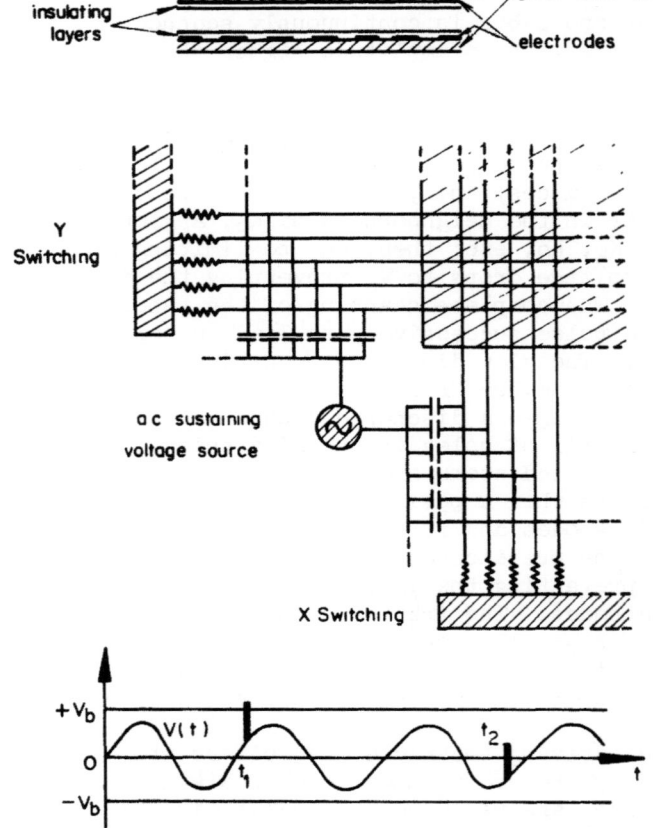

FIGURE 18. Principle of the a.c. gas discharge panel operating in storage mode; the a.c. voltage is continuously applied; firing and extinguishing of a discharge is obtained by the superposition of a pulse, applied between a row and a column, which is added to (instant t_1) or subtracted from (instant t_2) the a.c. component.

---

indefinite storage capability. The discharge can be extinguished using a pulse of the same polarity, coming from the same switching circuits, if this pulse is applied at an instant t_2 (Fig. 18), so that it subtracts from the a.c. component and turns off the discharge; indeed, if t_2 is chosen so that the last wall charge just compensates the previous one, the situation becomes again identical to that before the initiation of the discharge and the cell no longer fires. The characteristics of the addressing pulses are: a duration of about 2 μs and an amplitude of 200 V; 512×512 element panels have been made with an area of 22×22 cm^2; larger area will certainly be available soon.

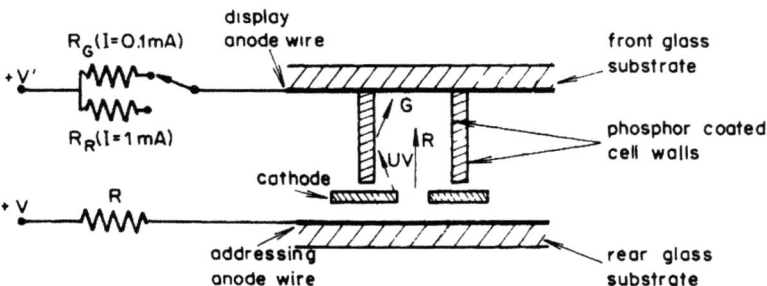

FIGURE 19. Dichrome discharge panel using the "Self-Scan" system; at low current (0.1 mA) neon emits chiefly U.V. radiations which induce the green emission of the phosphor layer; at high current (1 mA) neon emits chiefly red-orange radiations.

In the future, polychrome displays can be expected using stimulation of different phosphors incorporated in the displays, by ultraviolet emitting discharges [28], in a way similar to that used in a fluorescent lighting lamp filled with mercury vapor. Of course, for full TV color displays, the matrix will have to incorporate three times more elements. For data displays, a simple dichrome discharge panel has been proposed recently [29], combined with the "Self-Scan" system, as shown on Fig. 19. It makes use of the fact that, in neon filled cells, the light is emitted chiefly in the U.V. part of the spectrum at low current density, whereas it is in the red-orange part, at high current density. Then, by commutation of the limiting resistances R_G and R_R, the current in the display cell can be adjusted either at 0.1 mA, for U.V., or at 1 mA, for red-orange. The cell walls are coated with a green emitting phosphor whose emission is the only one visible at low current; at high current its emission is saturated and is exceeded by the red-orange discharge emission. In a manner similar to that used with "Penetron" tubes, the emitted light hue could be progressively changed from green to yellow and red-orange by adjusting the cell current, which makes this simple system very promising for color data displays.

PROJECTION DISPLAYS

Let us consider now the active projection displays and look at Table 1 once more. In the category of self-light-emitting displays the large brightness needed for projection limits the choice to CRT using electroluminescent phosphors excited by a high energy electron beam, which was the first high-resolution system

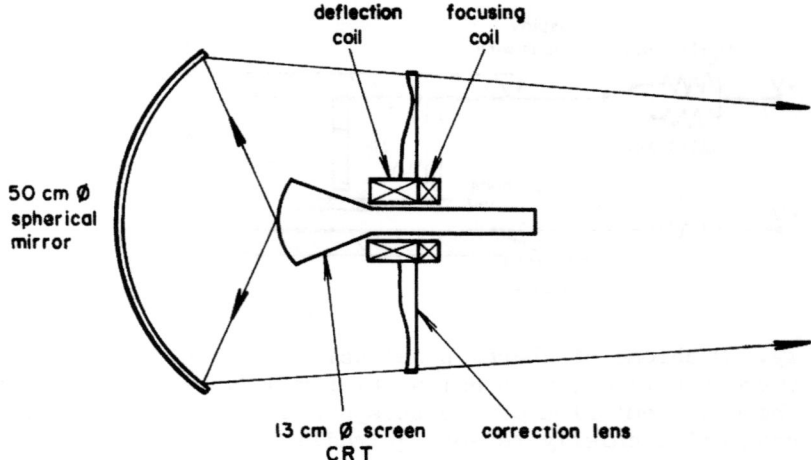

FIGURE 20. Projection display using a cathode-ray tube and a Schmidt optics.

---

used formerly for the projection of TV pictures. At the opposite end of the Table there are displays using the deflection of a laser beam which is well-known for its high brightness. The third and most densely populated category concerns image light-valves modulating a powerful cinema type lamp.

In this last category one could consider displays based on light absorption and using the same principles as those described for passive displays, such as photochrome, Xerographic or particle orientation displays. Nevertheless it must be emphasized that the average luminance of a picture is in general a third of the peak luminance which means that, in average, at least two thirds of the light power focused on the modulating target is dissipated in it; this raises the problem of the energy evacuation which is solved, in the case of cinema, by the movement of the film. Thus light absorption displays are not well suited for large screen projection and we will consider only, in the third category, image light valves operating either by a change of the direction or by a change of the polarization state of the light and using liquid, plastic or solid mediums. It can be noted that mediums such as liquid crystals and ferroelectric ceramics offer the two modulation process possibilities.

Projection Using Cathode Ray Tubes

The first TV projections using CRT date from 1948 [30]. Due to the large emitting aperture, corresponding approximately to a

Lambertian law, the most efficient collection-projection system is the Schmidt system using a spherical mirror and a correction lens, as shown on Fig. 20. Taking into account the useful aperture, the tube shadow effect and the other light losses, the efficiency of this system is of the order of 25%. In contrast with the other projection systems which will be described, the CRT has a consumption proportional to the mean luminance of the picture, instead of the peak luminance. In order to have a realistic comparison between the different system performances, we may assume that the ratio of peak to average luminance of the pictures is generally close to 3. In 13 cm diameter projection cathode ray tubes, using a 50 kV acceleration voltage [31], the maximum acceptable mean electrical dissipation of the screen is about 25 W with an air flow cooling. Assuming a luminous efficiency of 8 lm/W for a white projection tube phosphor, the average light flux can reach:

$$25 \times 8 \times 0.25 = 50 \text{ lm}$$

which would correspond, in comparison with other systems, to an equivalent peak output of about 150 lm. As with a non-directive screen, in a dark room, a peak screen brightness of 30 cd/m^2 is obtained with a peak illumination close to 100 lm/m^2, a projected area of 1.5 m^2 can be obtained with a CRT projector. Of course several square meters areas could be reached, using a directive screen, but the useful field of observation is then reduced and the number of possible viewers is not appreciably increased.

The color TV version makes use of three separated systems with red, green and blue phosphor tubes respectively. Registration is difficult and needs a dynamic correction of the trapezoidal distortions. The luminous efficiencies of the phosphors adapted to projection tubes are in the order of 5 lm/W for red, 16 lm/W for green and 2.5 lm/W for blue. The light flux limitation comes here from the red tube which has to give about 30% of the whole luminance. The mean output can reach:

$$\frac{25 \times 5 \times 0.25}{0.3} \approx 100 \text{ lm}$$

corresponding to an equivalent peak output of about 300 lm and a screen area of 3 m^2 with a non-directive screen.

These examples show that it is difficult to reach large areas using CRT, due to the limitation of the acceptable tube screen dissipation. Improvements have been proposed, such as the use of silica windows or the use of sufficiently heat conducting windows cooled by means of a water jacket. The use of such improvements results in an increase of the light output and thus of the projected picture area but reduces strongly the tube lifetime. On the other hand it must be emphasized that their is a danger of failure of the projection tube, when the displayed picture contains some bright straight lines on a dark background.

Laser Display

Lasers are high brightness light sources. The high directivity of the beam and its small cross-section allow an easy deflection at standard TV rates; light deflectors can be mechanical for instance, using vibrating or rotating mirrors, or acoustico-optical, using diffraction by ultrasonic waves in a liquid or a solid. The video signal can be applied to an electrooptic modulator using, for example, the Kerr effect in nitrobenzene or the Pockels effect in a potassium dihydrogen phosphate crystal (KDP). For large screen TV displays, however, lasers' possibilities don't seem very attractive, due to their low efficiencies. Indeed the yield of c.w. lasers emitting in the visible does not exceed about 10^{-3}, at present. Table 2 shows, for example, the yields η of a typical Argon-ion laser for the main lines emitted [32], with the indication of the luminous efficienty Y corresponding to each wavelength. The sum $\Sigma\eta$ gives an overall yield of 10^{-3} and the sum $\Sigma\eta Y$ gives a luminous efficiency of only 0.21 lm/W. Taking into account the maximum transmission of the system, at most 0.7, and the relative projection time, about 0.7 with the usual 20% of the time for line flyback and 10% for frame flyback, the overall efficiency drops to 0.1 lm/W. With a non-directive screen and a peak luminance of 30 cd/m^2 corresponding to an illumination of 100 lm/m^2, the specific consumption of the system is then about 1 kW per square meter of screen. The consumption would therefore be 10 kW for a 10 m^2 picture area.

For TV color displays, it can be seen, from Table 2, that the two most efficient rays of the Ar$^+$ laser can give green and blue primaries, at 514.5 and 488 nm wavelength. The red primary can be obtained using a Krypton-ion laser, at a 647 nm wavelength. Figure 21 shows the corresponding color triangle compared to that of the NTSC standard and of a typical shadow mask tube; it can be seen that the blue primary coordinates are far from those needed which means that the rendering of blue and purple hues will not be satisfactory. On the other hand, due to the excentric position of the NTSC white reference point C in the laser color triangle, its reproduction would make a very inefficient use of the green component of the Ar$^+$ laser resulting in a low overall efficiency; in order to increase the latter, the displayed white will probably be greenish; taking into account the consumption of the second laser and the necessary balance between the energies of the three color primaries, the overall consumption is about

LARGE AREA DISPLAY MATERIALS AND DEVICES

TABLE 2. Ar^+ Laser Main Lines.

	λ(nm)	η	Y(lm/W)	ηY(lm/W)
	476.5	0.7×10^{-4}	82	6×10^{-3}
B →	488	5.5×10^{-4}	120	66×10^{-3}
	498.7	0.8×10^{-4}	190	15×10^{-3}
G →	514.5	3×10^{-4}	410	123×10^{-3}

$\Sigma \eta = 10^{-3}$

$\Sigma \eta Y = 0.21$ lm/W

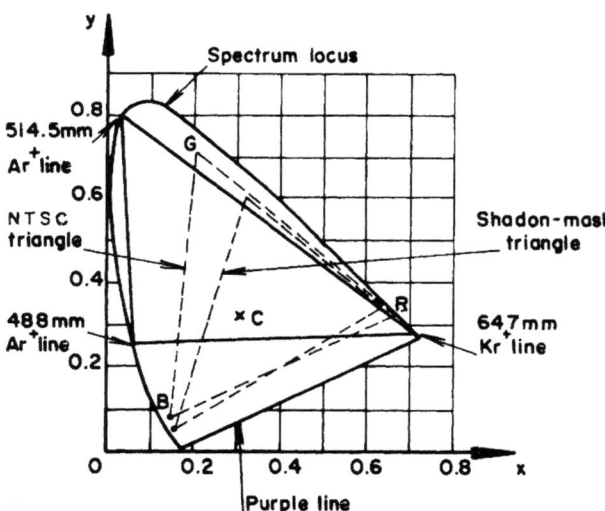

FIGURE 21. Color diagram showing three color triangles corresponding respectively to Kr^+ and Ar^+ laser lines, NTSC primaries and shadow-mask phosphors.

doubled compared to the single color projection; thus a maximum illumination of 100 lm/m² on the screen is obtained with a consumption in the order of 2 kW per square meter of screen.

Such low efficiencies tend to indicate that laser TV projection won't find a widespread use as long as a substantial improvement of the laser yields is not obtained. Nevertheless, in the case of data displays, where the bright areas represent one tenth or less of the whole surface, present laser efficiencies can be sufficient if a random-access scanning system is used; however the complexity of such scanning systems is, in general, much greater than that of standard saw-tooth scanning systems.

Image Light Valve Projection Displays

Image light valves are the most promising techniques for large screen projections when the modulation process does not involve a light absorption phenomenon, as mentioned above. In order to classify the various devices we will distinguish three main types of modulation processes:

- light deflection or diffraction on a surface;
- light scattering in a film or a plate;
- induced birefringence in a film or a plate.

(i) <u>Light deflection or diffraction devices</u>. In this modulation process, use is made of deflection or diffraction of the light induced by the deformation of the surface of a liquid or a plastic film or of a solid membrane. The well known example of such a system is the "Eidophor" whose operating principles were proposed by F. Fischer as early as 1939 and whose development was completed around 1950 [33,34]. In order to transform the light deflection into an amplitude modulation of the light, all these devices make use of "Schlieren optics" various forms of which are shown on Figs. 22 and 23 for transmission and reflection operation respectively.

On Fig. 22(a), the principle of operation is described: the transparent medium T, whose surface can be deformed by electrostatic forces induced by electric charges deposited by an electron beam EB, is placed between two lenses L_1 and L_2. The input light passes through a diaphragm D_1 placed in the front focal plane of L_1. In the absence of any deformation on T, the lenses L_1 and L_2 make, in the back focal plane of L_2, an image of the aperture of D_1 which coincides with a second diaphragm D_2 whose shape is the negative of D_1; the magnification is equal to the ratio of the focal lengths f_2/f_1, and the light is completely stopped by D_2. If deformations exist on some points of the surface of T, the light will be deflected while passing through T and the corresponding light rays will make an image of D_1 which does not coincide anymore with D_2. Thus a certain amount of light coming from these points passes through D_2 and can be focussed on a screen

LARGE AREA DISPLAY MATERIALS AND DEVICES

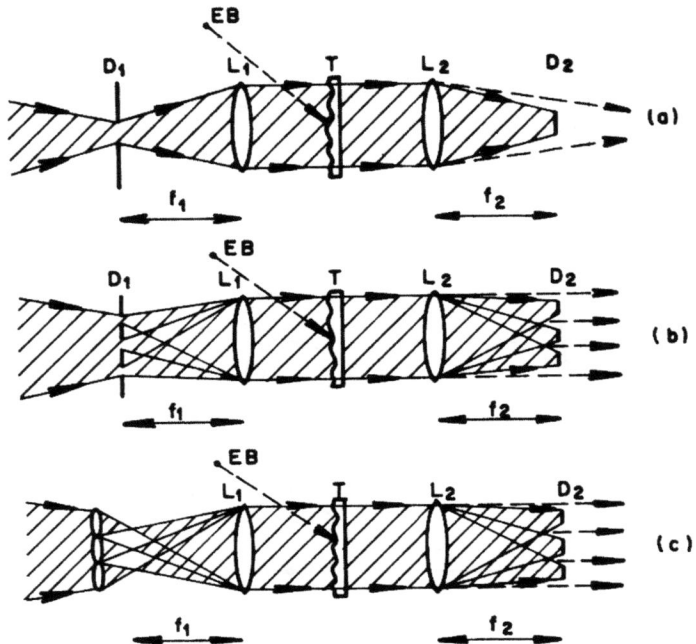

FIGURE 22. Transmission mode "Schlieren optics" system using a deformable medium T placed between two lenses and incorporating: (a) circular input D_1 (positive) and output D_2 (negative) diaphragms; (b) grating shaped diaphragms D_1 and D_2; (c) cylindrical lenses in place of D_1 in order to avoid a loss of the input light.

---

using an objective; this objective can be added to the system, after D_2, or can be placed in lieu of the lens L_2, if the distance between this lens and the target T is about equal to the focal length f_2. The deflection, and hence the amount of light projected on the screen, increases when the deformation is increased resulting in a reproduction of the picture including half tones.

Figure 22(b) shows an improvement in which both diaphragms have the shape of a grating resulting in an increased sensitivity of the modulation process. On Fig. 22(c), the first grating is replaced by a series of cylindrical lenses in order to avoid the loss of half of the incoming light, as would be the case with the set up of Fig. 22(b). It must be noted that the use of such lenses increases the aperture angle of the input light, resulting in a decrease by at least a factor of 2 of the luminance of the light in the input plane.

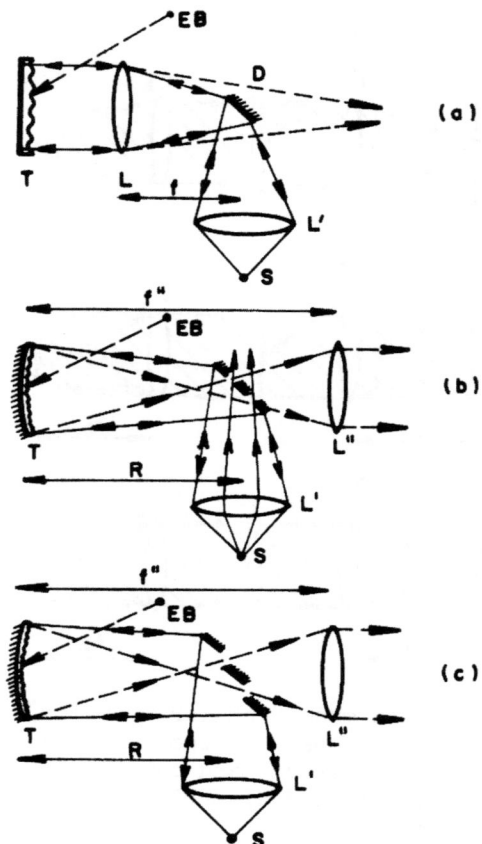

FIGURE 23. Reflection mode "Schlieren optics" system incorporating: (a) a simple mirror-diaphragm D, a lens L and a deformable medium T deposited on a plane mirror; (b) a bar-mirror and a deformable medium T deposited on a concave mirror; (c) a "Venetian blind" shaped bar-mirror in order to avoid a loss of the input light.

On Fig. 23(a) is described the principle of operation in reflection mode with the light reflected on a side of the deformable medium T, a single mirror D being used as diaphragm. In the absence of any deformation on T, the image of the mirror-diaphragm coincides with itself and the light is reflected toward the lamp, through the condensor lens L', and is lost. As previously, the lens L can act as an objective and project on a screen the images of the deformed points of the target.

On Fig. 23(b), the single mirror diaphragm is replaced by a bar-mirror in order to increase, as on Fig. 22(b), the sensitivity of the modulation process; besides the lens L is replaced by a concave mirror which holds the deformable film T and whose center of curvature coincides with the bar-mirror. The lens L" projects on the screen the image of the target T. On Fig. 23(c) the bar-mirror is "Venetian blind" shaped in order to avoid a loss of the input light. It must be noted that the cross section of the light beam is then increased by a factor of 2 resulting in a reduction by 2 of the luminance as in the case of Fig. 22(c). The use of a "Venetian blind" thus replaces the loss of half the input light by a reduction of the luminance which results in a need of increasing the numerical aperture of the optical system.

In the systems shown on Fig. 22(c) and 23(c) a maximum of about half the input light can pass through the diaphragm when large deformation exists on the whole surface of T. The maximum efficiency of the system is thus 50%.

The "Eidophor" apparatus (Gretag, Switzerland) is similar to that described on Fig. 23(c). The target is composed of an oil film deposited on a spherical mirror which rotates slowly in order to avoid permanent deformation. The signal is applied to an electron gun and acts on the cross section of the electron beam hitting the target. The acceleration voltage is of the order of 15 kV so that the secondary emission ratio is less than unity. The low level signal corresponds to a large unfocused beam and a uniform electric charge pattern which results in a uniform electrostatic field and in no deformation. The high level signal corresponds to a narrow focused beam which deposits a thin line of electric charges; these charges induce high electrostatic forces thus creating a sharp groove in the oil film which deflects the light.

In order to reproduce motion, the viscosity and the electric conductivity of the oil film are adjusted so that deformations of the film disappear before the next scanning. This is obtained by a proper choice of the oil and by a regulation of the operating temperature. Figure 24 shows the shape of the output characteristic versus time during the period T of the frame scanning. Due to the exponential decay of the deformations, the average duty cycle is in the order of 1/2 which represents the ratio between the hatched surface and the surface limited by the maximum ordinate L_m.

Let us calculate the consumption of the system, using a Xenon lamp with an efficiency of 40 lm/W, for a highlight illumination of the screen of 100 lm/m^2. The following efficiencies are assumed:

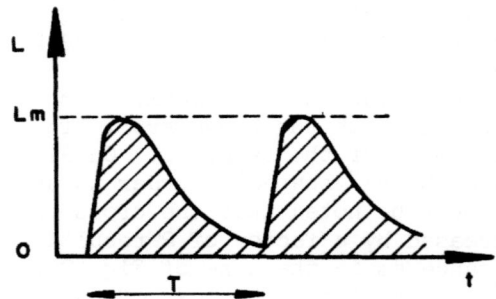

FIGURE 24. Light output versus time characteristic of the "Eidophor" during the period T of the frame scanning.

0.25 for the collecting optics;
0.5 for the output transmission of the bar-mirror;
0.5 for the duty cycle;
0.75 for the objective transparency;
0.7 for the transmission of the remaining optical elements.

The specific consumption per unity of screen area is then:

$$\frac{100 \text{ lm/m}^2}{0.25 \times 0.5 \times 0.5 \times 0.75 \times 0.7 \times 40 \text{ lm/W}} \approx 75 \text{ W/m}^2.$$

Due to the large size of the useful part of the target, 54×72 mm^2, it is possible to focus a large amount of light on it. With a Xenon lamp operating at about 3 kW, the highlight output can reach 4000 lm allowing a luminance of 30 cd/m^2 on a 40 m^2 non-directive screen. In the case of color TV projection, three systems similar to that of Fig. 23(c) are used in combination with only one Xenon lamp and two dichroic mirrors separating the white beam in three components, red, green and blue respectively. Due to the efficiency of the trichrome separation and of the elimination of the yellow components of the spectrum, the maximum output of the color "Eidophor" is in the order of 3200 lm with the same Xenon lamp.

Such light outputs are sufficient to cover practically all the large screen projection needs. Nevertheless the "Eidophor" exhibits a lot of drawbacks: it is expensive, bulky and not easy to operate; indeed, it is demountable to permit the replacement of the cathode, whose lifetime is limited by the presence of residual oil vapor pressure, and it incorporates a pumping system including a mechanical and a molecular pump. Therefore attempts have been made to build smaller size sealed-off versions.

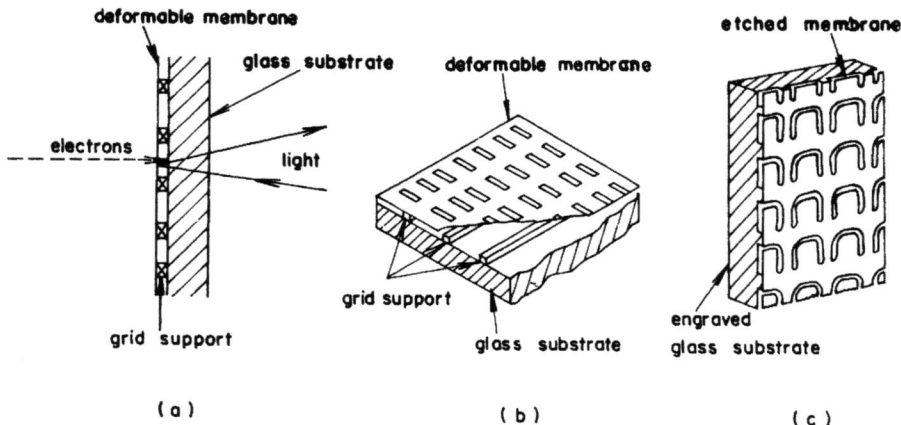

FIGURE 25. Examples of thin solid membranes used as deformable medium: (a) continuous membrane stretched on a grid support; (b) segmented membrane; (c) membrane etched in the form of tongues.

In one version [35], a rotating oil film is used, with a small target area and a light transmission mode similar to that described on Fig. 22(c). The overall system is small and sealed-off but its maximum light output is only in the order of 700 lm. In a color version using only one such system, a vertical grating is superimposed, on each point of the target area, to the normal horizontal grating parallel to the scanning lines. Using two separate gratings at the input, it is then possible to modulate the green component, with the horizontal grating, and simultaneously the red and blue components, with the vertical grating, using the first and the second order of diffraction. Such a system results in a simple and cheap color projection device, but the highlight output is then reduced by a factor of 2 in comparison with the black and white device: about 350 lm.

In another version [36], a mica sheet separates the deformable film chamber from the CRT's main vacuum chamber; the system operates in the reflection mode (Fig. 23(a)).

In a transmission mode version called "Lumatron" (CBS [37]), the deformable medium is a thermoplastic film which can be written on only when it is heated; by cooling the target after writing, the system acts as a storage device.

In all these systems erasing can be achieved using secondary emission with a reduced velocity electron beam, in a way similar to that described in the beginning, for passive displays (Fig. 2 to 4).

At the same image light valve display family can be connected devices using deformable thin solid membranes stretched on a grid support in front of a glass substrate, as shown on **Fig. 25**. The membrane thickness is in the order of 0.1 to 0.2 μm so that the electrons, with an acceleration voltage of about 20 kV, pass through it and deposit charges, on the glass substrate, which attract the membrane. Thus the membrane can be made of a metal, such as beryllium [38]; nickel and various alloys have also been proposed with a segmentation of the membrane, as shown on Fig. 25(b), in order to increase its flexibility [39]; for the same purpose, it is also possible to use a membrane etched in the form of tongues, as shown on Fig. 25(c) [40]. At present the light output obtained with such systems does not reach that given by the sealed-off light valve: 700 lm for black and white TV.

(ii) <u>Light scattering devices</u>. For large screen projection applications, using light scattering in films or plates, the same Schlieren optics set ups, as described on Fig. 22 and 23 with deformable mediums, can be used. The first usable materials exhibiting electronically induced scattering are nematic liquid crystals extensively described in the previous sections (Fig. 5 and 6). For addressing the display, a cross-bar system does not seem well adapted, due to the small dimensions of the targets used for projection. An electron beam addressing device has been proposed [41] using a CRT with a mosaic face plate which consists of a glass window in which wires are embedded; the latter serve to make transverse electrical connections between the inside and the outside of the CRT. In order to operate in reflection mode, a segmented metallic mirror is deposited on the face plate and the liquid crystal film is sandwiched between this mirror and a plate coated with a transparent conductive layer. The resolution of the system described is not very high and is limited by the mosaic window, by the segmented mirror and by the fact that these components are not registered with each other; the contrast ratio also is poor, in the order of 5 only. In the future, however, these characteristics could be probably greatly improved, as in other systems using liquid crystals.

A third addressing possibility is an optical one and makes use of a photoconductive layer. This possibility is under study in several laboratories. In a first device [42], a ZnS photoconductive layer, sensitive only in the UV and transparent in the visible, is placed directly in contact with the liquid crystal, as shown on Fig. 26(a), the projection system working then in transmission as in the cases described on Fig. 22. In two other devices [43,44], use is made of a more efficient photoconductive layer, such as selenium or selenium compound, sensitive in the visible, and the system operates in the reflection mode, as in the cases described on Fig. 23; a metallic segmented mirror is

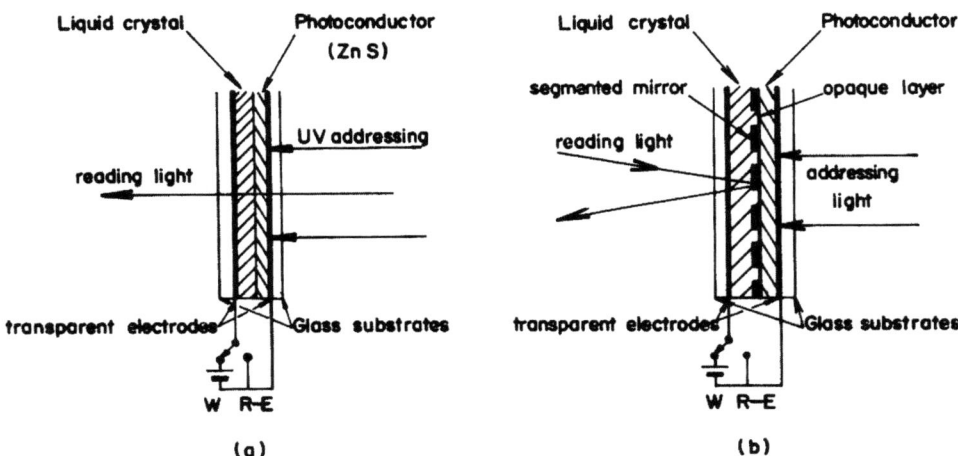

FIGURE 26. Nematic liquid crystal light-valve device involving the dynamic scattering effect and optically addressed by means of a photoconductive layer: (a) transmission operation using a ZnS photoconductive layer sensitive in the UV only; (b) reflection operation with a segmented mirror.

placed between the liquid crystal and the photoconductive layer, as shown on Fig. 26(b), with possibly the addition of a non-conducting opaque layer introduced in order to separate more efficiently the addressing side from the reading side of the system; the segmented mirror could be replaced by a dielectric mirror.

To write a charge pattern at the surface of the liquid crystal, an image is projected, in UV or visible light, on the photoconductor, thus inducing carriers proportionally to the exposure (i.e. to the product illumination x time); simultaneously a d.c. voltage is applied between the two transparent conductive layers deposited on the glass substrates, resulting in the transport of charges towards the liquid crystal. To erase, the whole photoconductor area is flooded with light, while the two electrodes are short circuited; then the motion of the induced carriers is controlled by the voltage pattern previously written on the liquid crystal and, if the amount of light is sufficient, the carriers completely discharge the liquid crystal cell.

These systems seem very promising but are, at present, at the laboratory stage only. The main difficulties occur from:

- lack of contrast and of speed, as in the case of the systems using liquid crystals described in the previous sections;

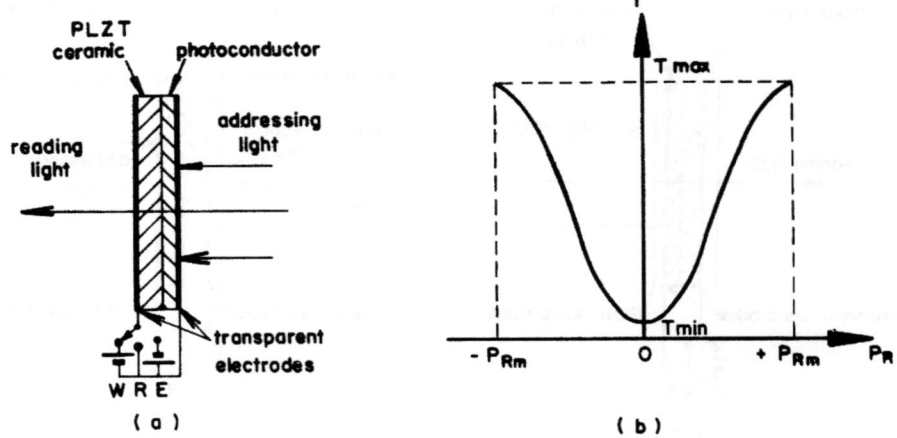

FIGURE 27. Light-valve device involving light scattering in a coarse-grained PLZT ceramic: (a) structure of the device optically addressed by means of a photoconductive layer; (b) transmission versus remanent electrical polarization characteristic curve of the ceramic.

- lack of modulation efficiency due to the difficulty of having an optics with a sufficient aperture to collect the major part of the scattered light;
- limitation of the reading light due to the difficulty to find an efficient optical screening between the reading and the addressing sides.

A second kind of material exhibiting electronically induced light scattering was recently proposed [45]: lead-lanthanum-zirconate-titanate (PLZT) ferroelectric ceramics which were previously used in induced birefringence devices, as we will see in the next paragraph. In these ceramics the proportion of zirconate and titanate is adjusted to values in the vicinity of 65% and 35% and 5 to 10% lanthanum are introduced in order to improve the transparency; the Curie temperature then ranges between 100 and 200°C. For scattering mode, a coarse-grained (grain size >3 μ) ceramic is better adapted than the fine grained (grain size ≈1 μ) ceramics previously used. The basic optically addressed device comprises a ceramic plate, of a thickness ranging between 0.1 and 0.3 mm, a photoconductive film and two transparent electrodes, as shown on Fig. 27(a).

Such ferroelectric ceramics exhibit domains which are randomly oriented, when the plate is on the zero remanence state. As the ferroelectric domains do not have the same index of refraction in all directions, this random orientation results in a large scattering of the light and a low transmission in the direction of the incoming light. When an electric field is applied, the domains are progressively oriented in the same direction, resulting in a decrease of the scattering and in an increase of the transmitted light, as seen on Fig. 27(b). The ratio of the maximum to the minimum transmission T_{max}/T_{min} can exceed 100 for a small angular aperture (about 2°). In practice, in large screen projection applications, the angular aperture has to be large and, in order to obtain a sufficient contrast, Schlieren optics systems would be perhaps better adapted, provided the minimum scattering, occurring at maximum remanent polarization P_{Rm}, is sufficiently low to give a satisfactory contrast.

The reading light is applied only when the writing light has been suppressed and, during reading, the electrodes are short-circuited. During the erasure, a voltage is applied with a sign opposite to that during writing, while the photoconductor is flooded with light, which results in the saturation of the polarization of the ferroelectric ceramic.

Due to the large number of problems still to be solved, it is too early to foresee what will be the future of such a system in the field of large screen projection. It must be noted, in particular, that the characteristic curve shown on Fig. 27(b) is reversible only because it is the remanent polarization P_R which has been chosen as the abscissa; the relation existing between this remanent polarization and the previous applied field is in fact hysteretic. Therefore such systems are probably better adapted to storage displays.

(iii) <u>Induced birefringence devices</u>. In display devices involving induced birefringence, use is made of a thin plate of a transparent material exhibiting an electro-optic effect when an electric field parallel to the direction of propagation of light is applied across the plate.

A first possibility is to use a crystal exhibiting the Pockels effect; this possibility was described as early as 1936 by L. S. Kaysie [46] and M. von Ardenne [47], the proposed material being a ZnS crystal; in fact, for a practical application, this material was not suitable, due to its low electro-optic sensitivity and to the difficulty of growing sufficiently large crystals, and the attempts to use it were unsuccessful.

During the 2nd world war large dimension electro-optic crystals were grown, of the family of potassium-dihydrogen-phosphate (KH_2PO_4 or KDP in abbreviated form); they were manufactured, in fact, for their interesting piezoelectric

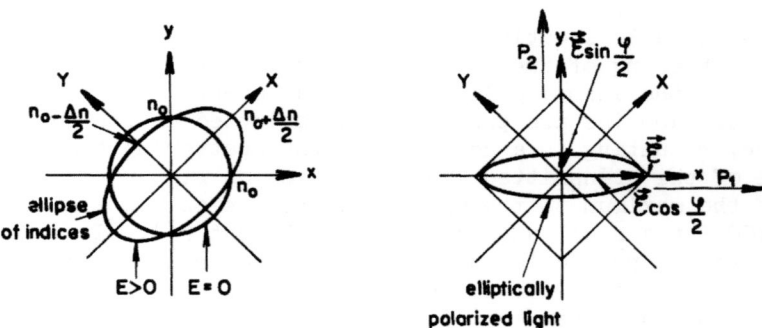

FIGURE 28. Left: optical indicatrix in the xy plane of a KD_2PO_4 crystal with an electric field E applied along the z axis. Right: elliptically polarized light emerging from the crystal, for an input light polarized parallel to the x axis.

---

properties; indeed the Pockels effect appears in the same kinds of crystals which exhibit the piezoelectric effect, that is in crystal classes which do not have a center of symmetry.

Figure 28 shows how the Pockels effect can be used in a KDP plate cut perpendicularly to its optical axis z. In the absence of an electric field, the crystal presents the same index of refraction n_o whatever the direction of the electric vector of the light $\vec{\mathcal{E}}$ in the xy plane. When an electric field E parallel to z is applied, a birefringence appears in the xy plane, the index being increased by $\frac{\Delta n}{2}$, for example, for a direction of the $\vec{\mathcal{E}}$ vector parallel to the first bisectrix X, and decreased by $\frac{\Delta n}{2}$ for a direction of $\vec{\mathcal{E}}$ parallel to the second bisectrix Y of the x and y crystallographic axes, as shown on the Fig. If the input light is polarized in the direction of the x axis, by means of a polarizer P_1, the light is then divided into two components propagating with different velocities; at the output of the plate of thickness l, these two components exhibit a phase difference:

$$\varphi = \frac{2\pi l \Delta n}{\lambda} \qquad (4)$$

which results in an elliptically polarized light; the transmittance, after a second polarizer P_2 crossed with the first one, is then:

$$T = \frac{r^2}{2} \sin^2 \frac{\varphi}{2} = \frac{r^2}{2} \sin^2 \frac{\pi l \Delta n}{\lambda} \qquad (5)$$

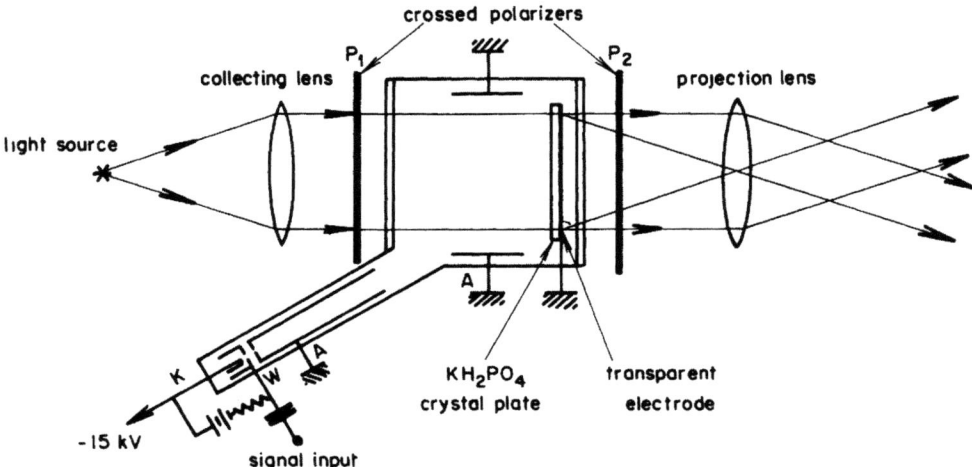

FIGURE 29. Transmission mode light-valve tube using a KH_2PO_4 plate operated at room temperature and addressed by means of a modulated electron beam.

where the factor $\frac{1}{2}$ comes from the polarization of the light and the factor r from the yield of each polarizer.

As the Pockels effect is linear, Δn is proportional to the electric field, that is to the ratio V/l in which V is the potential difference existing between the two faces of the plate. Thus φ is proportional to V and the transmittance can be written:

$$T = \frac{r^2}{2} \sin^2 kV, \qquad (6)$$

in which the constant k is independent of the thickness l of the plate; up to $\varphi = \frac{2\pi}{3}$, that is 75% modulation depth, the modulation characteristic versus V is essentially quadratic as in the case of most image tubes.

Since 1950 several laboratories have studied display devices using crystals of the KDP family, [48-55]. In the earlier works [48-53], use is made of crystals operating at room temperature and the set up is, in general, similar to that described on Fig. 29. The second side of the crystal plate is coated with a transparent conductive layer connected to the grounded anode. During the writing cycle, the cathode is connected to about -15 kV so that the secondary emission coefficient of the target is smaller than 1; thus the various points of the target get negatively charged proportionally to the electron beam current which is modulated by

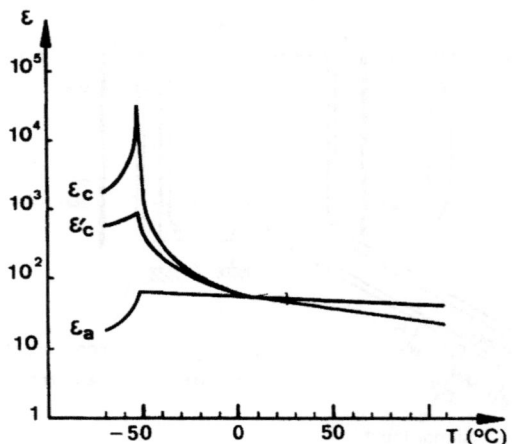

FIGURE 30. Variation, versus temperature, of the dielectric constants of KD_2PO_4: $\mathcal{E}_a$ in a direction perpendicular to the optical axis; $\mathcal{E}_c$ and $\mathcal{E}'_c$ in a direction parallel to the optical axis for a free crystal and a clamped crystal respectively.

---

the video signal. For erasing, the cathode could be connected to about -2 kV so that the secondary emission coefficient of the target would be greater than 1, as described in the beginning of this lecture (Fig. 2-4). In fact, this is not necessary because the discharge time constant of the crystal $\rho \mathcal{E} \mathcal{E}_o$ is less than 0.1 second at room temperature.

The results of these early works, in particular the uniformity and the resolution of the displayed pictures, were not very satisfactory. The main difficulty arises from the fact that the various points of the target have potentials which may differ by several kilovolts; indeed, a phase shift φ equal to $\frac{2\pi}{3}$ requires a potential difference of 5 kV, if ordinary KDP is used, and 2.4 kV, if deuterated KDP (KD_2PO_4 or DKDP) is used, the latter being the most sensitive commercially available crystal of the KDP family. With a double-pass operation, that is in reflection mode [52], these potential differences can be reduced by a factor of two, but this is still far from sufficient for obtaining good results.

In more recent work [54,55] use was made of the fact that the DKDP crystal is ferroelectric, with a Curie temperature near -50 C, and that, when the temperature varies, the induced birefringence is not only proportional to the electric field but also to the electric polarization, that is to the product $\mathcal{E}E$ when the crystal is operated above the Curie point. Figure 30 shows

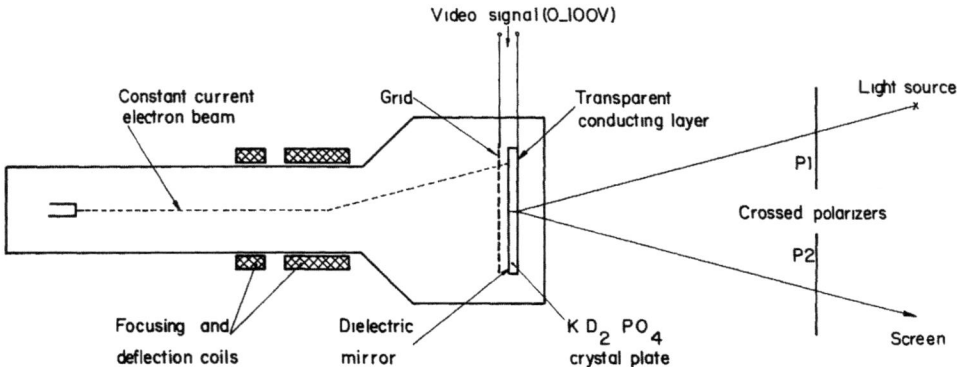

FIGURE 31. Reflection mode light-valve tube (Titus) using a KD_2PO_4 plate operated just above the Curie temperature and addressed by means of a constant current electron beam; the video signal is applied between a grid placed close to the crystal plate and a transparent conducting layer deposited on the other face of the plate.

---

the variations, with temperature, of the relative dielectric constant of the deuterated KDP for an electric field in a direction perpendicular to the optical axis ($\mathcal{E}_a$) and in a direction parallel to the optical axis; in the latter case, it is shown both for a free crystal ($\mathcal{E}_c$) and for a clamped crystal ($\mathcal{E'}_c$). Actually, as the Pockels effect in a free crystal is accompanied by a piezo-electric effect which might cause coupling between the different points of the picture, it is necessary to clamp the crystal plate, for instance by gluing it on a transparent substrate; one then observes that $\mathcal{E'}_c$ reaches a maximum of approximately 700 at a temperature near the Curie point. Besides, one observes that $\mathcal{E}_a$ is about 10 times lower than $\mathcal{E'}_c$, which makes it possible to choose a target thickness exceeding the distance separating two elements in the picture, without loss of resolution. Using the reflection mode, the voltage required for 75% modulation of the light (i.e. for $\varphi = 2\pi/3$) is reduced to about 100 V at -50°C instead of 1.2 kV at room temperature.

The discharge time constant $\rho \mathcal{E} \mathcal{E}_0$ of the crystal is about 1 hour at -50°C, whereas it is less than 0.1 second at room temperature. In the device designed at LEP [54,55], erasing and writing are simultaneously obtained by using the set-up shown on Fig. 31. The target is bombarded by an electron beam whose acceleration voltage is in the order of 500 to 800 V, so that the saturation secondary emission yield is greater than 1 (Fig. 3); a grid is placed in front of the target, at a distance of about 50 μm. The electron beam, which has a constant intensity, then

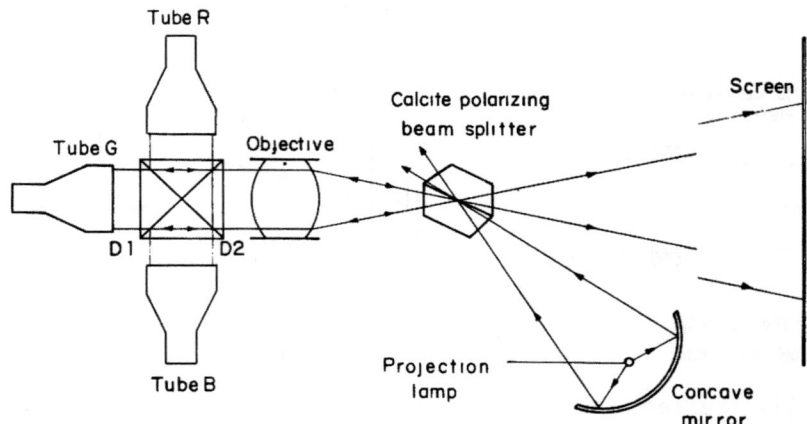

FIGURE 32. Projection system showing a possible arrangement, adapted to color TV, using three light-valve tubes, two dichroic mirrors (D_1, D_2) and a single optical equipment.

---

works practically as a flying short-circuit between the grid and the point of impact on the target; indeed, when the potential of the point impinged upon by the electron beam is lower than the potential of the grid, the secondary emission yield is higher than 1 and the potential of the point increases; in contrast, when this potential exceeds that of the grid, the secondary emission yield is lower than 1 and the potential of the point decreases.

It is therefore sufficient to apply the video signal between the grid and a transparent conducting layer, deposited on the second face of the crystal plate, to ensure that the various points of the target get charged to the corresponding video voltage when they are hit by the electron beam. The erasure and writing functions are therefore combined and the result is an entirely flicker-free operation; indeed, the charge at each point remains constant between two successive scans of the electron beam, as well as during these scans, for points which have not moved on the image. In addition, as a result of the fact that the voltage pattern stored on the target does not depend on the intensity of the electron beam, no line structure is apparent on the picture, without any loss in vertical resolution.

An experimental sealed-off tube, called Titus (for "Tube Image a Transparence Variable Spatio-temporelle"), has been constructed. Its dimensions are about 40 cm in length and 10 cm in the largest diameter; it uses a target with a 30×40 mm^2 area and a 0.25 mm thickness; the cooling is achieved with two stages of Peltier cells located in vacuum inside the tube. Figure 32 shows an experimental projection system using this tube. The light

from a Xenon lamp is condensed by means of a concave mirror onto
a calcite polarizing beam splitter which transmits to the tube one
of the polarized components of the light. The objective is placed
between the beam splitter and the tube so that the light beam
incident on the target has a mean direction which is normal to the
latter; when the light beam is reflected and passes through the
objective and the beam splitter again, only the second polarized
component of the light is transmitted to the screen. The figure
also shows a possible arrangement adapted to color television and
using only one light source and a single objective: R, G and B
are three identical tubes receiving red, green and blue components
respectively; D_1 and D_2 are two dichroic mirrors of the color
television camera type.

With the same Xenon lamp as that used for the Eidophor, black
and white television pictures have been recently projected with a
highlight output of 2500 lm, a contrast of 40 and a resolution of
700 elements per line. In comparison with the complexity of the
Eidophor this system seems very attractive. The study of the
black and white projection is near completion; the next stage will
be devoted to color projection; in this case the tolerances on the
uniformity of the projected pictures, which is related to the
quality of the crystal used, will probably be more stringent.

As in the case of nematic liquid crystal displays used with
polarized light (Fig. 10), it is possible with a single tube to
display data in color if the product of the birefringence Δn by
the optical path l in the crystal varies between 1.2 and 1.5 µm,
for instance, as shown on Fig. 12. This would correspond to a
video signal ranging between about 700 and 800 V. To avoid such
large signal voltages, it is sufficient to add in front of the
tube a birefringent plate, in mica for instance, which gives a
retardation, for a double pass, in the order of 1.2 or 1.5 µm;
this retardation adds algebrically with that given by the tube for
a signal varying from 0 to 100 V. Thus the addition of this
retardation plate converts the amplitude modulation of the light
into a hue modulation, the range being for instance green-yellow-
red or red-brown-green, depending on the value of the constant
retardation introduced.

Another device, the Phototitus, derived from the Titus tube
and using the same DKDP target cooled near -50°C, has been
demonstrated recently [56]. It is optically addressed by means
of a photoconductive layer deposited between the dielectric mirror
and a second transparent electrode, as shown on Fig. 33. During
the writing cycle, a voltage is applied between the electrodes
while a picture is projected on the photoconductive layer.
Erasure is accomplished by short-circuiting the electrodes and
flooding the photoconductor with light. The resolution of this

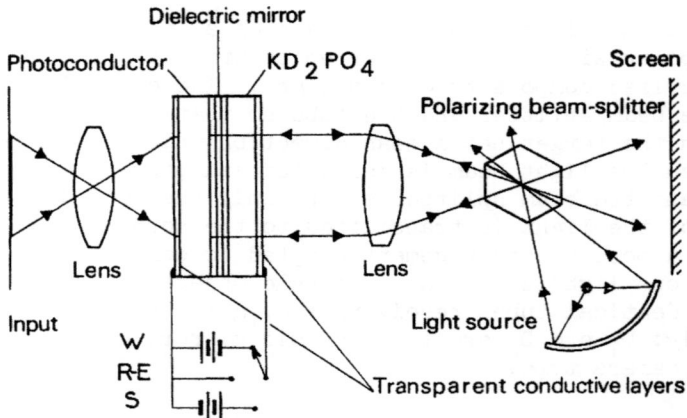

FIGURE 33. Reflection mode light-valve device (Phototitus) using a KD_2PO_4 plate operated just above the Curie temperature and optically addressed by means of an amorphous selenium photoconductive layer. Subtraction, from a first image, of a second one can be effected by inverting the voltage polarity during the writing of the latter (position S of the switch).

---

device reaches 1500 elements per line. Color data display with a single Phototitus can be obtained as for the Titus tube.

Due to the difficulty in optically separating efficiently the writing side from the reading side with the dielectric mirror, this device seems better adapted to medium size picture displays and to optical data processing in coherent or incoherent light. For this last application, it can be noted that subtraction, from an image, of a second one can be effected by inverting the voltage polarity during the writing of the latter.

Monocrystals exhibiting the Pockels effect are not the only usable materials for induced birefringence devices. Indeed one can use the same types of materials as those described for light scattering devices. The first ones are nematic liquid crystals used with polarized light and controlled by an a.c. voltage (with an about 10 kHz frequency), as described for the flat active displays. For projection devices, the most suitable system would probably use an optical addressing by means of a photoconductive layer, as discussed in the previous section. Nevertheless, in large screen projection systems using a powerful reading light, the adequate optical separation of the reading side from the writing side would probably be difficult.

The second kind of usable materials is the fine-grained PLZT ferroelectric ceramics [57], which can also be addressed by means

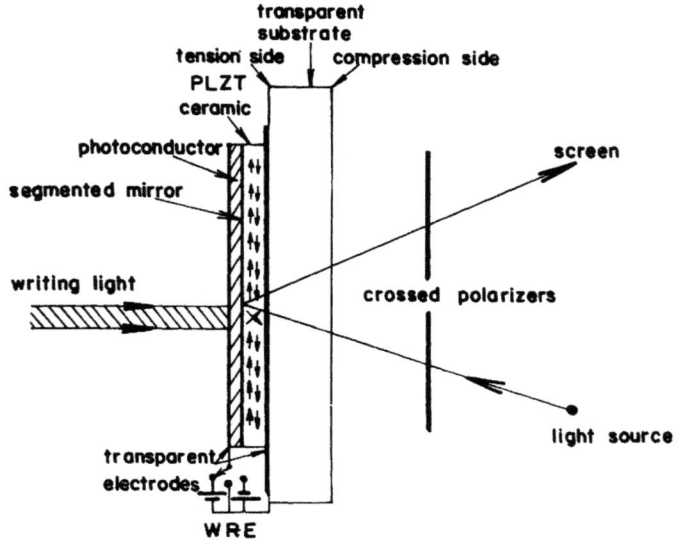

FIGURE 34. Reflection mode light-valve device making use of induced change of birefringence in a fine-grained PLZT ceramic and optically addressed by means of a photoconductive layer; the ceramic plate is cemented on a transparent substrate and mechanically "strain-biased".

of a photoconductive layer and which can work in the transmission as well as in the reflection mode [58-60]. Usually such polycrystalline materials only present a Kerr effect in which the birefringence is induced by an electric field perpendicular to the direction of propagation of the light; in order to obtain an effect with an electric field parallel to this direction, it is necessary to pole the material so as to render it anisotropic in the plane of the plate. The most convenient technique seems to use a "strain-biased" plate [59,60]; for this, a thin ceramic plate is cemented on a transparent substrate, and the assembly is placed under stress as shown on Fig. 34. The internal field created in the ceramic plate, due to the piezoelectric effect, produces an alignment of the ferroelectric domains along the tension axis, resulting in a uniform birefringence in the plate. If an area of the plate is illuminated while a voltage is applied between the transparent electrodes of the device, an electric field appears in the ceramic; then a certain number of domains are rotated; this results in a change of the birefringence of the area and a change in the transmission of the system if the plate is placed between two crossed polarizers.

As the birefringence varies around a mean value, without passing through zero, one observes a modulation of the hue of the transmitted light, when the system is used with a white reading light, in the same way as was previously described on Fig. 12. If an amplitude modulation is desired, it is necessary to operate with a monochromatic light or to use a compensator in order to cancel the initial birefringence of the strain-biased ceramic.

As in the case of the coarse-grained PLZT ceramics used in the scattering mode, it is too early to foresee what will be the future of such systems for display purposes. However it must be noted that, at present, the lifetime of strain-biased devices is not sufficient; indeed it is limited by the occurrence of cracks during operation of the device. If this effect cannot be prevented, the scattering mode of operation would probably be a better solution in the future.

CONCLUSION

It is difficult to conclude this brief survey of large area display materials and devices; indeed a great number of the devices described are still in the research state. In the class of large area flat displays, we will probably see, in the near future, an important development of gas discharge panels, whereas liquid crystal devices seem very promising in a more distant future. In the class of large screen projection systems, the Eidophor fills practically all the requirements but has several drawbacks, in particular its price, bulkiness and the necessity of a heavy maintenance; we will see probably, in the near future, less cumbersome sealed-off devices based on the same principle and using liquid mediums or membranes and other devices based on scattering or electro-optic effects in solids or in liquid crystals.

REFERENCES

[1] C. Feldman, J. Opt. Soc. Amer., 47, pp. 790-794 (1957).
[2] T. E. Sisneros, P. A. Faeth, J. A. Davis, E. H. Hilborn, Inform. Display, Vol. 7, no. 4, April 1970, pp. 33-37.
[3] J. Frost, Inform. Display, Vol. 6, no. 5, Sept-Oct. 1969, pp. 43-46.
[4] P. G. R. King, J. Inst. Elec. Engrs. (London) 93 A, pp. 171-172 (1946).
[5] P. M. Heyman, I. Gorog, B. Faughnan, IEEE Trans., Vol. ED-18, no. 9, pp. 685-691 (1971).
[6] H. L. Bjelland, Proc. 3rd Nat. Symp. Soc. Inf. Display, San Diego, California, Feb. 1964, pp. 286-299.

[7] C. K. Clauer, J. D. Kuehler, U.S. Patent no. 3,109,062, Oct. 1963.
[8] J. S. Donal, Proc. IRE, Vol. 31, pp. 195-208 (1943).
[9] J. S. Donal, D. B. Langmuir, Proc. IRE, Vol. 31, pp. 208-214 (1943).
[10] B. Kazan, D. P. Foote, D. G. Marlow, Proc. IEEE, Vol. 56, no. 3, pp. 338-339 (1968).
[11] G. H. Heilmeier, L. A. Zanoni, L. A. Barton, Proc. IEEE, Vol. 56, no. 7, pp. 1162-1171 (1968).
[12] G. Assouline, M. Hareng, E. Leiba, IEEE Trans., Vol. ED-18, no. 10, pp. 959-964 (1971).
[13] L. T. Creagh, A. R. Kmetz, R. A. Reynolds, IEEE Trans, Vol. ED-18, no. 9, pp. 672-679 (1971).
[14] B. J. Lechner, F. J. Marlowe, E. O. Nester, J. Tults, Proc. IEEE, Vol. 59, no. 11, pp. 1566-1579 (1971).
[15] J. R. Hansen, R. J. Schneeberger, IEEE Trans., Vol. ED-15, no. 11, pp. 896-906 (1968).
[16] M. Hareng, G. Assouline, E. Leiba, Revue Technique Thomson-CSF, Vol. 3, no. 4, pp. 781-783, Dec. 1971.
[17] Electronics, Vol. 44, no. 25, Dec. 6, 1971, pp. 7E-8E.
[18] A. Vecht, N. J. Werring, J. Phys. D., Appl. Phys. 1970, Vol. 3, pp. 105-120.
[19] M. Yoshiyama, Electronics, Vol. 42, no. 6, March 17, 1969, pp. 114-118.
[20] Electronics, Vol. 44, no. 25, Dec. 6, 1971, p. 9E.
[21] G. W. Taylor, IEEE Trans, Vol. ED-16, no. 6, pp. 565-575 (1969).
[22] Electronics, Vol. 43, no. 11, Mary 25, 1970, pp. 112-117.
[23] D. G. Thomas, IEEE Trans., Vol. ED-18, no. 9, pp. 621-627 (1971).
[24] F. Auzel, C. R. Acad. Sciences Paris, T. 262 B, pp. 1016-1019 (1966).
[25] F. Auzel, C. R. Acad. Sciences Paris, T. 263 B, pp. 819-821 (1966).
[26] Electronics, Vol. 43, no. 5, March 2, 1970, pp. 120-125.
[27] R. L. Johnson, D. L. Bitzer, H. G. Slottow, IEEE Trans., Vol. ED-18, no. 9, pp. 642-649 (1971).
[28] S. van Houten, R. N. Jackson, G. F. Weston, Proc. of the S.I.D., Vol. 13/1, First Quarter 1972.
[29] R. A. Cola, Electronics, Vol. 44, no. 15, July 19, 1971, pp. 66-69.
[30] P. M. van Alphen, H. Rinia, Philips Tech. Rev., Vol. 10, no. 4, pp. 69-78 (1948).
[31] T. Poorter, F. W. de Vrijer, Philips Tech. Rev., Vol. 19, no. 12, pp. 338-355 (1958).
[32] E. F. Labuda, E. I. Gordon, R. C. Miller, IEEE Jour. of Quant. Elec., Vol. QE1, no. 6, pp. 273-279 (1965).
[33] E. Baumann, Jour. of Brit. Inst. of Radio Engrs., Vol. 12, no. 2, pp. 69-78 (1952).

[34] E. I. Sponable, Jour. of SMPTE, Vol. 60, no. 4, pp. 337-343 (1953).
[35] W. E. Good, IEEE Trans., Vol. BTR 15, no. 1, pp. 21-24 (1969).
[36] Electronics, Vol. 45, no. 2, January 17, 1972, pp. 32-34.
[37] R. J. Doyle, W. E. Glenn, IEEE Trans., Vol. ED-18, no. 9, pp. 739-747 (1971).
[38] M. Auphan, L'Onde Electrique, Vol. 36, no. 357, pp. 1040-1045 (1956).
[39] J. A. van Raalte, Applied Optics, Vol. 9, no. 10, pp. 2225-2230 (1970).
[40] G. O. Langner, Electronics, Vol. 43, no. 25, Dec. 7, 1970, pp. 78-83.
[41] J. A. van Raalte, Proc. IEEE, Vol. 56, no. 12, pp. 2146-2149 (1968).
[42] J. D. Margerum, J. Nimoy, S. Y. Wong, Applied Physics Letters, Vol. 17, no. 2, pp. 51-53 (1970).
[43] D. L. White, M. Feldman, Electronics Letters, Vol. 6, no. 26, Dec. 31, 1970, pp. 838-839.
[44] M. Frappier, G. Assouline, M. Hareng, E. Leiba, Nouv. Rev. d'Optique appliquee, Vol. 2, no. 4, pp. 221-228 (1971).
[45] W. D. Smith, C. E. Land, Appl. Phys. Lett., Vol. 20, no. 4, pp. 169-171 (1972).
[46] L. S. Kaysie, Television and Shortwave World, May 1936.
[47] M. von Ardenne, Telegraphen-Fernsprech-Funk und Fernseh Technik, Vol. 27, pp. 518-524 (1938); Vol. 28, pp. 180-184 (1939); Vol. 28, pp. 403-407 (1939).
[48] E. Burstein, J. W. Davisson, P. L. Smith, J. E. Dehnel, Abstract J. Opt. Soc. Am., Vol. 41, no. 4, p. 228 (1951).
[49] R. W. Weeks, W. E. Dickinson, U.S. Patent, no. 2 983 824, May 1961 (filed May 1955).
[50] E. Lindberg, Electronics, Vol. 36, no. 51, Dec. 20, 1963, pp. 58-61.
[51] E. J. Calucci, Inform. Display, Vol. 2, no. 2, March-April 1965, pp. 18-22.
[52] D. H. Pritchard, R.C.A. Review, Vol. 30, no. 4, pp. 567-592 (1969).
[53] S. Rissmann, H. Vosahlo, Jenaer Jahrbuch, Vol. 1, pp. 228-244 (1960).
[54] G. Marie, Philips Res. Rept., Vol. 22, no. 2, pp. 110-132 (1967).
[55] G. Marie, Philips Tech. Rev., Vol. 30, no. 8-9-10, pp. 292-298 (1969).
[56] M. Grenot, J. Pergrale, J. Donjon, G. Marie, Appl. Phys. Letters, Vol. 21, no. 3, pp. 83-85 (1972).
[57] C. E. Land, P. D. Thacher, Proc. IEEE, Vol. 57, no. 5, pp. 751-768 (1969).
[58] A. H. Meitzler, J. R. Maldonado, D. B. Fraser, Bell Syst. Tech. Journ., Vol. 49, no. 6, pp. 953-967 (1970).

[59] J. R. Maldonado, A. H. Meitzler, Proc. IEEE, Vol. 59, no. 3, pp. 368-382 (1971).
[60] J. R. Maldonado, L. K. Anderson, IEEE Trans., Vol. ED-18, no. 9, pp. 774-777 (1971).

CHAPTER 15

MAGNETISM

C. Haas

Laboratory of Inorganic Chemistry
Materials Science Center of the University
Zernikelaan, Groningen, The Netherlands

INTRODUCTION

It is well-known that atoms or ions with partly filled electron shells can posses magnetic moments. The moments are due to the orbital motion of the electron around the nucleus and to the spin of the electron. In free ions the energy levels of the atoms are determined by interactions between the electrons, between the electrons and the nucleus and by spin orbit interactions. In complexes and in solids the surrounding of the magnetic ion by other atoms or ligands influences the energy levels strongly. These crystal or ligand field effects cause a further splitting of the ionic energy levels, and also result in many cases in a partial quenching of orbital magnetic momentum [1-3].

The energy levels and the magnetic susceptibility of isolated paramagnetic ions are dealt with in other texts. In these paragraphs we will consider the interactions between magnetic ions in solids and some of the consequences of these interactions for the magnetic properties. We will restrict the discussion to the magnetic properties of non-conducting materials.

INTERACTIONS BETWEEN MAGNETIC IONS

Introduction

The most important interaction between magnetic ions is the so-called exchange interaction, a quantum-mechanical effect which is a consequence of the Pauli-exclusion principle [4-5]. In this paragraph we will discuss two types of exchange interaction which occur in non-conducting solids. These are the direct interaction, caused by the direct electrostatic interaction and overlap of the orbitals of the magnetic electrons, and the superexchange interaction via the intermediate non-magnetic ions.

Consider two ions, with spins $\vec{S}_1$ and $\vec{S}_2$, and magnetic moments $\vec{\mu}_1 = g_1 \mu_B \vec{S}_1$, $\vec{\mu}_2 = g_2 \mu_B \vec{S}_2$ (g is the spectroscopic splitting factor, μ_B the Bohr magneton). Phenomenologically the exchange interaction can be written as

$$H' = -2J \vec{S}_1 \vec{S}_2 \qquad (1)$$

where J is the exchange interaction constant. For $J > 0$, the interaction is ferromagnetic, and tries to orient the two spins parallel to one another. An antiferromagnetic interaction $J < 0$ favors an antiparallel orientation of the spins.

The eigenstates of (1) are labeled by the total spin quantum number S. The operator for the total spin is $\vec{S} = \vec{S}_1 + \vec{S}_2$, so that $\vec{S}^2 = \vec{S}_1^2 + \vec{S}_2^2 + 2\vec{S}_1\vec{S}_2$. Because the eigenvalues of $\vec{S}^2$, $\vec{S}_1^2$, $\vec{S}_2^2$ are $S(S+1)$, $S_1(S_1+1)$ and $S_2(S_2+1)$, the eigenstates of H' are

$$E(S) = -J\{S(S+1) - S_1(S_1+1) - S_2(S_2+1)\} \qquad (2)$$

For the case of two atoms with spin $S_1 = S_2 = \frac{1}{2}$, we obtain a triplet state $E(S=1) = -\frac{1}{2}J$ and a singlet state $E(S=0) = +3/2\, J$. The energy difference between the two states is $2J$.

Expression (1) describes an isotropic exchange interaction. Experimental data show that in some cases the exchange interaction is anisotropic. It is given for example by an expression of the type

$$H' = D(\vec{S}_1 \times \vec{S}_2), \qquad (3)$$

which was proposed by Moriya [6]. This interaction tries to orient the spins $\vec{S}_1$ and $\vec{S}_2$ perpendicular to one another; it is responsible for the occurrence of so-called weak ferromagnetism [6,7] (a small magnetization due to a small angle between two sublattice magnetizations in a normally antiferromagnetic material).

The magnetic dipole - dipole interaction between two magnetic moments μ_1 and μ_2 is given by

$$H'' = \frac{\vec{\mu}_1 \cdot \vec{\mu}_2}{r^3} - \frac{3(\vec{\mu}_1 \vec{r})(\vec{\mu}_2 \vec{r})}{r^5} \quad (4)$$

where $\vec{r}$ is the vector between the two atoms. This type of interaction is usually much weaker than the exchange interaction.

Direct Exchange Mechanisms

In order to explain the mechanism of exchange interactions* we consider first the simple case of two atoms A and B, each with one electron in a single non-degenerate orbital. The Hamiltonian is written as

$$H = -\frac{\hbar^2}{2m}\nabla_1^2 + V(\vec{r}_1) - \frac{\hbar^2}{2m}\nabla_2^2 + V(\vec{r}_2) + \frac{e^2}{r_{12}} \quad (5)$$

The kinetic energy of the two electrons is represented by the terms $-\frac{\hbar^2}{2m}\nabla_1^2$, and $-\frac{\hbar^2}{2m}\nabla_2^2$, the potential energy by $V(\vec{r}_1)$ and $V(\vec{r}_2)$. The term e^2/r_{12} represents the electrostatic repulsion between the two electrons.

We will show that one can distinguish three contributions to the exchange interaction:

a) The electrostatic repulsion between the electrons leads to a ferromagnetic interaction (the same mechanism leads to Hund's rule for atoms: the state with the lowest energy has maximum S).

b) If there is overlap between the two orbitals φ_A and φ_B occupied by the electrons, an attractive potential in the region of overlap causes an antiferromagnetic contribution to the exchange interaction.

c) The mixing up of ionic configurations (i.e. states with two electrons on one atom) leads to an effective antiferromagnetic interaction.

The effects a) and b) are called potential exchange, c) is called kinetic exchange.

---

*A survey of the theory of exchange interactions is given in refs. 8 and 9.

We first discuss the effect of electrostatic repulsion. For two electrons 1 and 2, occupying orthogonal orbitals φ_A and φ_B, the possible wave functions are:

$$\psi(1,1) = \frac{1}{\sqrt{2}} \left\{ \varphi_A(1)\varphi_B(2) - \varphi_B(1)\varphi_A(2) \right\} \alpha_1\alpha_2$$

$$\psi(1,0) = \frac{1}{2} \left\{ \varphi_A(1)\varphi_B(2) - \varphi_B(1)\varphi_A(2) \right\} \left\{ \alpha_1\beta_2 + \alpha_2\beta_1 \right\} \quad (6)$$

$$\psi(1,-1) = \frac{1}{\sqrt{2}} \left\{ \varphi_A(1)\varphi_B(2) - \varphi_B(1)\varphi_A(2) \right\} \beta_1\beta_2$$

$$\psi(0,0) = \frac{1}{2} \left\{ \varphi_A(1)\varphi_B(2) + \varphi_B(1)\varphi_A(2) \right\} \left\{ \alpha_1\beta_2 - \alpha_2\beta_1 \right\} \quad (7)$$

The states are labelled by the quantum numbers S and M of the total spin $\vec{S} = \vec{S}_1 + \vec{S}_2$; the one-electron spin functions are α ($M = +\frac{1}{2}$) and β ($M = -\frac{1}{2}$). For the triplet state ψ (S,M) with $S = 1$, $M = 1, 0, -1$, the spins are parallel, and the orbital function changes sign if the two electrons are interchanged (Eq. (6)). For the singlet state with $S = 0$ and $M = 0$, it is the spin function which changes sign if the two electrons are interchanged (Eq. (7)). Thus all four functions fulfil the Pauli principle, which requires that the total wave function changes sign if two electrons are interchanged.

The energy is calculated from $E = \int \psi^* H \psi d\tau_1 d\tau_2$; the result is

$$E(S = 1) = 2E_0 + K_0 - J_0$$
$$E(S = 0) = 2E_0 + K_0 + J_0 \quad (8)$$

with

$$E_0 = \int \varphi_A^*(1) \left\{ -\frac{\hbar^2}{2m} \nabla_1^2 + V(\vec{r}_1) \right\} \varphi_A(1) d\tau_1 =$$

$$= \int \varphi_B^*(1) \left\{ -\frac{\hbar^2}{2m} \nabla_1^2 + V(\vec{r}_1) \right\} \varphi_B(1) d\tau_1 \qquad (9)$$

$$K_0 = \int \varphi_A^*(1) \varphi_B^*(2) (e^2/r_{12}) \varphi_A(1) \varphi_B(2) d\tau_1 d\tau_2$$

$$J_0 = \int \varphi_A^*(1) \varphi_B^*(2) (e^2/r_{12}) \varphi_B(1) \varphi_A(2) d\tau_1 d\tau_2$$

E_0 is the orbital energy of a single electron in φ_A or φ_B. The Coulomb integral K_0 represents the classical electrostatic repulsion between an electron in φ_A and an electron in φ_B. The exchange integral J_0 is a non-classical term which arises directly from the antisymmetry of the wave functions.

From this calculation we find an energy difference $2J_0$ between the singlet and the triplet state. It is easily seen that $J_0 > 0$, so that this term gives a ferromagnetic contribution to the exchange interaction constant J, defined earlier.

If the orbitals φ_A and φ_B are not orthogonal, i.e. if $S_0 = \int \varphi_A^*(1) \varphi_B(1) d\tau_1 \neq 0$, the wave functions (6) and (7) are not properly normalized, and the energy should be calculated from $E = \int \psi^* H \psi d\tau_1 d\tau_2 / \int \psi^* \psi d\tau_1 d\tau_2$. The result is

$$E(S = 1) = \frac{2E_0 + K_0 - J_0 - 2S_0 V}{1 - S_0^2} \qquad (10)$$

$$E(S = 0) = \frac{2E_0 + K_0 + J_0 + 2S_0 V}{1 + S_0^2}$$

with

$$V = \int \varphi_A^*(1) \left\{ -\frac{\hbar^2}{2m} \nabla_1^2 + V(\vec{r}_1) \right\} \varphi_B(1) d\tau_1 \qquad (11)$$

Because the potential energy $V(\vec{r})$ is due to the nuclei (and the core electrons) it is usually attractive; thus V will be negative, and it can cause an antiferromagnetic coupling.

Finally we consider the influence of ionic states, i.e. states with two electrons on one atom, such as $\varphi_A(1)\varphi_A(2)$ or $\varphi_B(1)\varphi_B(2)$. The spins of two electrons in the same orbital need to be antiparallel. Therefore the ionic states have S = 0.

$$\psi_{AA} = \frac{1}{\sqrt{2}} \varphi_A(1)\varphi_A(2) \{\alpha_1\beta_2 - \alpha_2\beta_1\} \tag{12}$$

$$\psi_{BB} = \frac{1}{\sqrt{2}} \varphi_B(1)\varphi_B(2) \{\alpha_1\beta_2 - \alpha_2\beta_1\}$$

Because these ionic states have the same spin as the unperturbed singlet state $\psi(0,0)$, it is possible to admix the ionic states into the singlet state. The new singlet state is given by*

$$\psi'(0,0) = \sqrt{1-2\lambda^2}\, \psi(0,0) + \lambda\psi_{AA} + \lambda\psi_{BB} \tag{13}$$

The coefficient λ is easily calculated with perturbation theory. The result is:

$$\lambda = \left(\frac{1}{U}\right) \int \psi_{AA}^* H\psi(0,0) d\tau_1 d\tau_2 = -\frac{b\sqrt{2}}{U} \tag{14}$$

The corresponding change of the energy is given by

$$\Delta E(S=0) = -\frac{4b^2}{U} \tag{15}$$

In these equations U is the energy difference between the ionic state and the singlet state. The so-called transfer integral b is given approximately by

$$b = \int \varphi_A^*(1) H\varphi_B(1) d\tau_1 \tag{16}$$

---

*For simplicity we assume here orthogonal orbitals φ_A and φ_B ($S_0=0$).

In this way it is found that the admixing of ionic states lowers the energy of the singlet state. A contribution of this type is not possible for the triplet state. Therefore the ionic states cause an antiferromagnetic interaction.

Direct Exchange Between Transition Metal Atoms

The quantitative discussion of the exchange interactions is a matter of great difficulty. Even for very simple systems it has not been possible to calculate in a satisfactory way (for example by ab initio computer calculations) the exchange interactions quantitatively. However, it is possible to derive qualitative rules for the strength and sign of exchange interactions between transition metal ions in ionic crystals (Goodenough-Kanamori rules [9-12].

For the discussion of the direct exchange interaction between transition metal atoms we have to take into account the spatial symmetry of the orbitals, and also the fact that on each atom there are five d orbitals. We assume, as is usually done, that the transfer contributions to the exchange interaction dominate. Only if the transfer effects are zero (for example by symmetry) the sign of the remaining weak interaction is determined by the potential exchange.

We consider first the contribution to the exchange interaction of a particular orbital of atom A with a particular orbital of atom B. The total interaction will consist of a sum of contributions of this type. The various cases are shown in Fig. 1. The total spin of all electrons occupying orbitals other than φ_A and φ_B are indicated by S_A and S_B arrows, respectively. The spin of an electron in φ_A is coupled to S_A by intra-atomic exchange interactions (Hund's rule); these are usually stronger than the interatomic exchange interactions. In each case the possibility of transfer of an electron between the orbitals φ_A and φ_B is considered for a parallel (left hand side) and an antiparallel (right hand side) orientation of S_A and S_B;* the transfer is indicated by the dotted arrows. The energy of the transfer state is U. In each case we give the contribution to the energy ΔE_F for

---

*The parallel or ferromagnetic orientation of the spins corresponds to the wave functions $\psi(1,1)$ and $\psi(-1,-1)$, the antiparallel or antiferromagnetic orientation to the wave functions $1/\sqrt{2} \{\psi(1,0) \pm \psi(0,0)\}$. These antiferromagnetic wave functions are not eigenfunctions of the hamilton operator H for two electrons.

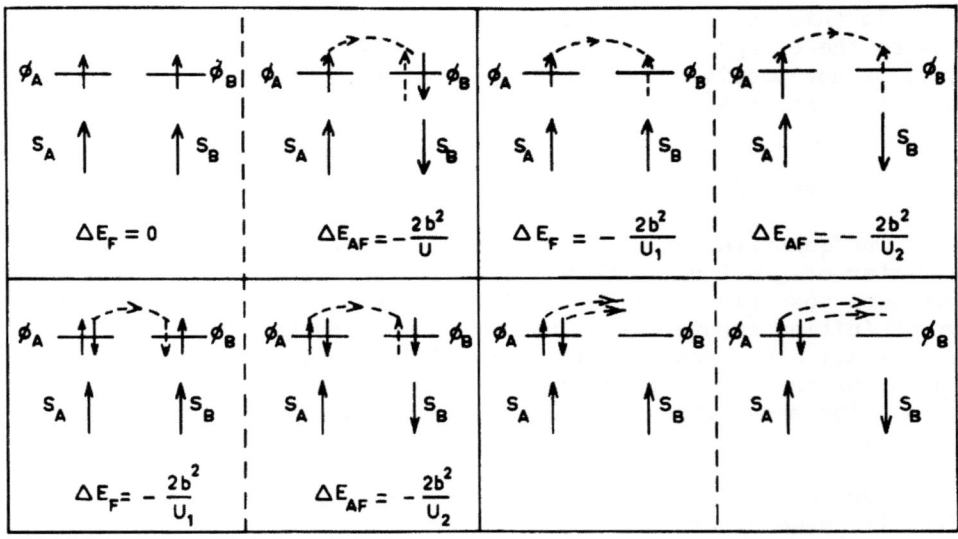

FIGURE 1. Contributions to the exchange interaction, caused by the transfer between orbitals φ_A and φ_B. 1) Two half-filled orbitals; strong antiferromagnetic interaction. 2) Half-filled + empty orbital; because $U_2 > U_1$ and $U_2 - U_1 \ll U_1$, there is a weak ferromagnetic interaction. 3) Filled + half-filled orbital; because $U_2 > U_1$ and $U_2 - U_1 \ll U_1$, there is a weak ferromagnetic interaction. 4) Filled + empty orbital; for both orientations, two types of transfer are possible; the net result is found to be a weak antiferromagnetic interaction.

---

the parallel, and ΔE_{AF} for the antiparallel orientation of the spins S_A and S_B, as a result of the transfer between orbitals φ_A and φ_B.

The magnitude of the direct exchange interaction between two cations depends strongly on the distance, but also on the symmetry of the d orbitals involved. As an example we consider transition metal ions in an octahedral coordination of anions. The anions produce a ligand field splitting, with different energies for the e_g type orbitals which point in the direction of the anions, and the t_{2g} type orbitals. From Fig. 2 it is immediately clear that the transfer integral between two t_{2g} orbitals is large, whereas the transfer between t_{2g} and e_g orbitals vanishes. Thus one expects that direct exchange is strongly antiferromagnetic between two cations with half-filled t_{2g} shells, i.e. between $Cr^{3+}(t_{2g}^3)$, or between Mn^{2+}, $Fe^{3+}(t_{2g}^3 e_g^2)$.

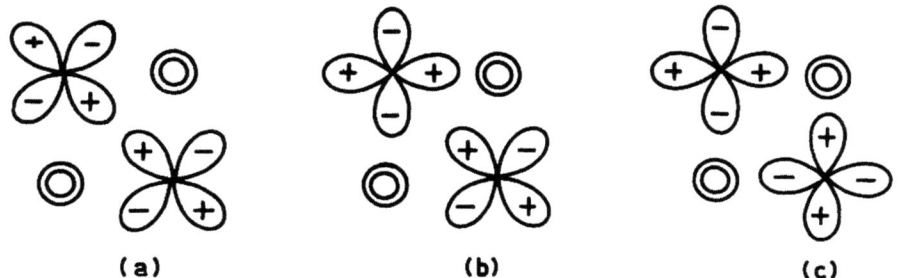

FIGURE 2. Transfer between d orbitals of cations in an octahedral coordination of anions. The anions are shown as double circles. a) large transfer between t_{2g} orbitals; b) transfer between e_g and t_{2g} orbitals vanishes; c) weak transfer between e_g orbitals.

Superexchange

The direct exchange is due to overlap of the orbitals of two cations, and will therefore strongly decrease with increasing distance between the cations. In many magnetic compounds the magnetic ions are separated from one another by non-magnetic atoms, so that the direct exchange must be very small. Nevertheless, in many cases the observed magnetic interactions are not at all weak. This is due to the so-called superexchange interaction between the magnetic ions via the intermediate anions [8]. This superexchange interaction is directly related to the covalency of the metal d orbitals.

The interaction of a magnetic ion and the surrounding ligand atoms causes a covalent mixing of cation orbitals and ligand orbitals [1]. If we consider for example a transition metal atom with an incomplete d shell, the magnetic electrons will occupy orbitals of the type $\varphi' = \varphi_d - \lambda \varphi_L$, where φ_L is a linear combination of ligand orbitals. Thus the magnetic electrons are no longer localized at the central atom, but have also a non-zero density at the ligand atoms. This effect is observed directly in electron paramagnetic resonance as the so-called superhyperfine coupling: the delocalization of the electron spin introduces a coupling with the nuclear spins of the ligands.

With the new orbitals which also extend over the ligand atoms, we can now discuss the superexchange interaction along the same lines as we did for direct exchange. For two magnetic ions A and B, the two orbitals are $\varphi'_A = d_A - \lambda_A \varphi_L$ and $\varphi'_B = d_B - \lambda_B \varphi_L$. Thus if the magnetic ions have a common ligand L, overlap and transfer effects between φ'_A and φ'_B are possible via the intermediate ligand orbital φ_L.

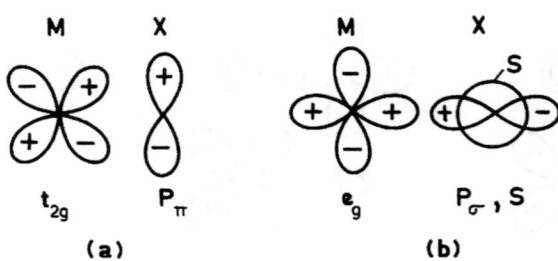

FIGURE 3. Covalency in octahedral coordination. a) Overlap of t_{2g} and p_π causes π-bonding: $\varphi'_{t_{2g}} = t_{2g} - \lambda_\pi p_\pi$. b) Overlap of e_g with p_σ and S causes σ-bonding: $\varphi'_{e_g} = e_g - \lambda_\sigma p_\sigma - \lambda'_\sigma s$.

As an example, we discuss the superexchange interactions between magnetic ions M in a crystal MX with the NaCl-structure [9]. In this crystal structure, all cations are surrounded by an octahedron of anions X. This causes a ligand field splitting of the d levels into t_{2g} and e_g levels. The e_g orbitals are involved in σ bonding with s and p ligand orbitals, the t_{2g} orbitals contribute only to π bonding. (See Fig. 3.)

In the NaCl-structure we distinguish between 180°-superexchange (ions M-X-M on one line) and 90°-superexchange (ions M-X-M form an angle of 90°) (Fig. 4). Just as for direct exchange it is usually the transfer effect which dominates. We expect transfer between two e_g orbitals to be proportional to λ_σ^2, transfer between an e_g and a t_{2g} orbital proportional to $\lambda_\sigma \lambda_\pi$, etc. Because $\lambda_\sigma > \lambda_\pi$, the transfer between e_g orbitals will be strong, between e_g and t_{2g} orbitals moderate and between two t_{2g} orbitals weak. Furthermore, the rules for superexchange are exactly the same as for direct exchange; for example, it is strongly antiferromagnetic between two half-filled orbitals, etc. (Fig. 1). However, we should also consider the geometry of the orbitals involved. It is easily seen that for 180°-superexchange the transfer between two e_g orbitals is large, whereas the transfer between e_g and t_{2g} vanishes by symmetry (Fig. 5).

Although considerations of this type are only qualitative, they have been and still are very useful for classifying and predicting exchange interactions in magnetic materials. A recent example [13] is the application to exchange interactions in the layer-type compounds $ACrS_2$ (A = Li, Na, K, Cu, Ag). The structures of these compounds are based on a cubic close packing of anions with Cr^{3+} (t_{2g}^3) ions occupying octahedral holes, so

MAGNETISM 381

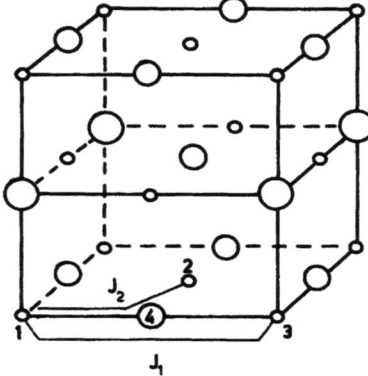

FIGURE 4. Superexchange in a MX crystal with NaCl structure; M = small circles; X = large circles; $J_1 = 180°$-superexchange $M_1 - M_3$; $J_2 = 90°$-superexchange $M_1 - M_2$ via X_4.

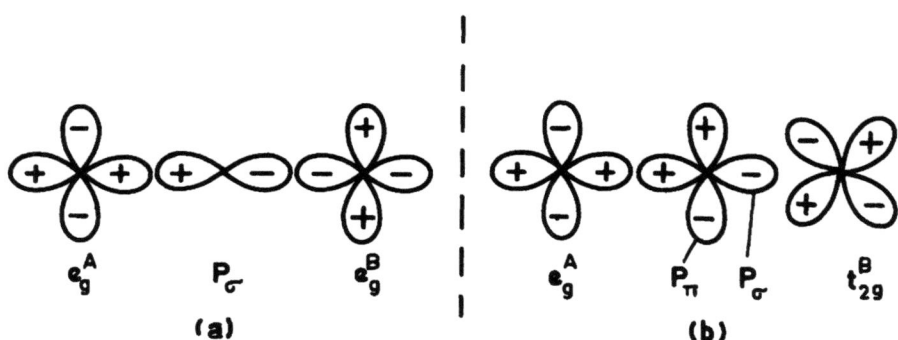

FIGURE 5. $180°$-superexchange between cations A and B. (a) Transfer between e_g^A and e_g^B is possible via p_σ. (b) Transfer between e_g^A and t_{2g}^B vanishes by symmetry.

---

that CrS_2 sandwiches are formed. A strong correlation is found between the asymptotic Curie temperature θ (which is a measure of the exchange interactions) and the Cr-Cr distances in the layers (Fig. 6). This correlation can be understood by considering the magnetic interactions between the spins of adjacent Cr^{3+} ions. In the first place, one expects a strongly antiferromagnetic direct exchange between the Cr^{3+} ions with the half-filled t_{2g} shell. Also the superexchange between Cr ions in the layer should be considered. Because the Cr-S-Cr angle is close to $90°$, this interaction involves an anion p orbital, a t_{2g} orbital (half filled) on one Cr and an e_g orbital (empty) on the other Cr (Fig. 7). Transfer between a half-filled and an empty orbital is

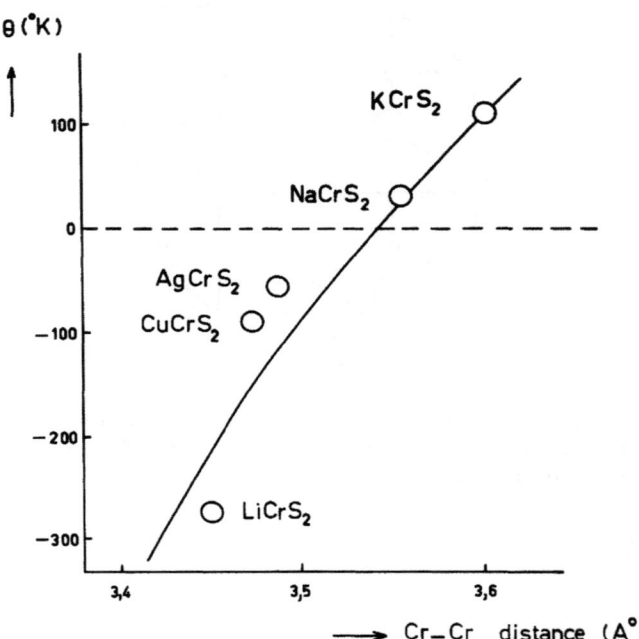

FIGURE 6. Asymptotic Curie temperature θ as a function of the Cr-Cr distance in compounds $ACrS_2$ (A = Li, Na, K, Cu, Ag).

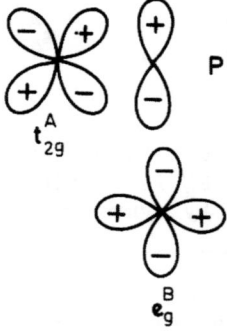

FIGURE 7. 90°-superexchange; transfer $t_{2g}^A - e_g^B$ is possible via ligand orbital p; the transfer is proportional to $\lambda_\sigma \lambda_\pi$.

expected to be ferromagnetic. Therefore one expects two competing interactions, a ferromagnetic superexchange and an antiferromagnetic direct exchange. The direct exchange decreases rapidly with increasing Cr-Cr distance. This is in complete agreement with the observed data (see Fig. 6).

FERROMAGNETISM

Molecular Field Theory of Ferromagnetism

In this paragraph we calculate the influence of the exchange interaction on the magnetic properties. We consider a simple solid, with one type of magnetic atoms of spin $\vec{S}_i$ and magnetic moment $g\mu_B \vec{S}_i$. It is assumed that each magnetic atom has a ferromagnetic exchange interaction J with z neighboring ions. The Hamiltonian of the spins in the presence of an applied magnetic field is:

$$H = - \sum_{ij}{}' J\vec{S}_i\vec{S}_j - \sum_i g\mu_B \vec{S}_i \vec{H} \qquad (17)$$

The sum $\sum'$ is only over neighboring spins i and j.

Even in the absence of an applied field, the lowest energy is obtained if all spins are oriented parallel to one another. At T = 0, the solid is in the ferromagnetic ground state, with magnetization $M_0 = Ng\mu_B S$ (N is the number of spins), and energy $E_0 = -NzJS^2 - Ng\mu_B SH$.

At higher temperature states with spins deviating from the perfectly ordered arrangement also occur. These excitations, which are called spin waves or magnons cause a decrease of the magnetization with increasing temperature. At a certain critical temperature, the Curie temperature T_C, the spontaneous magnetization M (for H = 0) vanishes, and above T_C the solid is paramagnetic.

The calculation of the statistical properties of an assembly of exchange-coupled spins is very difficult. The simplest method to treat such a system is the molecular field approximation [14,15]. In this theory the influence of the neighboring spins on the spin of a particular ion is treated as an effective exchange field:

$$-\vec{S}_i \sum_j 2J\vec{S}_j = - g\mu_B \vec{S}_i \vec{H}_i^e \qquad (18)$$

For the spins S_j we substitute the average value of all spins $\langle S \rangle$, and we obtain

$$\vec{H}_i^e = 2zJ \langle \vec{S} \rangle / g\mu_B \qquad (19)$$

This approximation implies that correlations between neighboring spins are neglected.

We now consider the states of a spin i under the influence of the total field $H + H_i^e$. The energy levels are given by $E_m = -mg\mu_B(H + H_i^e)$, where m is the magnetic quantum number for the component of S along the field ($m = -S, -S+1, \ldots, +S$). The probability for a state m to be occupied, is proportional to $\exp -(E_m/kT)$. Therefore the magnetization $M = Ng\mu_B \langle S \rangle$ is given by

$$M = N \frac{\sum_{m=-s}^{+s} mg\mu_B \exp\{(mg\mu_B/kT)(H + H_i^e)\}}{\sum_{m=-s}^{+s} \exp\{(mg\mu_B/kT)(H + H_i^e)\}} \quad (20)$$

The sums in (20) are easily evaluated; the result is

$$M = Ng\mu_B S\, B_s(x) \quad (21)$$

$$x = \frac{g\mu_B S}{kT}(H + H_i^e) \quad (22)$$

$$B_s(x) = \left(\frac{2S+1}{2S}\right) \coth\left(\frac{2S+1}{2S}\right)x - \left(\frac{1}{2S}\right)\coth\left(\frac{x}{2S}\right) \quad (23)$$

$B_s(x)$ is the so-called Brillouin function.

From these equations we can calculate the spontaneous magnetization $M(H = 0)$ (Fig. 8). The magnetization vanishes at the Curie temperature T_C, given by

$$T_C = \frac{2zJS(S+1)}{3k} \quad (24)$$

We also calculate the magnetic susceptibility χ for $T > T_C$. Consider the case of a weak magnetic field. For small x the Brillouin function can be approximated as $B_s(x) \simeq \left(\frac{S+1}{3S}\right)x$. We obtain for the magnetic susceptibility $\chi = M/H$:

$$\chi = \frac{C}{T - \theta} \quad (25)$$

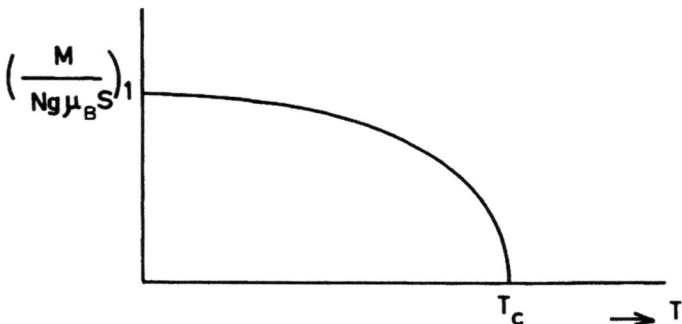

FIGURE 8. Spontaneous magnetization M of a ferromagnet as a function of the temperature.

The Curie constant C is given by $C = Ng^2\mu_B^2 S(S + 1)/3k$, the asymptotic Curie temperature θ is equal to T_C. Equation (25) is the well-known Curie-Weiss law.

As mentioned already, the molecular field method is an approximate method which neglects short range order. It can be shown by more accurate methods [16] (for example the Bethe-Peierls method, [17,18] which solves the Hamiltonian for a cluster of a central ion with its neighbors) that although the long range order vanishes at T_C, short range order persists above T_C. This has the effect of lowering T_C for a given strength of the exchange interaction. For spins $S = \frac{1}{2}$ on a b.c.c. lattice, the molecular field theory gives $kT_C = 4J$, the Bethe-Peierls method $kT_C = 2.91J$, whereas the exact result based on series expansion methods, is $kT_C = 2.64J$.

Spin Waves and Magnons

In this paragraph we discuss very briefly the collective excitations of the spin system [19-21]. We consider again the Hamiltonian [17] for an assembly of interacting spins $S = \frac{1}{2}$. The Hamiltonian for $H = 0$ can also be written as

$$H = -J \sum_{ij}{}' \{S_{zi} S_{zj} + \tfrac{1}{2} S_i^+ S_j^- + \tfrac{1}{2} S_i^- S_j^+\} \qquad (26)$$

where $S_i^{\pm} = S_{xi} \pm iS_{yi}$. The ferromagnetic ground state is $\chi_0 = (\chi_{\alpha 1}, \chi_{\alpha 2}, ---, \chi_{\alpha N})$; i.e. the state with all N spins in a state

α, with $m_s = \frac{1}{2}$. The effect of the operators $S^{\pm}$ and S_z on spin functions α and β is: $S^+\chi_\alpha = 0$; $S^+\chi_\beta = \chi_\alpha$; $S^-\chi_\alpha = \chi_\beta$; $S^-\chi_\beta = 0$; $S_z\chi_\alpha = \frac{1}{2}\chi_\alpha$; $S_z\chi_\beta = -\frac{1}{2}\chi_\beta$. For the energy of the ground state we obtain:

$$E_0 = \langle \chi_0 | H | \chi_0 \rangle = -1/4 \, NzJ \qquad (27)$$

We now consider an excited state, with the spin of atom n not in χ_α but in a state χ_β with $m_s = -\frac{1}{2}$: $\chi_n = (\chi_{\alpha 1}, \chi_{\alpha 2} \text{-------}, \chi_{\beta n}, \text{---------} \chi_{\alpha N})$. For the effect of the operator H on χ_n we find:

$$H \chi_n = (E_0 + zJ) \chi_n - \sum_{i=1}^{z} J \chi_{n+i} \qquad (28)$$

(the sum is over all neighbors n+i of atom i). Equation (28) shows that χ_n is not an eigenfunction of H. The proper eigenstates of H are linear combinations of the Bloch-type:

$$\psi_k = \sum_n e^{i\vec{k}\vec{R}_n} \chi_n \qquad (29)$$

These states are characterized by a wave vector $\vec{k}$; $\vec{R}_n$ is the position of atom n. For the energy of a state $\psi_{\vec{k}}$ we find $E_{\vec{k}} = \langle \psi_{\vec{k}} | H | \psi_{\vec{k}} \rangle$:

$$E_{\vec{k}} = E_0 + J \left\{ z - \sum_i e^{i\vec{k}\vec{b}_i} \right\} \qquad (30)$$

The vectors $\vec{b}_i$ are the vectors connecting the central atom with its neighboring atoms.

In the states $\psi_{\vec{k}}$ the excited spin is not localized at a particular atom, but is propagating through the crystal. Such a propagating spin deviation is called a spin wave (Fig. 9). The excitation energy is $\hbar w_{\vec{k}} = E_{\vec{k}} - E_0$; a single quantum of energy $\hbar w_{\vec{k}}$ and wave vector $\vec{k}$ is called a magnon. For small values of k we find approximately $\hbar w_{\vec{k}} \simeq \frac{1}{2} J \sum_i (\vec{k}\vec{b}_i)^2$, and for a simple body-centered cubic lattice this reduces to $\hbar w_{\vec{k}} = Jk^2 a^2$. (a is the unit-cell edge). The full magnon dispersion curves ($\hbar w_{\vec{k}}$ as a

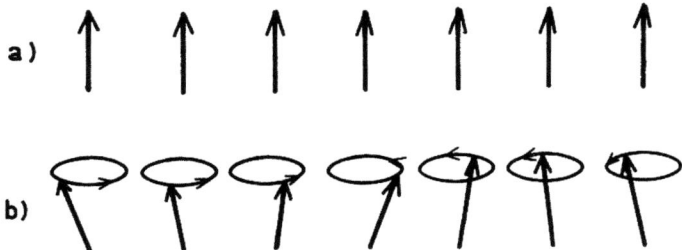

FIGURE 9. a) Ferromagnetically-ordered spins; b) propagating spin wave.

---

function of $\vec{k}$) can be determined experimentally from inelastic neutron scattering experiments [22].

Because a magnon corresponds to a single reversed spin, the magnetization of a crystal with $n = \sum_{\vec{k}} \bar{n}_{\vec{k}}$ magnons is simply given by $M = Ng\mu_B S - ng\mu_B$. In thermal equilibrium the average number of magnons $\bar{n}_k$ with wave vector $\vec{k}$ is $\{\exp(\hbar w_{\vec{k}}/kT) - 1\}^{-1}$. The number of states between k and k + dk per unit volume is $\left(\frac{1}{2\pi}\right)^3 4\pi k^2 \, dk$, so that

$$n = \sum \bar{n}_k = \left(\frac{1}{2\pi}\right)^3 \int_{\text{(Brillouin zone)}} \frac{4\pi k^2 dk}{e^{(\hbar w_{\vec{k}}/kT)} - 1} \quad (31)$$

At low temperatures, the integral can be approximated by an integral from zero to infinity. From a straight-forward calculation we obtain

$$M = M_0 (1 - A\, T^{3/2}) \quad (32)$$

This is the Bloch $T^{3/2}$-law for the change of the magnetization at low temperature.

Magnetic Anisotropy; Ferromagnetic Resonance

So far we have discussed only the orientation of the spins with respect to each other, as a result of the exchange interactions. There exist weaker interactions [23] which bind the spins to specific crystallographic directions. These are the magnetic dipole-dipole interactions (this term vanishes in cubic crystals) and interactions caused by spin-orbit coupling. The latter effect is a single ion effect. The orbital momentum of the electron is coupled to the crystallographic directions by means of the crystal field interactions, and the spin-orbit interaction of the spin to the orientation of the orbital momentum. These single-ion effects are important for ions with a large orbital momentum (for example Co^{2+} in an octahedron of anions). The magnetic anisotropy is observed directly if one measures the magnetization vs field curves of single crystals for various orientations of the field (Fig. 10).

The anisotropy energy is usually written as a power series in the direction cosines of the angles of the magnetization with the crystallographic axes. For a hexagonal crystal one finds for the anisotropy energy

$$E_A = K_1 \sin^2\theta + K_2 \sin^4\theta + \text{-------} \quad (33)$$

where θ is the angle between the magnetization M and the hexagonal c-axis. For cubic crystals one obtains

$$E_A = K_1(\alpha_1^2\alpha_2^2 + \alpha_1^2\alpha_3^2 + \alpha_2^2\alpha_3^2) + K_2\alpha_1^2\alpha_2^2\alpha_3^2 + \text{------} \quad (34)$$

where α_1, α_2, α_3 are the cosines of the angles of M with the x, y and z axis, respectively. From the experimental data one finds for Fe at 300°K, $K_1 = 4.2 \times 10^5$ erg/cm^3 and $K_2 = 1.5 \times 10^5$ erg/cm^3; in agreement with these values one finds indeed that the energy is lowest if M is parallel to the [100] direction (with $\alpha_1 = \alpha_2 = \alpha_3 = 1/\sqrt{3}$).

It is not at all obvious that a series expansion converges rapidly. In fact, in some metallic compounds one has observed anisotropy energies which depend on the orientation of the magnetization in a complicated way. The anisotropy in alloys depends critically on details of the Fermi surface.

The magnetic anisotropy can also be determined from ferromagnetic resonance measurements [24]. The classical equation of motion of magnetization M of the sample in an applied field H is:*

$$\frac{d\vec{M}}{dt} = \gamma(\vec{M} \times \vec{H}) \quad (35)$$

---

*The field $\vec{H}$ includes the demagnetizing field; a consequence is that the resonance frequency depends on the shape of the sample.

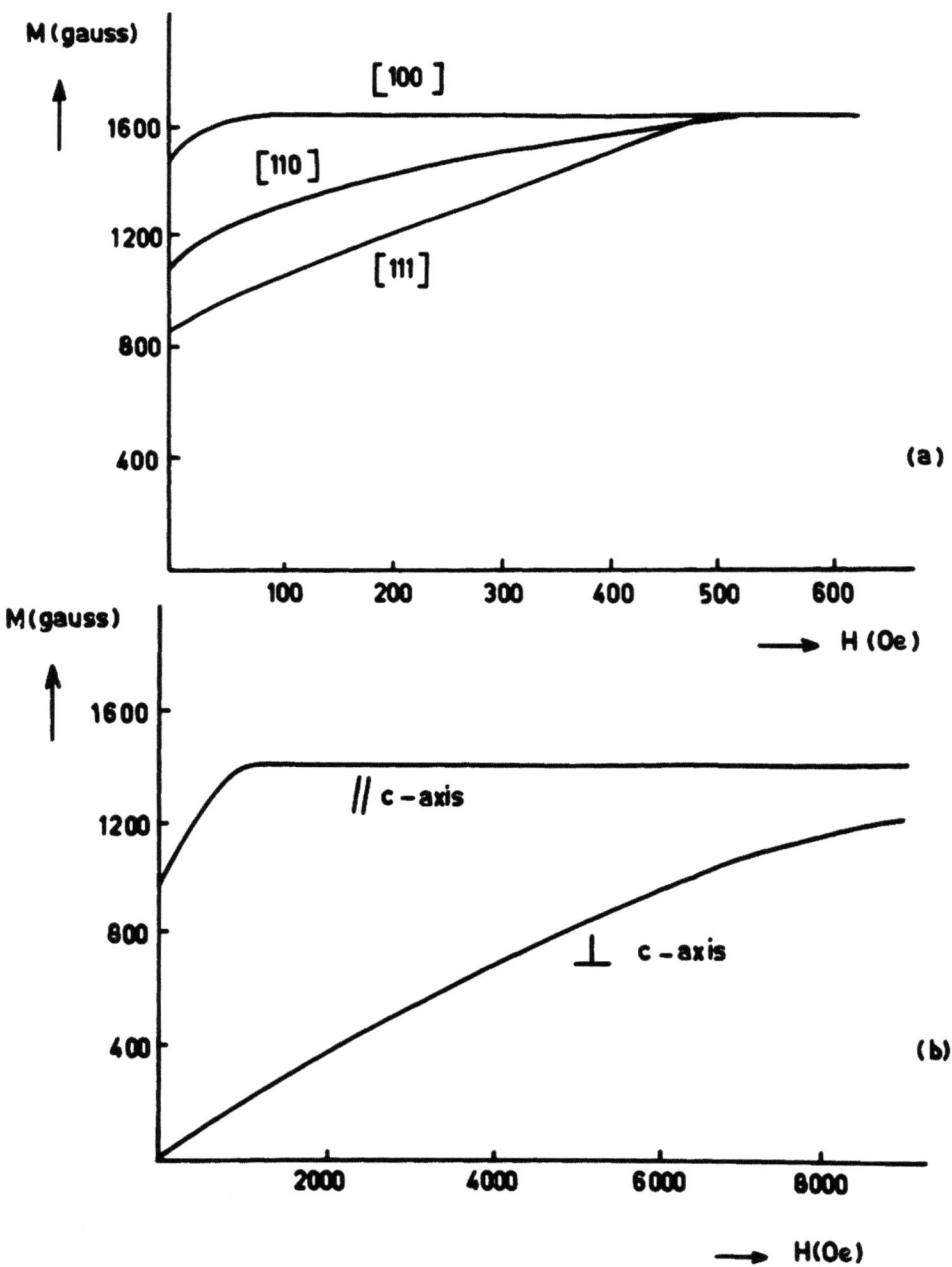

FIGURE 10. Magnetization curves of single crystals of (a) iron and (b) cobalt.

where $\gamma = eg/2mc$. Thus, for a field $H_0 \parallel z$, the equations of motion become

$$\frac{dM_x}{dt} = \gamma H_0 M_y; \quad \frac{dM_y}{dt} = -\gamma H_0 M_x; \quad \frac{dM_z}{dt} = 0 \qquad (36)$$

with solutions

$$M_x = a \sin \omega t; \quad M_y = a \cos \omega t; \quad M_z = \text{constant} \qquad (37)$$

with $\omega = \gamma H_0$. Thus the magnetization precesses around the applied field with a frequency ω. If the system is subjected to microwave radiation of frequency ω, a strong absorption occurs: this is the ferromagnetic resonance absorption.

Consider now the influence of the magnetic anisotropy of a uniaxial crystal on the ferromagnetic resonance. The energy of the sample in a field H_0 is $E = K \sin^2 \theta - MH_0 \cos \theta$, if H_0 is parallel to the axis, and the magnetization makes an angle θ with the axis. For small angles θ, E can be written approximately as $E = 2K - M(H_0 + H_A) \cos \theta$, with $H_A = 2K/M$. Thus the influence of the anisotropy for small angles is equivalent to the effect of an anisotropy field H_A along the axis. For resonance, the total field should be considered; the resonance condition is $\omega = \gamma(H_0 + H_A)$. Thus, from a determination of ω one can calculate directly the anisotropy field H_A.

Ferromagnetic Domains

The total energy of a piece of magnetized material also has a term due to the external magnetic field. This term is $\int H_e^2 dV$, and is always positive. In order to minimize the total energy, small magnetic domains are formed in the material [25] (Fig. 11).

The separation between two domains has a finite thickness; it is called a Bloch wall [26]. Inside the wall the orientation of the magnetic moments changes gradually from the orientation of the moments in one domain to that in the other domain.

The formation of Bloch walls costs energy because the spins in the wall are not parallel to one another (this costs exchange energy), and they are not along the direction of easy magnetization (which costs anisotropy energy). In equilibrium the energy of the external field plus the energy of the wall is a minimum. From

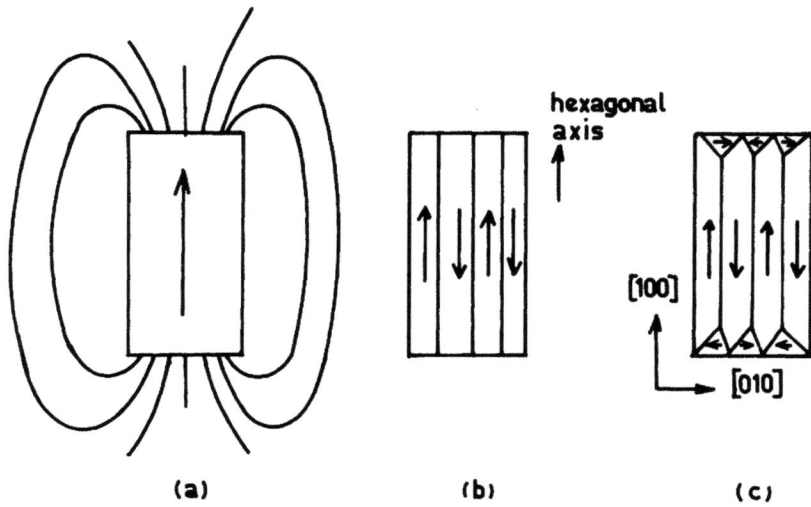

FIGURE 11. (a) Large external field for a single domain specimen. (b) Domains in hexagonal crystal (Co); easy magnetization along the hexagonal axis. (c) Domains in cubic crystal (Fe), with easy magnetization along [100] directions.

---

this condition it is possible to calculate the size of domains and the thickness of the Bloch walls; typical values are 10^{-2} and 10^{-5} cm, respectively.

The study of domains is of great importance for the understanding of magnetization processes, hysteresis, switching, etc. The domain structure depends on the geometry of the specimen. Of great interest (for memory applications in computers) is the study of domains in thin films. Recently cylindrical domains (bubbles) in thin films have been observed; these have quite fascinating properties [27,28].

We only mention very briefly the complicated phenomena which occur when a piece of ferromagnetic material is magnetized [29]. Let us start with a sample of zero magnetic moment, i.e. a sample where the magnetization of the domains compensate each other. In an applied field the following processes take place (Fig. 12):

a) For small fields H the Bloch walls between the domains begin to move in such a way that domains which have the magnetization M in the direction of H grow. Some of the Bloch walls move quite easily and reversibly, others are pinned at lattice defects. For these a certain minimum field is required

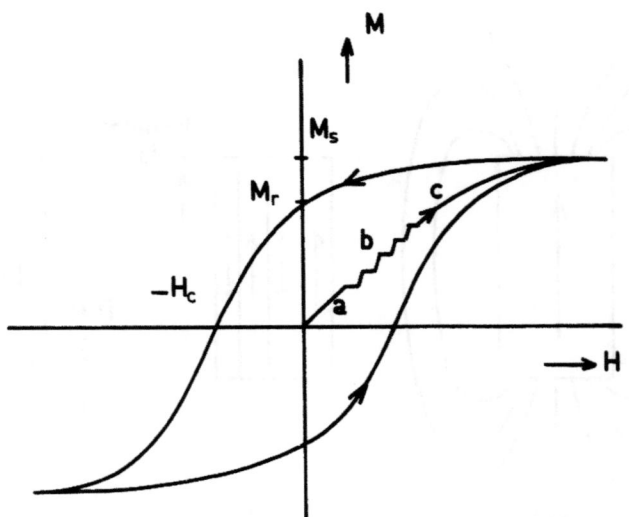

FIGURE 12. Magnetization process of a ferromagnetic material. (a) Reversible motion of Bloch walls. (b) Irreversible motion of Bloch walls (Barkhausen jumps). (c) Rotation magnetization. M_S = saturation magnetization; M_r = remanent magnetization; H_c = coercitive field.

---

before they begin to move. They move over a certain distance until they become blocked again at a different defect. Due to this irreversible motion of the Bloch walls, the magnetization increases jump wise (Barkhausen jumps).

b) At higher fields the magnetization in the domains rotates from the direction of easy magnetization to the direction of the field.

c) If the field is removed, the magnetization decreases. However, a certain remanent magnetization M_r remains even for H = 0, because some of the Bloch walls are pinned at defects. Thus, the magnetization vanishes only at a negative field, the coercitive field $-H_c$.

Magnetic materials show a wide range of coercitive fields [29]. For permanent magnets one needs so-called hard magnetic materials, with a high coercitive field H_c and a large saturation magnetization M_S; very large values have been obtained for rare-earth cobalt compounds Co_5Sm with H_c = 16000 Oe, M_r = 9000 Gauss [31]. For applications in transformers soft magnetic materials, with a large initial permeability $\mu = (dB/dH)_0$ and a small coercive force are required; examples are some Ni-Fe alloys (permalloys) with $\mu \sim 10^5$. In computers small ferrite rings are used as memory elements. Switching these elements consists of changing the magnetization from $+M_S$ to $-M_S$. For optimum performance a square hysteresis loop and a small coercive field ($H_c \sim 1$ Oe) is required. It was found recently in some materials that the magnetic properties (coercive field, permeability) can be changed by shining light on the specimen (photomagnetic effect [32]).

ANTIFERROMAGNETISM AND FERRIMAGNETISM

Molecular Field Theory of Antiferromagnetism

If the exchange interaction J between the spins is negative, the spins will try to orient themselves antiparallel to each other. This causes an antiferromagnetic ordering of the spins at low temperature [33].

Consider as an example a simple antiferromagnet with two equivalent magnetic sublattices [30,33]. All $\frac{1}{2}N$ spins on sublattice 1 are parallel, so that the sublattice magnetization is $M_1 = \frac{1}{2} N g \mu_B \langle S_1 \rangle$; for sublattice 2 $M_2 = \frac{1}{2} N g \mu_B \langle S_2 \rangle$. The exchange interactions are given by $H' = - \sum_{ij}' J_{ij} \vec{S}_i \vec{S}_j$. If a spin S_1 of sublattice 1 interacts with z_{11} spins of sublattice 1 (exchange constant J_{11}) and with z_{12} spins of sublattice 2 (exchange constant J_{12}), the exchange field acting on spin S_1 is:

$$H_1^e = 2z_{11}J_{11}\langle S_1 \rangle/g\mu_B + 2z_{12}J_{12}\langle S_2 \rangle/g\mu_B \tag{38}$$

Thus the exchange field can be written as

$$\begin{aligned} H_1^e &= \alpha M_1 + \alpha' M_2 \\ H_2^e &= \alpha' M_1 + \alpha M_2 \end{aligned} \tag{39}$$

$$\alpha = \frac{4z_{11}J_{11}}{Ng^2\mu_B^2} \quad \alpha' = \frac{4z_{12}J_{12}}{Ng^2\mu_B^2} \tag{40}$$

The average value $\langle S_1 \rangle$ of the component of the spin along the total field $H + H_1^e$ acting on it is calculated in the same way as for ferromagnets. The result is:

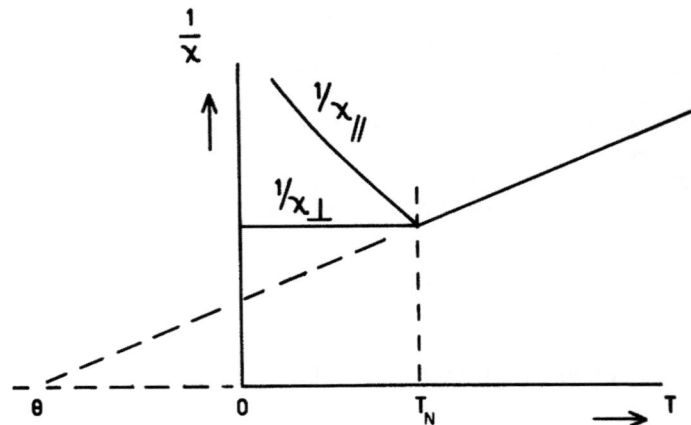

FIGURE 13. Reciprocal susceptibility for an antiferromagnet, as a function of the temperature.

$$M_1 = \tfrac{1}{2} Ng\mu_B S B_S \left\{ \frac{g\mu_B S}{kT} (H + \alpha M_1 + \alpha' M_2) \right\}$$

$$M_2 = \tfrac{1}{2} Ng\mu_B S B_S \left\{ \frac{g\mu_B S}{kT} (H + \alpha' M_1 + \alpha M_2) \right\}$$

(41)

For $H = 0$, we obtain for the antiferromagnetic solution $M_1 = -M$ the equation $M_1 = \tfrac{1}{2} Ng\mu_B S B_S (x)$ with $x = (g\mu_B S/kT)(\alpha - \alpha')M_1$. Thus, the temperature dependence of the sublattice magnetization of the antiferromagnet is precisely the same as that of a ferromagnet (Fig. 8). The sublattice magnetization vanishes above the Néel temperature T_N, given by

$$T_N = \tfrac{1}{2} C (\alpha - \alpha') \qquad (42)$$

C is the curie constant

$$C = \frac{Ng^2 \mu_B^2 S(S+1)}{3k} \qquad (43)$$

In the paramagnetic region $T > T_N$, and for small applied field, we obtain with the approximation $B_S(x) \simeq \left(\frac{S+1}{3S}\right) x$:

$$M_1 = \frac{C}{2T} (H + \alpha M_1 + \alpha' M_2)$$

$$M_2 = \frac{C}{2T} (H + \alpha' M_1 + \alpha M_2)$$

(44)

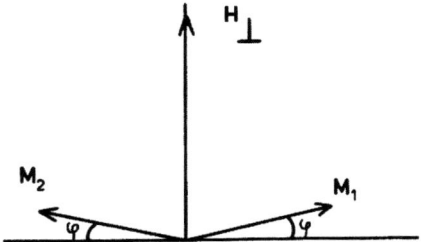

FIGURE 14. Change of the sublattice magnetizations of an antiferromagnet by a perpendicular field $H_\perp$.

The magnetic susceptibility $\chi = (M_1 + M_2)/H$ is (Fig. 13)

$$\chi = \frac{C}{T - \theta} \qquad (45)$$

$$\theta = \tfrac{1}{2} C (\alpha + \alpha') \qquad (46)$$

The susceptibility at T_N is $\chi(T_N) = -(1/\alpha')$. Below T_N, the susceptibility is anisotropic. For an applied field perpendicular to the sublattice magnetizations, the susceptibility $\chi_\perp$ is easily calculated. The field causes the orientation of the sublattice magnetization to change by a small angle φ (Fig. 14). The energy is

$$E = -\tfrac{1}{2} \alpha (M_1^2 + M_2^2) - \alpha' M_1 M_2 - (M_1 + M_2)H \qquad (47)$$

For small φ, this can be written as

$$E = -(\alpha - \alpha')M^2 - 2\alpha' M^2 \varphi^2 - 2MH\varphi \qquad (48)$$

if $|M_1| = |M_2| = M$. The angle φ will take the value for which the energy is a minimum; from $(\partial E/\partial \varphi) = 0$ we obtain $\varphi = -(H/2\alpha' M)$. The magnetization along the field is $M_\perp = 2M\varphi = -(H/\alpha')$, so that $\chi_\perp = -(1/\alpha')$.

The susceptibility $\chi_\parallel$ for fields parallel to the sublattice magnetizations is more difficult to calculate; it decreases at lower temperature (Fig. 13).

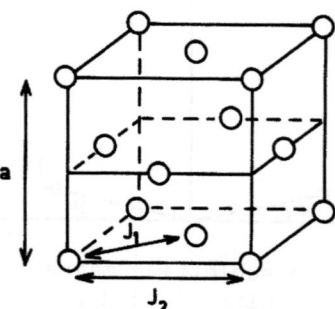

FIGURE 15. Exchange interactions in a fcc lattice of magnetic atoms (lattice constant a).

Types of Antiferromagnetic Order

For a given crystalline arrangement of ions, several different types of antiferromagnetic order are possible [30]. In this section we discuss as an example the ordering of spins on a face-centered lattice (Fig. 15). We consider only the exchange interactions of a spin with its 12 nearest neighbors (distance $\frac{1}{2} a \sqrt{2}$, interaction J_1) and the 6 next-nearest neighbors (distance a, interaction J_2). The paramagnetic Curie temperature θ is easily calculated with a molecular field model to be $\theta = 2S(S + 1)(12 J_1 + 6 J_2)/3k$. The Néel temperature depends on the type of magnetic order; it can be calculated with a molecular field model in the manner indicated in the preceding paragraph. The exchange energy E_0 at low temperature is calculated directly from $H = - \sum'_{ij} J_{ij} \vec{S}_i \vec{S}_j$; it is directly related to the Néel temperature by

$$E_0 = - \frac{3NSkT_N}{2(S + 1)} \qquad (49)$$

In Fig. 16 several types of magnetic order in the fcc structure are shown. The corresponding exchange energies are given in Table 1. The stability of the various types of order depends on the ratio (J_1/J_2); the ranges of stability are easily calculated from the energies given in Table 1. The result is shown in Fig. 17.

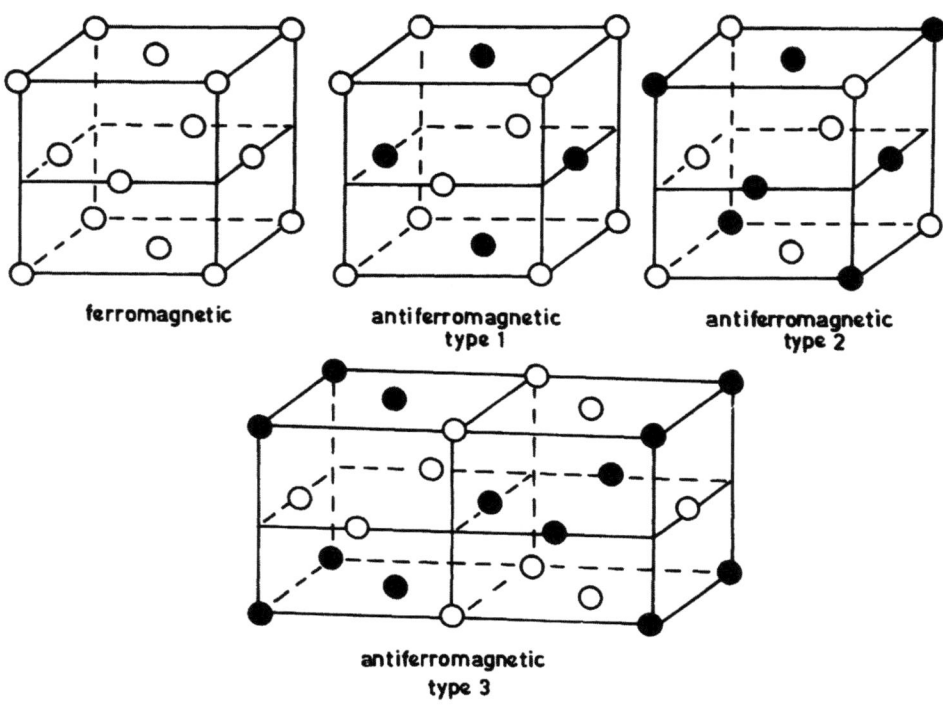

FIGURE 16. Types of magnetic order in a fcc lattice.

TABLE 1. Energies of Magnetic Structures in a fcc Lattice.

	neighbors at		E_0
	$\frac{1}{2}a\sqrt{2}$	a	
ferromagnetic	12 parallel	6 parallel	$-NS^2(12\ J_1 + 6\ J_2)$
antif. type 1	4 parallel + 8 antiparallel	6 parallel	$-NS^2(-4\ J_1 + 6\ J_2)$
antif. type 2	6 parallel + 6 antiparallel	6 antiparallel	$+NS^2\ 6\ J_2$
antif. type 3	4 parallel + 8 antiparallel	4 parallel + 2 antiparallel	$-NS^2(-4\ J_1 + 2\ J_2)$

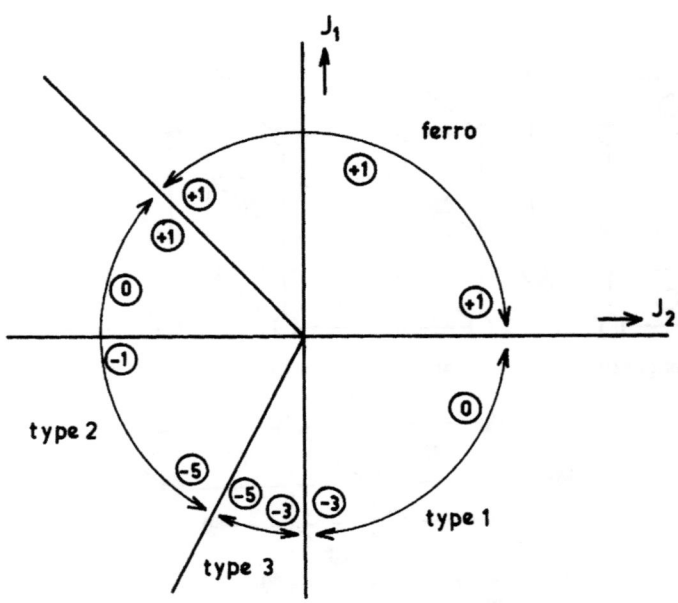

FIGURE 17. Stability ranges of various types of magnetic order for a fcc lattice of magnetic ions. The numbers along the circle indicate the ratio (θ/T_N).

In Table 2 experimental data of magnetic compounds with an fcc lattice of magnetic atoms are given. There is a reasonable agreement with the simple theory presented above.

TABLE 2. Experimental Data on Magnetic Compounds with an fcc Lattice of Magnetic Atoms.

Compound	Crystal Structure	$\theta(°K)$	T_N or $T_C(°K)$	θ/T_N	Type of Order
MnO	NaCl	-610	120	-5.0	antif. type 2
FeO	NaCl	-570	200	-2.9	antif. type 2
CoO	NaCl	-330	290	-1.1	antif. type 2
NiO	NaCl	-1300	520	-2.5	antif. type 2
MnS	NaCl	-400	130	-3.1	antif. type 2
MnS	ZnS	-980	160	-6.1	antif. type 3
MnSe	NaCl	-360	170	-2.1	antif. type 2
MnS_2	pyrite	-600	60	-10.0	antif. type 3
$MnSe_2$	pyrite	-480	100	-4.8	complicated
$MnTe_2$	pyrite	-530	80	-6.5	antif. type 1
EuO	NaCl	+76	76	1.0	ferrom.
EuS	NaCl	+19	16	1.2	ferrom.
EuSe	NaCl	+9	4.6	-	<2.8°K: ferrom. <4.6°K: antif.
EuTe	NaCl	-6	8	-0.8	antif.

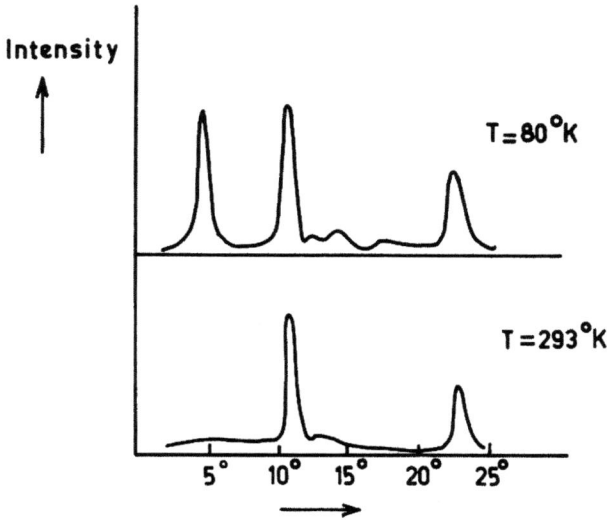

FIGURE 18. Neutron diffraction of MnO (T_N = 120°K); θ is the Bragg angle of scattering.

The magnetic structure can be determined from neutron diffraction investigations [22,34]. This method is similar to the well-known method of X-ray diffraction of crystals. Whereas the electronic charge density is responsible for the scattering of X-rays, the neutrons are scattered by the atomic nuclei and, by virtue of their magnetic moments, also by the magnetic moments of the atoms. In the paramagnetic region the nuclei contribute to the Bragg reflections; the disordered magnetic moments produce only a diffuse scattering. However, below T_N or T_C, the magnetic moments are ordered and will also contribute to the Bragg reflections. Because the symmetry and the size of the unit cell of the magnetic structure may differ from that of the crystallographic structure, it is possible that certain Bragg reflections, which are absent in the paramagnetic state, appear in the magnetically ordered state (Fig. 18). From a comparison of the diffraction patterns above and below the ordering temperature it is possible to deduce the magnetic structure.

Ferrimagnetism

If a magnetic material has two (or more) nonequivalent magnetic sublattices which are oriented antiparallel to each other, a net magnetization exists at low temperature; such a material is called ferrimagnetic [35].

We consider a simple ferrimagnet, with two sublattices 1 and 2. The exchange fields are

$$H_1^e = \alpha_{11} M_1 + \alpha_{12} M_2$$
$$H_2^e = \alpha_{21} M_1 + \alpha_{22} M_2 \tag{50}$$

The coefficients α_{12} and α_{21} are equal because both terms $\alpha_{12} M_2$ and $\alpha_{21} M_1$ arise from a single term $\alpha_{12} M_1 M_2$ in the expression for the total energy. However, $\alpha_{11} \neq \alpha_{22}$ because the two sublattices are not equivalent. We obtain in the usual way

$$M_1 = N_1 g_1 \mu_B S_1 B_{S_1}(x_1)$$
$$M_2 = N_2 g_2 \mu_B S_2 B_{S_2}(x_2) \tag{51}$$

with $x_1 = \dfrac{g_1 \mu_B S_1}{kT}(H + H_1^e)$; $x_2 = \dfrac{g_2 \mu_B S_2}{kT}(H + H_2^e)$.

N_1 and N_2 are the numbers of atoms of sublattices 1 and 2, with spin quantum numbers S_1 and S_2 and magnetic moments $g_1 \mu_B S_1$ and $g_2 \mu_B S_2$, respectively.

We first calculate the susceptibility in the paramagnetic region for weak fields. Using again the expression $B_S(x) \simeq (S+1) x/3S$, we obtain

$$M_1 = (C_1/T)(H + H_1^e)$$
$$M_2 = (C_2/T)(H + H_2^e) \tag{52}$$

with $C_1 = N_1 g_1^2 \mu_B^2 S_1(S_1+1)/3k$ and $C_2 = N_2 g_2^2 \mu_B^2 S_2(S_2+1)/3k$. Substituting the expressions for H_1^e and H_2^e gives:

$$(T - C_1 \alpha_{11}) M_1 - C_1 \alpha_{12} M_2 = C_1 H$$
$$-C_2 \alpha_{12} M_1 + (T - C_2 \alpha_{22}) M_2 = C_2 H \tag{53}$$

From these equations one can calculate M_1 and M_2, and from these the susceptibility $\chi = (M_1 + M_2)/H$.

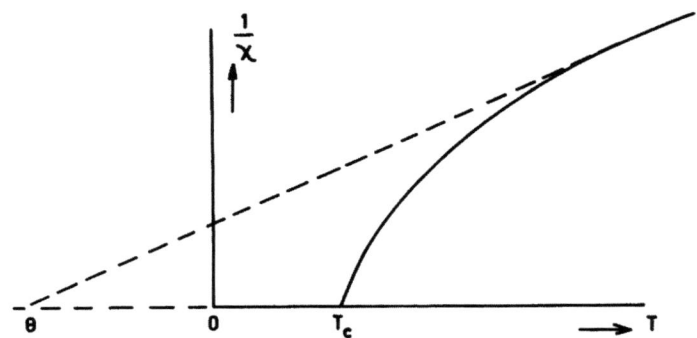

FIGURE 19. Reciprocal susceptibility of a ferrimagnet.

The result is

$$\frac{1}{\chi} = \frac{T - \theta}{C} - \frac{\sigma}{T - \theta'} \tag{54}$$

$$C = C_1 + C_2; \quad \theta' = \left(\frac{C_1 C_2}{C}\right)(\alpha_{11} + \alpha_{22} - 2\alpha_{12})$$

$$\sigma = \left(\frac{C_1 C_2}{C^3}\right)[C_1^2(\alpha_{11} - \alpha_{12})^2 + C_2^2(\alpha_{22} - \alpha_{12})^2 - \tag{55}$$

$$- 2C_1 C_2 \{\alpha_{12}^2 - (\alpha_{11} + \alpha_{22})\alpha_{12} + \alpha_{11}\alpha_{22}\}]$$

$$\theta = \frac{1}{C}(\alpha_{11} C_1^2 + \alpha_{22} C_2^2 + 2\alpha_{12} C_1 C_2).$$

These equations show that at high temperature $1/\chi$ approaches a straight line $1/\chi = (T - \theta)/C$, with an asymptotic Curie temperature θ (Fig. 19). The susceptibility becomes infinite at the Curie temperature T_C, which is given by

$$T_C = \tfrac{1}{2}(C_1\alpha_{11} + C_2\alpha_{22}) + \tfrac{1}{2}\{(C_1\alpha_{11} - C_2\alpha_{22})^2 + 4C_1 C_2 \alpha_{12}^2\}^{\tfrac{1}{2}} \tag{56}$$

The spontaneous magnetization ($H = 0$) can also be calculated from equations (51). Because $M_1 \neq M_2$, a net magnetization M results below T_C; this net magnetization may exhibit a curious temperature dependence. Some examples are shown in Fig. 20.

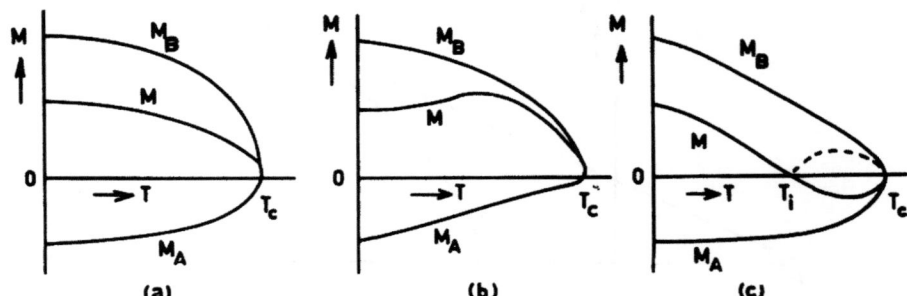

FIGURE 20. Temperature dependence of the sublattice magnetizations M_A and M_B, and of the total magnetization $M = M_A + M_B$, for a ferrimagnet. (a): the temperature dependences of M_A, M_B and M are equal. (b): weak molecular field at M_A. (c): weak molecular field at M_B (inversion point T_i).

Suppose that the saturation magnetization of sublattice B is larger than that of sublattice A. If the molecular field on sublattice A is small, M_A will decrease rapidly with increasing temperature. In that case the total magnetization $M = M_A + M_B$ may exhibit a maximum (Fig. 20b). If, on the other hand, the molecular field on B is small, M_B decreases more rapidly than M_A with increasing temperature, and the total magnetization may show an inversion point (Fig. 20c).

A large number of non-conducting magnetic oxides exhibits ferrimagnetism. Many of these materials have found application in the electronic industry as permanent magnets, as ferrite cores for memory elements in computers, in microwaves devices, etc.

An important class of materials are the magnetic oxides $A[B_2]O_4$ with the cubic spinel structure [36,37]. The oxygen ions form essentially a face-centered cubic lattice. The cations occupy the sites A and B; cations on site A are surrounded by a tetrahedron, cations on site B by an octahedron of oxygen ions. In a so-called normal spinel, the A sites are occupied by divalent ions (Cu^{2+}, Fe^{2+}, Mg^{2+}, Ni^{2+}, Mn^{2+}, etc.) and the B sites by trivalent ions (Fe^{3+}, Mn^{3+}, Cr^{3+}, etc.). In an inverse spinel the divalent ions occupy half of the B sites, the trivalent ions are distributed over the A sites and the remaining B sites.

In most of the spinels there is a strong antiferromagnetic superexchange interaction between the cations on A and B sites. As a consequence, most spinels are ferrimagnetic with two magnetic sublattices M_A and M_B. For example, in magnetite, which is an inverse spinel, the sublattice magnetizations are $M_A = 5 \mu_B$ (Fe^{3+}), $M_B = 9 \mu_B$ ($Fe^{2+} + Fe^{3+}$) and the saturation magnetization is (approximately, because orbital contributions were neglected) $4 \mu_B$.

The magnetic properties of spinel type oxides have been studied extensively [36,37]. It is possible in many cases to obtain a magnetic material of desired properties by making suitable solid solutions. For example the magnetic anisotropy of spinels can be varied strongly by doping with small amounts of Co^{2+} (due to the orbital momentum, this ion has a large single-ion anisotropy), and $Co_{0.03}Ni_{0.97}Fe_2O_4$ has essentially zero anisotropy.

A class of magnetic oxides with a very large magnetic anisotropy are the hexagonal ferrites; these ferrimagnetic materials are used as permanent magnets. An example is ferroxdure, with composition $BaFe_{12}O_{19}$ [37,38].

Finally we mention the magnetic oxides with the garnet structure [39-41]. These materials with formula $Z_3Fe_5O_{12}$ are cubic. The iron ions are all trivalent; two occupy octahedral, the other three occupy tetrahedral sites. Sites Z have a dodecahedral coordination; these sites can be occupied by Y^{3+}, and by rare-earth ions Gd^{3+}, Tb^{3+}, Dy^{3+}, Ho^{3+}, etc. The rare-earth garnets show an inversion point (Fig. 20c). The most important garnet is the so-called YIG ($Y_3Fe_5O_{15}$); of this material single crystals of good quality can be grown. This results in a very narrow line width for ferromagnetic resonance.

Other Types of Magnetic Order

In this paragraph we mention briefly some other types of magnetic order.

Experimental data on some magnetic spinels were found to be incompatible with a simple type of spin ordering. It was shown subsequently that in these materials a non-collinear spin ordering occurs [42]. Consider for example a spinel with a very strong antiferromagnetic exchange interaction between B-site cations, and a weaker antiferromagnetic interaction between A- and B-site cations. In such a situation the lowest energy is obtained for a triangular configuration of the spins: the B-site cations form two magnetic sublattices $\vec{M}_B$ and $\vec{M}_{B'}$ which make an angle with each other (Fig. 21). A spin structure of this type occurs in $CuCr_2O_4$ [43].

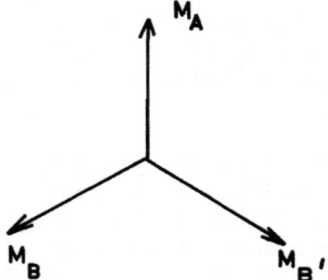

FIGURE 21. Triangular spin configuration.

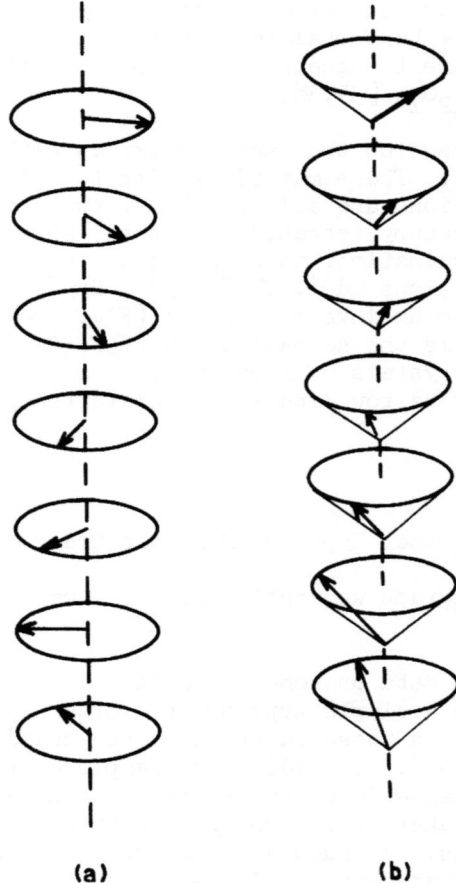

(a)   (b)

FIGURE 22. Antiferromagnetic (a) and ferrimagnetic (b) spiral.

It has been shown that the presence of competing exchange interactions can lead to magnetic spirals [44-46]. As an example we discuss a simple hexagonal lattice [47]. We assume that the interaction between spins in a plane perpendicular to the hexagonal c-axis is very strong and aligns all spins in the plane ferromagnetically. We assume that there is a ferromagnetic interaction between nearest neighbors along the axis ($J_1 > 0$), and an antiferromagnetic interaction between next-nearest neighbors along the axis ($J_2 < 0$). The exchange energy due to the interaction between the layers is

$$E = -2NS^2(J_1 \cos\theta + J_2 \cos 2\theta) \qquad (57)$$

if θ is the angle between spins in adjacent planes. Minimizing E with respect to θ gives $\cos\theta = -(J_1/4 J_2)$ as the stable configuration for $|4 J_2| > |J_1|$ (Fig. 22). This spiral configuration is antiferromagnetic, i.e. it has no net magnetization. Ferrimagnetic spirals are also possible (Fig. 22). Spiral spin configurations have been observed in rare earth metals [48], $CoCr_2O_4$ [49], etc.

REFERENCES

[1] Ballhausen, C. J., Introduction to Ligand Field Theory, Mc Graw-Hill Book Company, New York, 1962.
[2] Griffith, J. S., The Theory of Transition Metal Ions, Cambridge University Press, Cambridge, 1961.
[3] Van Vleck, J. H., The Theory of Electric and Magnetic Susceptibilities, Oxford University Press, Oxford, 1932.
[4] Heisenberg, W., Z. Phys. 38, 411, 1926.
[5] Dirac, P. A. M., Proc. Roy. Soc. (London) 112A, 661, 1926.
[6] Moriya, T., Magnetism, Vol. 1, Rado, G. T. and Suhl, H., Eds, Academic Press, New York, 1963.
[7] Dzyaloshinskii, I. E., Sov. Phys. JETP, 6, 621, 1958.
[8] Anderson, P. W., Solid State Physics, Vol. 14, Seitz, F. and Turnbull, D., Eds., Academic Press, New York, 1963.
[9] Goodenough, J. B., Magnetism and The Chemical Bond, Interscience Publ., New York, 1963.
[10] Goodenough, J. B., J. Phys. Chem. Solids, 6, 287, 1958.
[11] Goodenough, J. B., Phys. Rev., 100, 564, 1955.
[12] Kanamori, J., J. Phys. Chem. Solids, 10, 87, 1959.
[13] Engelsman, F. M. R., Van Laar, B., Wiegers, G. A. and Jellinek, F., to be published.
[14] Weiss, P., J. Physique, 6, 667, 1907.
[15] Smart, J. S., Effective Field Theories of Magnetism, W. B. Saunders Company, Philadelphia, 1966.
[16] Stanley, H. E., Critical Phenomena, W. A. Benjamin, New York, 1972.

[17] Bethe, H., Proc. Roy. Soc. (London), 150A, 552, 1935.
[18] Peierls, R. E., Proc. Roy. Soc. (London), 154A, 207, 1936.
[19] Bloch, F., Z. Phys., 61, 206, 1930.
[20] Walker, L. R., Magnetism, Vol. 1, Rado, G. T. and Suhl, H., Eds., Academic Press, New York, 1963.
[21] Keffer, F., Spin Waves, in Encyclopedia of Physics, Vol. 18/2, Wijn, H. P. J., Ed., Springer Verlag, Berlin, 1968.
[22] Izyumov, Y. A. and Ozevov, R. P., Magnetic Neutron Diffraction, Plenum Press, New York, 1970.
[23] Kanamori, J., Magnetism, Vol. 1, Rado, G. T. and Suhl, H., Eds., Academic Press, New York, 1963.
[24] Kittel, C., Phys. Rev., 76, 743, 1949.
[25] Dillon Jr., J. F., Magnetism, Vol. 3, Rado, G. T. and Suhl, H., Eds., Academic Press, New York, 1963.
[26] Bloch, F., Z. Phys., 74, 295, 1932.
[27] Bobeck, A. H., Fischer, R. F., Perneski, A. J., Remeika, J. P. and Van Uitert, L. J., IEEE Trans. Magnetics, MAG-5, 544, 1969.
[28] Thiele, A. A., Bell Syst. Techn. J., 48, 3287, 1969.
[29] Kneller, E., Ferromagnetismus, Springer, Berlin, 1962.
[30] Morrish, A. H., The Physical Principles of Magnetism, John Wiley and Sons, Inc., New York, 1965.
[31] Buschow, K. H. J., Luiten, W., Naastepad, P. A. and Westendorp, F. F., Philips Techn. Rev., 29, 336, 1968.
[32] Enz, U., Lems, W., Metselaar, R., Rynierse, P. I. and Teale, R. W., IEEE Trans. Magnetics, MAG-5, 467, 1969.
[33] Néel, L., Ann. physique, [10] 18, 5, 1932.
[34] Bacon, G. E., Neutron Diffraction, Oxford University Press, Oxford, 1962.
[35] Néel, L., Ann. physique, 3, 167, 1948.
[36] Gorter, E. W., Philips Res. Repts., 9, 295, 321, 1954.
[37] Smit, J. and Wijn, H. P. J., Ferrites, John Wiley and Sons, New York, 1959.
[38] Went, J. J., Rathenau, G., Gorter, E. W. and Van Oosterhout, G. W., Philips Techn. Rev., 13, 194, 1951.
[39] Bertaut, E. F. and Forrat, F., Compt. Rend. (Paris), 242, 382, 1956.
[40] Geller, S. and Gilleo, M. A., Acta Cryst., 10, 239, 1957.
[41] Pauthenet, R., Ann. physique, 3, 424, 1958.
[42] Yafet, Y. and Kittel, C., Phys. Rev., 87, 290, 1952.
[43] Prince, E., Acta Cryst., 10, 554, 1957.
[44] Yoshimori, A., J. Phys. Soc. Japan, 14, 807, 1959.
[45] Kaplan, T. A., Phys. Rev., 116, 888, 1959.
[46] Villain, J., J. Phys. Chem. Solids, 11, 303, 1959.
[47] Enz., U., J. Appl. Phys., 32, 22S, 1961.
[48] Elliott, R. J., Magnetism, Vol. 2A, Rado, G. T. and Suhl, H., Eds., Academic Press, New York, 1965.
[49] Dwight, K. and Menyuk, N., J. Appl. Phys., 40, 1156, 1969.

CHAPTER 16

MAGNETIC ALLOYS

Roberto L. Colombo

Centro Sperimentale Metallurgico

Rome, Italy

INTRODUCTION

Over the last 20 years, magnetic alloys have been the subject of much less scientific work than before, mostly because of the competition from ferrites, whose properties, unlike those of alloys, can be nearly custom-tailored to fit special needs, specially in the electronic industry.

However, alloys still make up for by far the larger proportion of magnetic materials produced every year. They are generally much cheaper than ferrites, have superior mechanical properties and can more easily be machined into complicated shapes. Moreover, some of their magnetic properties are still unique and cannot be matched by any ferrite, as it is shown in Table 1.

TYPES OF MAGNETIC ALLOYS

Magnetic alloys are usually distinguished in soft and hard. Soft materials are characterized by high saturation magnetization, high permeability, low coercivity and small a.c. losses, while hard materials show high coercivity, high remanence and high energy product. Among soft magnetic alloys we can list alloys for power engineering, for communication engineering, for loading or inductance coils, for device transformers, choke coils, relays, electromagnets and shields, and so on. All these, as it will be shown later, have rather basically different properties. Hard magnetic alloys, instead, though having different uses, are selected mostly according to their performance from the point of view of the same properties and, naturally, cost.

TABLE 1. Some Selected Magnetic Properties of Metals and Ferrites.

Property	Metals or Alloys	Ferrites
Highest saturation magnetization, T	2.158 (pure Fe) 2.46 (Fe-35% Co)	0.60 (magnetite) 0.60 (Co-ferrite)
Highest Curie temperature, C	1121 (pure cobalt)	670 (Li-ferrite)
Highest initial permeability	1.63×10^5 (Supermalloy)	4×10^3 (commercial Mn-Zn ferrite)
Highest maximum permeability	5×10^6 (single crystal Si-Fe) 10^6 (square-loop Perminvar, single crystal iron)	Unknown, but inferior
Lowest coercivity, $A\ m^{-1}$	1.3 (Extra-pure, zone melted iron) 0.16 (Supermalloy)	8 (commercial Mn-Zn ferrite) 4 ("dense" Mn-Zn-ferrite)
Highest coercivity, $A\ m^{-1}$	710×10^3 (Co-Cu-Sm alloy) 320×10^3 (Pt-Co alloy)	160×10^3 (Ba-ferrite)
Highest energy product, $J\ m^{-3}$	184×10^3 (Pr-Sm-Co alloys)	32 (experimentally 40) $\times 10^3$ (Sr-ferrite)

SOFT MAGNETIC MATERIALS

Table 2 lists typical uses and required properties of soft magnetic materials. It will be seen that while, broadly speaking, iron and dilute iron alloys are used for power engineering, iron-cobalt alloys are used for electromagnets and magnetic amplifiers, and nickel alloys cover the rest of the applications.

Alloys For Power Engineering

For power engineering, high saturation magnetization, high flux density (i.e. high permeability) at low magnetizing force

TABLE 2. Uses and Property Requirements for Soft Magnetic Materials (after Stanley, op. cit.).

Uses	Property Requirements
Power Applications	
Distribution and power transformers High-quality motors and generators	Low core losses; high permeability at low and medium induction; high saturation
Instrument Transformers	
Audiofrequency transformers	Low core losses; high permeability at low and medium inductions
Pulse transformers	High permeability
Inductance Coils	
Audiofrequency	Low hysteresis and high permeability
Carrier-frequency	Very low hysteresis and eddy-current loss
Radiofrequency	High permeability at low magnetizing force
Miscellaneous	
Relays Magnetic shielding	High permeability; low remanence and low coercive force. For alternating-current applications, core loss should be low.
Magnetic amplifiers	Rectangular hysteresis loops: low hysteresis

(usually defined at 800 or 2000 Am^{-1}) and low core losses, which imply low hysteresis loss and comparatively high electrical resistance, are essential. In fact, the total losses P_t are the sum of the hysteresis loss P_h and the Eddy current loss P_e

$$P_t = P_h + P_e \tag{1}$$

The hysteresis loss is proportional to the frequency, according to the formula

$$P_h = f A_h \tag{2}$$

where $A_h = \oint B\, dH$ is the area of the hysteresis loop, B is the induction and H is the magnetic field. The Eddy current loss in the case of an indefinite sheet of an isotropic magnetic material with constant permeability, assuming a uniform distribution of the flux, can be calculated to be

$$P'_e/f^2 = \pi^2 t^2 B_p^2/6\rho\sigma = p'_e \tag{3}$$

where t is the thickness, B_p the peak induction, ρ the electrical resistivity and σ the density.

If we put Eqs. (2) and (3) into Eq. (1) we find

$$P_t/f = A_h + p'_e f \tag{4}$$

As a consequence, losses should show in a plot against frequency as a straight line: the ordinate of the intercept would be the area of the hysteresis loop, i.e. a measure of the static losses, and the slope p_e a measure of the Eddy current loss (Fig. 1, curve A). On the contrary, the results of experiments are represented by curve B, where we can find both a curvature (the initial part) and a slope anomaly.

In fact, Eq. (3) is based on two erroneous assumptions, viz. that permeability is constant throughout the hysteresis cycle, and that the flux distribution is uniform inside the material. On the first count, Brailsford and Burgess [1] have proved that assuming a field-dependence of permeability does not alter enough the calculations to have them fitting experimental data. As for the second argument, it is known that the flux does not penetrate deeply inside the material at high frequency (skin effect), which is the reason why we use thin laminations to make transformers and rotating machines; however, for such laminations and the frequencies involved in power engineering, the influence of the skin effect is not important. But the flux cannot be uniformly distributed also because of the Weiss domain structure. Back in 1950, Williams, Shockley and Kittel [2] calculated the Eddy current loss in a theoretical sheet having a single Bloch wall running parallel to the applied field. Their results were extended later on by Pry and Bean [3], who used a more sophisticated model, having a uniform bar domain structure and plane 180° Bloch walls at equal 2L intervals; they found:

$$p_e = \frac{8t^2 B_p^2}{\pi\rho\sigma} \frac{2L}{t} A\left(\frac{2L}{t}, \frac{B_p}{B_s}\right) \tag{5}$$

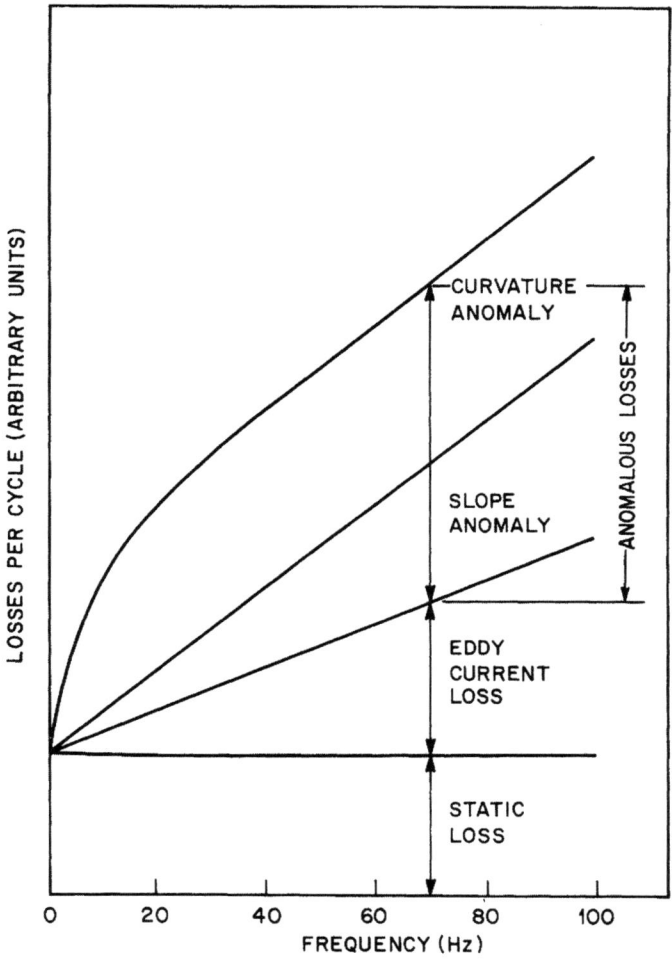

FIGURE 1. Losses per cycle vs frequency.

where A is a rather complicated function of the arguments. Equation (5) is a definite improvement in comparison to Eq. (3); however, p_e appears to be frequency independent, which does not explain the curvature anomaly. As a consequence, some people [4] find good agreement with experiments and some [5,6] find serious discrepancies. An improved treatment is due to Carr [7], who derived the total losses as the area of a dynamic cycle

$$P_t = \oint H \, dB = \oint H \, dB/dt \, dt \qquad (6)$$

where dB/dt is calculated considering the motion of the Bloch walls of a domain structure. At comparatively high frequencies, the result is the same as that of Pry and Bean, but at very low frequencies Carr states that the motion of a Bloch wall can be completed before the external field has increased enough to start the motion of the next one. In general, the number of walls N in motion at a given instant is a function of the frequency. An approximate solution for Eq. (6) for sinusoidal B and large domains is

$$P_t = \frac{\pi^2 B_p^2 f^2}{CB_s N(f)} + 4H_o B_m f \qquad (7)$$

where C is a constant and H_o a "friction field".

A common feature of the models considered up to now is that the walls remain plane during their motion. Brouwer [8] had already shown in 1955 that the condition is not valid but for low frequencies and weak fields. Experimental evidence has been gathered recently by Overshott, Preece and Thompson [9], Boon and Robey [10], and Hellmiss [11]. They all derive theoretical treatments, which contain some factors to be determined experimentally. Moreover, they assume that the wall velocity is a linear function of the applied field, which, according to later work by Boon and Robey [12], is not always true. As a consequence, it has become customary to introduce a coefficient of anomalous loss

$$\eta = (P_t - P_h)/P'_e = P_e/P'_e \qquad (8)$$

and derive it from experiments; η can take values from about 1 to about 4, depending mostly on frequency, but also on the material, its thickness, grain size and texture (these two latter influencing the domain structure).

However, texture is far more important in the context of the hysteresis loss, because of the superior magnetic properties of the b.c.c. iron single crystal along the $\langle 001 \rangle$ directions (Fig. 2). The obtainment of a sharp, favorable texture in magnetic sheets makes use of secondary recrystallization, so that an outline of recrystallization and grain growth is in order.

When a cold worked metallic matrix is heated, those of its properties, which have undergone changes in the deformation process, gradually recover their previous values. Upon heating at higher temperatures, new crystals appear and grow at the expenses of the rest of the material, until they consume all of it.

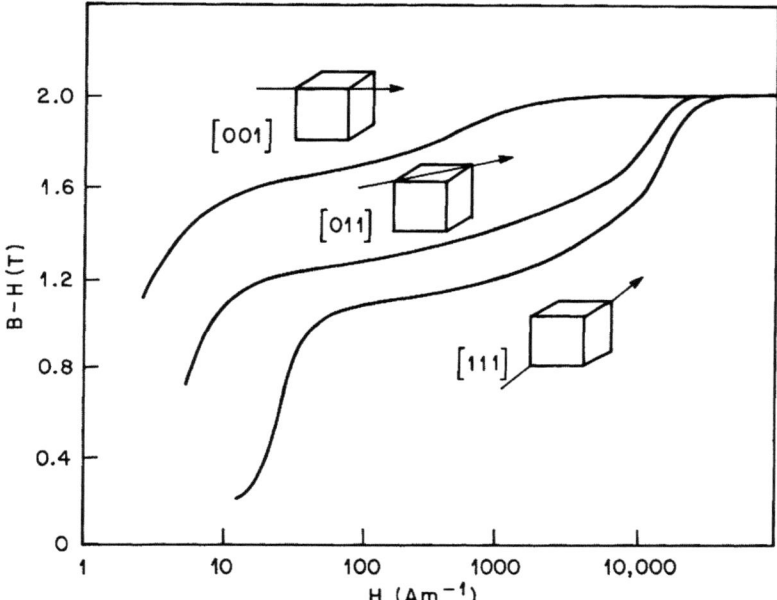

FIGURE 2. Dependence of magnetization on orientation for Alpha-iron single crystals.

The former process is called recovery, the latter primary recrystallization.

On prolonging the heat treatment, larger grains with more than six legs in a micrograph tend to grow at the expenses of the smaller ones [13,14] (grain coarsening or primary grain growth). The driving force for such a process is the minimization of the overall grain boundary energy. Following Detert [15] we can express the rate of grain growth by the relation

$$d\bar{D}/dt = \eta_1 M \gamma_B / \bar{D} \qquad (9)$$

where $\bar{D}$ is the average grain size, t is the time, M a mobility term, which depends on composition and texture (if existing), γ_B the average grain boundary energy per unit area and η_1 a numerical factor. Therefore the time exponent of continuous grain coarsening is 1/2.

Secondary recrystallization takes place when generalized grain coarsening is strongly impeded, and only a few grains, which act as nuclei, can grow. Then the initial growth rate is

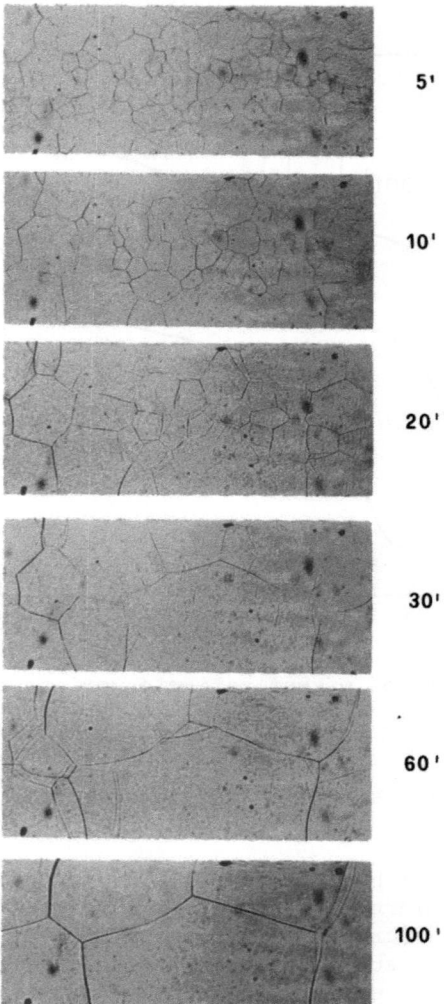

FIGURE 3. Progress of secondary recrystallization in 3% Si-Fe at 1000°C (section, HT microscopy, 10 X).

independent of time and the motion of the boundaries of the privileged grains is towards the center of curvature. Figures 3 and 4 show the progress of secondary recrystallization in sheet specimens of 3% silicon-iron.

Generalized grain coarsening can be inhibited by dispersed particles of a second phase. According to Zener [16], continuous grain growth must come to end when a limiting grain size D_{lim} has been reached, such that

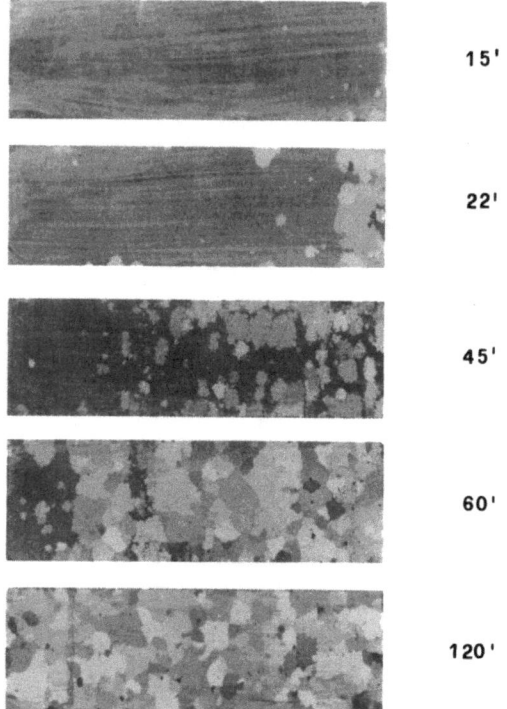

FIGURE 4. Progress of secondary recrystallization in 3% Si-Fe at 960°C (front view, etched).

$$D_{lim} = \eta' d_i / f \qquad (10)$$

where f is the volume fraction and d_i the average diameter of particles (considered spherical), and η' is a factor between 1 and 2. By heating the material in a temperature zone, where the particles very gradually begin to coarsen and dissolve, some selected grains can start to grow again. Burke and Turnbull [17] have shown that for this to happen the average grain diameter must be less than D_{lim} and those grains will grow with a size larger than average.

Another kind of hampering to grain coarsening is represented by a strong texture existing in the matrix [18], and Dunn [19] has observed that it can lead to secondary recrystallization. Within the textured fraction of the matrix only low-angle grain boundaries exist, which have a lower energy and are less mobile than large-angle boundaries. In general, a grain largely disoriented to the textured fraction would tend to shrink, unless its diameter D is considerably larger than average, and precisely unless

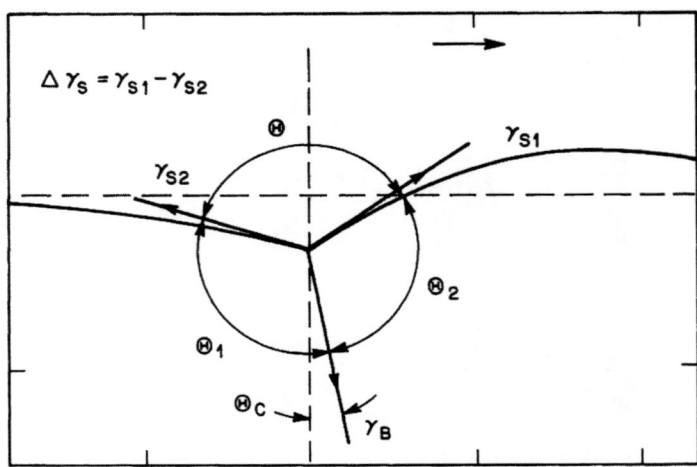

FIGURE 5. Diagram showing surface profile for grain boundary migration against thickness effect inhibition.

$$D > (\gamma_B/\gamma_{LAB})\bar{D} \qquad (11)$$

where γ_{LAB} is the energy of a low-angle boundary (in general 0.3 to 0.5 times γ_B). Therefore, when D is 2-3 times larger than $\bar{D}$, the grain will tend to expand and act as a nucleus for secondary recrystallization.

Thermal grooves formed during bright annealing of thin sheets (Fig. 5) can also prevent grain coarsening (thickness effect). Mullins [20] has shown that a matrix with an average grain diameter of $\bar{D} \geq 2t$, where t is the sheet thickness, is stabilized by them (thickness inhibition). To promote secondary recrystallization, some grains must act as nuclei [21]. This happens when a difference of surface energy $\Delta\gamma_S$ exists between an individual grain and the surrounding ones, and its diameter is such that

$$D \geq (2\gamma_B t)/(\gamma_B + 4\Delta\gamma_S) \qquad (12)$$

It must be born in mind that γ_S depends on impurity absorption, so in different atmospheres different textures can be produced.

Alloys for power engineering are all iron-based. A short outline of the fabrication procedure is given here. Traditionally, iron was melted in the open-hearth furnace, later on the electrical furnace came into the picture. Most promising today is the melting from molten pig iron and scrap in a basic oxygen furnace through the L-D process, so that carbon, manganese, sulfur and phosphorus are eliminated by reaction with oxygen and a suitable slag. In any case, the molten metal is tapped into a ladle and, if required, the alloying elements are added before pouring into molds to form ingots, which are hot rolled to a thickness of about 2 mm. Rolling to final thickness and annealings follow. Carbon is removed in thin strips by annealing in moist hydrogen.

Iron and Low-Carbon Steel

Abundance and excellent magnetic properties made iron the first soft magnetic material and justify its continued use. The highest permeability and saturation magnetization are obtained when impurities, specially carbon, nitrogen, oxygen and boron, are reduced to a minimum. Cioffi, Williams and Bozorth [22] have measured extremely high permeabilities in single crystals annealed for long times in pure hydrogen at 1300°C. Even better results have been obtained by Mager and Hillmann [23] in zone-refined iron. Obviously, such materials are too expensive to be used commercially.

ARMCO iron is melted under conditions, which remove carbon and manganese, but leave a lot of oxygen in the metal. However, oxide particles are large and have little effect on magnetic properties, and saturation magnetizations of 2,15 T are easily obtained. In forgings and castings, annealing only results in relieving the internal stresses, but in sheets it also increases the grain size and reduces the content of the harmful impurities. Optimum results could be obtained by annealing in dry hydrogen while in the γ phase, but the treatment is difficult and expensive, so that working at about 800°C in moist hydrogen and finishing at the same temperature in dry hydrogen is preferred. In this way, carbon and nitrogen, which are particularly harmful because they induce aging, are reduced to a minimum (below 50 ppm).

Cube-on-edge (COE), or {110} [001] textured iron has been produced by thickness [24] or sulphide particle [25] inhibition. However, most applications of iron are in rotating machines, in which the magnetic flux runs in more than one direction, and textured iron is not used.

Enormous quantities of cold-rolled, low carbon steel strip has been used as a cheap magnetic material, specially for small motors. The steel must be annealed by the user in order to be fully decarburized, but is often preferred to pure iron because of the higher manganese content, which provides a higher electrical resistivity and makes easier for it to recrystallize after cold work.

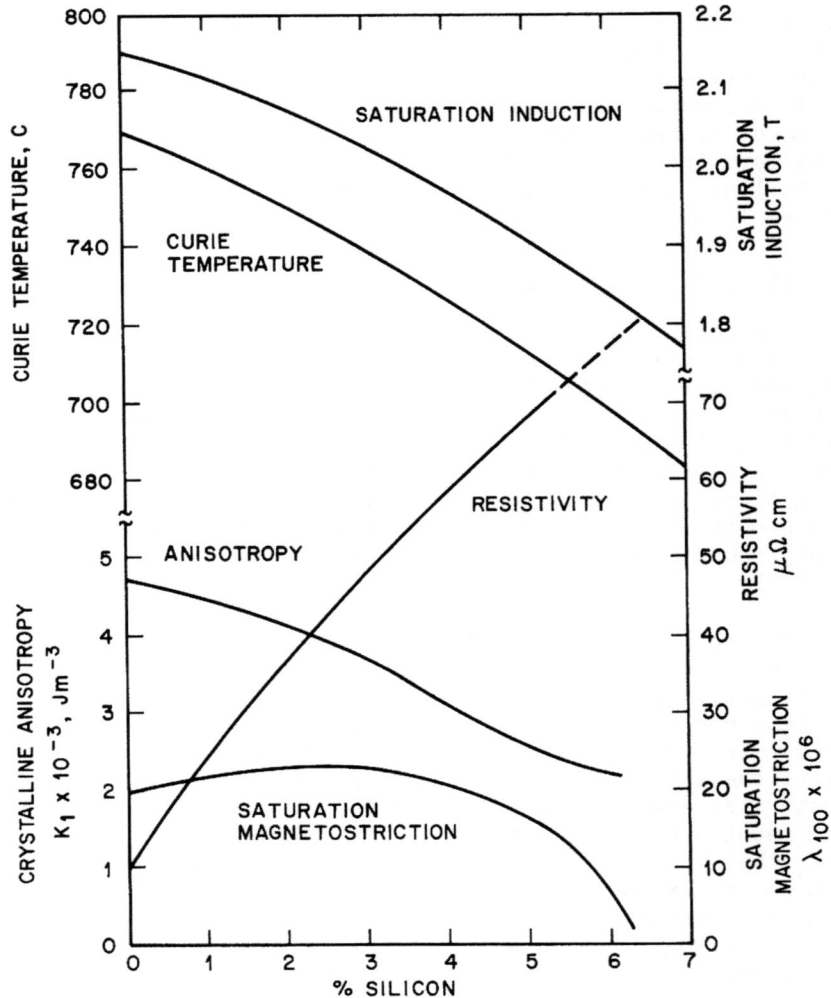

FIGURE 6. Properties of iron-silicon alloys (after Littman, op. cit.).

Silicon Iron

(i) <u>Intermediate grades</u>. A major contribution to magnetic steel was made about 1900 by the addition of silicon as performed by Hatfield and coworkers [26] in England. Because of the existence of a γ loop, alloys containing more than 2.15% silicon in weight and no carbon are in the α phase up to the melting point, but the presence of minimal amounts of carbon displaces the boundary of the loop to the right. Some of the relevant properties of the alloys are shown in Fig. 6. The better performances of silicon-iron with respect to mild steel or pure iron are largely due to its higher electrical resistance, but also to its inherently lower hysteresis loss. Thinner materials provide lower Eddy current loss, but the resulting lower packing factor limits the induction attainable with a given stack height. In practice, sheets from 0.28 to 0.63 mm are used for motors and small transformers operating at 50-60 Hz. Thinner material is used for higher frequencies.

Until 1930, silicon-iron alloys were usually pack hot rolled to final gauge. Cold rolling has come in use more recently since the development of welding has permitted to produce continuous coils from the hot rolled plates. Despite its good properties and the possibility of somewhat higher silicon contents (cold rolling above 3.5% silicon is not practical, because of brittleness), very little hot rolled silicon-iron is produced today, due to cost considerations. Cold rolled material is strip annealed to remove carbon and increase the grain size. If the fabrication procedures require later punching or shearing, a successive annealing at 700°C to remove stresses may be necessary.

Some typical properties of intermediate grade steels, which are usually called non-oriented because texture is not produced intentionally, are listed in Table 3.

TABLE 3. Properties of Some Non Oriented Electric Steels (Mild Steel in the Form of Sheet, Cast Iron of Ingots, Sheet Iron of 2 mm Thick Plate, the Rest 0.5 mm Thick Strip).

Material	ρ $\mu\Omega cm$	B_S, T	H_c, $A\ m^{-1}$	$\mu 800$	$P_{1.5,50}$ $w\ kg^{-1}$
Normalized mild steel (0.2% C)		2.14	320	1500	
Cast magnetic iron (0.015% C)	10.7	2.15	68	1575	
Sheet magnetic iron (as above)	10.7	2.15	89	1575	10.5
Decarburized steel (0.01% C)	12.5	2.14	72	1530	7.3
1.1% Si steel	25	2.10			8.2
1.6% Si steel	32				6.0
2.1% Si steel	42	2.04	36	1485	4.7
2.5% Si steel	46				4.0
3% Si steel	50	1.98	32	1450	3.1

ρ = resistivity
B_S = saturation induction
H_c = coercivity
$\mu 800$ = permeability at 800 A m^{-1}
$P_{1.5,50}$ = core losses at 1.5 T, 50 Hz

(ii) <u>Grain-Oriented Grades</u>. Virtually all the grain-oriented silicon-iron contains 3.2% silicon and exhibits the COE texture, so that the rolling direction corresponds to one of the easy magnetization axes.

The first to produce COE oriented silicon-iron, largely by chance, was Goss in the mid-thirties [27]. Essentially, his procedure consisted of cold rolling in two stages, with an intermediate anneal in a reducing atmosphere at 800-900°C, a hot rolled slab about 2 mm thick to a final thickness of 0,25-0,28 mm, and annealing it at a temperature up to 1100°C in a non-oxidizing atmosphere. Goss did not immediately realize that his material was textured, and the discovery was made by Bozorth [28] instead.

The starting material contains approximately 0.06-0.10% Mn, 0.03% C, 0.02% S and as little as possible of other elements, specially Al (which could form small nitride and oxide precipitates, very detrimental to magnetic softness). Ingots are hot rolled from temperature near 1300°C to 3-8 cm thick, reheated if necessary, and finally hot rolled to 0.15-0.3 cm thick. The resulting "hot rolled band" is annealed at 800-1000°C, pickled or sandblasted and cold rolled in two approximately equal steps to final thicknesses of 0.25-0.35 mm with intermediate annealing at 800-1000°C. The resulting strip is continuously decarburized in moist hydrogen near 800°C, covered with a slurry containing magnesia, then coil annealed for very long times at temperatures slowly rising up to about 1200°C. The final material has a carbon content of about 50 ppm and a sulphur content of 20 ppm or so. COE texture can be more than 90%, the grain size between 0.1 and 0.5 cm; the induction measured at 800 A m^{-1} is about 1.8 T (which means that the relative permeability is 1,800), the total loss at 1.5 T and 50 Hz is less than 1.1 w kg^{-1}.

The COE texture in silicon-iron has been studied for the last 30 years and has contributed more than everything else to our knowledge of secondary recrystallization; nevertheless, our understanding of it is still far from complete. May and Turnbull [29] were first to show that MnS particles stabilize the primary grain size, and coarsen or dissolve at higher temperature, when secondary recrystallization is effective. Why during secondary recrystallization COE grains, which are a minority in the primary matrix, grow almost alone at the expenses of virtually all the others, is still unclear.

Several modifications of the standard procedure have been proposed, of which only final rolling at temperatures slightly above R.T. is sometimes practiced, to take care of cold brittleness [30]. Progress aims at making the material cheaper, e.g. through continuous casting or cold rolling in only one step;

or at obtaining better flatness and thickness control in the laminations; or again at reducing the spread in the magnetic properties, which is now relevant.

A lot of effort has been spent in trying primary growth inhibitors other than MnS. A good deal of information can be found in the articles by Matsuoka [31]; practically, only few of the proposed inhibitors can be used, namely MnSe [32] and perhaps MnTe, and specially VN and AlN. VN was proposed by Fiedler [33], who was able to produce much thinner COE oriented strips than was possible at the times with MnS. AlN was found unsuitable by Fiedler [34]; however, Taguchi and Sakakura [35] showed that between 800 and 1000°C it precipitates in the form of needles 1 μm long, bearing fixed orientation relations with the iron lattice and able to interfere with grain boundary motion (viz., with primary growth and secondary recrystallization) only as long as this troubles said relations; by this mechanism, several textures can be induced by annealing cold rolled single crystals. This discovery has led to the invention of an industrial procedure [36], by which 2.9% silicon-iron containing Mn and S (possibly replaced or accompanied by Se and/or Te) as in the standard process and 100-600 ppm of Al is hot rolled, annealed, cold rolled in two steps (of which the latter not less than 80% R.A.) with intermediate anneal at 1100-1150°C, and coil annealed as usual. Lower core losses (specially at 1.7 T) and much higher relative permeability at 800 A m^{-1} (over 1,900) than in the standard material are claimed: apparently, the progress is due to a sharper texture (with consequent larger grain size). I shall mention that results just as good are claimed in a recent patent by Goss, Black and Stewart [37], which describes a particular melting procedure in the L-D converter and straight use of MnS alone as the growth inhibitor.

The standard process does not give good results for the thinner laminations which are used e.g. in magnetic amplifiers, because of the loss of sulfur through the surface before the secondary recrystallization is complete. Littmann [38] devised a process by which a conventional strip 0.25 mm thick is cold rolled to the desired thickness and annealed to obtain primary recrystallization: the resulting texture is {120} [100] rather than {110} [100], but still quite favorable. Straight COE texture is currently obtained through a method developed by Kohler [39], who applied a sulfur coating to the strip surface.

For years people have thought of making silicon-iron with the cube-on-face (COF), or {100} [001], texture. The first method is based on a discovery by Walter and Dunn [40], who showed that in a silicon-iron sample held at 1000°C, {110} grains will grow at the expenses of the other in dry hydrogen, while {100} will do the same in the presence of tiny amounts of oxygen or sulfur [41,42]. The effect is believed to be due to surface energy differences. Another method is credited to a thickness effect and in fact can be used only for thin laminations [21].

Finally, Taguchi, Sakakura and Yasunari [43] obtained the same aim by cold cross rolling a band containing AlN as the primary growth inhibitor, annealing at 850-1200°C, cold rolling to final thickness with 50-84% R.A. and stack annealing from 1000 to 1300°C.

As it turned out later, COF textured silicon-iron strips of the usual thicknesses do not have lower losses than COE when used in a transformer [44]; partly because of too large grain size and relatively poor alignment, partly, perhaps, because of unfavorable closure domain structure. The {100} [u,v,w] textured material, which is produced in mostly the same way, would be good for rotating machines. Unfortunately, high cost restricts its use to special cases.

Soft Magnetic Alloys For Uses Other Than Power Engineering

It has been shown that uses other than power engineering are almost innumerable. On the contrary, types of alloys to fill them are comparatively few. As a consequence, I think it reasonable to treat all of them in a single chapter.

Nickel-Iron Alloys

Iron-nickel alloys containing less than 20% Ni are entirely α (b.c.c.) at R.T., the transformation from the high-temperature γ (f.c.c.) phase being irreversible and martensitic, while alloys from about 20 to 30% are a mixture of $\alpha + \gamma$ phases. On cooling to sub-zero temperatures, the residual γ-phase can transform to α-martensite. The nature of Fe-Ni austenite is open to questions. Weiss [45] has postulated the existence of two spin states for γ - Fe, separated by a small energy gap of about 0.037 eV, one antiferromagnetic with a Néel temperature of 8°K and the other ferromagnetic, the former having the lower energy. The energy gap is supposed to vary with composition in a f.c.c. Fe-base alloy and might go to zero with sufficient alloying, so that the ferromagnetic state becomes stable at all temperatures. Since the b.c.c. phase in iron is stabilized by the ferromagnetic transition, ferromagnetic Fe-Ni austenites do not transform to the α-phase.

The last important feature of the Fe-Ni phase system is the existence of long-range order corresponding to the Ni_3Fe coherent compound.

Nickel-iron alloys (and those derived from them) are usually produced by melting in an arc furnace (sometimes in an induction furnace) under an alkaline slag. Small amount of manganese and aluminum are added for desulfuration and deoxidation respectively, while carbon is kept below 0.1%. Alternatively, powder metallurgy can be used. Ingots are heated at a very high temperature (typically, 1200°C), then hot rolled without reheating and cold rolled with intermediate anneals, if necessary, in neutral or reducing atmospheres.

Commercial interest in Fe-Ni alloys is limited to three main compositions: about 35% Ni there are the Invar alloys, which are characterized by a negligible expansion coefficient in certain ranges of temperatures, while about 50% and 80% Ni there are Permalloys, whose main property is a very high initial and maximum permeability.

(i) _Permalloys_. From domain theory, rotational or wall-motion processes lead to the initial permeability

$$\mu_o \propto I_s^2 / K_{eff} \tag{13}$$

where I_s is the saturation magnetization and K_{eff} is an effective anisotropy constant, covering all sources of anisotropy. The existence of anisotropy means simply that the magnetic energy of a crystal depends on the direction of spontaneous magnetization. The energy is a minimum along the so-called "easy" axes. Types of anisotropy to be considered here are magnetocrystalline, magnetostrictive, thermomagnetic and slip-induced in origin.

Magnetocrystalline anisotropy tends to be determined by composition alone and to be insensitive to heat treatments (except for 75% Ni-Fe alloys). On the other side, magnetostrictive anisotropy results from the existence of elastic stresses in the material. As a consequence, those compositions, which minimize magnetocrystalline anisotropy and magnetostriction, should offer the highest initial permeability. The anisotropy constants pass through zero at about 70% Ni and magnetostriction does the same at about 80% Ni (Fig. 7), so that the preferred composition is often 78% Ni. If the alloy is given the so-called "double treatment" (i.e., annealing for 1 h or longer at 900°C, slow cooling and re-heating to 600°C, followed by quenching at least to 300°C) the initial permeability is about 10,000 and the maximum permeability 100,000. Permeability, however, is not the only choice criterion for an alloy of this type, because it just helps to reduce the coil weight, while saturation magnetization limits the core weight. If higher saturation and resistivity are desired (Fig. 8), the alloy containing 45% Ni is preferred: it has an initial permeability

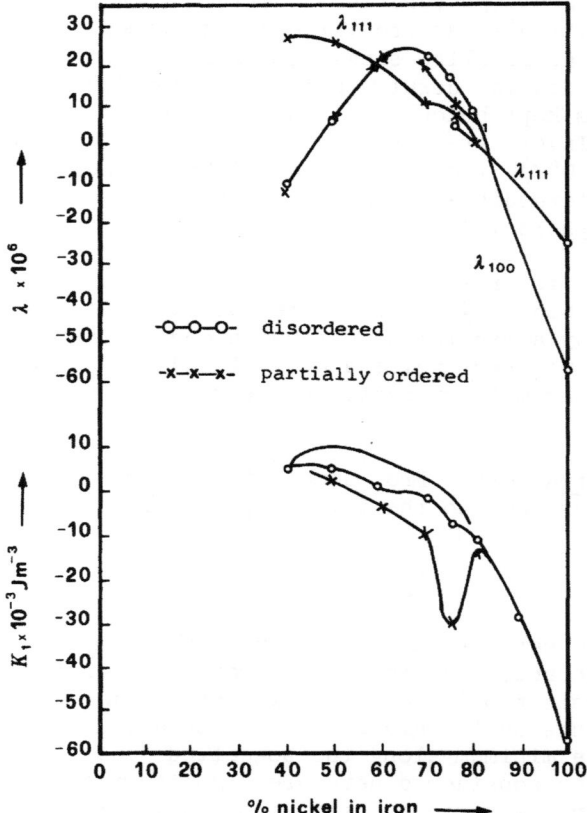

FIGURE 7. Anisotropy and magnetostriction constants in Fe-Ni alloys. -O-O-O- disordered; -x-x-x- partially ordered.

of less than 5,000 and a maximum permeability of the order of 20,000. A similar alloy (50% Ni), has its maximum permeability raised to about 80,000 by annealing in hydrogen at 1000°C, which treatment probably just removes such non-metallic impurities as oxygen, carbon, nitrogen and sulphur.

The alloys with composition near 50% Ni are often produced in the form of COF textured strip, which shows a nearly square hysteresis loop when tested in a field applied along the rolling direction (Fig. 9). These materials are ideally suited to magnetic amplifier cores: in fact, the square loop provides a sharply defined saturation, with minimal permeability above it and nearly infinite (differential) permeability in the insaturated region, virtually independent of the resetting field for minor loops. COF textured strip is produced by a process studied in Germany [46]

MAGNETIC ALLOYS

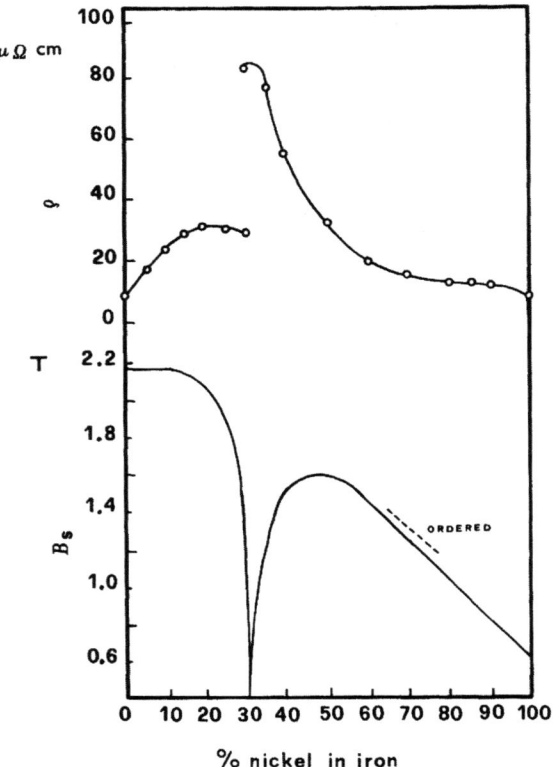

FIGURE 8. Resistivity ρ and saturation induction B_s of iron-nickel alloys.

---

and the Netherlands [47] in the 1930's. This consists of severe cold rolling (about 85-95%, but no more) and annealing in hydrogen at 900-1000°C (at higher temperatures secondary recrystallization, which results in a not-so-sharp {120} [001] texture, would occur [48]). The COF texture develops in a cold worked matrix by a process of oriented growth [48-50], whereby {100} nuclei grow at the expenses of the others. For this to occur, virtually each of the {100} nuclei must be surrounded by all the components of the rolling texture, and therefore a fine-grained starting material and no second phases are required [49].

I have so far neglected thermomagnetic and slip-induced anisotropy. Thermomagnetic anisotropy is a uniaxial anisotropy developed by annealing the alloys below the Curie temperature. If the annealing is performed in the presence of an applied magnetic field, the hysteresis loop squares up; if the applied field is absent, the hysteresis loop becomes constricted or

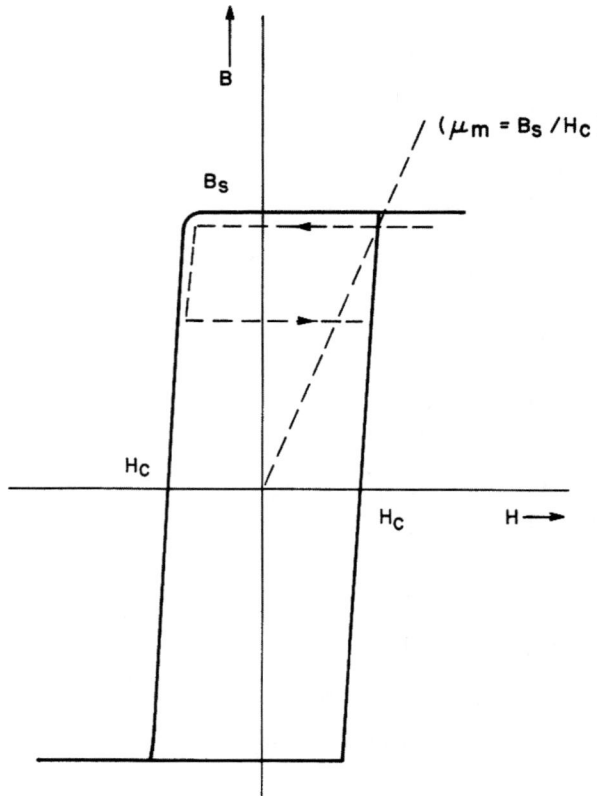

FIGURE 9. Square hysteresis loop (schematic).

wasp-waisted (since the effect is more pronounced in Perminvar alloys, see below, it is usually called "Perminvar effect"). The origin of this induced anisotropy is considered to be due to short range order [51,52], which is necessary to produce atom pairs like Ni-Ni or Fe-Fe rather than Ni-Fe and is distinguished from long range order associated with the Ni_3Fe coherent compound. In a magnetic field at temperatures lower than the Curie temperature, but high enough so that vacancy-assisted diffusion is operative (about 400°C for nickel), the atom pairs are aligned to the field and quenching to R.T. freezes this situation. In the absence of an applied field, the anisotropy becomes aligned randomly by the local magnetization of the domains. Thermomagnetic anisotropy is comparatively low ($1-3 \times 10^4$ J m^{-3}). Directional order, however, can be also established by severe plastic deformation through sheet rolling or wire drawing, the rolling direction or the wire axis becoming in the case of anisotropic material the easy axis of

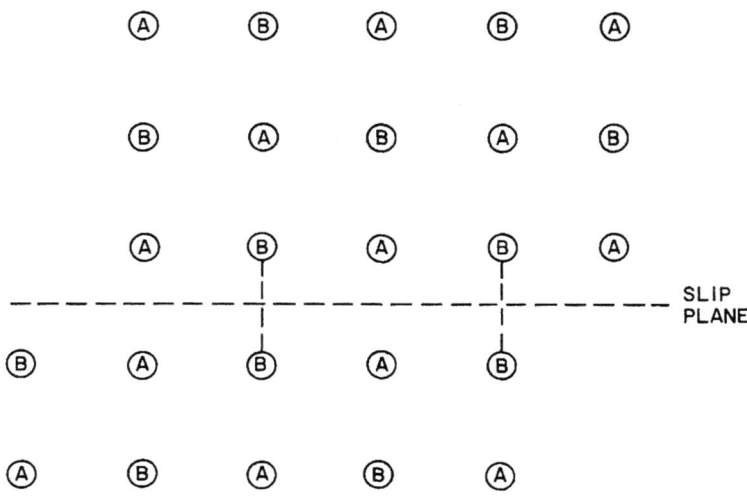

FIGURE 10. Mechanism of slip-induced magnetic anisotropy: displacement of one atomic plane by slip forming pairs of like atoms [53].

magnetization. The mechanism [53-55] is best explained bi-dimensionally in Fig. 10, and the effect is quite considerable (up to the order of 10^6 J m^{-3}). If a COF textured sheet is cold rolled, the rolling direction becomes a hard axis and the direction normal to it in the rolling plane becomes the easy axis. The alloy containing 50% Ni made in this way is interesting, because when the field is applied along the rolling direction the hysteresis loop has a very small area and the permeability is nearly constant (if not high: about 1800 max., with a coercive field of about 500 A m^{-1}). It is used for the cores of loading coils, which are high frequency inductors used in telephone lines.

(ii) <u>Ternary and quaternary Permalloy-related alloys</u>. Variations in the composition of soft ferrites, which improve their permeability at low frequencies, usually impair their high-frequency properties.

This is not the case with Mo-Permalloys, in which 4-5% Fe is replaced by Mo. The initial permeability increases (mostly due to a lower anisotropy) by a factor of 10 with respect to 78% Ni Permalloy, the resistivity increases and the double heat treatment is no more necessary to achieve the optimum properties, because the alloy must now be in a state of partial order, brought about by hydrogen annealing at 1300°C, followed by cooling at 100°C/h [56]. The alloy, submitted to bombardment by 2 MeV electrons in the presence of an applied field at the temperature of about 100°C, develops the most nearly ideal square loop behavior [57]. This "radiomagnetic order" is brought about by the vacancies introduced by the electron bombardment, which can migrate at temperatures very low as compared to those necessary to thermally produce them. The alloy containing also a small amount of Mn is called Supermalloy, and shows the highest recorded initial permeability.

The addition of Cu spreads the range of compositions over which reasonably high initial permeabilities can be achieved, though Cu alone does not increase the initial permeability very much. Better results are obtained by the simultaneous presence of Cu and Cr, with a little Mn (Mumetal and Satmumetal), or of Cu and Mo (1040 alloy). Copper is also important in that it greatly increases ductility, thus allowing the production of strip down to 10 µm in thickness for the fabrication of low loss cores. Table 4 shows some magnetic properties of several ternary and quaternary Permalloy-related alloys.

TABLE 4. Properties of Some Permalloy Related Alloys.

Alloy	Chemical Composition, %	Saturation magnetization, T	Coercivity $A\,m^{-1}$	max. permeability	resistivity $\mu\Omega\,cm$
Supermalloy	16 Fe, 5 Mo, bal. Ni	0.79	1	10^6	60
Mumetal	17 Fe, 5 Cu, 2 Cr, bal. Ni	0.80	2.5	10^5	62
Satmumetal		0.70	1	2.4×10^5	45
Supermumetal		0.80		2.8×10^5	55
1040	11 Fe, 14 Cu, 3 Mo, bal. Ni	0.60	1	10^5	62

I have so far left aside Fe-Co-Ni alloys, which constitute the Perminvar alloys. The Perminvar effect has already been mentioned in the context of Permalloys: in fact, the diagram of Fig. 11, derived from work by Elmen [58] and Masumoto [59], shows that most Permalloys lie in the so-called "Perminvar region". A typical hysteresis loop for a 43% Ni, 23% Co Perminvar is shown in Fig. 12 (from William and Goertz [60]). The characteristics of Perminvars annealed in the absence of an applied field, which

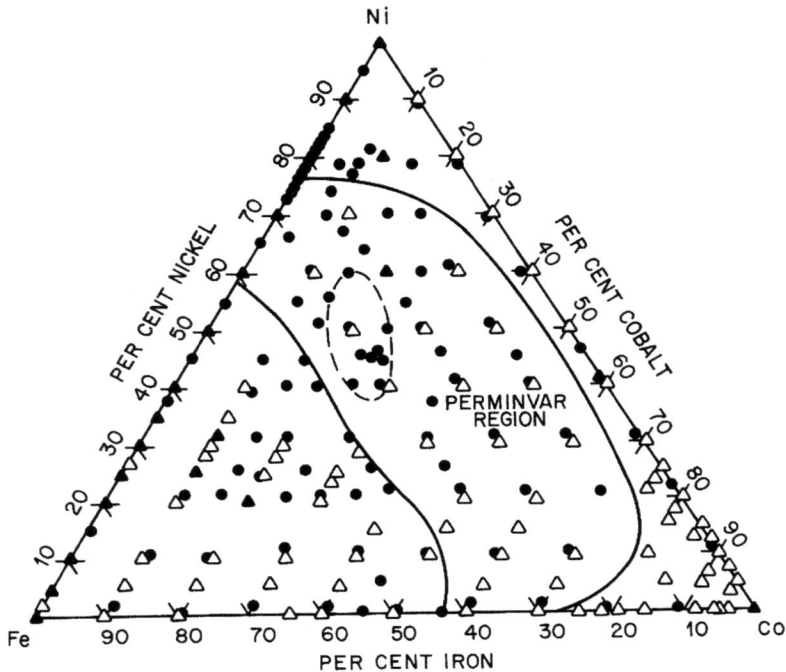

FIGURE 11. The iron-nickel-cobalt phase diagram (after Bozorth, op. cit.).

make them technically attractive, are: a) a constricted hysteresis loop and, in consequence, b) a permeability nearly constant in weak fields up to about 100 A m^{-1} and sharply rising at higher fields, c) a very low remanence, and d) a comparatively high coercivity. Perminvar annealed in an applied magnetic field will show, in the direction of the field, a square hysteresis loop, with low loss and a maximum permeability up to 415,000, and a very low constant permeability at weak fields in the perpendicular direction. The explanation for the Perminvar effect lies in the complex domain structure and the nature of the restoring forces, due to wall stabilization, by the applied field on wall displacement [61,62]: however, it is too intricate to be treated here.

(iii) <u>Invar alloys</u>. The invar alloys are important, not because of their behavior in the electromagnetic field, but because their large volume magnetostriction and spread of the Curie temperature result in an extremely low expansion coefficient below this. Figure 13 is a plot of the volume magnetostriction and magnetization at 3.44×10^5 A m^{-1} versus temperature in 30.2 and 35.9% Ni-Fe alloys, both annealed and 80% R.A. swaged, as

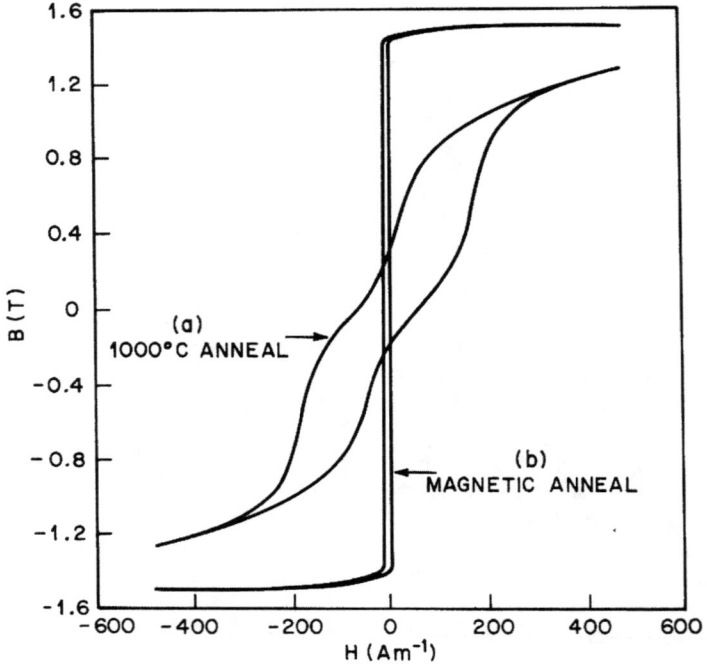

FIGURE 12. Influence of magnetic anneal on the hysteresis loop of Perminvar [60].

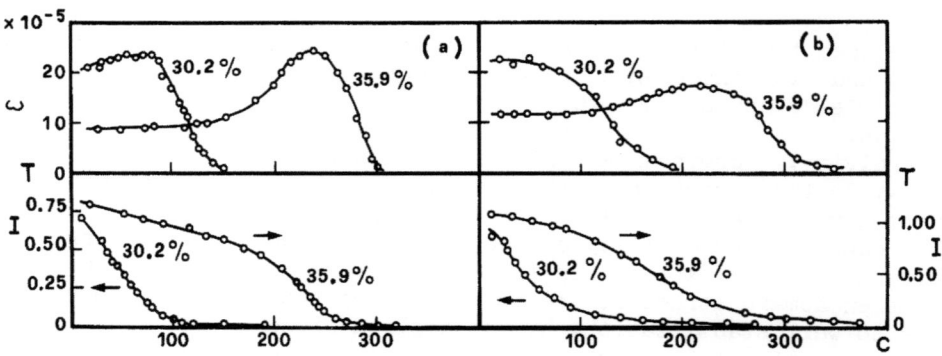

FIGURE 13. Volume magnetostriction ω at 554,200 A^{-1} and magnetization I at 344,000 A m^{-1} for 30.2 and 35.9% Ni-Fe Invar alloys annealed 24 h at 1000°C (a) and 80% R.A. swaged (b) [63].

---

measured by Tino and Maeda [64]: it will be easily seen that the most attractive alloy is the one with 36% Ni (sometimes with minor additions of other elements).

The "Invar anomaly" is poorly understood. It has been associated either with ordering phenomena in the spin alignment [64,45,65,66] or with the presence of two (or three) magnetically different phases (e.g., one paramagnetic or antiferromagnetic and the other ferromagnetic) with temperature-dependent boundaries and strong coupling between them [63,67-70]. The increase in the anomaly obtained by cold work (or quenching) could be due to stress-induced phase transformations, or to the large dependence of the Curie temperature on pressure, associated to the inhomogeneous stress system left over by rolling or swaging [71]. I just need say that it is possible, by the same token, to make alloys with controlled (and low) expansion coefficients, to be used, e.g., as electric leads into sealed vessels of certain glasses. Kovar and Fernico (31% Ni, 15% Co), Konel (70% Ni, 20% Co, 2.8% Ti) and Ferrichrome (30% Ni, 25% Co, 8% Cr) are among them.

Iron-Cobalt Alloys

The Fe-Co phase diagram [72] shows complete solubility of up to 76% Co in the b.c.c. α-phase of iron. Long range order is present over a composition range about 50%.

Possibly the most interesting property of the Fe-Co alloys is the saturation magnetization, which is higher than that of iron and reaches a maximum of 2.46 T at 35% Co. The alloy with 50% Co, with a saturation magnetization only a little lower, has higher maximum permeability (37,000 when melted in hydrogen and annealed at 940°C), and this led Elmen [73] to invent "Permendur", which is used in magnetic circuits of electromagnets and permanent magnets at high inductions, where it can save space and weight. Unfortunately, Permendur is very brittle and difficult to fabricate, but additions of vanadium or chromium, while somewhat deteriorating the magnetic properties, might improve the ductility. Among the alloys so derived we can quote 2V Permendur (49% Co, 1.8% V) and Hyperco (35% Co, 0.5% Cr).

Other Alloys

Because of the existence of the ordered Fe_3Al compound, it is possible to impart directional properties to Fe-Al alloys in the 13-16% Al range [74-77], so that they can show a comparatively high permeability (up to 53,000) and be used for relays in place of Ni-Fe alloys. The alloys with 16% Al are called Alperm or Alfenol, the one with 13% Al, which has a high magnetostriction [78], is called Alfer.

An alloy containing 9.5% Al and 5.6% Si has very good d.c. properties (maximum permeability up to 110,000) and high

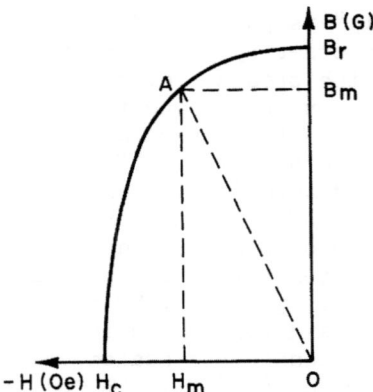

FIGURE 14. Demagnetization curve for a permanent magnet alloy (schematic).

resistivity (more than $100\,\mu\Omega$ cm, almost double than Permalloy) so that losses are lower than that of Permalloy over a wide range of frequencies. The alloy is called Sendust, and is very brittle, but sheets can be produced by slip casting or sintering, in which latter case the properties are worse. An ingenious method has been devised by Helms and Adams [79]: an alloy is first melted with a 25% deficiency of iron; because of its extreme brittleness, it is easily reduced to a fine powder, to which iron powder to restore the normal composition is added; the soft iron acts as a bonding agent, permitting direct rolling to a thin sheet, which is annealed at 1200°C for a short time and then cold rolled to the final thickness and annealed for a long time to homogenize the alloy by solid-state diffusion.

HARD MAGNETIC ALLOYS

Needless to say, hard magnetic materials are used to make permanent magnets. These materials are characterized by the part of the hysteresis loop which is contained in the second quadrant (Fig. 14). If we consider a ring-shaped magnetic circuit made of a magnet of length l_m and an air gap of length l_g and having a constant cross section A, we find, letting the indexes m and g refer to the magnet and the gap respectively,

$$B_m = B_g \qquad (14)$$

and, from the well known equation, if the circuit is saturated and then the current is switched off,

MAGNETIC ALLOYS

$$\oint H \, dl = 0$$

$$H_m l_m + H_g l_g = H_m l_m + \frac{B_m}{\mu_o} l_g = 0 \quad (15)$$

whence

$$-\frac{B_m}{H_m} = \mu_o \frac{l_m}{l_g} \quad (16)$$

It is so seen that the ratio of the induction in the magnet to the demagnetizing field depends only on its length to that of the air gap. Such ratio is the slope of the "load line" indicated in Fig. 14. By considering the energy stored in the circuit, we find

$$W = \frac{1}{8\pi} B_g H_g l_g A \quad (17)$$

and, for a specified length of the magnet and the air gap, with the aid of Eq. (15), that W is a maximum when the product $B_m H_m$ is a maximum too. That is why a magnet is often characterized by the energy product $(BH)_{max}$, which is the stored energy per unit volume and the area of the largest rectangle to be inscribed in the said quadrant of the hysteresis loop.

There are three cases in which magnetic materials don't work in the $(BH)_{max}$ condition. One is the geographic magnet, where the needle works near the remanence (which it does because of its shape). The second one concerns the focussing permanent magnet: since the magnetic field in a tubular circuit is nearly equal to the demagnetizing field inside it, one chooses conditions such that the circuit operates near the coercive field and uses materials of high coercivity. The last case is that of the hysteresis motor, which exhibits a couple independent of the load and is used in record players, gyroscopes and so on: here materials are used, which show the largest possible area of the hysteresis loop.

Let us consider a permanent magnet operating at (load) point A (Fig. 15). If a demagnetizing field is applied, the point shifts from A to B. If now the demagnetizing field is suppressed, point B is displaced along a straight line, called "recoil line" (BB') and the new load point will be the intersection A_1 of it with the load line OA. The slope of BB' is called "recoil permeability", which is, for a family of alloys, the smaller, the

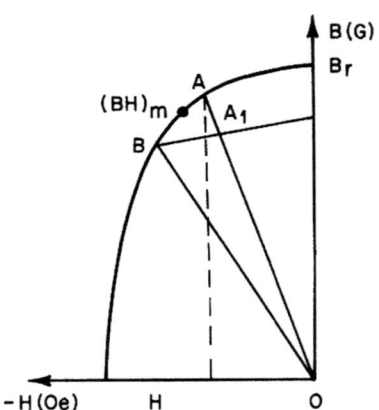

FIGURE 15. Stabilization of a magnetic circuit (schematic).

higher the coercivity (e.g., it is 1.20 for Pt-Co alloys and near 4 for some Alnicos). If another demagnetizing field is now applied, which is not larger than the earlier one, the load point will shift along the A_1B line rather than along the hysteresis loop and the magnetic circuit is called "stabilized". The magnetic recoil is important for dynamic operating conditions.

High coercivities may arise by three different mechanisms: a) the nucleation of domains may require very high fields, b) there may be energy barriers which hinder the motion of the domain walls, and c) the material may exist in the form of particles so fine, that domains cannot form and magnetization processes are only through spin rotation. The first mechanism is of no interest to us, because the only technical material concerned is yttrium orthoferrite, which is not an alloy. Barriers considered in the second mechanism can be inclusion or periodic strains. The first one to conceive of a theory for inclusions was Kersten [80], but Chikazumi [81] has shown his argument to be at fault. In fact, assuming equispaced particles of radius r and interdistance l, one finds:

$$(H_c)_{max} = 1.5 \times 10^5 \frac{I_s r^2}{l^2} \tag{18}$$

For iron, I_s = 2,158 T, and assuming $r = 10^{-5}$ and $l = 10^{-4}$ m, one calculates $(H_c)_{max} = 3 \times 10^3$ A m^{-1}. Periodic strains of maximum amplitude σ produce a maximum coercivity derived by Kersten [82]

$$(H_c)_{max} = \frac{3\lambda\sigma}{I_s} \tag{19}$$

where λ is the magnetostriction constant. If we assume $\lambda = 10^{-5}$, $\sigma = 100$ kg mm^{-2} (10^9 N m^{-2}) and $I_s = 1$ T, we find $(H_c)_{max} = 3\times10^4$ A m^{-1}. Fine particles can contribute to the coercivity through their crystal and shape anisotropy. If K is the relevant anisotropy coefficient of a crystal, the coercivity is

$$H_c = 2K/I_s \tag{20}$$

This formula would lead to values of the coercivity much higher than are actually met by real materials (at least of an order of magnitude for those having a very high anisotropy, like the Co-rare earth alloys). Part of the explanation lies in the fact that many of the particles will have sizes in excess of that, below which no walls are nucleated. If the particles are very anisotropic in shape and are properly aligned, a coercivity will appear given by the relation

$$H_c = (N_b - N_a)I_s \tag{21}$$

where N_a and N_b are the demagnetizing factors parallel and normal to the axis of saturation, i.e. corresponding to the major and minor axis respectively. For iron needles, a coercivity of about 10^6 A m^{-1} is easily calculated.

Remanence B_r can be calculated theoretically as a fraction of the saturation induction in the case of single-domain grains randomly oriented and uniformly magnetized. In this case, for uniaxial anisotropy $B_r/B_s = 0.5$, for cubic anisotropy and $K_1 > 1$ (six easy directions) $B_r/B_s = 0.832$, and for cubic anisotropy and $K_1 < 1$ (eight easy directions) $B_r/B_s = 0.866$. In practice, domain nucleation and other phenomena tend to reduce remanence, while orientation and induced anisotropy tend to increase it.

At the beginning of this century the only material used for permanent magnets was martensitic steel, and only in 1932 Alni, Remalloy and Silmanal, the first alloys inherently different from steel, were invented. Figure 16 shows the progress achieved in this century in terms of coercivity and energy product.

Though there are some materials having peculiar uses, for which others are not suitable, it is customary to classify hard magnetic alloys according to the mechanism by which they magnetically harden.

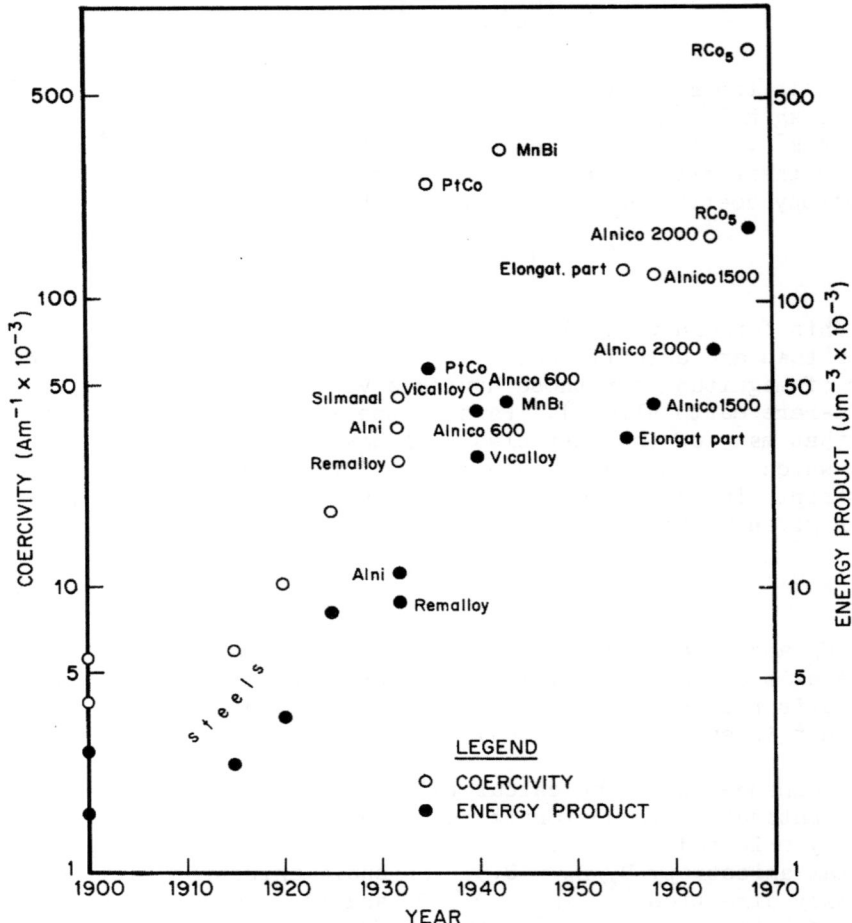

FIGURE 16. Progress in the properties of permanent magnet alloys in the course of this century. ○ coercivity ● energy product.

Martensitic Steels

Plain carbon (up to 1%) steel in form of hot worked, quench-hardened and untempered bars and strip is known since a very long time. It has a very high remanence, but a comparatively low coercivity.

Tungsten was introduced after 1855, and W (5-6%) steels containing 0.6-0.9% C were popular during the First World War. When tungsten became difficult to obtain, chromium was substituted, but Cr-steels had to be aged 24 h at 100°C before use. Both W- and Cr-steels retain the high remanence of carbon steel, but the coercivity is much higher. Unfortunately, the slow decomposition of martensite make them unstable over long times.

In 1920 Honda and Saito [83] introduced the KS (or cobalt) magnet steels. A wide range of compositions (0.7-1% C, 1-8% W, 1-6% Cr and 3-42% Co) have been used.

Table 5 lists the composition and relevant properties of some martensitic steels used for permanent magnets.

TABLE 5. Composition and Properties of Some Martensitic Steels.

Composition %	Remanence T	Coercivity $A\ m^{-1}$	$(BH)_{max}$ $J\ m^{-3}$
0.65 C, 0.85 Mn	1.00	3,150	1,450
1 C, 0.50 Mn	0.90	4,100	1,600
0.7 C, 5 W	1.05	5,600	2,650
0.6 C, 1 Cr, 0.45 Mn	0.95	4,150	1,850
0.9 C, 2 Cr, 0.45 Mn	0.93	4,800	2,100
1.1 C, 6 Cr, 0.4 Mn	0.95	5,900	2,400
0.9 C, 9 Co, 1.25 W, 5 Cr	0.78	9,750	3,300
0.7 C, 17 Co, 8.25 W, 2.5 Cr	0.90	13,600	5,200
0.8 C, 36 Co, 3.75 W, 5.75 Cr	0.96	18,250	7,450
0.7 C, 40 Co, 5 W, 4.25 Cr	1.00	19,350	8,250

Diffusion Hardening Alloys

Alni Alloys

Alni of composition near stoichiometric Fe_2NiAl (12% Al, 25% Ni, balance Fe) was invented by Mishima [84] in 1932. Alloys of this type have far better magnetic properties (B_r = 0.69 T, H_c = 36×10^3 A m^{-1}; $(BH)_{max}$ = 11,050 J m^{-3}) than even the best Co steels. On the other side, they are extremely hard and brittle, and very difficult to work and machine, which has resulted in the manufacturers producing generally finished components.

Though Alni alloys today have been generally replaced by the better performing Alnicos, a study of them can help to clarify a certain number of more general problems in permanent magnet alloys.

From the investigations of Bradley and Taylor [85] and Kiuti [86] it is known that at high temperatures the only phase present is the b.c.c. α-phase, which on cooling spinodally decomposes into the two phases α_1 (ordered Fe_3Al) and α_2. At R.T. the alloy is made up of monodomain α_1 particles embedded in the Ni- and Al-rich α_2 matrix. A third phase, γ, corresponding to the $Ni_3Al(Fe)$ composition, does not form in AlNi, although it can do in Alnicos.

It is worth noting that the decomposition of the α phase takes place at a temperature higher than the Curie temperature, therefore there is no use of any thermomagnetic treatment and Alnis are magnetically isotrope. Néel [87] has in fact proved that their coercivity results from the dispersion of monodomain particles in a slightly ferromagnetic matrix.

Alnicos

Ruder [88] and Mishima [89] discovered independently that the addition of cobalt and an appropriate thermal treatment greatly increase the coercivity of Alnis. Further additions are titanium, copper etc.

Though the phase diagram has never been established, Koch, Van der Steeg and de Vos [90] have shown that the situation in Alnicos differs from that in Alnis in that the γ phase exists at equilibrium and can form if the cooling down from high temperatures is not fast enough, while the decomposition of the α phase into α_1 and α_2 occurs at temperatures lower than the Curie temperature. The ferromagnetic α_1 particles are in the form of ellypsoids, which grow preferentially along the $\langle 001 \rangle$ axes of the matrix: this fact allows the devising of thermomagnetical treatments to induce shape anisotropy. The γ phase, which on cooling transforms into the b.c.c. α_3 phase, is detrimental to the magnetic properties and care is taken in order to avoid its formation.

Alnicos are usually prepared by melting in the induction furnace in a magnesia crucible, and care is taken to prevent contamination, particularly from carbon and phosphorus. Casting temperature is between 1600 and 1800°C. Since, as pointed out before, both hot and cold workings are impossible, castings are produced as near to the final shapes as it is possible, and then only given a final grinding. Powder metallurgy is another possibility. Most metal magnets of comparatively small size (loudspeakers, electric motors and generators, instrument needles, microelectronic devices etc.) are made of Alnicos.

As it is easy to conceive, columnar rather than equiaxial crystallization in cast parts can be an advantage [91] and special techniques have been devised to meet this end.

(i) <u>Titanium Free Alnicos</u>. During the cooling in the mould, a considerable amount of γ phase is formed in these alloys, so that an annealing at 1300°C is necessary to dissolve it. The treatment is performed in a moly resistors furnace with a hydrogen atmosphere, then the parts are cooled in still air down to about 900°C, introduced in a magnetic field of some 1.2×10^6 A m^{-1}, where their temperature reaches 600°C in 15-20 min, and finally cooled completely in non-controlled conditions. After this "magnetic quenching", the material is tempered for 6 h at 650°C and 24 h at 550°C.

It has been found later [90] that 0.2-0.4% Si delays the formation of the γ phase, so that parts as cooled in the mould are free of it. In this case, instead of the annealing at 1300°C, a simpler homogenization treatment of 20-30 min at 920-940°C is needed prior to the magnetic quench. Tempering takes place at 640°C for 1 h and 520 for 4 h. Table 6 lists composition and magnetic properties of some popular equiaxed Alnicos.

TABLE 6. Composition and Magnetic Properties of Some Equiaxed Alnicos

Chemical Composition, %	B_r T	H_c A m^{-1}×10^3	$(BH)_{max}$ J m^{-3}×10^3
17.5 Al, 12 Ni, 24 Co, 3 Cu, 0.2 Si, bal. Fe	1.40	40	40
18 Al, 14 Ni, 24 Co, 3 Cu, 0.4 Si, bal. Fe	1.27	48-50	40-42
8 Al, 4.5 Ni, 24 Co, 3 Cu, 2.3 Nb, bal. Fe	1.14	64	32-34

In order to obtain a columnar rather than equiaxed crystallization, the alloy is cast in a mould, which is placed upon a cooler. To avoid that part of the heat escape from the side walls and start a parasitic crystallization, several practices are available, the simplest of which are by pre-heating the mould at 1000-1200°C or making it in an exothermic material, which is set afire just before pouring. Continuous [92] and electroslag [93] casting have also been used with success. In such ways, alloys with coercivities up to 70×10^{-3} A m^{-1}, and energy products of $56-60 \times 10^3$ J m^{-3} are produced.

(ii) <u>Titanium Bearing Alnicos</u>. Quite early it was found that titanium is beneficial to the magnetic properties of Alnicos.

However, the thermomagnetic treatment becomes slightly more complicated. Alloys are annealed at 1250°C for 30 min, cooled in still air down to 900°C, then in a magnetic field down to 600°C and again in still air down to R.T. Then they are given a magnetic anneal at 800-810°C for 10 min (not longer, least precipitates coarsen) with cooling in a magnetic field down to 600°C in 15 min (much less if the titanium content exceeds 6%), and finally a double temper of 6 h at 650 and 24 h at 550°C.

Alloys of this type show remanences up to 0.84 T, coercivities up to 168×10^3 A m^{-1}, and energy products up to 48×10^3 J m^{-3} respectively [94].

Of the other additions, niobium [95,96] is particularly noticeable.

Titanium presence in Alnicos originates the phenomenon called "constitutional suprafusion", which results in a fine-grained structure and prevents columnar crystallization in any of the usual ways. Harrison and collaborators [97] have shown that the addition of 0.2% sulfur effectively prevents constitutional soprafusion. By this way energy products of 80×10^3 J m^{-3} are easily obtained.

For certain applications, parts with multipolar magnetization are required, so that magnetically isotropic materials are produced. Table 7 lists composition and properties for some of them.

TABLE 7. Composition and Magnetic Properties of Some Isotropic Alnicos.

Chemical composition, %	B_r T	H_c A m^{-1}×10^3	$(BH)_{max}$ J m^{-3}×10^3
11 Al, 22 Ni, 2.2 Co, 7 Cu, 0.7 Ti, bal. Fe	0.65	42	11.2
10.5 Al, 25 Ni, 4.5 Co, 4 Cu, 0.8 Ti, bal. Fe	0.63	50	12
9.8 Al, 21.5 Ni, 12 Co, 1.8 Cu, 0.8 Ti, bal. Fe	0.70	55	12.2
8.2 Al, 19.5 Ni, 19.5 Co, 5 Cu, 5.5 Ti, bal. Fe	0.57	80	15.2

Permanent magnets for car speedometers are often of Alnicos incorporated in a thermosetting resin. They are produced by grinding isotropic Alnicos to a fine powder and adding to this a few percent of bakelite powder, which is then hot compacted. Of course, remanence and energy product are less than for the bulk material.

Order-Disorder Transformation Hardening Alloys

Platinum-Cobalt Alloys

The equiatomic Pt-Co alloy has been known for a comparatively long time. Despite its high price, it has found some application (watch-making, inertial navigation, focussing in progressive-wave tubes) because of its very good magnetic properties and ductility.

The magnetic hardening mechanism has been studied, among others, by Brissoneau and Coll [98]. With a proper thermal treatment, it is possible to obtain in the alloy the formation of the ordered phase PtCo in the shape of quadratic platelets perpendicular to the $\langle 110 \rangle$ axes of the disordered f.c.c. matrix. These platelets, which group themselves to minimize the strain energy, so that in a single crystal one of the equivalent orientations is twice as frequent as the other two, have a very large crystal anisotropy (about 5×10^6 J m^{-3}), and impart a maximum coercivity when their size is of the order of 0.1 mm [99].

The alloy is generally prepared by vacuum or inert atmosphere melting in zircon or alumina crucibles. Ingots are either hot rolled, or cold rolled after annealing at 1000-1100°C. Magnetic properties are acquired after quenching from 1000°C (to freeze-in the disordered state) and ordering 30 min at 650°C. In this condition the material is magnetically isotropic, with remanence 0.64 T, coercivity 320×10^3 A m^{-1} and energy product 72×10^3 J m^{-3}.

Precipitation Hardening Alloys

Cunife and Cunico

Cunife (60% Cu, 20% Ni and 20% Fe) and Cunico (50% Cu, 22% Ni and 28% Co) have been invented in 1937 by Neumann and Coll [100], and in 1938 by Dannöhl and Neumann [101] respectively.

At high temperature these alloys are costituted by a single f.c.c. solid solution, while at a lower temperature there exist two f.c.c. phases, one rich in Cu and the other in Ni and Fe or Co.

The alloys are cast into ingots, which are water quenched from 1000°C or 1100°C, cold rolled and aged a few hours at 500-700°C. An anisotropy in the rolling direction is so acquired. Cunife shows a remanence of 0.52 T, a coercivity of 42×10^3 A m^{-1} and an energy product of 80×10^3 J m^{-3}; for Cunico the figures are 0.35, 53 and 61 respectively. Both are no more in use in Western Europe, where they have been replaced by Vicalloys.

Remalloys or Comalloys

These alloys, mostly Fe(70-80%)MoCo, have been invented in 1932 by Köster [102] and Seljeaster and Rogers [103].

They are prepared in the electric arc or induction furnace, under a slag, which prevents the oxidation of Mo. Only hot rolling is practicable, but after quenching from 1300°C in boiling water, machining is possible. Magnetic hardening is performed by aging at 700°C. The best alloy so far produced exhibits 1.03 T remanence, 27×10^3 A m^{-1} coercivity and 8.8×10^3 J m^{-3} energy product.

Vicalloys

The Vicalloys, discovered by Nesbitt and Kelsall [104] in 1940, retain to these days a considerable importance for microelectronics, instruments and hysteresis motors (watchmaking, gyroscopes for inertial navigation etc.). The chemical composition is 52% Co, 9-13% V, bal. Fe. The main advantage is that Vicalloys can be easily cold worked and machined.

These alloys are melted in the electric furnace, slightly hot worked, quenched from 1100°C, heavily (90% R.A. and more) cold rolled and tempered at 500-600°C. At this point, the alloy richest in vanadium shows a remanence of 0.9 T, a coercivity of 48×10^3 A m^{-1} and an energy product of 30×10^3 J m^{-3}.

Magnetic properties of Vicalloys still present a major problem [105]. In the as-cast condition, an alloy with 13% V is entirely in the non-magnetic or scarcely ferromagnetic f.c.c. α phase. This transforms into the ferromagnetic b.c.c. β phase either by liquid nitrogen quenching or by cold rolling (which is actually practiced). The rolling results in an induced magnetic anisotropy in the rolling direction (L), but coercivity is only 3-4000 A m^{-1} and remanence very low (0.1-0.2 T): the easy axis is along the direction normal to the sheet plane (N). By heating below 300°C, the anisotropy vanishes and the three directions, L,T and N, become equivalent. This shows that the important $\langle 110 \rangle$ fiber texture is not responsible for the anisotropy. Above 300°C a new anisotropy arises, whereby N becomes the difficult axis, the rolling plane the easy plane, with the L direction slightly favored over T. Apparently this is linked to the formation of a FeCo superlattice, that is to ordering, but little more is known, and attempts to pick up sources of anisotropy (crystalline, shape or strain) have so far failed. At temperatures above 500°C, the $\alpha \rightarrow \gamma$ transformation starts and the coercivity sharply rises; the electron microscope has revealed that the particle size is about 0.1 µm, which would induce an imperfect monodomain behavior.

Heusler Alloys

Heusler alloys are made out of non-ferromagnetic elements, of which one is Mn, to which others such to enlarge the lattice parameter (Bi, Al, Ag) are added. They are, in fact, a beautiful confirmation of the Heisenberg and Bethe theories.

Silmanal

In 1932 Potter [106] showed that he could increase the coercivity of the classic Heusler alloy (Cu_2MnAl) by replacing Cu with Ag. The alloy named Silmanal contains 86.75% Ag, 8.80% Mn and 4.45% Al. It is melted with the induction furnace in a graphite mould, and the ingots are water quenched from 760°C and cold rolled (70% R.A.), so that the material acquires a magnetic anisotropy. After a final annealing of 48 h at 250°C, B_r is about 0.06 T, H_c is about 46×10^3 A m^{-1} and $(BH)_{max}$ is 680 J m^{-3}. As it is seen, the only interesting property is the coercivity and the alloy is used to make needles for magnetometers.

Fine Particles

The first prediction of the coercivity of fine powder agglomerates is contained in a long forgotten Russian paper [107]. A little later, Dean and Davis [108] made a permanent magnet by compressing fine powders of iron and other metals prepared by electrodeposition on a mercury electrode. In France pure Fe and 30% Co-Fe spherical particle powder magnets have been industrially produced [109].

Progress has been achieved later on, taking advantage of an idea by Néel [110]. Mendelsohn, Luborsky and Plaine [111] proposed a method to prepare monodomain particles having a high shape anisotropy (viz., elongated particles). An iron or cobalt-iron plate is electrolitically dissolved with a mercury cathode, so that an amalgam is obtained, from which elongated particles are magnetically extracted. The metal powder is added with a lead alloy, then pressed in a magnetic field and vacuum annealed to eliminate residual mercury. The resulting ingots are pulverized again and compacted either in or without a magnetic field. Table 8 lists the properties of some materials.

TABLE 8. Properties of Fe and Co-Fe Elongated Particle Powder Magnets.

Metal Volume %	Remanence T		Coercivity $A\ m^{-1} \times 10^3$		Energy Product $J\ m^{-3} \times 10^3$	
	isotr.	anis.	isotr.	anis.	isotr.	anis.
Iron						
0.25	0.310	0.375	72.0	80.8	7.2	11.2
0.30	0.390	0.500	64.8	72.8	9.6	13.6
0.35	0.460	0.600	57.6	64.8	9.6	16.0
0.40	0.540	0.700	48.0	53.6	10.4	17.2
0.45	0.620	0.790	38.4	44.8	8.8	17.6
Iron-Cobalt						
0.25	0.335	0.450	91.2	106.0	9.6	20.8
0.30	0.410	0.570	84.0	100.0	11.2	24.8
0.35	0.485	0.690	77.2	92.0	12.8	27.2
0.40	0.562	0.790	65.6	80.0	12.8	28.0
0.45	0.650	0.900	54.4	68.0	12.0	28.8

RCo_5 Compounds

The RCo_5 compounds (where R is Y or a lanthanide element) are the ultimate in terms of magnetic properties, so that fabrication difficulties and cost have not discouraged those interested in their development.

The first work on the magnetic properties of these compounds has been published in 1959 [112] and the first work on the crystalline structure appeared in the same year [113]. Both subjects have been reviewed and implemented by Lemaire in 1966 [114].

It was found that:

a) the R-Co phase diagram is very complicated, the RCo_5 are only some of the phases present, and above 1100°C decompose into R_2Co_7 and R_2Co_{17};

b) the RCo_5 compounds form by peritectic reactions;

c) all the compounds are hexagonal, the easy direction being along the hexagonal axis;

d) the RCo_5 compounds exhibit a very large crystalline anysotropy energy;

e) saturation magnetization and Curie temperature for all the compounds are elevated;

f) all the R metals have a very high affinity for oxygen, which they would displace from most refractory crucibles when in the molten state; as a consequence, they must be prepared by arc melting or by induction melting in recrystallized alumina crucibles; if absolute purity is required, only by levitation melting;

g) magnetic properties are stable only if a high enough density is obtained.

The molten element mixture is rapidly cooled, then the product is annealed 48 h at 1100°C and ground. The powder is compacted in a magnetic field of 2.5-8×10^6 A m^{-1}, then either hydrostatically compacted [115], or sintered at 1100°C [116], or both [117].

RCo_5 compounds are envisaged for microwave applications. Their properties have been discussed recently by Martin and Benz [118]. As an example, we can quote remanence up to 1 T, coercivities up to 710×10^3 A m^{-1} and energy products up to 184×10^3 J m^{-3} for $Pr_{0.5}Sm_{0.5}Co_5$. Mischmetal-Co compositions, though not nearly as good from the magnetic point of view, are attractive on reasons of lesser cost.

It is worth noting that by replacing a part of the Co with Cu, Fe or both [119,120], the RCo_5 phase can be precipitated in the metal matrix and grinding and following steps can be omitted. This advantage is obviously very great, since the magnetic properties remain quite attractive, if not as impressive as the above.

Semi-Hard Magnetic Materials

Recently a new class of magnetic materials, having coercivities in the range 1000-5000 A m^{-1} and very high remanence (1.5 T and more), has been developed to fill in needs of high performance hysteresis motors and magnetic memories for computers.

The first of these materials, an alloy containing 45% Co, 6% Ni, 4% V, bal. Fe, has been invented by Betts [121]. Nesbitt and Gyorgy [122] have shown that Au additions to a Permalloy water quenched from 900°C, cold rolled and aged at 650°C, cause the precipitation of an Au-rich phase and the consequent hardening.

Chin and Coll [123] have obtained similar results with a Co-rich Co-Fe alloy containing Ti: since Ti adversely affects rolling properties, the most promising composition appears to be 85% Co, 12% Fe, 3% Ti.

BIBLIOGRAPHY

Excellent books are available, which cover the field of magnetic materials. Below are listed just a few of them in cronological order.

[1] R. M. Bozorth, Ferromagnetism (Van Nostrand, Princeton, 1951).
[2] F. Pawlek, Magnetische Werkstoffe (Springer-Verlag. Berlin, 1952).
[3] K. Hoselitz, Ferromagnetic Properties of Metals and Alloys (Oxford University Press, London, 1952).
[4] F. Brailsford, Magnetic Materials, III Edition (Methuen, London, 1960).
[5] J. K. Stanley, Electric and Magnetic Properties of Metals (A.M.S., Metals Park, Ohio, 1963).
[6] J. C. Anderson, Magnetism and Magnetic Materials (Chapman & Hall, London, 1966).
[7] R. S. Tebble and D. J. Craik, Magnetic Materials (Wiley-Interscience, London, 1969).
[8] R. E. Berkowitz and E. Kneller, eds, Magnetism and Metallurgy (Academic Press, New York, 1969).

I have also heavily preyed on the following review articles:

[1] M. F. Littmann, "Iron and Silicon - Iron Alloys", IEEE Trans. Magn., $\underline{7}$ (1971) 48.
[2] G. Y. Chin, "Review of Magnetic Properties of Fe-Ni Alloys", IEEE Trans. Magn., $\underline{7}$ (1971), 102.
[3] C. Bronner, "Alliages pour Aimants Permanents", Conference given to the Association Belge pour l'Etude, l'Essai et l'Emploi des Materiaux (Dewarichet, Bruxelles, 1971).

REFERENCES

[1]   F. Brailsford and J. M. Burgess, Proc. IEE, $\underline{108C}$ (1961) 458.
[2]   H. J. Williams, W. Shockley and C. Kittel, Phys. Rev., $\underline{80}$ (1950) 1090.

[3] R. H. Pry and C. P. Bean, J. App. Phys., 29 (1958) 532.
[4] D. A. Leak and W. E. Duckworth, JISI, 201 (1963) 588.
[5] M. Hu and G. Wiener, J. App. Phys., 30 (1959) 86 S.
[6] D. A. Wycklendt and R. M. Kay, J. App. Phys. 32 (1961) 368 S.
[7] W. J. Carr, Jr., J. App. Phys., 30 (1959) 90 S.
[8] G. Brouwer, J. App. Phys. 26 (1955) 1297.
[9] K. J. Overshott, I. Preece and J. E. Thompson, Proc. IEE, 115 (1968) 1840.
[10] C. R. Boon and J. A. Robey, Phys. Stat. Sol., 33 (1969) 617.
[11] G. Hellmiss, Z. angew. Phys., 28 (1969) 24.
[12] C. R. Boon and J. A. Robey, J. Phys. D: App. Phys., 3 (1970) 327.
[13] P. Feltham, Acta Met., 5 (1957) 97.
[14] M. Hillert, Acta Met., 13 (1965) 227.
[15] K. Detert, "Secondary Recrystallization" in: F. Aaessmer, ed., Recrystallization of Metallic Materials (Dr. Rieder Verlag GMBH, Stuttgart, 1970) Ch, 5, p. 109.
[16] C. Zener, quoted by C. S. Smith, Trans. AIME, 175 (1948) 15.
[17] J. E. Burke and D. Turnbull, Progr. Met. Phys. 3 (1952) 220.
[18] P. A. Beck and P. R. Sperry, Trans. AIME, 190 (1949) 163.
[19] C. G. Dunn, Acta Mec., 1 (1953) 163; 2 (1954) 173.
[20] W. W. Mullins, Acta Met., 6 (1958) 414.
[21] F. Assmus, K. Detert and G. Ibe, Z. Metallic., 48 (1957) 344.
[22] P. P. Cioffi, H. J. Williams and R. M. Bozorth, Phys. Rev., 51 (1937) 1609.
[23] A. Mager and M. Hillmann, Z. angew. Phys., 13 (1961) 171.
[24] C. G. Dunn and J. Walter, Trans. AIME, 221 (1960) 413.
[25] D. M. Kohler, J. App. Phys., 38 (1967) 1176.
[26] W. F. Barrett, W. Brown and R. A. Hatfield, Sci. Trans. Roy. Dublin Soc. 7 (1900) 67; J.I.E.E. 31 (1901) 674.
[27] N. P. Goss, U.S. Pat. 1,965,559 (1934); Trans. ASM 23 (1935) 511.
[28] R. M. Bozorth, Trans. ASM, 23 (1935) 1107.
[29] J. E. May and D. Turnbull, Trans. AIME, 212 (1958) 769; J. App. Phys., 30 (1959) 210 S.
[30] G. H. Cole, R. L. Davidson and V. W. Carpenter, U.S. Pat. 2,307,391 (1943).
[31] T. Matsuoka, Tetsu-to-Haganè, 52 (1966) 1635; Trans. I.S.I.J. 7 (1967) 19; ibid, 238.
[32] F. A. Malagari, U.S. Pat. 3,556,873 (1971).
[33] H. C. Fiedler, Trans. AIME, 221 (1961) 1201; 227 (1962) 776.
[34] H. C. Fiedler, J. App. Phys., 38 (1967) 1067.
[35] S. Taguchi and A. Sakakura, Acta Met. 14 (1966) 405; A. Sakakura, J. App. Phys., 40 (1969), 534; A. Sakakura and S. Taguchi, Metall. Trans., 2 (1971) 205.
[36] S. Taguchi and A. Sakakura, J. App. Phys., 40 (1969) 1539.
[37] N. P. Goss, J. Black and W. J. Stewart, U.S. Pat. 3,438,820 (1969).
[38] M. F. Littmann, U.S. Pat. 2,473,156 (1949).

[39] D. M. Kohler, U.S. Pat. 3,333,993 (1967).
[40] J. L. Walter and C. G. Dunn, Trans. AIME, $\underline{215}$ (1959) 465; Acta Met. $\underline{8}$ (1960) 497.
[41] D. M. Kohler, J. App. Phys., $\underline{37}$ (1960) 408 S.
[42] G. W. Wiener, I. App. Phys., $\underline{35}$ (1964) 856.
[43] S. Taguchi, A. S. Sakakura and T. Yasunari, U.S. Pat. 3,163,564 (1964).
[44] M. F. Littmann, J. App. Phys. $\underline{38}$ (1967) 1104.
[45] R. J. Weiss, Proc. Phys. Soc., $\underline{82}$ (1963) 281; L. Kaufman, E. V. Clougherty and R. J. Weiss, Acta Met., $\underline{11}$ (1963) 323.
[46] O. Dahl and J. Pfaffenberg, Z. Phys., $\underline{71}$ (1931) 93; O. Dahl and F. Pawlek, Z. Phys, $\underline{94}$ (1935) 504.
[47] J. L. Snoek, Physica, $\underline{2}$ (1935) 403.
[48] F. Pawlek, Z. Metallk., $\underline{27}$ (1935) 160.
[49] P. A. Beck, Phil. Mag. Suppl., $\underline{3}$ (1954) 245; P. A. Beck and H. Hu, in H. Margolin, ed., "Recrystallization, Grain Grow and Textures" (A.S.M., Metals Park, Ohio, 1966) p. 461; H. Hu, in: L. Himmel, ed., "Recovery and Recrystallization of Metals (Interscience, New York, 1966).
[50] L. Néel, Compt. Rend., $\underline{237}$ (1953) 1613; J. Phys. Rad., $\underline{15}$ (1954) 225.
[51] S. Chikazumi and T. Oomura, J. Phys. Soc. Japan, $\underline{10}$ (1955) 842.
[52] S. Tamiguchi and M. Yamamoto, Sci. Repts. Res. Tohoku Univ. $\underline{A6}$ (1954) 330; S. Taniguchi, ibid., $\underline{A7}$ (1955) 269.
[53] S. Chikazumi, K. Suzuki and H. Iwata, J. Phys. Soc. Japan, $\underline{12}$ (1957) 1259.
[54] H. J. Bunge and H. G. Müller, Z. Metallk, $\underline{45}$ (1957) 26.
[55] G. Y. Chin, J. App. Phys., $\underline{36}$ (1965) 2915; Mat. Sci. Eng. $\underline{1}$ (1966) 77, G. Y. Chin, E. A. Nesbitt, J. H. Wernick and L. L. Vanskike, J. App. Phys., $\underline{38}$ (1967) 2623.
[56] O. L. Boothby and R. M. Bozorth, J. App. Phys., $\underline{18}$ (1947) 173.
[57] R. S. Sery and D. I. Gordon, J. App. Phys., $\underline{36}$ (1965) 1221.
[58] G. W. Elmen, J. Franklin Inst., $\underline{206}$ (1928) 317.
[59] H. Masumoto, Sci. Repts. Tohoku Imp. Univ., $\underline{18}$ (1929) 409.
[60] H. J. Williams and M. Goertz, J. App. Phys., $\underline{23}$ (1952) 316.
[61] M. Yamamoto, S. Taniguchi and K. Aoyagi, Phys. Rev., $\underline{102}$ (1956) 1295.
[62] L. Néel, J. App. Phys., $\underline{30 S}$ (1959) 3.
[63] Y. Tino and T. Maeda, J. Phys. Soc. Japan; $\underline{24}$ (1968) 729.
[64] C. Zener, Trans. AIME, $\underline{203}$ (1955) 619.
[65] M. Shiga, J. Phys. Soc. Japan, $\underline{22}$ (1967) 539; M. Shiga and Y. Nakamura, ibid, $\underline{26}$ (1969) 24.
[66] D. A. Colling and M. P. Mathur, J. App. Phys., $\underline{42}$ (1971) 5699.
[67] M. Shimizu and S. Hirooka, Phis. Lett. $\underline{27 A}$ (1968) 530; $\underline{30 A}$ (1968) 133.

[68] S. Kachi, H. Asano and N. Nakanishi, J. Phys. Soc. Japan, 25 (1968) 285; 909; S. Kachi and H. Asano, ibid., 27 (1969) 536; H. Asano, ibid., 542.
[69] A. Katsuki and K. Terao, J. Phys. Soc. Japan, 26 (1969) 1109, K. Terao and A. Katsuki, ibid., 27 (1969) 321, 326.
[70] W. F. Schlosser, J. App. Phys., 42 (1970) 1700; J. Phys. Chem. Solids, 32 (1971) 939.
[71] A. Ferro, private communication.
[72] W. C. Ellis and E. S. Greiner, Trans. ASM, 29 (1941) 415.
[73] G. W. Elmen, U.S. Pat. 1,739,752 (1959).
[74] R. M. Bozorth, H. J. Williams and R. J. Morris, Phys. Rev., 58 (1940) 203.
[75] H. Masumoto and H. Saito, unpublished report (1945), quoted by R. M. Bozorth, op. cit.
[76] M. Sugihara, J. Phys. Soc. Japan, 15 (1960) 1456.
[77] E. Adams, J. App. Phys., 33 (1962) 1214.
[78] K. Honda, H. Masumoto, Y. Shirokawa and T. Kabayshi, unpublished report (1945), quoted by R. M. Bozorth, op. cit.
[79] H. H. Helms and E. Adams, J. App. Phys., 35 (1964) 871.
[80] M. Kersten, Z. angew. Phys., 7 (1956) 313; 8 (1956) 496.
[81] S. Chikazumi, Physics of Magnetism (English edition, J. Wiley, New York, 1966) p. 287.
[82] M. Kersten, in R. Becker, ed., Probleme der technischen Magnetisierungkurve (Verlag Julius Springer, Berlin, 1938) p. 42.
[83] K. Honda and S. Saito, Sci. Repts. Tohoku Imp. Univ., 9 (1920) 417; Phys. Rev. 16 (1920) 495; U.S. Pats. 1,338,132 - 1,338,134 (1920).
[84] T. Mishima, Ohm, 19 (1932) 353; U.S. Pat. 2,027,996 (1936).
[85] A. J. Bradley and A. Taylor, Proc. Roy. Soc. (London) 166 A (1938) 353.
[86] S. Kiuti, Japan Nickel Rev., 9 (1941) 78.
[87] L. Néel, Compt. Rend., 225 (1947) 109.
[88] W. E. Ruder, U.S. Pat. 1,968,569 (1934).
[89] T. Mishima, U.S. Pat. 2,027,994 (1936).
[90] A. J. J. Koch, M. G. Van der Steeg and K. J. de Vos; Proc. of the 2nd Conf. on Magnetism and Magnetic Materials (IEE, New York, 1957) 173; Berichte der Arbeitgemeinshaft Ferromagnetismus (Verlag Stahleisen, Düsseldorf, 1959) p. 130.
[91] J. E. Gould, Cobalt, 23 (1964) 1.
[92] C. P. Marks, Z. angew. Phys. 21 (1966) 83.
[93] A. Hruska, Z. angew. Phys. 21 (1966) 85.
[94] C. Bronner, E. Planchard and J. Sauze, Cobalt, 31 (1966) 63; 32 (1966) 124.
[95] W. Wright, Cobalt, 48 (1971) 115.
[96] C. Bronner, J. P. Haberer, E. Planchard and J. Sauze, Cobalt, 46 (1970) 15; C. Bronner, Trans. AIEEE, 6 (1970) 301.

[97] J. Harrison, British Pats. 987,636 and 999,523 (1962);
D. Hadfield and J. Harrison, British Pat. 1,085,934 (1963);
J. Harrison, Z. Angew. Phys., 21 (1966) 101, J. Harrison
and W. Wright, Cobalt, 35 (1967) 63.
[98] P. Brissoneau, A. Blanchard and H. Bartolin, Trans IEEE
on Magnetics, 2 (1966) 479.
[99] Ya. S. Shur, L. M. Magat, G. V. Ivanova, A. J. Mitsck,
A. S. Yermolenko and G. V. Ivanov, Fizhika Metallov i
Metallovedenie, 26 (1968) 241.
[100] H. Neumann, A. Büchner and H. Reinboth, Z. Metallk., 29
(1937) 173.
[101] W. Dannöhl and H. Neumann, Z. Metallk., 30 (1938) 95.
[102] W. Köster, Arch. Eisenhuttenw., 5 (1932) 12; Z. Electroch.
28 (1932) 549; Stahl u. Eisen, 53 (1933) 849.
[103] K. S. Seljeaster and B. A. Rogers, Trans Am. Soc. Steel
Treating 19 (1932) 553.
[104] E. A. Nesbitt and G. A. Kelsall, Phys. Rev., 58 (1940) 202.
[105] H. Fahlenbrach, Cobalt, 49 (1970) 196.
[106] H. H. Potter, Phil. Mag., 13 (1932) 233.
[107] J. Antik and T. Kubyshkina, Wiss. Ber. Univ. Mosk., 11
(1934) 143.
[108] R. S. Dean and C. W. Davis, U.S. Pat. 2,239,144 (1941).
[109] Société Ugine, Brit. Pat. 596,875, (1948).
[110] L. Néel, Compt. Rend., 224 (1947) 1550.
[111] L. J. Mendelsohn, F. E. Luborsky and T. O. Paine, J. App.
Phys., 26 (1955) 1274; F. E. Luborsky, L. J. Mendelson and
T. O. Paine, ibid., 28 (1957) 334.
[112] E. A. Nesbitt, J. N. Wernick and E. Corenzwit, J. App.
Phys., 30 (1959) 365.
[113] J. H. Vernick and S. Gelber, Acta Cryst., 12 (1959) 662.
[114] R. Lemaire, Cobalt, 32 (1966) 132 and 201.
[115] K. H. J. Bushow, W. Luiten, P. A. Naastepad and
F. F. Westendorp, Philips Tech. Rev., 29 (1968) 336;
K. H. J. Buschow, P. A. Naastepad and F. F. Westendorp,
J. App. Phys., 40 (1969) 4029.
[116] D. K. Das, IEEE Trans. on Magnetics, 53 (1969) 214.
[117] D. L. Martin and M. G. Benz, Cobalt, 50 (1971) 11.
[118] D. L. Martin and M. G. Benz, IEEE Trans. on Magnetics, 7
(1971) 285.
[119] Y. Tawara and M. Senno, Japan. J. App. Phys., 7 (1968) 966;
H. Senno and Y. Tawara, ibid., 8 (1969) 118.
[120] E. A. Nesbitt, G. Y. Chin, R. C. Sherwood and J. H. Wernick,
J. App. Phys., 40 (1969) 4006; E. A. Nesbitt, R. H. Willens,
R. C. Sherwood, E. Buchler and J. H. Wernick, App. Phys.
Lett., 12 (1968) 361.
[121] J. C. Betts, Electrical Manufacturing, 3 (1959) 101.
[122] E. A. Nesbitt and E. M. Gyorgy, J. App. Phys., 32 (1961)
1305.
[123] G. Y. Chin, E. A. Nesbitt, J. H. Wernick and D. R. Mendorf,
J. App. Phys., 40 (1969) 760.

MICROWAVE FERRITES AND APPLICATIONS

D. Treves

The Weizmann Institute of Science

Rehovot, Israel

INTRODUCTION

It has been well known for many years that electromagnetic wave propagation in magnetized media can lead to nonreciprocal effects. The classical case is that of the optical Faraday effect in which the plane of polarization rotates clockwise or counter-clockwise, with respect to the propagation direction, according to whether the wave propagates parallel to the magnetizing field or in the opposite direction.

However not until after the second world war was there a material suitable for the microwave frequency range. The discovery of ceramic magnetic materials (Ferrites) with specific resistivities of the order of 10^8 ohm cm changed the situation drastically. Whereas in the case of a conductor an electromagnetic wave is essentially reflected, in the case of a magnetic insulator the wave can penetrate into the material and interact strongly with the atomic magnetic moments. In addition to being insulators these materials have low losses, and resonances in the microwave frequency range, thus enabling the construction of very useful resonant devices.

However, the biggest contribution of ferrites to the microwave technology, is due to their gyromagnetic properties that enabled the development of compact passive nonreciprocal devices such as the isolators and circulators which are today used in every modern radar or microwave communication system.

The ferrite materials most commonly used belong to the spinel and garnet families. For very high frequency operation hexagonal ferrites or antiferromagnets may be used. Their usefulness is to a large extent due to the possibility of tailor-making materials with magnetic properties such as saturation magnetization, resonant frequency and line width specifically required for particular applications. For example, attenuators require high loss materials, while most other devices require a low loss material. Narrow line width is a bonus for band pass filters and limiters, but for high power devices, broad line widths are a must. For millimeter wave operation very high magnetocrystalline anisotropy such as found in the hexagonal magnetoplumbites is an advantage, while for low frequency operation one takes advantage of the low magnetic moment and low losses of the garnets.

In this short chapter we shall briefly describe the structure and magnetic properties of the major families of ferrites. A few illustrative examples will be given on how, through proper understanding of the magnetochemistry of these ferrites, one can control those properties that are of interest to the microwave engineer.

In the third section, the behavior of magnetic materials under microwave fields will be examined, and in the last paragraph examples of a few devices will be given. These devices are the basic kinds and have been chosen to give an idea of the principles involved. Practical designs are usually elaborations and combinations of the simple structures shown.

The purpose of this writeup, is to make the layman aware of the principles involved in microwave ferrites and devices. For more detailed information the reader is referred to the many excellent books and review articles on the subject. The material in this chapter was drawn mostly from a few major publications, and mainly these will be referred to here. References to the original works, which are extremely numerous, can be found in the mentioned references.

MATERIALS

Ferrite materials consist in general of cations and oxygen. As with other polycrystalline ceramic materials they are usually prepared by reacting oxides of the cations at high temperature in the solid state, (sintering) [1]. The oxides are intimately mixed and prefired at about 1000°C. Chemical reaction occurs already in this stage. The material is then reground, pressed into the final shape at pressures of hundreds of atmospheres, and sintered at 1100-1400°C. For intimately mixing and homogeneity sometimes more sophisticated methods are used such as coprecipitation of the oxide from solution of the metal nitrates by mixing with ammonium hydroxide.

For special applications requiring very low loss or narrow line width, single crystals are used. These can be grown from the melt from supersaturated solutions or by hydrothermal methods [2]. More recently, with the development of miniature microwave devices there is a need for relatively thin ferrite films. These can be prepared by vapor deposition, a technique that may also be suitable for the preparation of single crystal epitaxial layers [3].

Once a chemical composition has been chosen for a particular use, one has to remember that many properties of interest to the electrical engineer depend to a large extent on technological parameters which must be controlled and optimized. Typical examples are porosity and grain size, which can have a large effect on line width and magnetic losses.

Some important references on ferrite materials are the books by Smit and Wijn [4] and by Morrish [5], the review by Hudson [6] and the encyclopedic work of Boxer et al. [7].

Spinels

This family of cubic materials with general formula $S^{2+}T_2^{3+}W_4^{2-}$ is named after the mineral $MgAl_2O_4$ with which they are isomorphic. The large oxygen anions form an approximate face centered cubic lattice. The smallest cubic unit cell contains eight formula units. The cations are located in eight of the 64 tetrahedral interstices (A sites) and in sixteen of the 32 octahedral interstices (B sites). The crystal structure is described in Fig. 1. If the A sites are all occupied by the divalent (trivalent) ions the spinel is called a normal (inverted) spinel. The notation for a partially inverted spinel would be:

$$S_\delta T_{1-\delta} [S_{1-\delta} T_{1+\delta}] W_4$$

where the ions on the tetrahedral sites are given in front of the square brackets and the octahedral ions between the brackets. This notation will be used throughout this chapter.

In essentially all spinels of practical value in microwave devices W is oxygen and the majority of the cations are iron. Sometimes cations with a valency different than two or three can be present; charge compensation is then maintained by proper ratios of the other ions. A classical example is $Fe[Li_{0.5} Fe_{1.5}]O_4$.

The interaction that induces magnetic ordering in the spinels is superexchange between two cations via an intermediate oxygen ion. This interaction, which induces an antiparallel spin

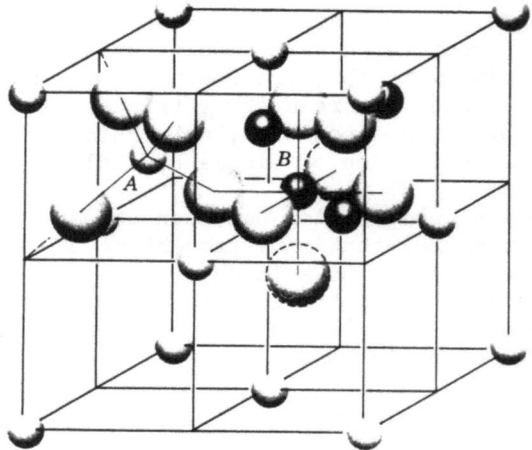

FIGURE 1. The unit cell of spinels. The light small sphéres are ions in tetrahedral coordination, the dark ones in octahedral coordination and the large spheres represent oxygen ions.

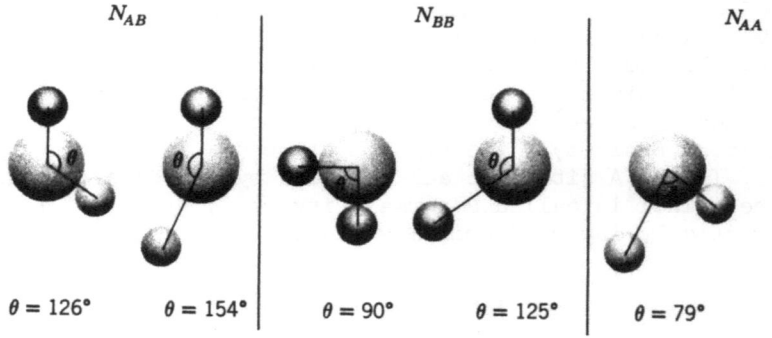

FIGURE 2. Important ion pair geometries that occur in spinels.

---

ordering, is strongest if the three ions are collinear, and decreases rapidly with separation [8]. Figure 2 shows qualitatively the more important configurations of ion pairs occurring in the spinel lattice. It can be seen that the most favorable geometrical configuration exists between ions in A and B sites. It is indeed found experimentally that the superexchange interaction is dominant. The net result is that in almost all cases the A site magnetic cations have their spins antiparallel to the B site magnetic cations. One can speak then of the A and B sublattices. The macroscopic magnetization is given by the difference of the magnetization of the A and B sublattices. The Curie point is determined by the strength of the AB superexchange interaction and by the number of such bonds. Thus for example $Fe^{3+}[Li^+_{0.5}Fe^{3+}_{1.5}]O^{2-}_4$ has a high Curie point but a relatively low saturation moment. The latter can be estimated from the spin only values: in the A sites each Fe^{3+} ion has $5\mu_B$ and in the B sites Li^+ is diamagnetic so that one is left with $7.5\mu_B$ due to $Fe^{3+}_{1.5}$. The balance is then $2.5\mu_B$, quite close to the experimental value of $2.6\mu_B$.

Spinels of practical value from the point of view of microwave applications are magnesium, nickel and lithium spinels. While the second is an inverted spinel the first one is only partially inverted with the general formula:

$$Mg^{2+}_\delta Fe^{3+}_{1-\delta} [Mg^{2+}_{1-\delta} Fe^{3+}_{1+\delta}]O^{2-}_4$$

The degree of inversion is a function of the thermal history of the material δ being given by [9]

$$\frac{\delta(1+\delta)}{(1-\delta)^2} = e^{-E/kT}$$

where k = Boltzman's constant,
T = quenching temperature
and E corresponds to the energy required to transfer a magnesium ion from a B site to an A site. The experimental value of E/k has been found to be approximately $1200°K$ [10]. It is therefore possible to control the saturation magnetization of this spinel by an appropriate heat treatment: a lower quenching temperature results in a lower δ, a higher degree of inversion and thus a lower saturation moment.

There is another mechanism whereby the magnetic moment, a very important parameter for microwave ferrites, can be controlled: Diamagnetic ions can be substituted for magnetic ones.

Tetrahedral interstices are smaller than octahedral ones. Therefore one would expect smaller ions to prefer such sites. Thus, Ga^{3+} will replace tetrahedral Fe^{3+} in an inverted spinel and will increase the net moment (at least for small Ga^{3+} concentrations), while Ti^{3+} will prefer octahedral sites and cause the net moment to decrease. In addition to ionic dimension, site preference is to a large degree determined by the electronic configuration of the ion involved [11]. For example, although Al^{3+} is smaller than Fe^{3+} ($r = 0.5A°$ vs $r = 0.67A°$) it has a strong affinity for octahedral coordination and aluminum in

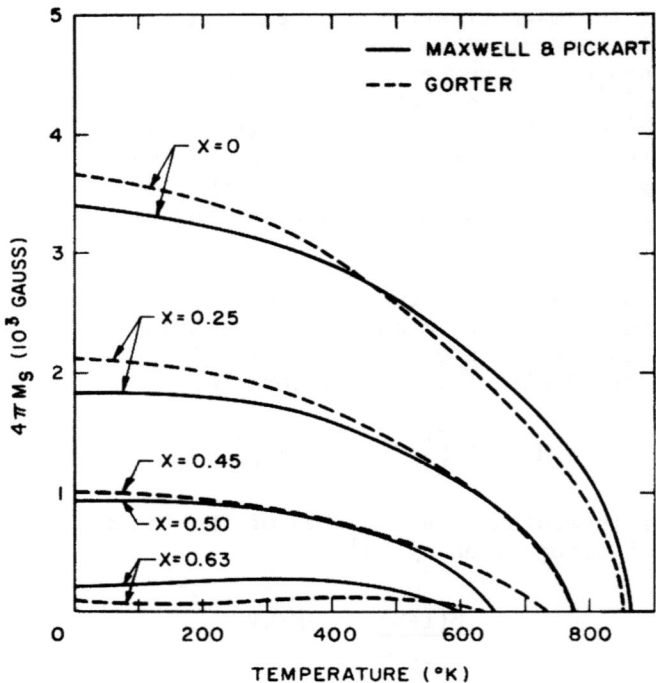

FIGURE 3. Saturation magnetization of polycrystalline $NiAl_xFe_{2-x}O_4$ as a function of temperature.

nickel, lithium or magnesium spinel replaces predominantly Fe^{3+} ions in the octahedral sites. This results in a rapid decrease of magnetic moment, without affecting too greatly the Curie temperature. In $NiFe_{2-x}Al_xO_4$ for example, the net magnetization actually passes through zero at approximately x=0.6, while T_c is still of the order of 600 K. Figure 3 shows the temperature dependence of the magnetization for several values of x [12,13]. A similar situation occurs in lithium spinel at $\delta \approx 1$ [14]. Cr^{3+} and Ti^{3+} ions behave similarly to Al^{3+} [15].

Diamagnetic Zn^{2+} has a strong preference for tetrahedral sites. Its substitution for the divalent ion in an inverted spinel like nickel spinel will transfer Fe^{3+} ions from A to B sites where they will replace the substituted Ni^{2+}. The net magnetic moment will increase because of the decrease in Fe^{3+} ions in A sites, and their increase in B sites at the expense of Ni^{2+} ions that possess a lower moment. See Fig. 4 [16]. Once the magnetization and Curie temperature of a required material have been determined by a major substitution in a simple spinel, additional important parameters such as magnetocrystalline

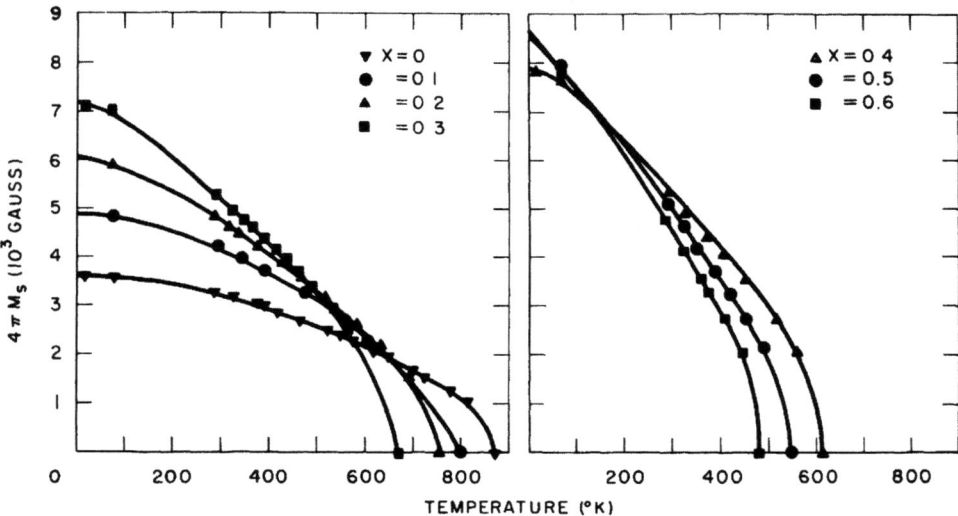

FIGURE 4. Saturation magnetization of polycrystalline $Ni_{1-x}Zn_xFe_2O_4$ as a function of temperature.

anisotropy, resistivity, porosity and grain size, which have a very strong influence on loss mechanisms and line width, can be controlled by minor additions. A low value of resistivity is often due to electron hopping associated with the simultaneous presence of ferrous and ferric ions on equivalent lattice sites. This may occur even in stoichiometric spinels such as $Fe[NiFe]O_4$ through a dissociation of the form

$$Ni^{2+} + Fe^{3+} \rightleftarrows Ni^{3+} + Fe^{2+}$$

Another possible explanation of the low resistivity of such spinels could be that it is very difficult to make perfectly stoichiometric materials: For example, excess oxygen, often associated with a low sintering temperature, results in the appearance of Ni^{3+} and oxygen deficiency, associated with a high sintering temperature, will induce Fe^{2+}. To insure high resistivity small amounts of an additional ion that can exist in two valence states are added. Again taking nickel spinel as an example, if the added ion in its low valence state has a greater affinity to oxygen than Ni^{2+}, and in its high valence state a lower affinity to oxygen than Fe^{3+}, then its presence will inhibit the formation of Ni^{3+} and Fe^{2+}. Such an addition is manganese or cobalt. As the amount of the multivalent additive is small conductivity through electron hopping between the multivalent ions does not

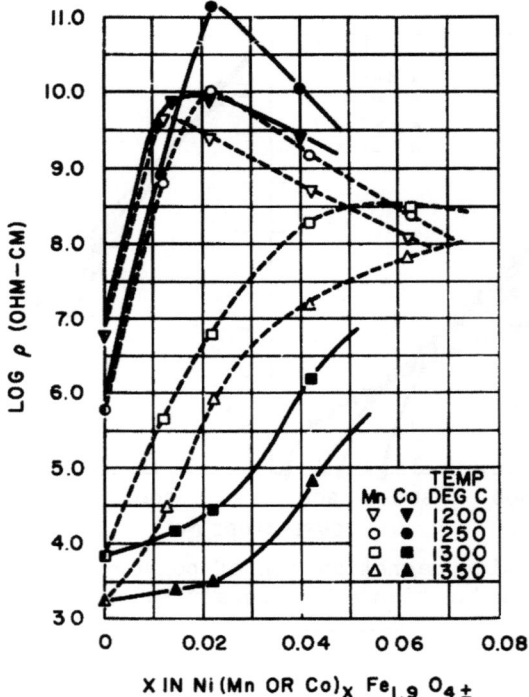

FIGURE 5. Dependence of the resistivity of nickel spinel on manganese and cobalt additions for various firing temperatures.

occur. Figure 5 shows the dramatic increase in resistivity that can be obtained. These curves [17] indicate also the above mentioned dependence of the resistivity on the sintering temperature. A similar increase in resistivity with addition of small amounts of manganese occurs in magnesium spinel [18].

The magnetocrystalline anisotropy of all spinels is negative and small, except for those that contain a large amount of Co^{2+}, in which case the anisotropy is positive and large [19]. Thus a small substitution of Co^{2+} in nickel or manganese spinel can reduce the anisotropy to very low values. This will usually reduce the resonance line width of polycrystalline microwave ferrites.

Porosity, which also strongly affects the line width, can be reduced by increasing the sintering temperature. However too high a temperature causes a dissociation and loss of such volatile components as lithium. Sintering aids, such as copper oxide, increase the reactivity of the starting materials, act as a flux

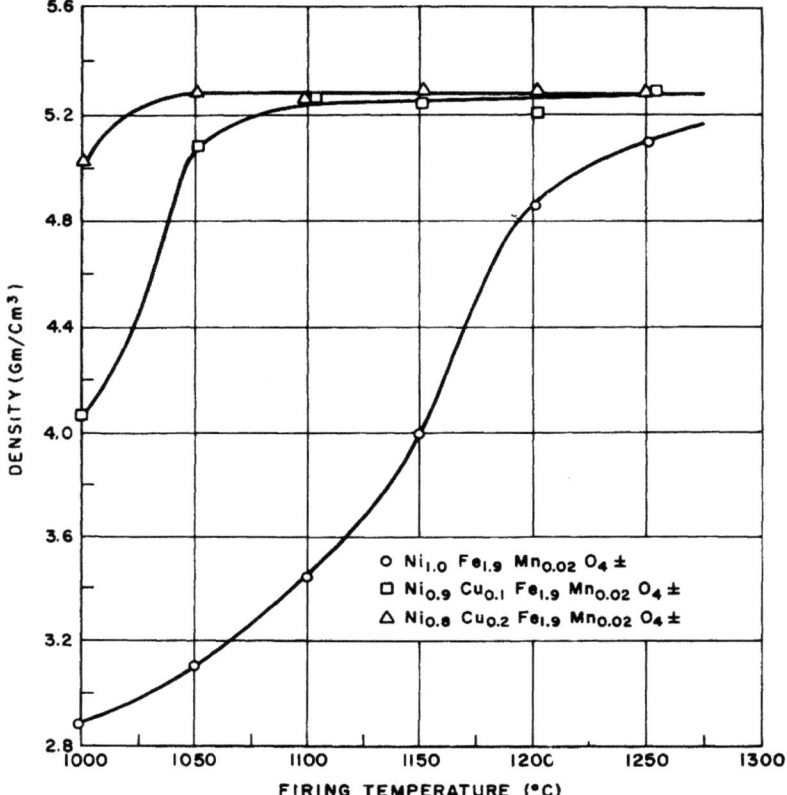

FIGURE 6. Density as a function of firing temperature for nickel spinels with copper additions.

and increase the diffusion rates during sintering. One can therefore obtain very low porosity at relatively low sintering temperatures [20] as shown for nickel spinel in Fig. 6.

Garnets

These materials have the general formula $R_3Fe_3[Fe_2]O_{12}$ where R is yttrium or a rare earth. The crystal is cubic with eight formula units per unit cell. Fe^{3+} occupies three tetrahedral d sites and two octahedral a sites. The R^{3+} ions are in dodecahedral c sites. While the R^{3+} ions are relatively large, (of the order of 1 Å) the dodecahedral sites are relatively small and therefore simple garnet occur only with the heavier (smaller) elements samarium to lutetium (atomic number 62-71). As in the spinels, the dominant magnetic interaction is superexchange between octahedral and tetrahedral sites. The R^{3+} ions are (if magnetic) coupled relatively weakly antiparallel to the tetrahedral ferric ions.

The most important simple garnets are $Y_3Fe_5O_{12}$ (YIG) and $Gd_3Fe_5O_{12}$ (GdIG). They owe their excellent microwave properties to the fact that all ions are trivalent and have zero orbital angular momentum. Therefore resistivity and spin lattice losses are very low. Furthermore, unlike the spinel, the crystal structure of the garnets is such that all the dodecahedral sites are occupied by all of the R^{3+} ions, and all the tetrahedral and octahedral sites are occupied by the Fe^{3+} ion. There is therefore an intrinsic degree of order that is lacking in the spinels. Due to these three effects, microwave losses are very low and resonance line widths very narrow. Their main disadvantages are perhaps the relatively high price of yttrium and the rare earth oxides, and their low Curie temperatures, (approximately $550°K$).

Yttrium is diamagnetic and the net magnetic moment in YIG is therefore due only to the excess Fe^{3+} ion in the tetrahedral sites. Since all magnetic ions are here of the same species, the temperature dependence of the net magnetic moment is similar to that of a single ion and it decreases smoothly to zero at the Curie point.

In GdIG the Gd moment is opposite that of the net moment of the ferric sublattices. It is only weakly coupled to the iron moment and it therefore decreases rapidly with temperature. On the other hand at very low temperature the full moment of $7\mu_B$ per Gd^{3+} ion is developed, so that a total of $3\times 7-5=16\mu_B$ per formula unit is expected, in a direction opposite to the net ferric moment. At some intermediate temperature ($290°K$) there is a compensation point where the net moment is zero, (accidental antiferromagnetism). A similar effect occurs for many of the garnets with magnetic rare earths, the compensation temperatures all being below that of GdIG.

Many substitutions are possible in garnet: By introducing more than one rare earth in the c sublattice one can control the magnetization and its temperature derivative without essentially affecting the Curie temperature. However in practice only Y^{3+} and Gd^{3+} are used as the other ions posses orbital angular momentum which will introduce heavy spin-lattice losses. Sometimes however small amounts of other rare earth ions are purposefully introduced in order to broaden line widths and increase the power handling capability of the material.

As in spinels, Fe^{3+} can be substituted by nonmagnetic ions. In the garnets Al^{3+} and Ga^{3+} prefer tetrahedral sites and thus reduce the net ferric moment, while In^{3+} and Sc^{3+} prefer octahedral sites [21]. In both cases the Curie temperature decreases rapidly with substitution.

Another useful family of garnets is derived for example from YIG by substituting Y^{3+} by Ca^{2+} and Fe^{3+} by V^{5+} [22]. Charge compensation is maintained by the proper ratio of substituents. Vanadium replaces Fe^{3+} exclusively in tetrahedral sites so that the formula of these materials, (denoted by YCaVIG) is

$$Y_{3-2x} Ca_{2x} Fe_{3-x} V_x [Fe_2] O_{12}$$

The magnetization decreases linearly with x and passes through zero at x = 1. On the other hand the Curie temperature decreases only very slowly with x, as shown in Fig. 7. Another advantage of these materials is that less of the expensive yttrium is used. The remaining yttrium can be replaced by bismuth [6].

The versatility and the excellent microwave properties [23] of the garnet family are such that the present trend seems to be that these materials are slowly replacing spinels in microwave devices.

Hexagonal Ferrites

In addition to the spinels and garnets, other materials have been investigated, in particular for the purpose of increasing the frequency range of devices. The most important ones have a hexagonal structure and therefore relatively large magnetocrystalline anisotropy. The best known material of this family is $BaFe_{12}O_{19}$, which is extensively used as a permanent magnet. Four fifths of the crystal structure of this material is identical to the spinel. The (111) axis of the latter coinciding with the hexagonal c axis. The last fifth is a hexagonally packed oxygen layer in which one oxygen atom is replaced by barium.

The order of stacking of the spinel and hexagonal layers along the c axis can be varied and as usual many substitutions are possible. One can therefore tailor make materials with the easy magnetization direction either along the c axis or in the basal plane [24]. Anisotropy fields can vary between a few kOe [25] to 50 kOe [26].

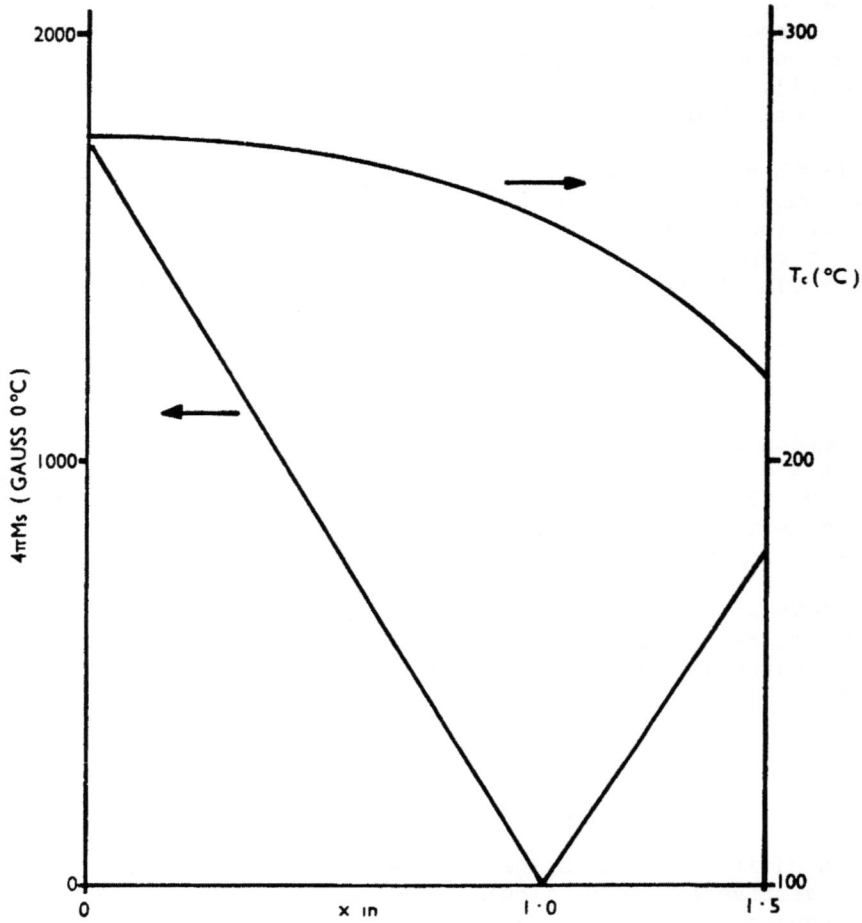

FIGURE 7. Room temperature magnetization and Curie temperature for $Y_{3-2x}Ca_{2x}Fe_{5-x}V_xO_{12}$ as a function of x.

BEHAVIOR OF MAGNETIC MATERIALS IN RF FIELDS

In order to understand the operation of microwave devices and the role of ferrites in them, it is necessary to briefly analyze the interaction of an electromagnetic field with magnetic materials. For a detailed discussion of this subject see for example [27,28,29,30 and 1].

The behavior of the magnetization M is completely described by the equation of motion

$$\vec{\dot{M}} = \gamma \vec{M} \times \vec{H}_{eff} - \alpha(\vec{M} \times \vec{\dot{M}})/M \tag{1}$$

Here $\vec{H}_{eff}$ is an effective field, the dot denotes the time derivative, $\gamma = ge/2mc$ where g is the gyromagnetic ratio, and α is a damping coefficient. Other forms of the damping term have been used [31]. If $\vec{H}_{eff}$ consists only of a constant field, a solution of (1) is a uniform precession of $\vec{M}$ around the direction of $\vec{H}_{eff}$ with an angular frequency

$$\omega_r = -\gamma H_{eff} \tag{2}$$

For a free electron (g=2) $\gamma = -1.76 \times 10^7$ sec^{-1} oe^{-1}; thus a free electron precessed at a frequency of 2.8 MHz in the field of 1 oe. Most ferrites have effective g factors close to 2. Interesting exceptions are ferrites with compensation points. The effective gyromagnetic ratio is the ratio of the net magnetic moment to the net angular momentum. Thus in a simple two sublattice description

$$\gamma_{eff} = (M_1 - M_2) / \left(\frac{M_1}{\gamma_1} - \frac{M_2}{\gamma_2} \right)$$

if $\gamma_1 \neq \gamma_2$ the nominator and denominator will go through zero at different temperatures and γ_{eff} may attain abnormally large values. This effect has indeed been found experimentally in many cases [32]. The effective field is in general a vector sum of steady and rf applied fields, shape and magnetocrystalline anisotropy fields, and any other effective fields that may be present. The magnetic moment $\vec{M}$ has a constant component $\vec{M}_o$ and a time dependent component $\vec{M}_{rf}$ (the latter at the frequency of the rf driving field) in the quasistationary state. It is assumed that the sample is magnetically saturated, and for the time being that the exchange forces are such that the magnetic moments throughout the sample remain essentially parallel to each other. The effect of breaking up into domains will be mentioned later.

The steady state solution of (1) yields the equilibrium conditions. Even this relatively simple problem can be solved only for a very limited number of simple situations. The dynamical solution is carried out under the approximation that the deviations around the equilibrium state are small, that is, the problem is linearized in the components of the time dependents magnetization. In the analysis of high power nonlinear effects these higher order terms must of course be retained.

The resonance frequency is found from the condition that $\vec{M}_{rf}$ has a nontrivial solution when the driving field $\vec{H}_{rf}$ is zero and neglecting damping; one obtains Eq. (2). For the case of negligible anisotropy, the following are a few examples often encountered in devices:

For a flat plate with the applied field, $\vec{H}$, in the plane of the specimen

$$H_{eff} = [H(H + 4\pi M_o)]^{1/2} \qquad (3)$$

For $\vec{H}$ applied perpendicular to the plate

$$H_{eff} = H - 4\pi M_o \qquad (4)$$

For a long circular cylinder with $\vec{H}$ along its axis

$$H_{eff} = H + 4\pi M_o \qquad (5)$$

In all these cases shape anisotropy with its associated demagnetizing fields plays a very important role, as shown by the appearance of M_o in the expression of H_{eff}.

For a sphere, shape anisotropy vanishes and if $\vec{H}$ is parallel to the symmetry axis of a crystal with uniaxial anisotropy or to the [100] direction of a cubic crystal

$$H_{eff} = H + 2K_1/M_o \qquad (6)$$

If the constraint that all magnetic moments rotate in phase under $\vec{H}_{rf}$ (k = 0, uniform precession mode) is relaxed, one can obtain from (1) all the spin wave modes [33]. H_{eff} will then be a function also of k dependent magnetostatic and exchange terms. In ferrimagnets and antiferromagnets there are more degrees of freedom than in a ferromagnet and there are therefore more than one k = 0 mode [34]. One is similar to the ferromagnetic mode, others are associated with a motion that involves an angle between sublattices and therefore exchange forces. These are usually highly energetic modes and their resonance frequency is therefore much higher than that of the ferromagnetic mode.

The time dependent components $\vec{M}_{rf}$ as a function of the steady state conditions and $\vec{H}_{rf}$ are also found from (1). The general solution has the form

$$\vec{M}_{rf} = \chi \vec{H}_{rf} \qquad (7)$$

where χ is the susceptibility tensor. For the case of a strong applied field in the z direction, and $\vec{H}_{rf}$ only perpendicular to it, χ will have the form

$$\chi = \begin{pmatrix} \chi_{xx} & \chi_{xy} & 0 \\ -\chi_{xy} & \chi_{yy} & 0 \\ 0 & 0 & 0 \end{pmatrix} \quad (8)$$

For the simple case of a sphere with zero anisotropy

$$\chi_{xx} = \chi_{yy} = \frac{\gamma M_o (\omega_r + i\omega\alpha)}{(\omega_r + i\omega\alpha)^2 - \omega^2} \quad (9)$$

$$\chi_{xy} = \frac{-i\omega\gamma M_o}{(\omega_r + i\omega\alpha)^2 - \omega^2} \quad (10)$$

where ω is the frequency of H_{rf}.

The microwave engineer is usually concerned with the permeability tensor μ usually written as

$$\mu = \begin{pmatrix} \mu_{xx} & -iK & 0 \\ iK & \mu_{yy} & 0 \\ 0 & 0 & 1 \end{pmatrix} \quad (11)$$

which is related to the tensor χ by

$$\begin{aligned} \mu_{xx} &= 1 + 4\pi\chi_{xx} \\ \mu_{yy} &= 1 + 4\pi\chi_{yy} \\ iK &= -4\pi\chi_{xy} \end{aligned} \quad (12)$$

When the precessing angle is relatively large, as at resonance and under high rf driving fields and if the motion of the magnetization is elliptical, there will be a significant rf component of the magnetization M_{2z} at twice the resonance frequency, in the equilibrium direction z.

The component M_{2z} is calculated from (1) by retaining second order terms in the rf components of the magnetization. For the optimum case of a linearly polarized effective rf field h, and resonance conditions M_{2z} is given by [35].

$$M_{2z} \approx \pi \gamma M_o h^2/\omega \Delta H$$

where ΔH is the small H_{rf} resonance line width.

As we have seen in the above discussion, the resonance frequency can be determined by the strength of the applied field H. However one has to remember that this field has to be large enough to saturate the magnetic material, and to prevent it from breaking up into magnetic domains. Thus it has to be larger than the demagnetizing fields which are of order $4\pi M_o$. For relatively low microwave frequency operation (1-10 KMHz) it is therefore imperative to reduce M_o either by using substituted ferrites as described in the section on materials or by using materials near their Curie temperatures.

For frequencies above 30 KMHz, impractically high field would be required. One can then use such materials as the magnetoplumbites that possess very high anisotropy. The latter contribution to H_{eff} is dominant, and the role of H is then mainly to saturate the sample and to fine-tune the resonance frequency.

The line width H of a resonance curve of a magnetic material is related to the damping parameter by

$$2\alpha\omega = \gamma \Delta H \qquad (13)$$

A detailed study of the relaxation mechanisms that control the line width [36] are beyond the scope of this presentation, and only some salient features related to the materials will be mentioned.

Experimentally, line widths of useful materials vary from a small fraction of an oersted for the best single crystal YIG samples to hundreds of oersteds for polycrystalline spinels and rare earth doped YIG.

The width of the line is a measure of the rate at which energy is dissipated from the uniform precession mode to either the lattice or to other spin wave modes degenerate with the uniform mode, and hence to the lattice. Direct spin lattice relaxation is caused mainly by the presence of ions with unquenched angular momentum such as the rare earths, or Fe^{2+} [37]. Thus the introduction or the avoidance of such ions constitute one method for controlling line widths.

The overlap of the spin wave spectrum with the uniform mode is a function of the shape of the sample, the magnetocrystalline anisotropy and the applied field. The coupling to these modes is through inhomogeneities in the material. Therefore in order to reduce this relaxation mechanism below the direct one, it is important to achieve a high degree of perfection, on both a macroscopic and a microscopic scale, something that can be achieved only in single crystal garnets.

In polycrystals it is probable that inhomogeneous broadening due to the porosity P and the crystal anisotropy are the dominant effects. Pores and inclusions cause the local field to vary from point to point in the material. This has two effects: first one may say that the resonance frequency varies from point to point, effectively increasing ΔH, and secondly higher spin wave modes may be excited, increasing the relaxation rate.

An order of magnitude estimate of the first effect [38] yields

$$\Delta H \approx P 4\pi M_o \qquad (14)$$

where P is the volume fraction of the voids.

For YIG this would be approximately 20 Oe/% porosity, in good agreement with measurements [38]. This contribution decreases with M_o and therefore it is not a serious problem in the low M_o materials used for low frequency operation.

Similar arguments concerning the anisotropy lead to the estimate [6]

$$\Delta H \approx K/M_o \qquad (15)$$

The difference here is that K is associated with spin orbit coupling that provides also a mechanism of direct spin lattice relaxation. The broadening described by (15) is particularly harmful for substituted ferrites or those with compensation points and low M_o used for low frequency applications. The advantage of these materials is that, in spite of their low M_o, they have high T_c and therefore good temperature stability. However their anisotropy fields are high.

Since K decreases rapidly as T_c is approached, one could use materials near T_c to obtain both low M_o and K, this however is at the expense of poor thermal stability. Obviously a compromise between these two effects and the reduction of K by small substitutions such as In in YCaVIG [39], can yield an optimum solution.

If the applied field is small, the magnetic material will not be saturated. It will break into magnetic domains, and inhomogeneous broadening will result. Additional dissipation mechanisms associated with domain wall motion come into play and spurious resonances at relatively low frequencies will appear [40]. These are attributed to restoring forces that pin the walls near imperfections in the magnetic material.

At high microwave powers dissipation mechanisms nonlinear in the rf field come into play. This very interesting subject can easily cover by itself a whole chapter in a treatise on magnetism [41], but will only be touched upon very briefly here.

At low power levels, due to the overlap of the spin wave spectrum and the uniform precession mode, energy is transferred from the latter to the spin wave at a rate related to ΔH_k, the spin wave line width. At higher powers, the transferred energy is enough to overcome the spin wave dissipation, and these will start to grow exponentially. The system is analogous to a parametric amplifier which will break up into spontaneous oscillation when a critical pump level is reached. From this point energy is transferred at a rapidly increasing rate from the uniform mode to the spin wave manifold, and the energy of the uniform mode increases very slowly as pump power is increased. As a first approximation, conservation of energy and momentum require the excited spin waves to have half the frequency of H_{rf}. The critical pump field for the onset of spin wave instability is proportional to the product of ΔH, the low field resonance line width, and some function of $\Delta H_k/4\pi M_o$ [42], and is thus very small, (of the order of millioersteds in YIG [43]).

In the usual configuration used for the excitation of the uniform precession mode, H_{rf} is perpendicular to the applied field H. If the frequency ω of H_{rf} is kept constant, and absorption spectra are derived by varying H, one finds that, as H_{rf} is increased, in addition to a broadening of the main resonance line, there appears often an absorption peak at values of H lower than those required for resonance. See Fig. 8. This subsidiary peak is due to the fulfillment of resonance conditions of spin waves of frequency $\omega/2$. The smallest threshold of fields for spin wave instability is obtained when, by proper design, the subsidiary peak is made to coincide with the uniform precession resonance peak.

In the parallel pump configuration H_{rf} is parallel to H. The uniform precession mode is not excited, and indeed, there is no absorption until a threshold for spin wave instability is reached. Therefore the measurement of threshold fields using this configuration is a convenient method for the study of spin wave line widths [44].

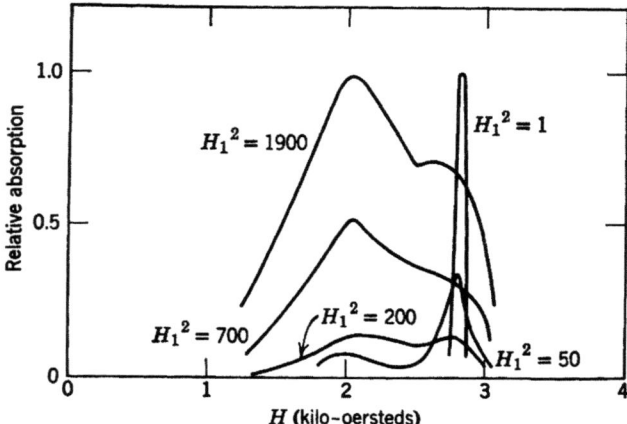

FIGURE 8. Absorption spectra of a nickel spinel single crystal for various microwave power levels. The ordinates of the curves for $H_1^2 = 1$ and 50 Oe^2 differ from those of the other curves.

DEVICES

A full analysis of microwave devices containing magnetic materials requires a simultaneous solution of the equation of motion and Maxwell's equations with appropriate boundary conditions that involve the shape of the magnetic specimen and their location in the waveguide. However, some insight can be gained by some very simple situations. A full analysis of devices can be found in [27,29,30].

For a sample very small compared to the wave length of the electromagnetic field (EMF) in it, propagation effects can be neglected. Note that the dielectric constant of ferrites in the microwave region is of the order of 10 [45]. In devices two main situations are encountered: in one the direction of propagation of the EMF is parallel to the static magnetization and the applied field, and in the other, perpendicular to it.

In the first situation the normal modes of the EMF for the simple case of an isotropic sample ($\chi_{xx}=\chi_{yy}=\chi_o$) are circularly polarized. This can be readily verified by substituting circularly polarized field $H_o(1_x + i1_y)$ and $H_o(1_x - i1_y)$ rotating counterclockwise and clockwise respectively about the z direction into (7) and calculating the appropriate $M_{\pm rf}$ components. The resulting permeability tensor for the circularly polarized components is

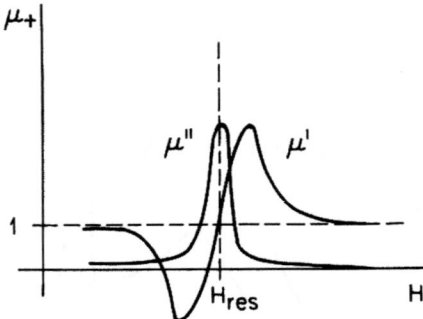

FIGURE 9. Typical behavior of the real and imaging part μ' and μ'' of the permeability near resonance.

$$\mu = \begin{pmatrix} \mu_o + K & 0 & 0 \\ 0 & \mu_o - K & 0 \\ 0 & 0 & 1 \end{pmatrix} \quad (16)$$

where $\mu_o = 1 + 4\pi\chi_o$. From Eqs. (9)-(12) one finds for the sphere

$$\mu_\pm = 1 + \frac{4\pi\gamma M_o}{(\omega_r + i\omega\alpha) \mp \omega} \quad (17)$$

Thus the permeabilities of the two modes are unequal. For the H_- case there is no resonance condition, and there is only a slow dependence of μ_- on frequency and the losses are generally low. For the H_+ case resonance occurs at $\omega = \omega_r$.

Figure 9 shows the general form of μ'_+ and μ''_+, the real and imaginary part of the permeability as a function of H. The losses are proportional to μ'' which peaks at $\omega = \omega_r$ and is small away from the resonance frequency.

There is some confusion in the literature about nomenclature: The magnetic moment M precesses in an effective field H as shown in Fig. 10. Thus resonance occurs for a circularly polarized H_{rf} rotating in the same direction. In the literature this polarization is sometimes called clockwise [27] and sometimes counter-clockwise [1]. A reversal of H reverses the sign of ω_r, and from (17) it is evident that the roles of $\mu\pm$ are now interchanged. Therefore the reversal of H is equivalent to the reversal of the circular polarization from one type to the other.

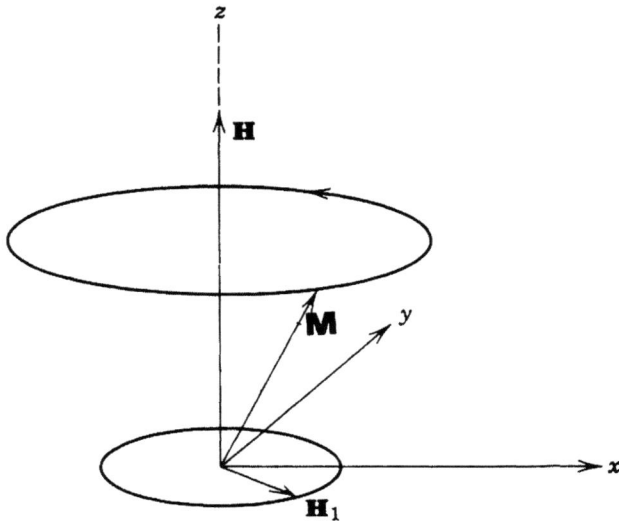

FIGURE 10. Direction of precession of the magnetization M about the field H.

In the other situation often encountered in devices, the direction of propagation is perpendicular to the applied field. The normal modes are fields linearly polarized parallel and perpendicular to the static magnetic field. For the first $\mu_\| = 1$, as there is no interaction with the magnetic moment, and for the second $\mu_\perp = (\mu_o^2 - K^2)/\mu_o$ [46], thus the material is birefringent. The behavior of $\mu_\perp$ is similar to that depicted in Fig. 9.

Several of the longitudinal devices utilize the Faraday rotation. A linearly polarized plane wave is a sum of left and right circularly polarized waves. According to (17) the permeability of these differ, and away from resonance, losses can be ignored. The velocity $v_\pm = c(\varepsilon \mu_\pm)^{1/2}$ differs for the two modes, resulting in a net rotation θ of the linear direction of polarization given by [47]

$$\theta = \frac{\ell \omega \varepsilon^{1/2}}{2c} \left[\left(1 + \frac{4\pi\gamma M_o}{\gamma H - \omega}\right)^{1/2} - \left(1 + \frac{4\pi\gamma M_o}{\gamma H + \omega}\right)^{1/2} \right] \quad (18)$$

where c is the light velocity in free space and ℓ is the length of the magnetic material in the direction of propagation.

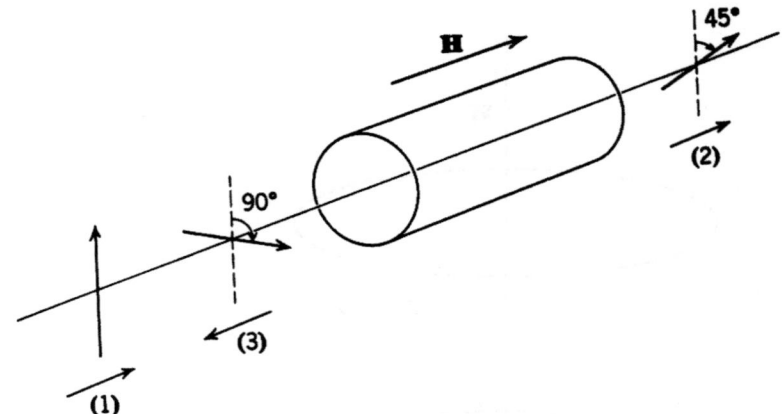

FIGURE 11. The Faraday rotation isolator.

The Faraday rotation isolator, schematically described in Fig. 11 consists of a device in a circular wave guide that produces a rotation $\theta = \pi/4$. Microwave energy is fed in and out of the circular section via rectangular wave guides rotated 45° so that flow in the direction of H is permitted. A field propagating in the opposite direction will also be rotated by $\theta = \pi/4$ in the original direction as indicated by (3) in Fig. 11 and will not be accepted by the rectangular guide at (1).

The Faraday rotation gyrator shown in Fig. 12 utilizes a $\theta = 90°$ section. A gyrator is a device that has zero relative phase shift in one direction and a phase shift of π in the other.

Figure 12 is almost self explanatory: the Faraday rotation causes the same rotation of 90° with respect to an external coordinate system, irrespective of the direction of propagation. On the other hand the 90° twist causes a rotation in opposite directions for the two directions of propagation. Therefore for energy flowing from left to right the rotations cancel while in the opposite direction they add to 180°.

The circulator is a multiport device with ports A,B,C... such that energy entering A exists at B, entering at B exists at C and so on. It can be built with the use of a gyrator. A four port device is shown in Fig. 13. The $\vec{\pi}$ symbol stands for the gyrator. The hybrid junction is a reciprocal device in which signals appearing in phase at the side arms, are due to a field at port 1, while if they are out of phase by 180° they are due to a field at port 3. The operation of the device is then self evident.

MICROWAVE FERRITES AND APPLICATIONS 473

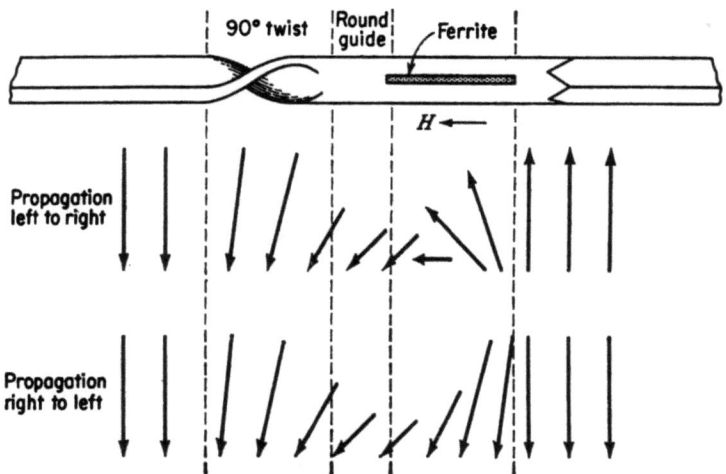

FIGURE 12. The Faraday rotation gyrator.

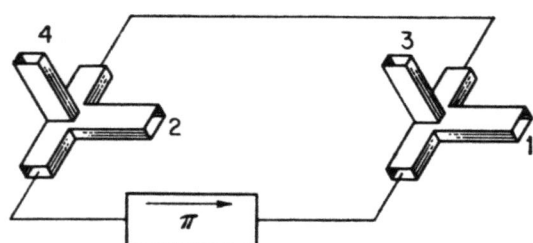

FIGURE 13. A four port circulator. The $\vec{\pi}$ symbol stands for a gyrator.

The configuration of Fig. 11 can be used as an amplitude modulator or switch: the Faraday rotation is a function of the applied magnetic field through the dependence of ω_r on the field. Therefore modulating the applied field will modulate the fraction of the component accepted by the rectangular wave guide at (2) in Fig. 11.

Through the dependence of the propagation velocity on the applied field, there exist various types of variable phase shifters. If the field is swept in time, the phase will be time dependent, and frequency modulation will result.

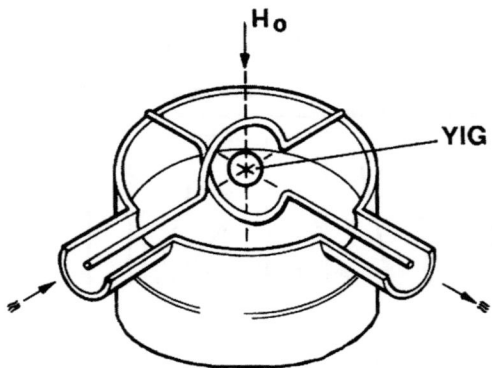

FIGURE 14. Typical configuration of a resonant transverse field device.

An example of a device with a transverse field configuration is shown in Fig. 14. This is a band pass filter; only fields at the resonance frequency are coupled by the YIG sphere between the two ports. No coupling exists at other frequencies because the coupling loops are orthogonal. A band stop filter is obtained by inserting a sphere in a transmission line: at resonance the strong absorption of the magnetic material drastically attenuates the transmitted power.

If the magnetic sample in such devices is in the region where the microwave magnetic field is linearly polarized, they will be reciprocal. If the field is circularly polarized they will behave in a nonreciprocal manner.

The geometrical arrangement shown in Fig. 14 is also used in power limiters. These make use of the nonlinear effects described in the paragraph on relaxation effects. The operating point of the device can be at the subsidiary peak, the main resonance peak or such that both coincide. The power at which limiting occur is of the order of tens of watts for the first case down to microwatts for the coincident case [48].

Ferrites under high driving field can also be used to generate frequency doubled microwave power. The second harmonic component of magnetization, M_{2z}, discussed in section III generates the second harmonic field. Magnetic materials are particularly useful in this respect at very high frequencies where diodes become progressively inefficient [1,35].

And finally, a few words about another device that utilizes nonlinear effects; the ferromagnetic amplifier [35]. This is essentially a parametric amplifier, where the ferromagnetic material behaves like a variable inductance modulated by a pumping field at ω_p. Coupled to this inductance are two tuned circuits at frequencies ω_1 and ω_2 known as idler and signal circuits. The frequencies obey the relation

$$\omega_1 + \omega_2 = \omega_p$$

As the energy at ω_p reaches a threshold which depends on the losses in the various circuits, oscillations will start in the idler and signal circuits and power will be drained from the pump. This is exactly the situation for the nonlinear loss effect described in section III, where ω_1 and ω_2 are essentially degenerate spin wave modes.

In the ferromagnetic amplifier the pump power is a little below threshold; thus an injected signal will be amplified at low noise. It is required that at least the signal tuned circuit be coupled to the microwave guide. Therefore either an external tuned circuit or low k magnetostatic modes are used. This device is being replaced by semiconductor amplifiers and varactor diode parametric amplifiers [49] that have, in recent years, improved tremendously in frequency coverage and noise figure.

REFERENCES

[1] R. F. Soohoo, IEEE Trans Magnetics, MAG-4, 118 (1968).
[2] J. W. Nielsen and E. F. Dearborn, The Phys. & Chem. of Solids, 5, 202 (1958).
[3] G. R. Pulliam, Chemical Vapor Growth of Single Crystal Magnetic Oxide Films, North American Aviation Inc., November 1966.
[4] J. Smit and H. P. J. Wijn, Ferrites, John Wiley & Sons, New York, 1959.
[5] A. H. Morrish, The Physical Principles of Magnetism, John Wiley & Sons, New York, 1965.
[6] A. S. Hudson, The Marconi Review, 33, 21 (1970).
[7] A. S. Boxer, J. F. Ollom and R. F. Rauchmiller, Properties of Ferrimagnetic Materials for Microwave Applications - Attachment to Technical Documentary Report No. ML-TDR-64-224, Bell Telephone Laboratories Inc., New Jersey, 1964.
[8] See ref. 5, p. 464.
[9] See ref. 4, p. 142.
[10] R. Pauthenet and L. Bochirol, J. Phys. Radium, 12, 249 (1951).
[11] J. B. Goodenough, in Magnetism and the Chemical Bond, Interscience, New York, 1963, p. 163.

[12] E. W. Gorter, Saturation Magnetization and Crystal Chemistry of Ferrimagnetic Oxides, Thesis, University of Leyden, 1954. Reprinted in Philips Res. Rev., 9, pp. 295-320, 321-365, and 403-443.
[13] L. R. Maxwell and S. J. Pickart, Phys. Rev., 92, 1120 (1953).
[14] A. Vassiliev, Ferrospinelles Comprenant l'Ion Li^+ et Contribution a ℓ'Etude de leurs Proprietes Magnetiques, Thesis, University of Paris, 1962.
[15] P. D. Baba, G. M. Argentina, W. E. Courtney, G. F. Dionne and D. H. Temme, IEEE Trans. Magnetics, MAG-7, 351 (1971).
[16] R. Pauthenet, Ann. Phys., 7, 710 (1952).
[17] L. G. Van Uitert, J. Chem. Phys., 24, 306 (1956).
[18] L. G. Van Uitert, J. Chem. Phys., 23, 1883 (1955).
[19] See ref. 7, p. 162.
[20] L. G. Van Uitert, J. Appl. Phys., 27, 723 (1956).
[21] See ref. 7, p. 73.
[22] W. R. Wilson, L. R. Hodges, Jr., G. P. Rodrigue and G. R. Harrison, J. Appl. Phys. 38, 1405 (1967).
[23] G. Winkler, IEEE Trans. Magnetics, MAG-7, 773 (1971).
[24] See ref. 5, p. 521.
[25] S. Dixon Jr., M. Weiner and R. Au Coin, J. Appl. Phys., 41, 1357 (1970).
[26] K. J. Button and T. S. Hartwick, in G. T. Rado and H. Suhl (Eds.), Magnetism, Vol. 1, Academic Press, New York, 1963, p. 649.
[27] A. G. Fox, S. E. Miller and M. T. Weiss, Bell System Tech. J., 34, 5 (1955).
[28] N. Bloembergen, Proc. IRE, 44, 1259 (1956).
[29] B. Lax and K. J. Button, Microwave Ferrites and Ferrimagnetics, McGraw Hill Book Company Inc., New York, 1962.
[30] K. J. Button and T. S. Hartwick, in G. T. Rado and H. Suhl (Eds.), Magnetism, Vol. 1, Academic Press, New York, 1963.
[31] See ref. 5, p. 549.
[32] J. S. van Wieringen, Phys. Rev., 90, 488 (1953).
[33] L. R. Walker, in G. T. Rado and H. Suhl (Eds.), Magnetism, Vol. 1, Academic Press, New York, 1963, pp. 299-381.
[34] See ref. 5, p. 607-624.
[35] See ref. 30, p. 658.
[36] C. W. Haas and H. B. Callen, in G. T. Rado and H. Suhl (Eds.), Magnetism, Vol. 1, Academic Press, New York, 1963, pp. 449-549.
[37] J. K. Galt and E. G. Spencer, Phys. Rev., 127, 1572 (1962).
[38] P. E. Seiden, J. Appl. Phys., 34, 1606 (1963).
[39] A. S. Hudson, IEEE Trans. Magnetics, MAG-5, 610 (1969).
[40] See ref. 5, p. 599.
[41] R. W. Damon, in G. T. Rado and H. Suhl (Eds.), Magnetism, Vol. 1, Academic Press, New York, 1963, pp. 551-620.

[42] H. Suhl, Proc. IRE, <u>44</u>, 1270 (1956).
[43] R. C. Le Craw, E. G. Spencer and C. S. Porter, J. Appl. Phys., <u>29</u>, 326 (1958).
[44] See ref. 5, p. 585.
[45] See ref. 4, p. 240.
[46] See ref. 5, p. 596.
[47] See ref. 5, p. 594.
[48] G. T. Rado and H. Suhl (Eds.), Magnetism, Vol. 1, Academic Press, New York, 1963, p. 656.
[49] L. E. Dickens, Proc. IRE, <u>60</u>, 328 (1972).

CHAPTER 18

CRYSTAL GROWTH

N. B. Hannay

Bell Laboratories, Murray Hill, New Jersey 07974

The importance of crystal growth for electronic materials is enormous, both because it provides a basis for research on the physics and chemistry of solids and because it is essential to the application of electronic materials. This generalization applies to all classes of electronic materials covered in this book - magnetic materials, semiconductors, and optical materials.

CRYSTAL GROWTH PROCESS [1]

Three basic factors govern the rate of crystal growth: diffusion, the surface energy, and the interface structure. The first two of these are already well understood in principle and in practice, and great progress is currently being made in the understanding of the effect of interface structures on crystal growth.

A number of diffusion processes are relevant to the growth process. One set has to do with the heat flows. There are external sources of heat, there is heat released in the freezing process, and there are various conduction processes, involving both the solid and the medium from which the crystal is being grown. These are illustrated in Fig. 1 for Czochralski growth from the melt.

There are also important mass flow considerations. The interface is the sink for the major component. In growth from solution the laws of diffusion describe the movement of atoms to this interface. Under steady-state conditions, Fick's first law

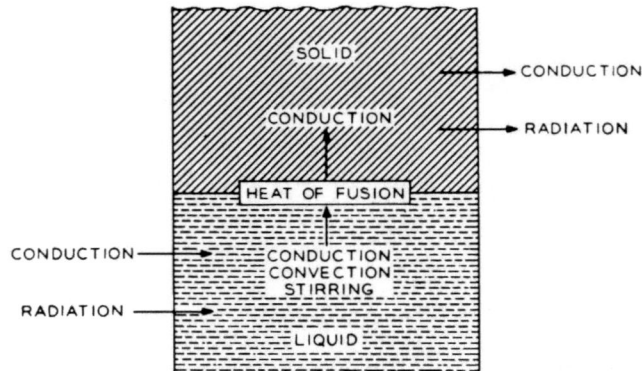

FIGURE 1. Heat flow in Czochralski crystal growth (after Tanenbaum [2]).

is useful and in nonsteady-state conditions the second law applies.

$$J = -D \frac{\partial c}{\partial x} \qquad (1)$$

$$\frac{\partial c}{\partial t} = D \nabla^2 c \qquad (2)$$

Fick's laws have been solved for most geometries involved in crystal growth. The more common geometries are diffusion toward a plane and toward a sphere.

The diffusion processes, both heat and mass, can be important if other processes are rapid and in this event they can limit the growth process. However, if the intrinsic growth rate of the crystal is slow, diffusion will not be limiting. Diffusion is also very important in the incorporation of impurities in the crystal, as will be discussed later in this chapter.

The second factor mentioned is that of surface free energy. Changes are usually small in comparison with the other factors, except for very small crystallites, and they rarely dominate the crystal growth process.

Finally, we must take into account the structure of the interface and its influence over growth rate. In **Fig. 2** an idealized model is shown of the interface. Atoms are added at sites which are most favorable energetically. In the case of a covalently bonded crystal, the most favored site is where the most bonds are formed by addition of the atom. In the case of an

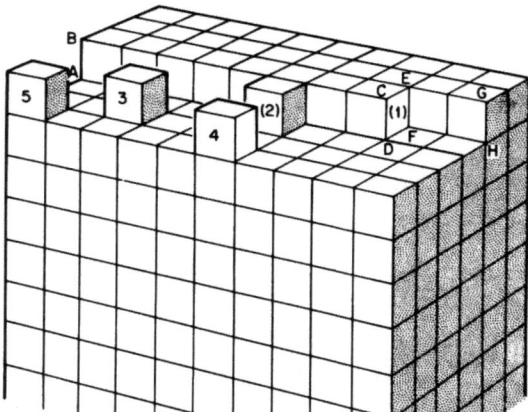

FIGURE 2. Model of interface structure showing various kinds of sites for addition of atoms to a growing crystal (after Laudise [1]).

---

ionically bonded crystal there will also be repulsive forces from nearby ions of like charge. The kink, site 1, is the most favorable site for a covalently bonded crystal, because three bonds are formed with the addition of the atom. A step, site 2, is the next best, because two bonds are formed. The least favored situation is the lone atom on a flat surface, as only a single bond is formed in this case, and three possibilities are shown for this. However, the second nearest neighbor effects lead to differences in these, as they also contribute to bond formation; thus site 3 is favored over site 4, which in turn is favored over site 5, because there are four, three and two next nearest neighbor atoms, respectively. In the ionic crystal (e.g. A^+B^-) both the repulsive forces from the next nearest neighbors and the attractive forces from the nearest neighbor and third nearest neighbor positions must be considered. Again, site 1 is most favored and site 2 is next, but for sites 3, 4 and 5 the order is reversed from the covalent case, i.e. site 5 is favored over 4, which in turn is favored over 3.

We next consider the kinetics of growth. We shall look at the interface on a microscopic scale to see what can be said about the theory of the interfacial structure and the growth rate [3].

The interfacial reaction, looked at microscopically, is described in the Wilson-Frenkel theory, as follows. The growth rate r is given by the difference between the rates for atoms arriving at, and leaving, the surface

$$r = r_A - r_L \tag{3}$$

where

$$r_A = k_A \exp(-\Delta E_A/RT)$$
$$r_L = k_L \exp(-\Delta E_L/RT) \tag{4}$$

The k's are rate constants, the E's activation energies. At equilibrium the rate of arrival of atoms and the rate at which they leave the surface are equal, so that $r = 0$. From this it follows that

$$k_A/k_L = \exp(-L/RT_E) \tag{5}$$

where $\Delta E_L - \Delta E_A = L$

L is the latent heat of fusion. If we let $\Delta T = T_E - T$ and $k_L = a\nu$, where a is the interatomic spacing and ν is the vibration frequency, then

$$r = a\nu \left(\frac{L\Delta T}{RT_E T}\right) \exp\left(\frac{-L}{RT_E}\right) \exp\left(-\frac{\Delta E_A}{RT}\right) \tag{6}$$

This shows that the growth rate depends upon the mobility of atoms at the surface $(a\nu)$, on the net flux of atoms to the surface $\left(\frac{L\Delta T}{RT_E T}\right)$, and on a Boltzman factor containing the activation energy for the freezing process.

The theory can be modified by assuming that growth will occur only on certain sites, for example, on screw dislocations, or as a result of a surface nucleation process. The whole expression in Eq. (6) is then multiplied by a factor which represents the fraction of the surface sites that are available for growth. In order to do this it is necessary to make explicit assumptions about a model for the surface and the number of sites available for growth, in order to calculate the growth rate, and the theoretical growth rate will be correct if the assumptions about the surface structure are correct. No general theory of this kind that is independent of the model assumed is possible, however.

It is apparent that in order to develop a true theory it is necessary to calculate the actual roughness rather than to assume some particular model. The simplest approach to this has been

CRYSTAL GROWTH

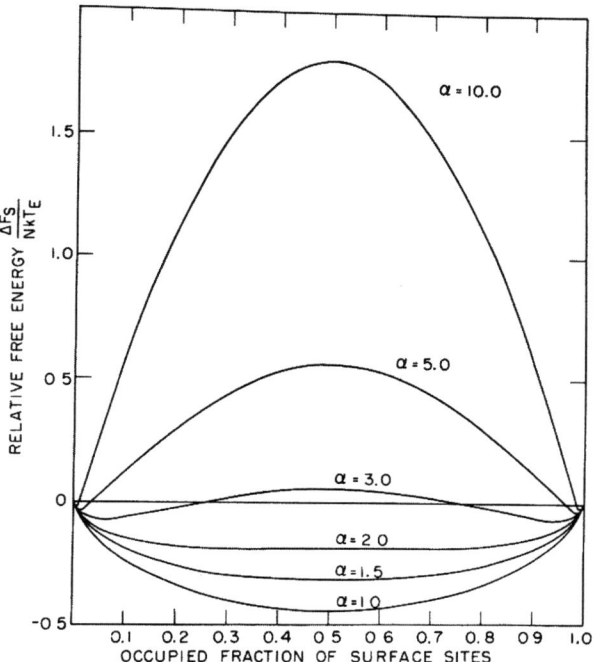

FIGURE 3. Free energy as a function of filling factor in a one-layer Bragg-Williams model. Curves represent different values of α (see text) (after Jackson [3]).

the Bragg-Williams model used by Jackson [3]. This is a calculation by statistical means of the random filling of a new layer on an initially plain surface. Changes in the free energy as the additional layer is added are calculated. The configurational entropy for the partially filled layer is found by considering the number of ways the atoms in the partially filled layer can be arranged; from this the free energy is obtained

$$\frac{\Delta G_s}{RT_E} = \alpha x(1-x) + x \ln(x) + (1-x) \ln(1-x) \qquad (7)$$

$$\text{where } \alpha = \left(\frac{L}{RT_E}\right) \xi$$

L/T_E is the entropy change associated with the transformation and ξ is a geometrical factor describing the packing on the surface. Thus it is seen that the controlling factor is α, a quantity which depends on ξ and on the entropy of crystallization. Figure 3 shows

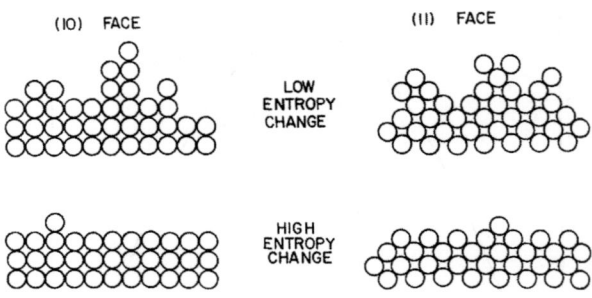

FIGURE 4. Multilayer interface structure (after Jackson [3]).

the free energy change plotted for various values of α and it is seen that the minimum free energy occurs when the layer is just half-filled, for α values less than 2; the minimum free energy occurs near x = 0 and 1 (which are equivalent since both represent completely filled layers) for $\alpha > 2$. Thus, rough surfaces correspond to low α, and atoms or molecules can be added readily as the roughness can be maintained with this addition. The roughness is independent of crystal face, so the growth will be isotropic. Rapid growth is expected for low α, therefore. Growth is slowest for flat surfaces, which correspond to large α, because it is unfavorable energetically to start new layers. For a given material, different crystal faces will represent different inherent degrees of roughness, and anisotropic growth may be expected.

The multilayer-model interface structure (Fig. 4) illustrates more clearly this roughness, as well as the dependence upon the α factor. It may be seen that when the entropy change is small, all crystal faces are rough, and growth is rapid and isotropic. When the entropy change is large, all faces are rather smooth, but certain crystal faces are more favorable for addition of atoms than others because more bonds are formed. Thus ξ becomes important. Growth is slow and anisotropic in this case. The detailed statistical mechanical calculation for the multilayer situation confirms the essential correctness of the simpler one-level calculation. Jackson assumed a rate of arrival independent of site, and a rate of departure determined by the number of nearest neighbors; surface jumps of atoms were allowed. He determined the various possible configurations for the surface and how one of them goes over into another as atom movements occur. A multilayer Bragg-Williams calculation was performed; the results have been confirmed by more sophisticated mathematical approaches that also take clustering into account, as these were found to give essentially the same answer. The results are shown in

CRYSTAL GROWTH

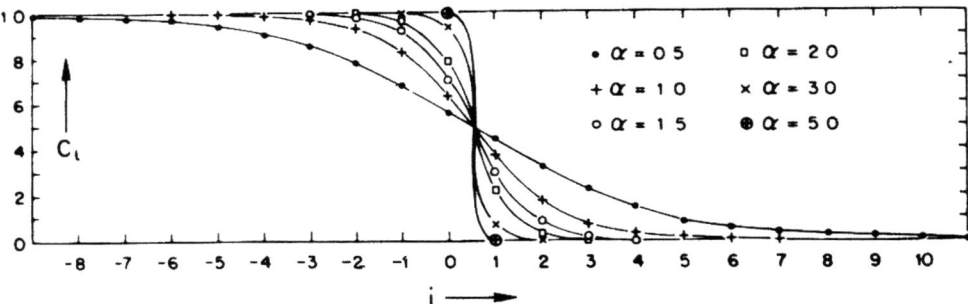

FIGURE 5. Structure of the "diffuse" interface, as shown by multilayer Bragg-Williams calculation. Fractional filling, C_i, of the ith layer is shown for values of i measured in and out from the nominal interface. Curves are for different values of α (after Jackson [3]).

Fig. 5. This shows that the partial filling extends over a number of layers and that the spreading out, or the diffuseness, of the interface is strongly dependent upon α.

The free energy for the multilayer structure is shown in Fig. 6. The abscissas go from 0 to 1, corresponding to the addition of one additional complete layer; clearly the interface can maintain any degree of roughness as this layer is added, depending upon the equilibrium configuration for the particular value of α, as shown in Fig. 5. It is seen that the curves are essentially flat to $\alpha=2$. At low α's there is no free energy barrier to the addition of more layers and growth will be rapid; the surface remains rough as atoms are added. However, at high α values there is a barrier to the addition of atoms, and an increase in free energy is required, initially. Growth will be slow in this case.

A quantitative theory of the growth rate has also resulted. At each point along the $\frac{n}{N}$ axis (Fig. 6) the equilibrium, or most probable, configuration is calculated. Having done this for one particular point, an atom is added to the system and the calculation is performed again. At high α's this addition results in an increase in the free energy. From the free energy change the exponential factor in the growth rate is determined, and the coefficients involve only the appropriate statistical multiplying factors. A sample result is shown in Fig. 7, which shows growth rate as a function of supercooling. For $\alpha=2$, with an entropy change of 6 and $\xi = 3$ (corresponding to the second most closely packed face), only a small supercooling is required for crystal growth. This is to be expected from Fig. 6 which showed that there was no barrier to the addition of layers on the basis of free

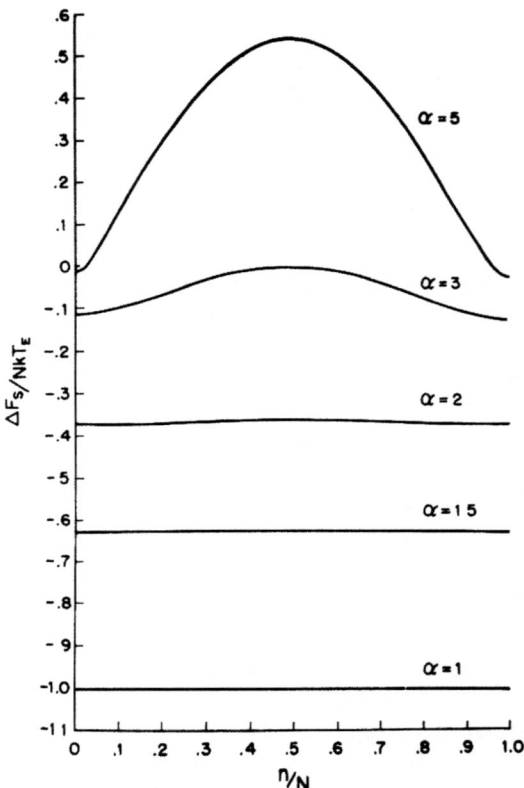

FIGURE 6. Free energy changes as new layers are added to the multilayer structure, for different values of α (after Jackson [3]).

energy calculations, for $\alpha = 2$. The situation is very different, however, for $\alpha = 3$, with an entropy change of 6 and $\xi = 2$ (corresponding to the closest packed face). In this case there is a free energy barrier, as shown in Fig. 6, and considerable supercooling is required in order for growth to occur at an appreciable rate. In this case there is a critical value of supercooling required for growth to occur. In Table 1 numerical results from the theoretical calculation of growth rate are shown and it can be seen that different rates are expected for various values of α and for the different crystal faces, in accordance with the foregoing.

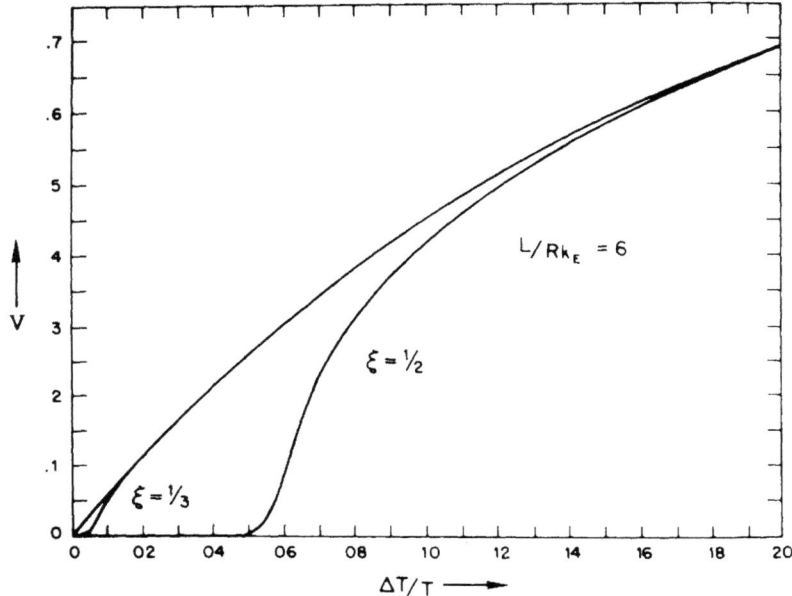

FIGURE 7. Growth rate as a function of supercooling for $\alpha = 2$ and $\alpha = 3$ (after Jackson [3]).

TABLE 1. Relative Growth Rates for Simple Cubic Crystals.

Entropy Change	Crystal Face (100)	(110)	(111)
1	.53	.55	.51
10	.01	.09	.51

CRYSTAL MORPHOLOGY

The theory of crystal growth outlined in the previous section provides the key to the origin of the morphologies in crystals [4]. Low α materials will grow isotropically and rapidly whereas high α materials will grow anisotropically and the rate will be limited by the slowest growing faces. Entropies of crystallization for several kinds of materials are shown in Table 2. From the entropies we can identify the conditions that control morphologies. We repeat - the only important factor is the value of α, and there is no additional dependence on the chemical nature of the solid. Experiments have been carried out on a variety of transparent

TABLE 2. Entropy Change on Crystallization [3].

$\frac{L}{kT_E}$	
1	Metals from Melt
3	Si, Ge, Sb, Bi, Ga from Melt
6	Most Organic Compounds from Melt
10	Metals from Vapor
20	Complex Molecules
>100	Polymers

organic and inorganic liquids. Their use permits the observation of growth morphology in a convenient fashion; thin layers were observed, through the freezing point, on a microscope slide, using transmitted illumination. A series of compounds were studied and some of these are shown in Fig. 8. Thus, carbon tetrabromide (a), with no impurities, shows a flat interface along the isothermal line, with growth rapid and completely isotropic. The entropy change for CBr_4 is 0.8. The addition of a small amount of impurity slows the growth through introduction of diffusion barriers in the melt, near the interface (next section). This leads to instabilities at the interface, as shown in (b), and causes the interface to break up into a cellular pattern of the kind commonly observed in reasonably pure metals. When an additional concentration of impurities is added, (c), this effect is increased and the growth morphology is dendritic. This diffusion-limited growth is similar to that observed in less pure metals with approximately the same entropy change. An example is shown in (d); this is tin, for which L/T_E = 1.34. At a higher entropy change, in trichloroacetic acid (entropy change = 2.14), the dendrites begin to show faceting (e). In this case growth is partly diffusion-limited and partly interface-limited. This is a condition that essentially reproduces the growth morphology of bismuth (f), which has an entropy change of 1.99. At a still higher entropy change (g), benzil, with an entropy change of 6, shows a growth morphology that is completely faceted, and quite anisotropic. In (h) the very anisotropic growth of calcium nitrate is illustrated; here the entropy change is 12.7. Finally, in (i), tristearin, with an entropy change of 63, shows conditions like those in crystal growth in semicrystalline polymers. Here the crystallization is spherulitic, high supercoolings (up to 50°) are common and there is a strong tendency toward spontaneous nucleation in the melt.

CRYSTAL GROWTH

FIGURE 8. Crystal morphologies for materials with different values of α: (a) pure CBr_4; (b) CBr_4 with a small amount of dye added; (c) CBr_4 with a higher concentration of added dye; (d) tin; (e) trichloroacetic acid; (f) bismuth; (g) benzil; (h) calcium nitrate; (i) tristearin (after Jackson and Miller [4]).

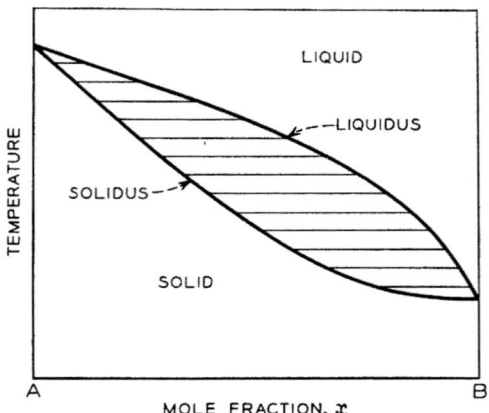

FIGURE 9. Binary solid solution illustrating distribution coefficients less than unity (B in A) and greater than unity (A in B) (after Thurmond [2]).

IMPURITY CONTROL IN CRYSTAL GROWTH

Impurity effects are extremely important in the growth of electronic crystals. A parameter that describes the incorporation of impurities into the crystal is the distribution coefficient [2]. This is illustrated in Fig. 9, which shows the phase diagram for a binary solid solution where there is a complete range of solid solubility over the whole range of composition. In general, the composition of the liquid and that of the solid will be different. The distribution coefficient is

$$k = C_S/C_L \qquad (8)$$

where C_S and C_L are the concentrations of impurity in the solid and the liquid. This may be less than, equal to, or greater than unity, depending upon the nature of the phase diagram. At the left side of Fig. 9, the liquid phase is richer in B than is the solid phase, hence the distribution coefficient for B in A is less than unity. On the right side of Fig. 9, the liquid phase contains a lower concentration of A than does the solid, and the distribution coefficient for A in B is greater than unity. When the solidus and liquidus coincide, $k = 1$. Usually the observed, or effective distribution coefficient, k_{eff}, is somewhat different from that expected under equilibrium conditions, k_{equil}. The latter may be observed, however, under very slow growth rates. At faster rates, it often happens that the impurity is concentrated or depleted in the region near the interface, as shown in Fig. 10(a). If the impurity is rejected at the interface by the solid, i.e. if the distribution coefficient is less than unity, then k_{eff} is greater

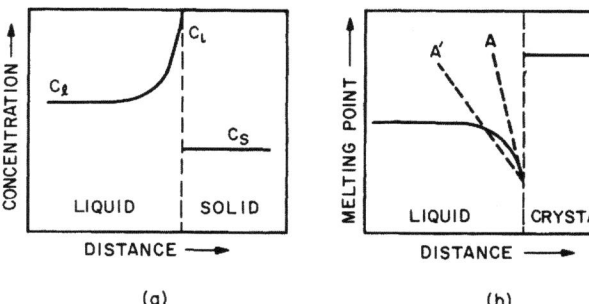

FIGURE 10. Impurity distribution through interface region for normal growth rates and k < 1 (a). The heavy line in (b) shows the corresponding melting points. Also shown are two possible temperature profiles in the liquid phase, one (A) a stable configuration and the other (A') leading to constitutional supercooling (after Tanenbaum [2]).

than k_{equil}. If k > 1, an analogous situation develops, with a depletion in the liquid near the interface. The details have been treated quantitatively [5]. k can also depend upon other impurities.

Still another factor of great important to crystal growth is constitutional supercooling [2], as shown in Fig. 10. In (a) the concentration profile is shown, and in (b) the melting points corresponding to this concentration profile are illustrated. Also shown in (b) are two possible situations depending upon the actual temperature gradient in the system. One of these is stable and the other unstable. The latter situation arises when the temperature gradient is so flat that the temperature in the liquid phase falls below the melting point near the interface. In this case the interface becomes unstable, and the cellular or dendritic growth which results causes an inhomogeneous distribution of impurities in the crystal. At very high doping levels constitutional supercooling can be extremely important, particularly when diffusion is slow and when the liquid and solid compositions are very different. Constitutional supercooling can be reduced by stirring, which has the effect of bringing k_{eff} closer to k_{equil}.

Another aspect of impurity control in solids that is often important is retrograde solid solubility [2], as shown in Fig. 11. As the temperature decreases the solubility of component B in A decreases, and under these conditions precipitation can occur if the kinetics permit.

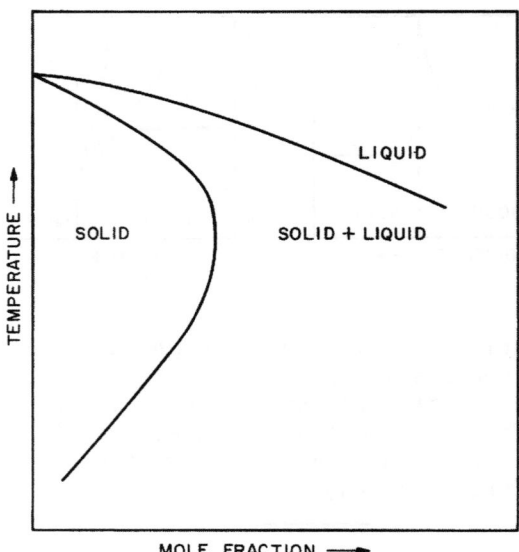

FIGURE 11. Retrograde solid solubility; solubility of B in solid A phase is limited and decreases with decreasing temperature (after Thurmond [2]).

Growth processes may be classified into conservative and nonconservative [2]. In the former no material is added to or subtracted from the system and the only change is the crystallization process itself. Examples of this are the Bridgman and Czochralski methods. Any control of impurities that is to be achieved is obtained by cropping of the crystal after successive stages of growth. Impurity distributions in such a situation are shown in Fig. 12.

In nonconservative processes, material can be added to the nonsolid phase during growth or removed by a process other than freezing. An example of this is zone refining, as material is added to the melt continuously during the crystallization process, as the zone moves. In Fig. 13 the impurity concentration in the solid, for different values of k, is shown as a function of zone lengths solidified. Zone-refining is widely used to purify solids, as repetitive operations lead to higher and higher purities without loss of material.

PRACTICAL GROWTH METHODS [1]

A great many methods are used to grow crystals and it is not practical to attempt a review of these in this chapter. The reader is referred to extensive discussions of this subject

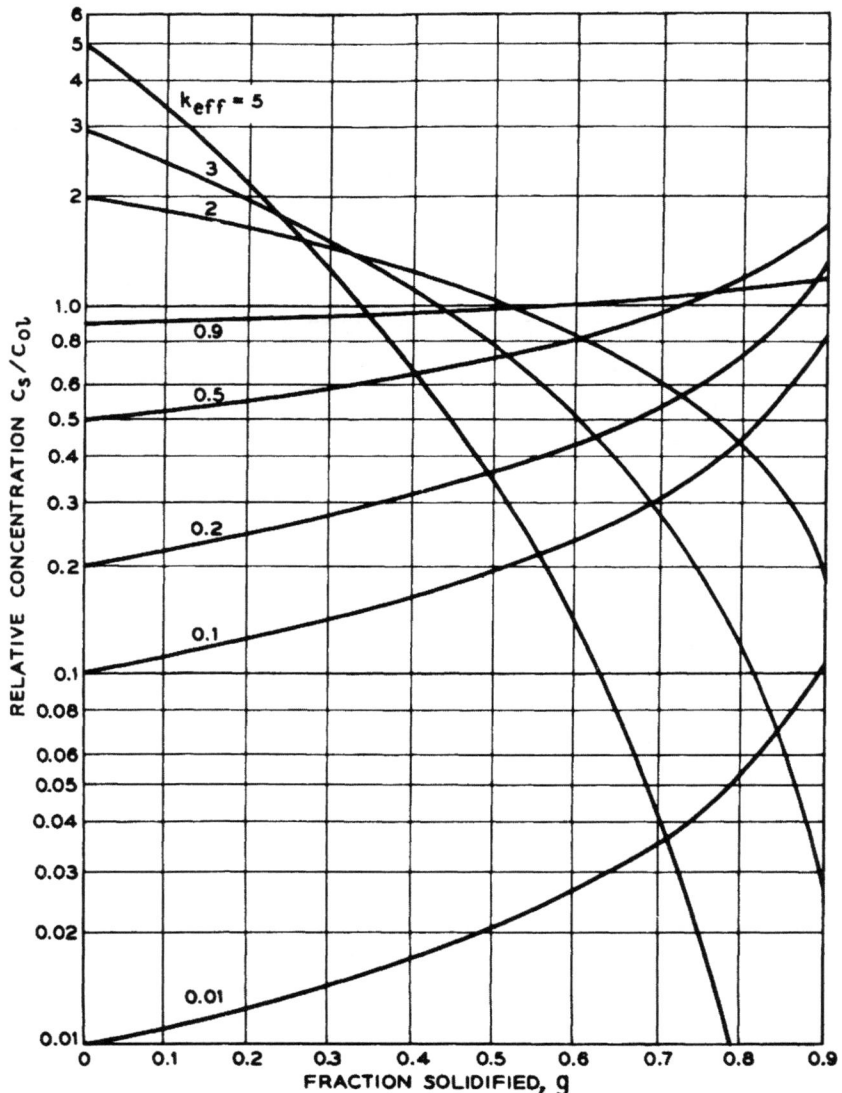

FIGURE 12. Impurity concentration as a function of fraction of the material crystallized, for different distribution coefficients (after Thurmond [2]).

available elsewhere. Growth methods may be monocomponent or polycomponent, in a chemical sense. Growth may be from the solid, liquid, or vapor phase. And growth may, or may not, involve a chemical reaction.

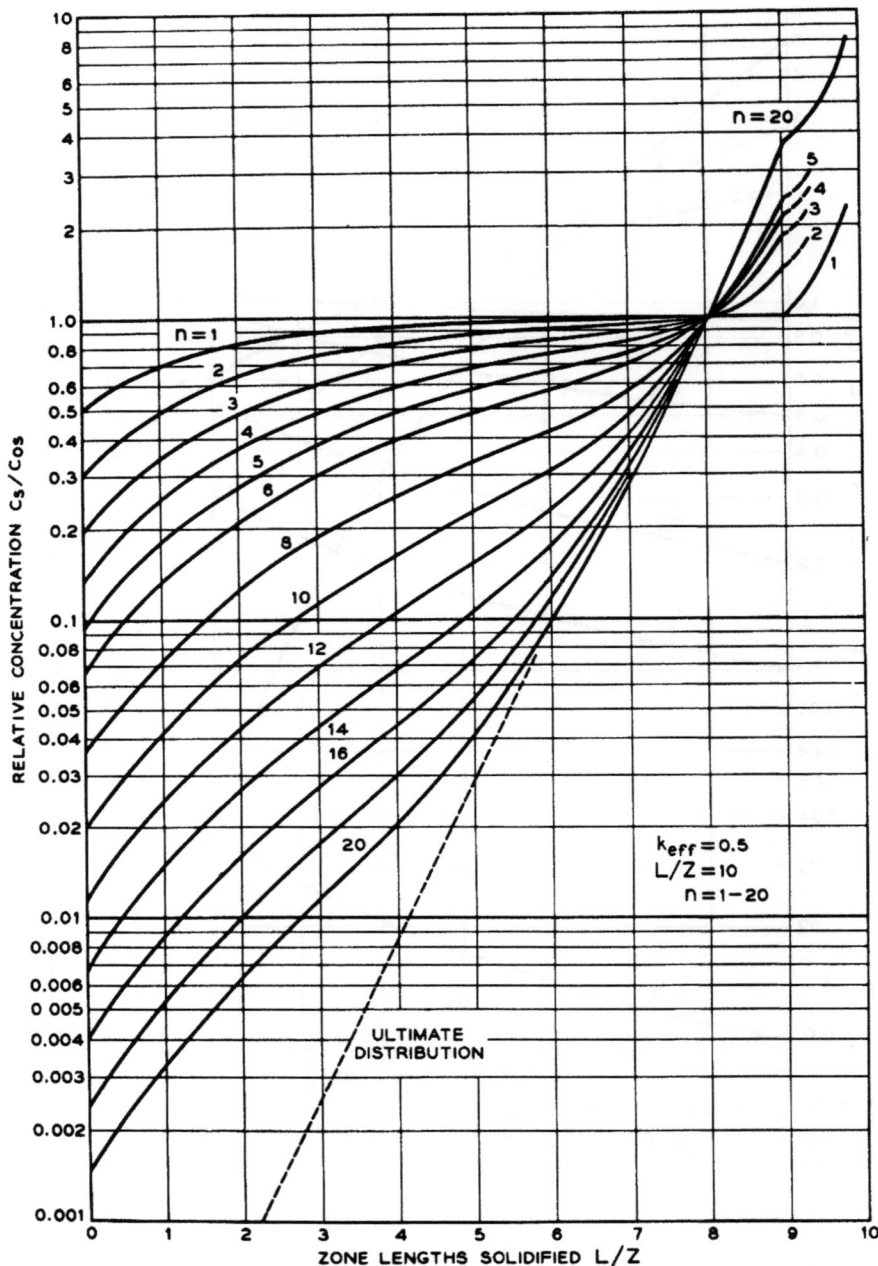

FIGURE 13. Impurity concentration as a function of zone lengths solidified in zone-melting process, for different distribution coefficients (after Thurmond [2]).

CRYSTAL GROWTH

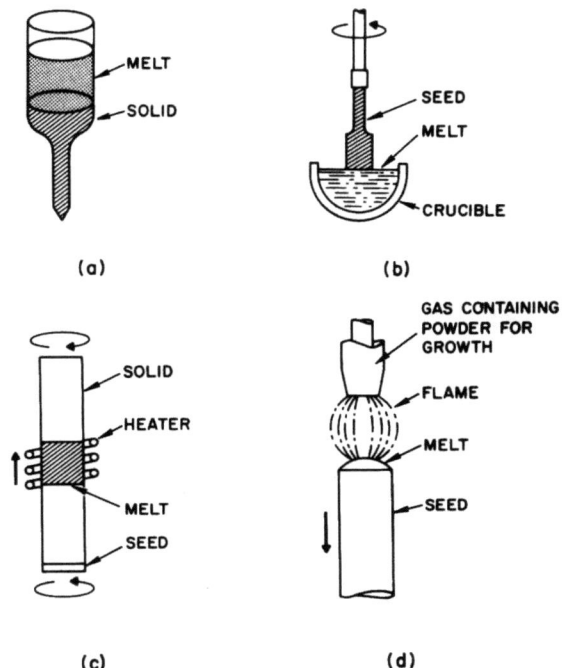

FIGURE 14. Crystal growth from the melts: (a) Bridgman-Stockbarger, (b) Czochralski, (c) floating zone, (d) Verneuil.

Among the most commonly used methods is the Bridgman-Stockbarger technique, which utilizes an open or closed, vertical or horizontal, container, and involves the directional freezing of a melt. A typical geometry is shown in Fig. 14(a). The advantages of the method are its simplicity and applicability to many materials; disadvantages include the possibilities of crucible contamination and uncontrolled nucleation. Table 3 summarizes some of the systems for which the Bridgman-Stockbarger technique has proved useful.

Also very widely used for growth from the melt is the Czochralski method, Fig. 14(b). This provides much better control over spurious nucleation, as the solid-liquid interface does not come into control with the crucible wall, and large crystals may be easily grown. There remains the hazard of crucible contamination. The Czochralski method provided the first single crystals of Ge and Si, and its importance for these led to exhaustive studies of the growth conditions and their effect on crystal quality. Two variations of the Czochralski method are of

TABLE 3. Some Representative Crystals Grown by Bridgman-Stockbarger Method [1].

Compound	Formula	Melting Point (°C)	Crucible Material	Cooling or Lowering Rate	Atmosphere
Silver bromide	AgBr	434	Pyrex, quartz, Pt	1-5 mm/hr	Cl_2(HCl or HBr pre-evacuate)
Argon	Ar	-189.4	Glass	1 mm/min	Ar
Copper	Cu	1083.2	Graphite	5-~20 cm/hr	Vacuum
Nickel	Ni	1455	Recrystallized Al_2O_3	0.1-0.2 mm/hr	Vacuum
Lithium	Li	179	Steel, stainless steel	2°-30°/hr	Ar
Fluorite	CaF_2	1392	Ta, Fe, Ni, or C	10 mm/hr	Vacuum

TABLE 4. Some Representative Crystals Grown by Crystal-Pulling [1].

Compound	Formula	Melting Point (°C)	Crucible Material	Pulling Rate	Atmosphere
Germanium	Ge	937			
Silicon	Si	1412			
Zinc	Zn	419	Pyrex	1.2cm/min	N_2
Gallium arsenide	GaAs	1240	Vitreous silica		As
Potassium chloride	KCl	770	Pt or porcelain		Air
Water	H_2O	0	Glass		Air
Calcium tungstate	$CaWO_4$	1535	Rh	0.5-2cm/hr	Air
Lithium niobate	$LiNbO_3$	1260	Pt	0.5-2cm/hr	Air
Sapphire	Al_2O_3	2050	Ir	0.5-2cm/hr	Air
Yttrium-aluminum garnet	$Y_3Al_5O_{12}$	~1900	Ir	0.5-2cm/hr	Air

particular interest. In one the seed is necked down, to grow out any dislocations, and then allowed to expand to full size; this has permitted the growth of larger dislocation-free crystals. In the other, a liquid flux, such as B_2O_3, is used to coat the melt and crystal surface in the growth of crystals with high decomposition pressures. The vessel is then pressurized with an inert gas. The B_2O_3 liquid encapsulant effectively prevents diffusion out of the decomposition products and avoids the complications arising from their chemical reaction with the vessel walls. Table 4 lists some examples of Czochralski-grown crystals.

Another important method of growth from the melt is the floating zone, Fig. 14(c). Surface tension and levitation from the magnetic field of the R.F. coil used to keep the zone molten overcome gravitation forces, thus leading to a stable zone. The great advantage of the method is the complete avoidance of crucible contamination. Examples of its use are listed in Table 5.

TABLE 5. Some Crystals Grown by Zone-Melting Technique [1].

Material	Melting Point (°C)	Boat
Germanium	942	Vitreous silica, graphite-coated
Silicon		Floating zone
Gallium arsenide	1240	Vitreous silica or silica coated with pyrolytic graphite
Tungsten	3370	Water-cooled copper reactor and float zone

Another method for growth from the melt that has proved useful is the Verneuil, Fig. 14(d). It is especially useful for refractory materials, for which high temperatures are needed; feed material, in powder form or from gaseous reactions, is supplied to the molten tip of a boule that is heated by flame, radiation (arc image, or laser), or plasma torch. Again no crucible is required, but crystal quality is usually inferior, although purity may be high. Examples are listed in Table 6.

Vapor-solid methods are also very useful. The simplest of them depend on the sublimation and condensation of a solid with a

TABLE 6. Some Crystals Grown by Verneuil and Arc-Image Techniques [1].

Material	Melting Point (°C)	
Al_2O_3	2040	Corundum, sapphire
Al_2O_3:Cr		Ruby
$CaWO_4$	1530	Scheelite
TiO_2	1830	Rutile
$NiFe_2O_4$	Above 1200	Nickel ferrite

reasonably high vapor pressure. A difficulty is the tendency toward nucleation of many crystals. Examples of this method are given in Table 7.

TABLE 7. Representative Crystals and Films Grown By Sublimation-Condensation [1].

Material	Sublimation Temperature (°C)	Condensation Temperature (°C)	Gas
Cd	320-330 (open or closed tube)	250-290	Vacuum, Air
Zn	375-475 (open or closed tube)	350-~380	Ar, He
CdS	1150-1200 (open or closed tube)	1100	Ar
ZnS	1550-1600 (open or closed tube)	1475-1500	Ar
SiC	>2500 (open or closed tube)	<2500	Inert atmosphere

A useful class of vapor-solid methods involves vapor transport with chemical reaction. The source may be a mixture of gases, which react chemically to produce the deposit, or it may be a solid over which reacting gases are passed, with reversal of the reaction at the seed crystal. A problem is that the chemical reactions must be tailored to each particular crystal and in case of compound crystal growth several vapor-solid reactions must be controlled simultaneously. Again, spurious nucleation must be avoided. Good results can be obtained when the chemistry is under control, and the epitaxial growth of Si from the vapor, widely used in the fabrication of integrated circuits, is an outstanding example of its use. Some examples of the application of chemical vapor deposition are given in Table 8.

TABLE 8. Crystals Grown by Vapor Reaction [1].

Crystal	Reaction
Iron	$Fe_{(s)} + 2HCl_{(g)} \underset{730°C}{\overset{1000°C}{\rightleftarrows}} FeCl_{2(g)} + H_{2(g)}$
Copper	$Cu_{(s)} + HI_{(g)} \underset{650°C}{\rightleftarrows} CuI_{2(g)} + H_{2(g)}$
Silicon	$Si_{(s)} + 4HCl_{(g)} \underset{1200°C}{\rightleftarrows} SiCl_{4(g)} + 2H_{2(g)}$
Gallium arsenide	$2GaAs_{(s)} + 6HCl_{(g)} \rightleftarrows 2GaCl_{3(g)} + As_{2(g)} + 3H_{2(g)}$
Gallium phosphide, GAP	$2GaP_{(s)} + H_2O_{(g)} \underset{1050°C}{\overset{1100°C}{\rightleftarrows}} Ga_2O_{(g)} + P_{2(g)} + H_{2(g)}$
Gallium arsenide, GaAs	$GaAs_{(s)} + \frac{3}{2}I_{2(g)} \underset{900°C}{\overset{1100°C}{\rightleftarrows}} GaI_{3(g)} + As_{(g)}$
	$2GaAs_{(s)} + 3ZnCl_{2(g)} \underset{900°C}{\overset{1100°C}{\rightleftarrows}} 2GaCl_{3(g)} + 2As_{(g)} + 3Zn_{(g)}$

Solution growth has been used for many water-soluble materials (Table 9). The method is simple, but is limited to water-soluble materials and it is slow. In the case of materials with low solubility, high temperatures and pressures, in a closed vessel, offer the possibility of useful growth rates. This "hydrothermal" method has been successfully used for the routine growth of large, high quality quartz crystals, on a commercial basis.

Solution growth can also be carried out in nonaqueous solvents. In this case it is often called "flux growth". A variety of fluxes have been used and crystals of a number of materials have been grown. Some examples are given in Table 10.

TABLE 9. Representative Crystals Grown from Aqueous Solvents [1].

Material	Formula	Method
Potassium alum	$KAl(SO_4)_2 \cdot 12H_2O$	Rotary crystallizer
Chrome alum	$KCr(SO_4)_2 \cdot 12H_2O$	Rotary crystallizer
Rochelle Salt (potassium-sodium tartrate)	$KNaC_4H_4O_6 \cdot 4H_2O$	Rotary crystallizer
Rock salt	$NaCl$	Evaporation

TABLE 10. Typical Crystals Grown from Molten-Salt Solvents [1].

Material	Formula	Solvent	Method
Yttrium-iron garnet (YIG)	$Y_3Fe_5O_{12}$	PbO	Slow-cooling
Barium titanate	$BaTiO_3$	KF	Evaporation and cooling
Barium titanate	$BaTiO_3$	TiO	Pulled from melt
Yttrium-aluminum garnet (YAG)	$Y_3Al_5O_{12}$	$PbO-PbF_2$	Slow-cooling
Sapphire or gallia	Al_2O_3 Ga_2O_3	PbF_2	Slow-cooling

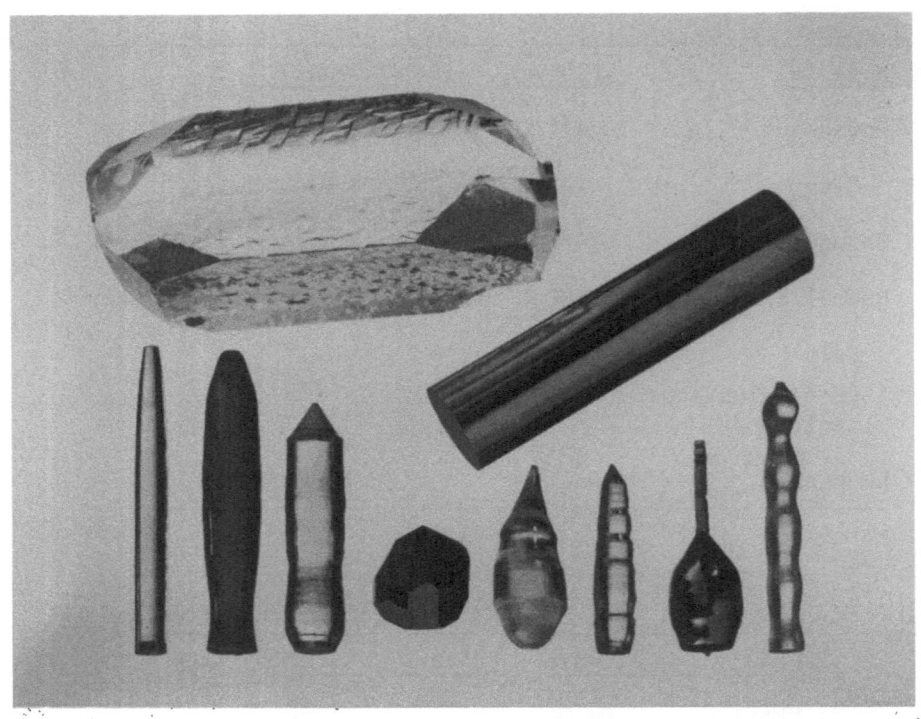

FIGURE 15. Some crystals of materials of interest in electronics.

A largely unexplored method of crystal growth that offers interesting practical possibilities is gel growth. The reacting ions are allowed to diffuse through the gel, with crystal growth occurring where they meet, within the gel medium. The gel controls the diffusion, and hence the growth rate.

Figure 15 shows some examples of crystals grown by some of the methods described.

REFERENCES

[1] R. A. Laudise, "The Growth of Single Crystals", Prentice-Hall, Englewood Cliffs, N. J. (1970).
[2] M. Tanenbaum (Chapter 3) and C. D. Thurmond (Chapter 4) in "Semiconductors", N. B. Hannay, Ed., Reinhold Publishing Co., New York (1959).

[3]   K. A. Jackson, in Proceedings of the Robert A. Welch Foundation, Conferences on Chemical Research, XIV Solid State Chemistry; in Proceedings of Interfaces Conference, Melbourne, Australia (1969); Chapter 12 in Materials Science Research, Vol. 4, Plenum Press, New York (1969).
      H. J. Leamy and K. A. Jackson, J. Appl. Phys. $\underline{42}$, 2121 (1971).
      R. A. Laudise, J. R. Carruthers and K. A. Jackson, Annual Review of Materials Science, Vol. 1 (1971).
[4]   Jackson, ref. 3, and C. E. Miller and K. A. Jackson, private communication.
[5]   J. A. Burton, R. C. Prim and W. P. Slichter, J. Chem. Phys. $\underline{21}$, 1987 (1953); J. A. Burton, E. D. Kolb, W. P. Slichter and J. D. Struthers, J. Chem. Phys. $\underline{21}$, 1991 (1953).

CHAPTER 19

SOLID STATE CHEMISTRY

N. B. Hannay

Bell Laboratories, Murray Hill, New Jersey 07974

One of the primary aims of materials science and engineering is to relate the macroscopic properties of solids to their structure. The structure includes both the electronic and the atomic structure.

THE PERFECT SOLID - CHEMICAL BONDING

In Chapter 1 of this book the modern band theory of solids has been briefly reviewed. This, of course, has provided a basis for our understanding of many of the electronic properties of solids. A chemical bond approach has also been exceedingly useful over the years and it adds a great deal of insight into the properties of solids. This began with concepts due to Pauling over 40 years ago [1]. He was the first to describe the bonding in solids in terms of a partial ionic character, recognizing that the bond contains both a covalent and an ionic component. He used the heats of formation of binary compounds to derive an electronegativity scale, and this in turn allowed him to estimate the degree of ionic character in a bond from the difference in electronegativities of the constituent atoms. Taking ΔH_{AA}, ΔH_{AB}, and ΔH_{BB} as the heats of formation of AA, AB, and BB bonds, the electronegativities are then derived from

$$(X_A - X_B)^2 \sim \Delta H_{AB} - \tfrac{1}{2}(\Delta H_{AA} + \Delta H_{BB}) \qquad (1)$$

From the electronegativities the ionicity f_i is defined as

	IA				IB	
	Li	Na	K	Rb	Cu	Ag
F	6	6	6	6	4	6
Cl	6	6	6	6	4	6
Br	6	6	6	6	4	6
I	6	6	6	6	4	4

	IIA			IIB				
	Ca	Sr	Ba	Be	Mg	Zn	Cd	Hg
O	6	6	6	4	6	4	6	6
S	6	6	6	4	6-4	4	4	6
Se	6	6	6	4	6-4	4	4	4
Te	6	6	6	4	4	4	4	4

FIGURE 1. Distribution of fourfold and sixfold coordination in $A^N B^{8-N}$ compounds, for N = 1,2 [2].

$$f_i = 1 - \exp[-(X_A - X_B)^2/4] \quad (2)$$

Until very recently there has been no essential improvement in the Pauling theory since it was first formulated. However, Phillips and Van Vechten [2] have now provided a reformulation, using spectroscopic data rather than heats of formation as a basis. They have applied their theory to all of the binary compounds $A^N B^{8-N}$ with 8 s-p valence electrons per bond. Thus, they include in their formulation the Group IV semiconductors as well as III-V, II-VI, and I-VII compounds. This is by far the largest group of homologous structures. In Fig. 1, the coordination in these compounds is shown for N = 1 and 2. This shows that some of the compounds fall into the rocksalt structure with sixfold coordination, and the rest in the tetrahedral coordination of the wurtzite or zincblende structure. Diamond is the prototype of the covalent structure and tetrahedral coordination is typical of predominantly covalent compounds such as Si, Ge, and the III-V compounds. Likewise, NaCl is the prototype of predominantly ionic structures and the sixfold coordination of the rocksalt structure is typical of mainly ionic compounds.

A clear test of any theory of ionicity would be a successful prediction of structure, that is, a prediction of fourfold or sixfold coordination. The full quantum-mechanical calculation of the cohesive energy is beyond present capabilities. It has been done only for a few ionic compounds - alkali-halides - with any degree of success, and it is unable at the present time to take into account covalency effects. Accordingly, a goal of the Phillips-Van Vechten theory is the prediction of structure.

In the Phillips theory, it is recognized also that the AB bond has a homopolar or covalent part, and an ionic or heteropolar part. The covalent part is identified with the symmetric wavefunction, and the ionic part with the antisymmetric wavefunction. The

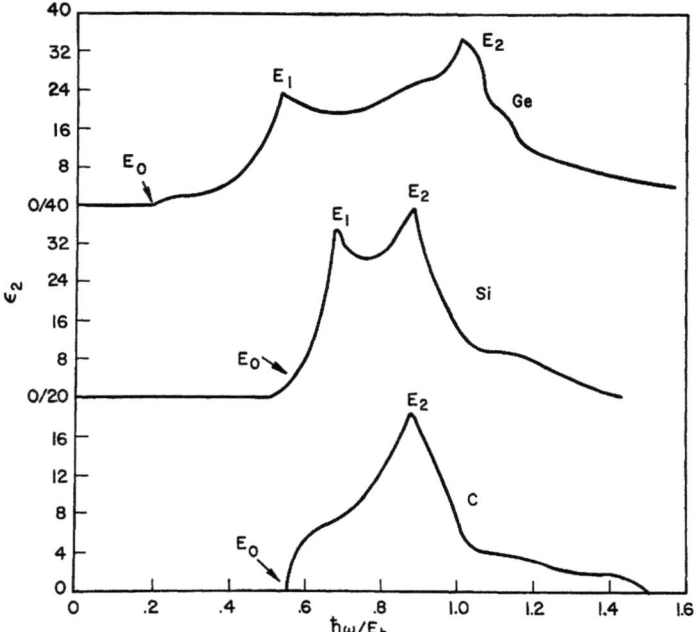

FIGURE 2. Out-of-phase dielectric constant, ε_2, for diamond, silicon, and germanium [2].

bonding states are predominantly on the more electronegative atom and they point toward the nearest neighbor atoms; these are the lower energy states. The antibonding higher energy states are predominantly on the more electropositive atom and point away from the nearest neighbor atoms.

Phillips defines the covalent and ionic parts of the bond spectroscopically, rather than through the heats of formation. To get the bond energies, he averages over all optical transitions, instead of doing the more difficult average over energy bands. The out-of-phase part of the dielectric constant, ε_2, is the part that corresponds to absorption; examples are shown in Fig. 2. Absorption peaks are commonly designated as E_0, E_1 and E_2. ε_1 can be derived from ε_2 through the Kramers-Kronig dispersion relation, and the total energy gap E_{total} is taken from the low frequency limit of ε_1 (i.e. the square of the refractive index), so that

$$\varepsilon(\omega) = \varepsilon_1(\omega) + i\,\varepsilon_2(\omega) \tag{3}$$

$$\varepsilon_1(o) = 1 + 4\pi Ne^2\hbar^2/m\,E_{total}^2 = n_o^2 \tag{4}$$

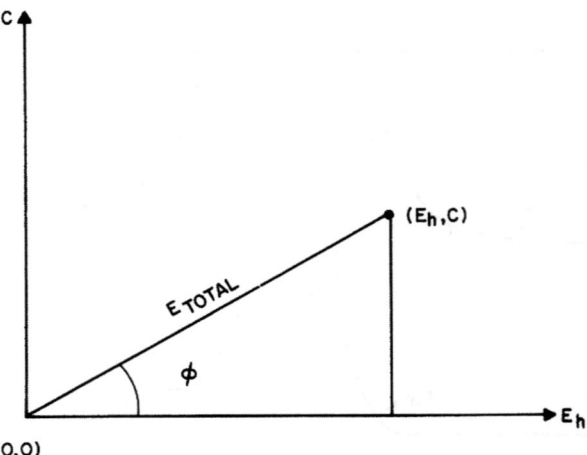

FIGURE 3. Relation between total energy gap E_{total}, homopolar gap E_h, and ionic gap, C [2].

In this N is the number of valence electrons per cm^3 and e and m are the charge and mass of the electron. E_{total} is a weighted isotropic average, and it measures the average energy required to excite a bonding electron from the valence band to an antibonding state in the conduction band. Figure 2 shows that in diamond E_2 is responsible for almost all of the absorption, and the average gap is approximately the same as the minimum gap because only a small part of the conduction band contributes importantly. In higher members of the Group IV elements, however, new peaks appear which correspond to additional transitions, and E_0 and E_1 become increasingly important. This is called metallization (dehybridization).

Next, Phillips considers how to separate E_{total} into the homopolar (covalent) part and the heteropolar (ionic) part. The homopolar part should depend only on the bond length. In the $A^N B^{8-N}$ compounds this is essentially unchanged across a series of compounds formed from the same row of the Periodic Table, that is, the Group IV element has the same bond length as the III-V, II-VI, and I-VII compounds in that row. This suggests that the completely covalent Group IV semiconductor, in which the total energy gap is determined entirely by the homopolar gap, will also give the homopolar part for all members in the series of compounds in that row. The homopolar part does vary greatly, however, from one row to the next; thus, it is 13.6 eV in diamond, and decreases to 3.1 eV in gray tin, using the refractive index for these materials and Eq. (4).

To get the ionic part we note that there is a 90° phase difference between the covalent and the ionic terms (Fig. 3) so

FIGURE 4. Energy gaps for germanium and zinc selenide [2].

that we have the relation

$$E_{total}^2 = E_h^2 + C^2 \qquad (5)$$

This defines a dimensionless ionicity

$$f_i = C^2/E_{total}^2 = \sin^2\varphi \qquad (6)$$

These quantities are shown in Fig. 4 for germanium and zinc selenide. Typical ionic and homopolar energy gaps are shown in Table 1. It is of interest to compare the Phillips ionic energy

TABLE 1

Crystal	E_h(eV)	C(eV)	Crystal	E_h(eV)	C(eV)
BN	13.1	7.71	InP	3.93	3.34
BeO	11.5	13.9	CdS	3.97	5.90
AlP	4.72	3.14	AlSb	3.53	3.10
GaAs	4.32	2.90	GaSb	3.55	2.10
ZnSe	4.29	5.60	ZnTe	3.59	4.48
InSb	3.08	2.10	InAs	3.67	2.74
CdTe	3.08	4.90	CdSe	3.61	5.50
ZnO	7.33	9.30	CuI	3.66	5.50
AlAs	4.38	2.67	CuBr	4.14	6.90
GaP	4.73	3.30	CuCl	4.83	8.30
ZnS	4.82	6.20			

gap with Pauling's electronegativity difference, as Eqs. (2) and (6) show that these should be linearly related. This is shown in Fig. 5; a roughly linear relationship is observed with deviations due to the approximate nature of the Pauling scale.

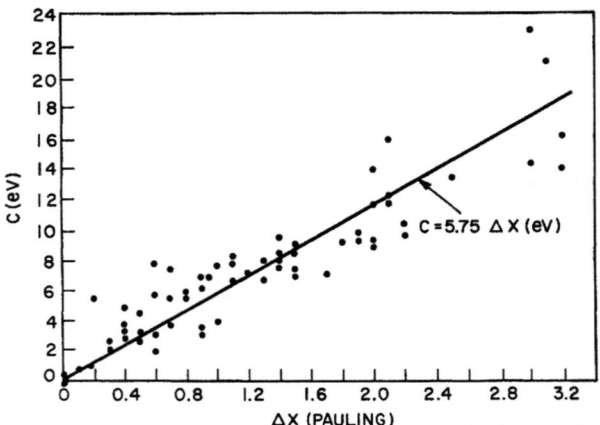

FIGURE 5. Comparison of Phillips ionic energy gap C and Pauling electronegativity difference for binary compounds [2].

Finally, these results can be used to predict structures. This is shown in Fig. 6. All of the $A^N B^{8-N}$ compounds are shown and it is seen that a line can be drawn which separates all of the sixfold and the fourfold coordination structures, with no mistakes. This line corresponds to an ionicity $f_i = 0.785$. Thus, whenever the ionicity is higher than this critical ionicity, the compound assumes the ionic sixfold coordination, that is the rocksalt structure, and whenever the ionicity is below this critical value, the compound assumes the covalent fourfold coordination of the diamond-zincblende structure. The Pauling ionicity scale applied to the same set of compounds predicts most of the structures correctly, but is inaccurate in eight cases. Borderline cases, in particular magnesium sulfide and magnesium selenide, which fall on the critical ionicity line, are found to be metastable in both coordination configurations. Silver iodide, which is slightly below the critical ionicity, is normally fourfold but is easily transformed to a sixfold rocksalt structure by the application of a few atmospheres of pressure.

The preceding discussion has applied to the average isotropic E_{total}. The theory has been extended to other energy gaps also [3]. Particular transitions corresponding to specific energy gaps of interest that are associated with energy states at certain important k values in the Brillouin zone, i.e. symmetry points, have been calculated. Both direct and indirect gaps were determined. The results include the lowest energy interband transitions, the quantities that are usually called the energy gap in semiconductors. In calculating these additional transitions it was assumed that

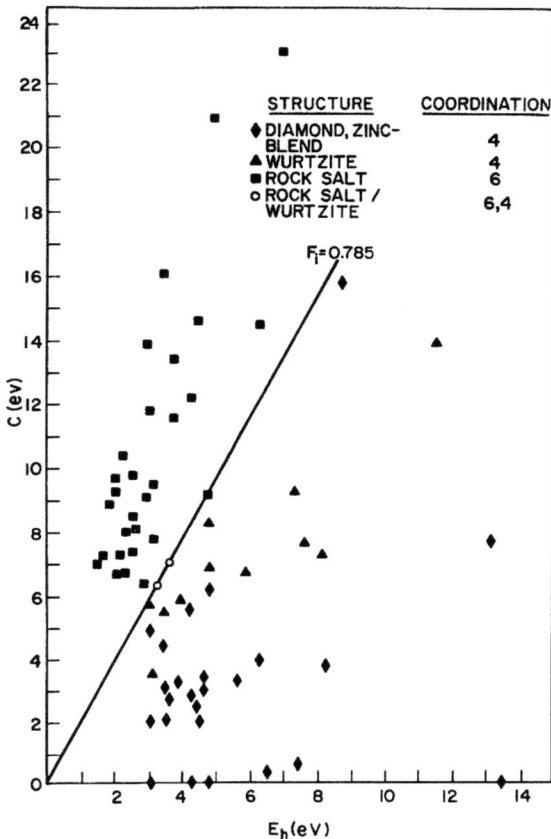

FIGURE 6. Prediction of structures; line corresponding to critical ionicity divides experimentally observed structures into groups with fourfold coordination (above line) and sixfold coordination (below line) [2].

the ionic gap, C, is constant for all transitions. This was determined by the procedure that has been described above, i.e. by deriving E_{total} from $\varepsilon_1(0)$, taking E_h from the homopolar semiconductors, and calculating C. C is also given by

$$C = be^{-kR}\left(\frac{Z_\alpha}{r_\alpha} - \frac{Z_\beta}{r_\beta}\right)e^2 \qquad (7)$$

where $b \approx 1.5$

where Z is the valence, r the covalent radius, e the electronic charge, and the exponential factor is the Thomas-Fermi screening factor. The term in the parentheses corresponds to the difference in electronegativities. The constant b cannot be calculated, but it is determined from the values of C derived from the Phillips-Van Vechten theory to be approximately 1.5. The C determined in this way for a particular material is added to the E_h for each transition. E_h varies from one transition to another; it is calculated first for all the transitions in the homopolar Group IV semiconductors using the relevant optical data, such as were shown in Fig. 2. E_h is then scaled for the other binary compounds using only the bond length. This procedure leaves no adjustable parameters in the calculation of E_h for a particular transition. This method has been applied to 19 tetrahedrally coordinated semiconductors with excellent results.

Another aspect of the Phillips-Van Vechten theory has been a consideration of the cohesive energy and heat of formation of solids [4]. An ultimate goal of bond theory would be the calculation from first principles of cohesive energies and heats of formation, as these quantities are fundamental to both physics and chemistry of solids. This remains a basic unsolved problem, although the spectroscopic theory of Phillips and Van Vechten can be used to approach these quantities.

The cohesive energy, ΔG_s, is the free energy per atom for complete atomization, while the heat of formation, ΔH_{AB}, is the enthalpy for the formation of the solid starting from standard states for the constituent elements. The quantities are directly related, through ΔG_{AB}, since

$$\Delta G_s = \Delta G_{AB} + \Delta G_A + \Delta G_B$$
$$\Delta G_{AB} = \Delta H_{AB} - T\Delta S_{AB} \quad (8)$$

where the subscript AB refers to formation of the compound from the standard states, and the subscripts A and B refer to cohesive energies for the standard states. The heat of formation is assumed in the theory to be proportional to the ionicity; this is similar to the Pauling approach, in which electronegativities and ionicities are derived directly from this proportionality. A phenomenological theory then gives the following expression for the heat of formation (at 298°K):

$$\Delta H_{AB} = \Delta H_o \left(\frac{a_{IV}}{a_{AB}}\right)^s \left[1 - b\left(\frac{2E_2}{E_0+E_1}\right)^2\right] f_i \quad (9)$$

ΔH_0 is an overall scaling factor. The metallization (dehybridization) factors are included in two ways, corresponding to the other terms in the equation. E_0, E_1, and E_2 were identified previously. s is determined from experiment and is found to give good agreement with experiment when it is assumed to have the value 3, as shown in Table 2.

TABLE 2

Crystal	$-\Delta H$(spectro.)	$-\Delta H_{expt}$
BN	63.6	60.8
BeO	128.5	143.1
AlP	22.2	39.8
GaAs	16.3	17
ZnSe	39.8	39
InSb	9.0	7.3
CdTe	25.1	22.1
ZnO	72.5	83.2
AlAs	17.3	27.8
GaP	24.8	24.4
ZnS	42.1	49.2
InP	20.3	21.2
CdS	35.8	38.7
AlSb	20.3	
GaSb	9.5	10.0
ZnTe	25.9	28.1
InAs	11.5	14.0
CdSe	30.9	32.6
CuI	14.8	16.2
CuBr	22.7	25.0
CuCl	34.9	32.8

The cohesive energy, ΔG_s, is found to correlate remarkably well with the ionicity, Fig. 7. Lines are drawn through data points for compounds corresponding to a particular row, or combination of rows, of the Periodic Table. The difference from one row to the next is due to the change in lattice constants, i.e. bond length. Most of the binding energy is associated with itinerant delocalized valence electrons. The slope, as the ionicity increases, suggests that ΔG_s is reduced with increasing f_i because energy is required for dielectric screening.

Electronic charge densities of bonds have been calculated from pseudopotential band structure theory [5]. A sample calculation for germanium is shown in Fig. 8(a) and for gallium arsenide in Fig. 8(b). It is seen that there is a charge shift in gallium arsenide, corresponding to a decrease in the excess

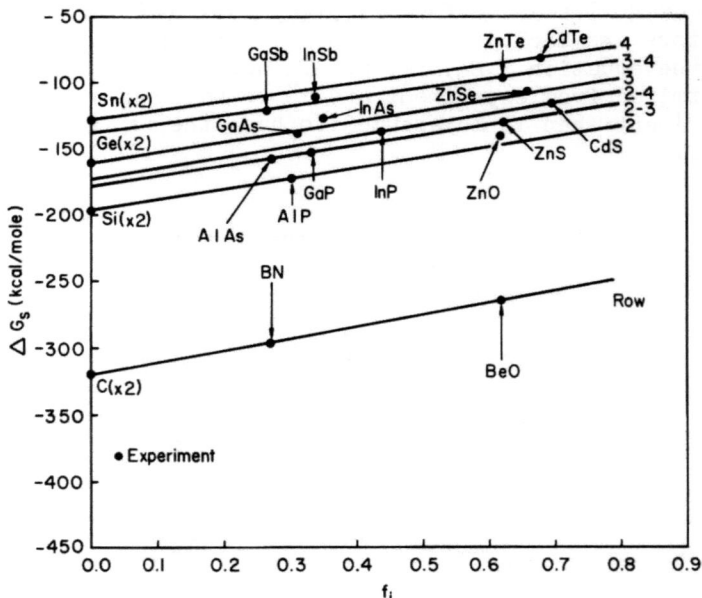

FIGURE 7. Cohesive energy for compounds of different ionicities [2].

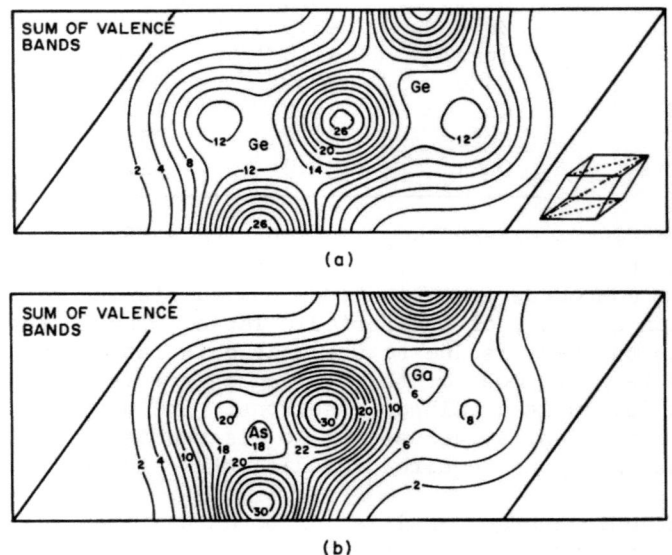

FIGURE 8. Calculated electronic charge densities of bonds in: (a) germanium, (b) gallium arsenide [5].

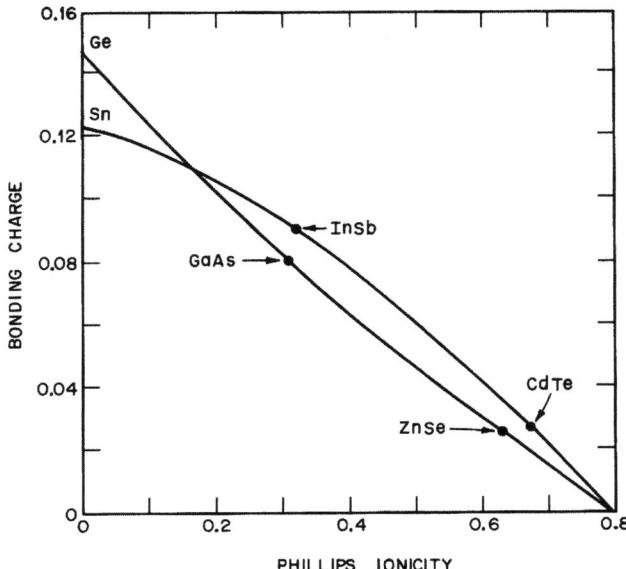

FIGURE 9. Electronic charge densities of bonds in binary compounds compared with Phillips ionicity [5].

covalent bond charge residing between the tetrahedrally coordinated atoms. It would be expected that when this excess covalent bond charge drops to zero, the structure will no longer remain tetrahedral, as it can lower its energy by shifting to a sixfold coordination. Therefore by comparing the calculated charge density with ionicity, as shown in Fig. 9, the extrapolation to a zero bond charge should give the critical ionicity. The value obtained in this way is in remarkably close agreement with that obtained by Phillips and Van Vechten, as discussed earlier.

THE IMPERFECT SOLID

Many of the properties of solids are determined by the overall chemical composition of the lattice and its geometry. However, a great many properties depend importantly on defects, and these may of course be either chemical or structural in nature. We thus turn to the consideration of some of the solid state chemistry of defects.

Concentration of Defects

First, we calculate the concentration of native structural defects [6,7]. We start with the vacancy, which corresponds to the removal of an interior atom to the surface; this is often called Schottky disorder. The free energy change is

$$\Delta G = \Delta E - T\Delta S = N_v E_v - T(N_v S_v - S_{cf}) \tag{10}$$

where E_v is the energy for the formation of the vacancy and there are N_v vacancies present. The entropy change from a perfect to an imperfect crystal has two terms. One is thermal, $N_v S_v$, and this results from the change in the ways the vibrational energy can be distributed over the vibrational modes, arising from the introduction of the vacancy. The other entropy term, S_{cf}, which is generally the more important one, is the configurational or mixing term; this results from the number of ways, W, that the vacancies can be distributed in the solid. For temperatures greater than $0°K$ there will always be a finite concentration of intrinsic defects at equilibrium, because S_{cf} is greater than 0. It is given by

$$S_{cf} = k \ln W = k \ln \left[\frac{(N+N_v)!}{N! \, N_v!} \right] \tag{11}$$

where k is Boltzmann's constant and N is the number of lattice sites. At equilibrium the free energy is at its minimum, with respect to changes in N_v, so that (neglecting S_v)

$$\left(\frac{\partial G}{\partial N_v} \right)_T = E_v - kT \ln \left(\frac{N+N_v}{N_v} \right) = 0 \tag{12}$$

Thus,

$$N_v = N \exp(-E_v/kT) \tag{13}$$

More generally, the right hand side is multiplied by a factor $\exp(S_v/k)$. These results are shown in Fig. 10. It is seen that at low concentrations of N_v, the free energy decreases because the entropy term predominates; at higher concentrations, however, the energy term outweighs the gain resulting from increasing the entropy, and the free energy begins to rise. The equilibrium situation corresponds to the minimum in G. If E_v is approximately 1 eV, which is a typical value, then the concentration of vacancies is approximately 10^{-5} at $1000°K$. Concentrations higher than the equilibrium concentration can be achieved by quenching the solid from some higher temperature.

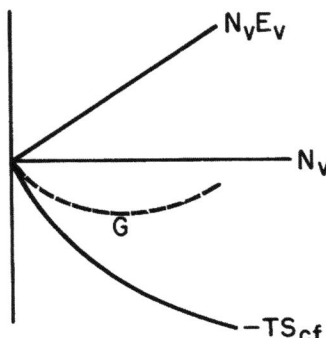

FIGURE 10. Free energy change with increasing vacancy concentration; energy and entropy terms are also shown [6].

Similar calculations can be carried out for Frenkel disorder, which results from the formation of a vacancy and an interstitial atom through the displacement of a lattice atom to an interstitial position, with the following result:

$$N_v \equiv N_I = (NN_{Int})^{\frac{1}{2}} \exp(-E_I/2kT) \qquad (14)$$

where E_I is the energy required to form the interstitial and N_{Int} is the number of interstitial sites. Similarly, Schottky-Wagner defects in ionic compounds, i.e. separated vacancies of opposite kinds in equal numbers, would have a concentration given by

$$N_v = N \exp(-E_p/2kT) \qquad (15)$$

where E_p is the energy to form the separated pair of vacancies.

Expressions (13), (14) and (15) are all Law of Mass Action expressions, with the equilibrium constants evaluated explicitly by statistical considerations.

Nonstoichiometry, which results from defects in compounds, is an exceedingly important manifestation of native defects in compounds [8]. It results from the equilibrium of both the M and the X atoms between the solid and the external phase. In addition, all of the above-mentioned defect modes can be active. One could work out in principle the concentration of defects, taking all of these equilibria into account, but usually only certain species and reactions are important and simplifications can be made. We will consider a simple example. We will assume no interstitials, but will assume that vacancies of either kind can form. The solid will be in equilibrium with the gas phase, and there will be no interaction between vacancies in the solid. We then need to write down a set of chemical reactions and consider the Law of Mass Action equations for these reactions. The chemical equilibria and LMA equations are

$$\left.\begin{array}{ll} M_M \rightleftharpoons M(g) + V_M & [V_M]P_M = K_M \\ X_X \rightleftharpoons X(g) + V_X & [V_X]P_X = K_X \\ \tfrac{1}{2}X_{2(g)} \rightleftharpoons X(g) & P_X/P_{X_2}^{\tfrac{1}{2}} = K_g \\ M(g) + X(g) \rightleftharpoons MX & P_M P_X = K_{MX} \end{array}\right\} \quad (16)$$

We add also the ionization equilibria and the hole-electron equilibrium, which is the LMA expression for the formation and breaking of the normal lattice bonding

$$\left.\begin{array}{ll} V_M \rightleftharpoons V_M^- + h^+ & \dfrac{[V_M^-][h^+]}{[V_M]} = K_V \\ V_X \rightleftharpoons V_X^+ + e^- & \dfrac{[V_X^+][e^-]}{[V_X]} = K_{V'} \end{array}\right\} \quad (17)$$

$$0 \rightleftharpoons h^+ + e^- \qquad [h^+][e^-] = K_i \qquad (18)$$

Finally, we have the charge neutrality condition and site balance condition

$$[e^-] + [V_X^-] = [h^+] + [V_M^+] \qquad (19)$$

$$[M_M] + [V_M] + [V_M^-] = [X_X] + [V_X] + [V_X^+] \qquad (20)$$

If we knew all of the equilibrium constants we could then calculate all the concentrations. In general, however, we do not have this information. K is an exponential function of the temperature, as in the previous expressions where we used a statistical approach to evaluate K. Generally, $\dfrac{P_M}{P_X}$ can be adjusted over a large range

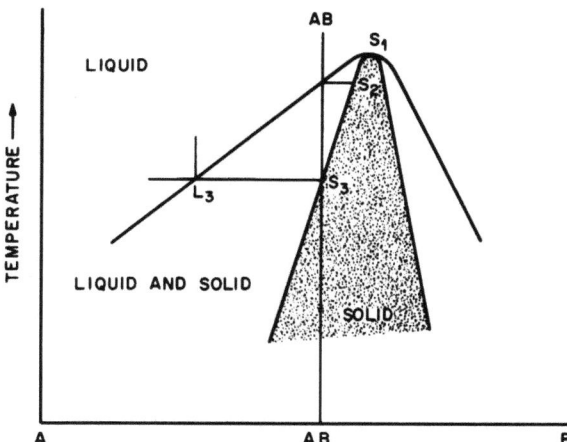

FIGURE 11. Phase diagram (schematic) showing how nonstoichiometry arises. The maximum melting point, corresponding to the minimum in free energy, is off-stoichiometry [6].

of values. High P_M will lead to fewer V_M and more V_X, and vice versa. It is usually difficult to determine experimentally the data that are required to solve completely for defect concentrations in even the simple binary compounds.

Nonstoichiometry can be treated statistically in much the same manner used above for the much simpler cases of defect formation [9]. Since the energy of formation for vacancies will usually differ between the cation and the anion, one species of defect will generally predominate.

The minimum free energy corresponds to the maximum in the melting point. The phase diagram, Fig. 11, will in general show a maximum congruent melting point that is off stoichiometry [6], reflecting the fact that the minimum free energy generally corresponds to a composition of the solid with a predominance of defects of one kind. There may be a range of solid solubility corresponding to different compositions within the solid phase. The liquid compositions in equilibrium with a given stoichiometry of the solid show that large differences between solid and liquid compositions will be normal.

In many cases it is important to add to the equilibria that involve the native defects in the solids, those which control the concentration of foreign impurities, since the doping of solids is an important practical matter. These equilibria are also subject to the Law of Mass Action and can be calculated in principle through consideration of additional equilibria with the external phase, added to those already discussed. The equilibrium constant describing the distribution of an impurity between the solid and the external phase is usually called the distribution coefficient (see Chapter 18).

Defect Interactions

Defects may interact and these interactions are important in solid state chemistry [6-8]. Interactions may be classified into two general categories, indirect and direct. In the former case, the coupling between defects is through the Law of Mass Action, and there is no need for the defects to be near each other in the lattice. In the case of direct interactions we will be concerned with instances in which two defects are found to be situated very close to each other within the lattice and a direct interaction of some sort develops.

Considering first the indirect interactions, one that has been studied quantitatively has to do with the solubility of impurities in a solid. A particularly convenient example for the study of this was lithium in silicon [10,6,8,9]. Lithium is a donor and its introduction into silicon is described by the following equations:

$$Li(ext) \rightleftharpoons Li(s) \rightleftharpoons Li^+ + e^- \qquad (21)$$

where $Li(ext)$ and $Li(s)$ are Li in the external and solid phases respectively, and e^- is an electron resulting from the ionization of $Li(s)$ in the semiconductor. The LMA expressions are

$$K_1 = \frac{[Li(s)]}{[Li(ext)]} \qquad (22)$$

$$K_2 = \frac{[Li^+][e^-]}{[Li(s)]} \qquad (23)$$

We next consider the effect of boron in the solid; it is an acceptor and ionizes to give holes, h^+

$$B \rightleftharpoons B^- + h^+ \qquad (24)$$

The electron and hole concentrations are connected through the hole-electron equilibrium (18). The lithium and boron ionizations will thus interact.

SOLID STATE CHEMISTRY

$$Li(ext) \rightleftarrows Li(s) \rightleftarrows Li^+ + e^-$$
$$+$$
$$B \rightleftarrows B^+ + h^+ \qquad (25)$$
$$\Updownarrow$$
$$(e^- h^+)$$

Solution of the Law of Mass expressions, and adding the charge neutrality condition

$$[B^-] + [e^-] = [Li^+] + [h^+] \qquad (26)$$

yields the result

$$[Li^+] = \frac{[B^-]}{1+Y} + \left\{ \left(\frac{[B^-]}{1+Y}\right)^2 + [Li^+]_o^2 \right\}^{\frac{1}{2}} \qquad (27)$$

where $Y = [1 + K_i(2ni/[Li^+]_o)^2]^{\frac{1}{2}}$ and $[Li^+]_o$ is the solubility of Li in undoped Si. Thus boron operates through the hole-electron equilibrium to increase the solubility of the lithium.

This solubility effect has an analogy in aqueous solutions, in which the ionization of the medium itself ($H^+ + OH^- \rightleftarrows H_2O$) occurs to approximately the same degree as the hole-electron ionization in the case of germanium and silicon. The corresponding solubility effect in aqueous solution would be the increase in solubility of a slightly soluble base, such as magnesium hydroxide, by the addition of a strong acid. The simple theory resulting in Eq. (27) is shown in Fig. 12, along with experimental data. It is seen that theory and experiment are in excellent agreement.

Another example of indirect defect interactions is the distribution of an atom between possible sites in a solid lattice. An example of this is the doping of gallium arsenide with silicon [11,6,9]. The silicon can enter on either a gallium or an arsenic site. In the former case it is a donor and in the latter case it is an acceptor. At low concentrations almost all of the silicon goes onto gallium sites, thus making the material n-type (see Fig. 13). The equilibria are:

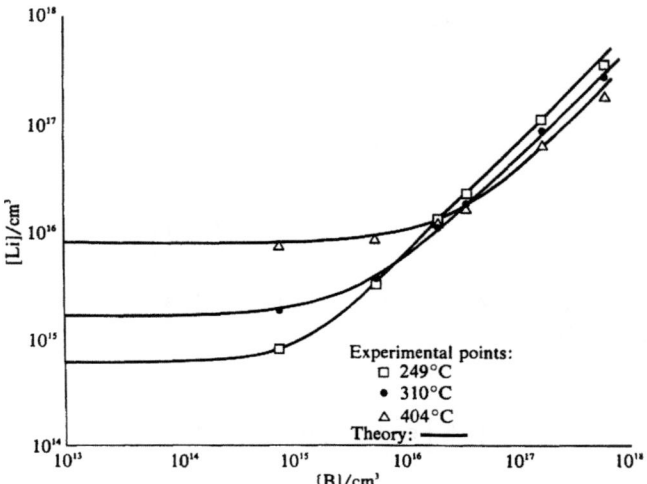

FIGURE 12. Comparison of Law of Mass Action theory (solid lines) for increase in solubility of lithium in silicon with added boron, and experiment (points) [10].

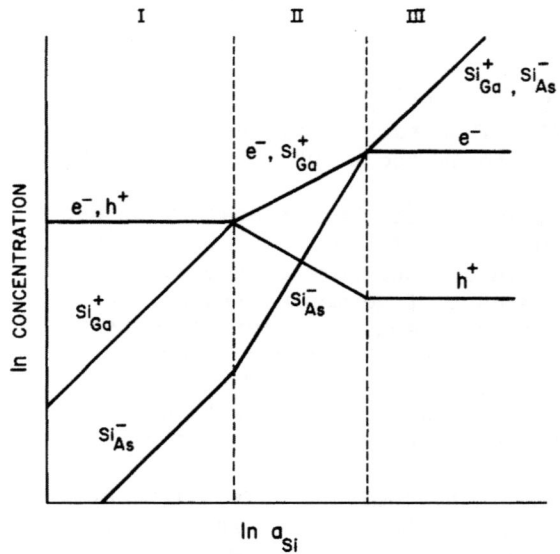

FIGURE 13. Law of Mass Action results (schematic) for distribution of Si on Ga and As sites in GaAs and for hole and electron concentrations, as a function of activity of Si in the external phase [12].

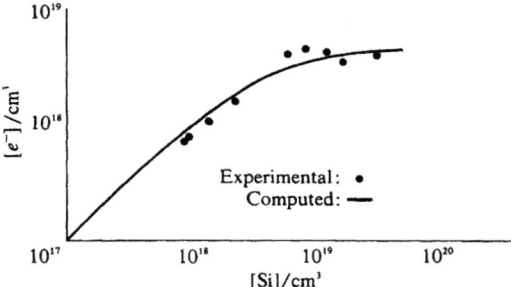

FIGURE 14. Comparison of Law of Mass Action theory (solid lines) for difference in electron and hole concentration as Si is added to GaAs, and experiment (points) [11].

$$Si_{Ga} \rightleftharpoons Si_{Ga}^{+} + e^{-}$$
$$\Updownarrow \quad +$$
$$Si_{As} \rightleftharpoons Si_{As}^{-} + h^{+} \quad (28)$$
$$\Updownarrow$$
$$(h^{+}e^{-})$$

Because the silicon favors gallium sites at low concentrations, the left-hand equilibrium is up; this is the situation in region II (Fig. 13). As additional silicon is added, both horizontal equilibria are affected, but the concentration of e^- is the larger. However, when $[e^-] - [h^+]$ becomes comparable to the intrinsic concentration of e^- and h^+, then the hole-electron equilibrium operates in such a way as to keep the difference approximately equal to the intrinsic hole and electron concentration (region III). Thus, further additions of silicon lead to a somewhat different balance. The top equilibrium begins to shift to the left, the left-hand equilibrium shifts down, and the lower horizontal equilibrium shifts to the right. This is "self-compensation". Law of Mass Action expressions can be written for this situation and from these the theoretical expression for $[e^-] - [h^+]$ shown in Figs. 13 and 14 is obtained. It is seen that experimental results are in good agreement with this. This clearly shows that at low concentrations almost all of the silicon acts as a donor but at high concentrations the electron concentration saturates corresponding to an approximately even distribution of silicon between sites, as a result of the self-compensation.

The correctness of this model can be corroborated by the addition of another donor. This would be expected to affect the equilibria because it independently adds electrons, and this shifts the various equilibria in predictable ways. The experimental results show the expected shifts.

We now consider direct interactions. One of the ways that this can happen is through the formation of ion pairs [10,6,8]. These are well known in aqueous solutions. Once again, lithium in silicon is a favorable experimental situation, because lithium has a high diffusion coefficient in silicon and it is easy to achieve equilibrium. The relevant equilibria are

$$Li^+ + A^- \rightleftharpoons \{Li^+A^-\} \qquad K_p = \frac{[P]}{[Li^+][A^-]} \qquad (29)$$

where A is a fixed (immobile) acceptor, [P] is the pair concentration, and [Li^+] is unpaired Li^+. The fraction of ions paired at a given temperature has been determined (Fig. 15). A number of effects resulting from ion pairing have been studied. Among these are the enhancement of solubility, because the formation of ion pairs is an additional mechanism for the removal of Li^+ from solid solution and therefore will increase the solubility of lithium in silicon. Another effect [10,6,8] is shown in Fig. 16, where the diffusion coefficient of lithium in silicon is shown as it is affected by the presence of acceptors. At high temperatures no pairs are formed and the normal diffusion coefficient is observed. However, at lower temperatures an appreciable fraction of the lithium is paired and the diffusion coefficient drops, because paired Li does not contribute to the flux of diffusing ions. The effective diffusion coefficient is related to the normal diffusion coefficient for lithium by:

$$D^* = D \left[\frac{[Li^+]}{[Li]_{total}} \right] \qquad (30)$$

The fraction paired is known, hence D^*/D can be calculated. The top line in Fig. 16 shows D, and the lower line the calculated D^*.

The case of zinc is interesting. This is a doubly charged acceptor, so that pairs with zinc would be expected to be more tightly bound, and also persist at higher temperatures. Thus zinc should have a larger effect on the diffusion coefficient and this in fact is observed.

The effect of direct interactions between impurities on luminescence spectra has also been studied and this of particular

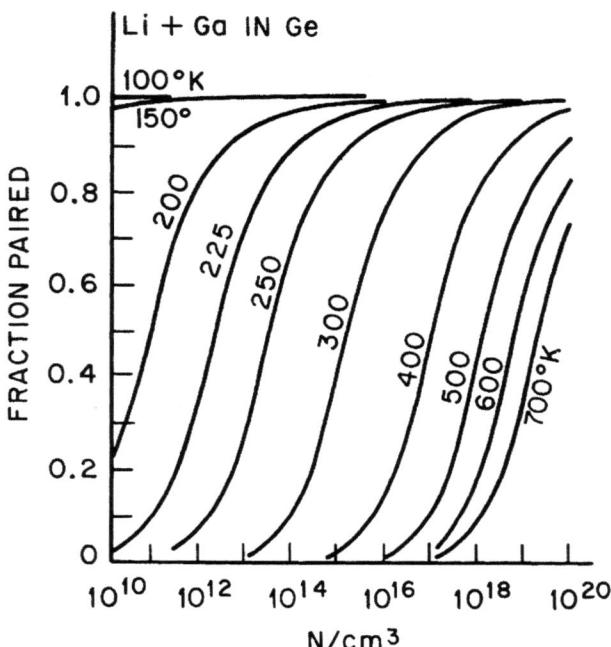

FIGURE 15. Fraction of ions paired for the system Li + Ga in germanium, where the lithium and gallium concentrations are equal [10].

---

importance in III-V and II-VI compounds (see Chapter 2). The achievement of substantial chemical control over gallium phosphide has permitted this system to be studied in detail, and the many lines of its luminescence spectra have been explained on the basis of impurity interactions. One mode of interaction that is of particular importance is emission due to the recombination of electrons on donors and holes on nearby acceptors [13]. These give rise to lines in both the red and the green part of the luminescent spectra and arise from impurity combinations, such as zinc plus oxygen or carbon plus sulfur. The energy of the radiation is modified by the separation of the donor and acceptor. Thus

$$E = (E_c - E_v) - (E_A + E_D) + \frac{e^2}{\varepsilon r} \qquad (31)$$

where $(E_c - E_v)$ is the energy gap, E_A and E_D are the acceptor and donor ionization energies, e is the electronic charge, ε is the dielectric constant, and r is the donor-acceptor separation

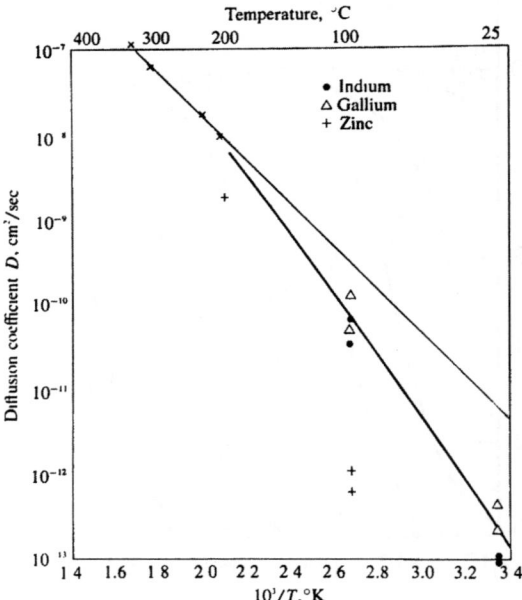

FIGURE 16. Effect of ion-pairing on the diffusion coefficient of Li in Ge [10].

distance. The allowed separations are given by the fixed geometry of the lattice. Consequently, one would not expect to see a continuous luminescent spectrum, but rather a series of discrete lines corresponding to the actual distances of separation available in the crystal lattice. This in fact is what is observed; moreover, the calculation of the relative intensities (counting number of possible pairs for a given separation distance in the lattice) and energies of the emission lines according to Eq. (31) is in excellent agreement with the observations, as shown in Fig. 17. The luminescence of GaP is discussed in greater detail in Chapter 2.

Solutions [14]

We now consider another way in which a solid can deviate from perfection, and this is through the random distribution of atoms in a solid solution. We first consider the conditions of formation of a solid solution and the factors that govern its stability. We are concerned with these matters because of the importance of solids containing small concentrations of impurities, as well as of alloy systems in metals, semiconductor compounds, complex magnetic oxide systems, and many other electronic solids.

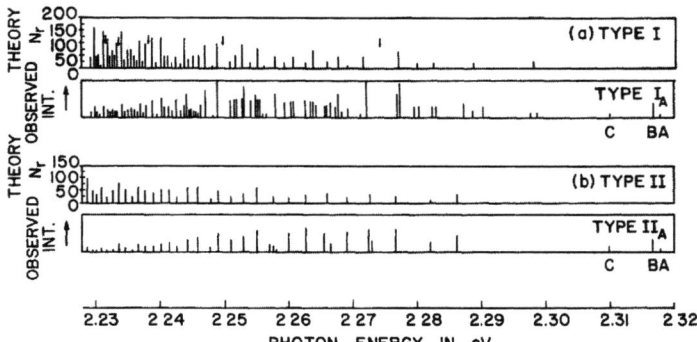

FIGURE 17. Comparison of theory and experiment for luminescence resulting from hole-electron recombination in GaP (after Thomas [13]).

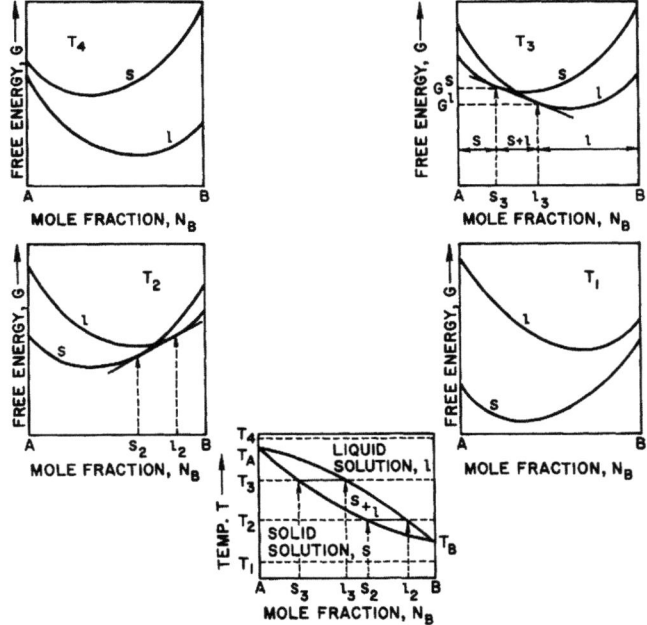

FIGURE 18. Relation of free energy curves for solid and liquid phases in binary system where A and B form a continuous series of solid solutions [14].

We first consider the system in which components A and B can form a complete range of solid solutions. The free energy of the solid and the liquid will have the shape shown in Fig. 18 for reasons similar to those explained earlier, in connection with defect formation. As a small amount of one component is added to the other, the free energy first decreases because of the gain in entropy, but at higher concentrations the free energy can increase again because the increase in internal energy outweighs the entropy term. As the temperature increases the free energy curve for the solid (S) moves down and that for the liquid (L) moves up, relatively. At temperature T_4, S is everywhere higher so that the system is liquid across the whole range of compositions. This may be seen in the phase diagram at the bottom. At T_3 the curves cross to the left of S_3, the solid curve is lower and so the system is solid. To the right of composition L_3, the liquid curve has a lower free energy and the system is liquid. Between the compositions S_3 and L_3 the whole system can lower its free energy by forming two phases, a solid and a liquid phase. The reason is that the chemical potential for each component is the same in the two phases, hence they can coexist. This occurs when a common tangent can be drawn to the curves since the chemical potential is directly related to the slope of the free energy curve. The free energy of the whole system thus lies on this tangent line, which is below both the solid and liquid curves. Again it can be seen that this situation is reflected in the phase diagram. The T_2 curves also cross, but are shifted more in the direction of the solid. Finally, at T_1, the solid free energy curve is always lower and we have the continuous series of solid solutions seen in the phase diagram.

The calculation of the free energy for a solid solution can next be considered [6]. We assume that nearest neighbor interactions only are important and let E_{AA}, E_{AB}, E_{BB} be the energies for the interaction of nearest neighbor pairs. Then we have for the energy and entropy the following expressions:

$$E = \frac{(Zf_A)(f_A N)}{2} E_{AA} + \frac{[Z(1-f_A)][(1-f_A)N]}{2} E_{BB}$$

$$+ (Zf_A)[(1-f_A)N]E_{AB}$$

$$= \frac{ZN}{2}\left[f_A E_{AA} + (1-f_A)E_{BB} + 2f_A(1-f_A)\left(E_{AB} - \frac{E_{AA}+E_{BB}}{2}\right)\right] \quad (32)$$

$$S = k \ln W = k \ln \frac{N!}{(f_A N)![(1-f_A)N]!} \quad (33)$$

Where Z is the coordination number and there are N total atoms, of which f_A are A. The first term in the energy is for AA

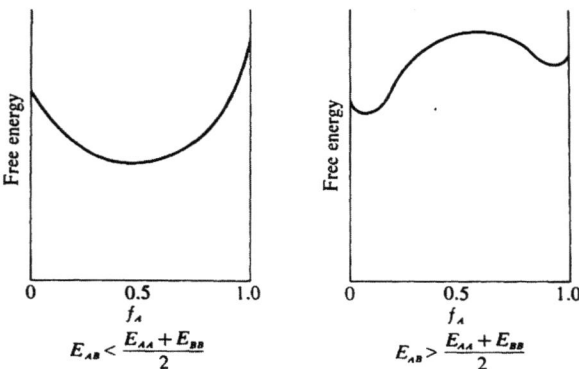

FIGURE 19. Free energy vs composition assuming: (a) the interaction energy of AB pairs is less than the mean of the interaction energies of AA and AB pairs, and (b) it is greater [6].

interactions, the second for BB and the third for AB. The multiplying factors are the number of atoms of a particular kind on a given site and the number of atoms of a particular kind in the nearest neighbor shell surrounding that site. Specific heat terms and strain energy have not been included in the energy expression, and the entropy is considered only to be configurational (mixing) entropy, as the thermal vibration entropy is less important. These results are shown in Fig. 19. Small impurity additions decrease the free energy because the entropy dominates, and at larger concentrations it depends upon whether the AB bond energy is greater or less than $1/2\ (E_{AA}+E_{BB})$. At the left we see that a continuous range of solid solutions will be formed, and at the right there will be two separate phases, each with a small solubility of one component in the other.

Another important situation in solids is spinodal decomposition, as shown in Fig. 20. Here we assume that the free energy curves at temperatures T_1, T_0 and room temperature are as shown at the left. At T_1 a continuous range of solid solutions will form, but as the temperature drops, a hump forms at intermediate compositions (as described earlier), so that a common tangent can be drawn at compositions S_0' and S_0'' and a mixture of phases is stable at this temperature. Consider now the inflection points, or spinodes, at T_0. Between these any small compositional fluctuation will lead to two subvolumes with a lower total free energy and therefore this is an unstable situation. However, between each spinode and the nearby minimum in the free energy curve a small fluctuation would lead to an increase in the free energy of the system. Hence, this is a metastable situation. Of course, larger fluctuations would lead to a decrease in the free energy. The corresponding phase diagram is shown in the right-hand part of the figure.

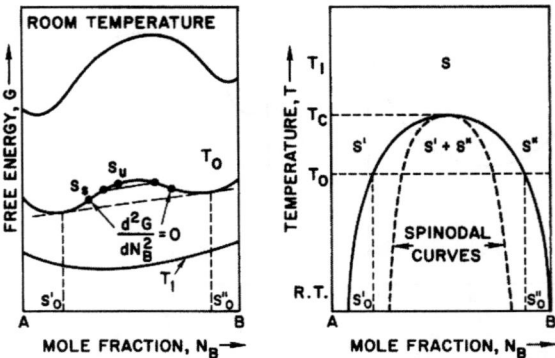

FIGURE 20. Spinodal decomposition: (a) assumed free energy curves for three temperatures, and (b) the corresponding phase diagram [14].

Finally, let us suppose that two phases for the solid exist. Each will have its free energy curve, and if we include also the free energy curve for the liquid, we will have situations such as those illustrated in Fig. 21(a). This illustrates how eutectics and peritectics, Fig. 19(b), result from particular free energy curves. In the case of the eutectic, the minimum for the liquid lies between the two solid phase minima, and in the case of the peritectic, it lies outside these compositions.

Finally we turn to the question of the theoretical analysis of the stability of solutions and of their properties. An extension [15] of the Phillips-Van Vechten theory to alloys has been carried out for III-V systems, because of their importance as electronic materials. An example is $GaAs_{1-x}P_x$. The basic equations

$$\Delta G = \Delta E - T\Delta S < 0 \qquad (34)$$

where

$$\Delta S = -R[(1-x) \ln (1-x) + x \ln x] \qquad (35)$$

and

$$\Delta E = -(1-x)\Delta H_{AB} - x\Delta H_{AC} + \Delta H_{AB_{1-x}C_x} \qquad (36)$$

show that the alloy is stable when the free energy change for formation is less than zero. Since ΔS is immediately obvious, it is seen that the problem reduces to a calculation of the heats of formation. This was discussed earlier in this chapter, Eq. (9).

SOLID STATE CHEMISTRY 531

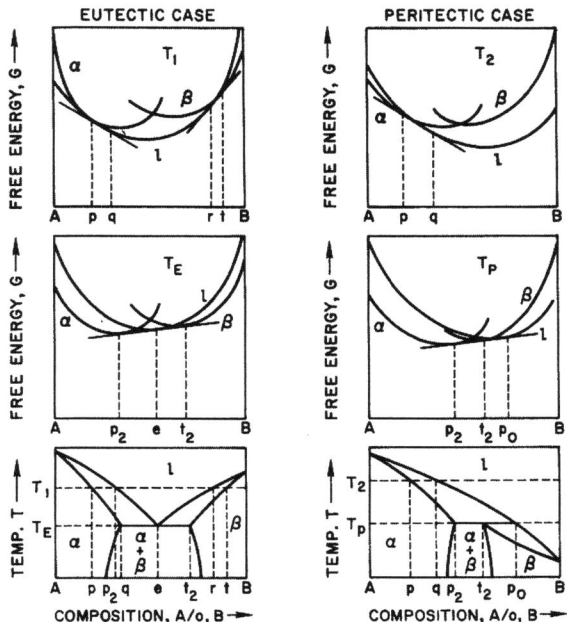

FIGURE 21. Free energy curves (a) when two solid phases exist and the minimum in the free energy curve for the liquid lies between or outside the minima in the curves for the solid; the corresponding phase diagrams (b), showing eutectic and peritectic compositions [14].

Thus it is possible to calculate the free energy of formation as a function of composition for a given alloy system, starting with a knowledge of the dependence of the band structure on x. The dependence is not linear, and in fact, experimental observation of bandgap shows that there is a quadratic dependence

$$E_o(x) = a + bx + cx^2 \qquad (37)$$

The bowing parameter, c, describes this deviation from linearity; the bowing is less for E_1 and E_2 than it is for E_0. The dielectric theory has been used [15] to calculate the bowing parameter. From this it was possible then to determine the heat of formation, as shown in Fig. 22, for several alloy systems.

This type of calculation has been extended to include all of the III-V compounds [16]. The calculation of the III-V phase diagrams follows almost immediately from the free energy

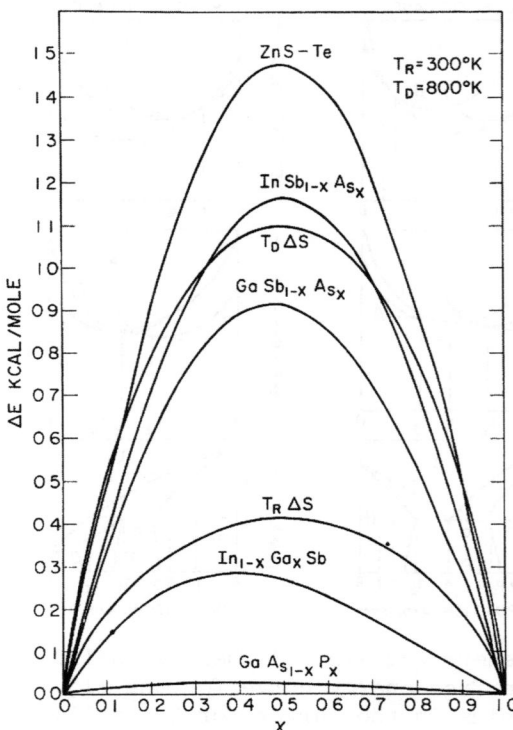

FIGURE 22. Enthalpy of mixing as a function of composition in alloys of III-V compounds (after Van Vechten and Bergstresser [15]).

determination; the free energy results, together with simple assumptions about the thermodynamic nature of the liquid solutions and the temperature and entropy of fusion, were used to calculate the phase diagrams.

The determination for solid solutions of the E_0, E_1 and E_2 symmetry point energy gaps provides a theoretical understanding for experimentally determined curves of energy gap vs composition. Figure 23 shows the GaAs-GaP system where the minimum energy for the direct gap and that for the indirect gap cross at a composition of 45% GaP. Data of this kind are not available for all alloy systems, and the theoretical curves for the gaps as a function of composition can therefore be used to estimate the composition for the transition from direct to indirect gap. A summary of the results for several III-V compound alloy systems is shown in Fig. 24, where both minimum direct and indirect gaps are indicated for the whole range of compositions, along with the gap at the crossover point.

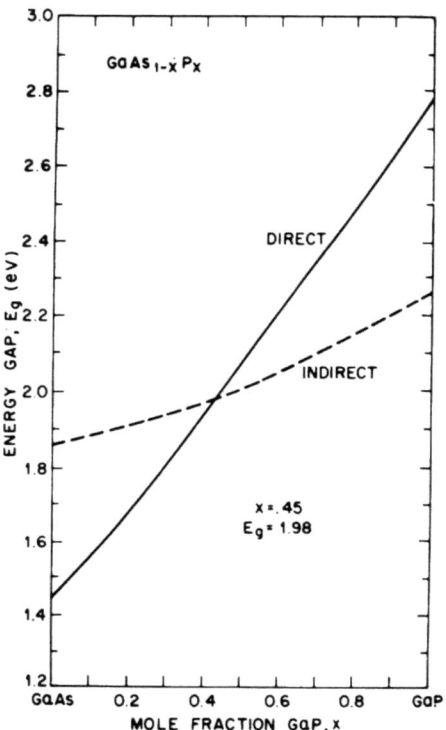

FIGURE 23. Direct and indirect gaps for $GaAs_{1-x}P_x$ [17].

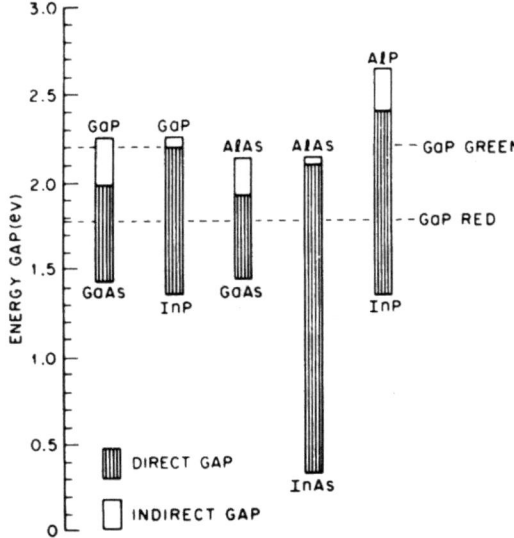

FIGURE 24. Energy ranges for minimum gaps over all values of x in several III-V compound alloy systems, showing crossover point from direct to indirect gap [17].

REFERENCES

[1] L. Pauling, "Nature of the Chemical Bond", 3rd edition, Cornell Univ. Press, Ithaca, New York (1960).

[2] J. C. Phillips, Physics Today, February 1970; Science 169, 1035 (1970); "Bonds and Bands in Semiconductors", Academic Press, New York (to be published); Rev. Modern Physics 42, 317 (1970).

[3] J. A. Van Vechten, Phys. Rev. 187, 1007 (1969).

[4] J. C. Phillips and J. A. Van Vechten, Phys. Rev. B, 2, 2147 (1970).

[5] J. P. Walter and M. L. Cohen, Phys. Rev. Letters 26, 17 (1971).

[6] N. B. Hannay, "Solid-State Chemistry", Prentice-Hall, Englewood Cliffs, N. J. (1967).

[7] R. A. Swalin, "Thermodynamics of Solids", John Wiley & Sons, New York (1962). F. A. Kröger, "Chemistry of Imperfect Crystals", Interscience, New York (1964). G. G. Libowitz (Chapter 7, Volume 2), R. A. Swalin and R. D. Weltzin (Chapter 6, Volume 2) and R. F. Brebrick (Chapter 5, Volume 3) in "Progress in Solid-State Chemistry", H. Riess, Ed., Pergamon Press, Oxford (1965 and 1967). A. D. Franklin, in "Point Defects in Solids", J. H. Crawford, Jr. and L. M. Slifkin, Eds., Plenum Press, New York (1972).

[8] C. S. Fuller, in "Semiconductors", N. B. Hannay, Ed., Reinhold Publishing Corp., New York (1959).

[9] H. Reiss, in "Chemical and Mechanical Behavior of Inorganic Materials", A. W. Searcy, D. V. Ragone and U. Colombo, Ed., Wiley-Interscience, New York (1970).

[10] H. Reiss, C. S. Fuller and F. J. Morin, Bell System Tech. Jl. 35, 535 (1956).

[11] J. M. Whelan, Proceedings of The International Conference on Semiconductor Physics, Prague (1960).

[12] F. A. Kröger, "Chemistry of Imperfect Crystals", Interscience, New York (1964).

[13] D. G. Thomas, Physics Today, February 1968; J. J. Hopfield and D. G. Thomas, Phys. Rev. Letters 10, #5, 162 (1963).

[14] A. G. Guy, "Introduction to Materials Science", Mc-Graw Hill, New York (1972).

[15] J. A. Van Vechten, International Conference on Physics of Semiconductors, IUPAP, Cambridge, Mass. (1970); J. A. Van Vechten and T. K. Bergstresser, Phys. Rev. B, 1, 3351 (1970).

[16] G. B. Stringfellow, J. Phys. Chem. Solids 33, 665 (1972).

[17] H. C. Casey and F. A. Trumbore, Materials Science and Engineering 6, 69 (1970).

CHAPTER 20

FILM TECHNIQUES

D. G. Thomas

Bell Laboratories

Murray Hill, New Jersey 07974

INTRODUCTION

This chapter is concerned with film techniques and epitaxy. From the practical point of view these are important topics. There are few electronic devices which do not use thin films in one form or another. Semiconductors make extensive use of epitaxial films; thin metal films are used for resistors and capacitors, and integrated circuits also make wide use of metal films, as do interconnection circuits. Thin insulating films are vital components of semiconductor devices often in active parts of the circuits where their properties dominate the device characteristics. The perfection of these insulating films can control the economics of manufacture of the devices. Photoelectric, and many optical devices also depend on thin films. Acoustic devices often involve fine patterns in thin films. Many memory devices of both the magnetic and semiconductor varieties depend on thin films. Furthermore, several superconducting devices are built around thin insulating films.

From the many possibilities, a selection of three topics has been made for this paper:

1. Certain corrosion properties of multilayer thin metallic films.
2. The increasing importance of Liquid Phase Epitaxy as a method for growing crystalline films.
3. Certain possible new applications of thin films for optical communications systems.

The first topic concerns certain manufacturing and reliability problems connected with the use of thin metallic films. One need not apologize for stressing such apparently unglamorous topics, for manufacturing problems dictate costs, and reliability more and more will determine customer acceptance. In these times when the results of technology are being judged more by their ability to make money than by their intellectual beauty, costs and customer acceptance are in the forefront of the minds of a good many people.

The problems to be discussed are often associated with the use of multilevel metallization schemes which are used simply to interconnect electrical components. For instance transistors and so forth may be interconnected on a single chip of silicon to form a silicon integrated circuit, in which case the metal is deposited on an SiO_2 film. Or silicon integrated circuits may be interconnected on a piece of ceramic to form a hybrid integrated circuit, perhaps with thin film resistors and capacitors. In this case the metal is deposited on a pure alumina substrate.

Gold has much to be said for it for interconnection purposes, since it is soft, it bonds easily to itself, it has high electrical conductivity and it is chemically very stable. However, just because it is stable, it does not stick very well to insulating substrates. Frequently, therefore, one uses a "glue" layer under the gold and this is often titanium. Presumably because of its high chemical reactivity, it provides exceptionally strong bonding to SiO_2, silicon nitride, ceramic substrates, and so forth. Furthermore, titanium alone is normally very resistant to corrosion because of the formation of a passivating surface layer. It was logical, therefore, to use a Ti-Au metallization system. After some difficulties, it was concluded that this pair of metals was incompatible with economical manufacture of high reliability devices. In particular, there were adhesion failures found in the finished devices, and unless rigid acceptance tests were imposed, there was the threat of long term reliability problems. A possible source of trouble was the interdiffusion of Au and Ti with undesirable compound formation between these elements, and so a diffusion barrier of Pt was placed between the Ti and Au. Indeed this improved the situation enormously.

Recent work has shown, however, that the benign influence of the Pt layer - and, in fact, for convenience of etching and deposition, Pd is often used in place of Pt - is, in most real situations, to be ascribed not to the diffusion barrier effect, but to much more subtle causes.

To determine failure rates, one often performs accelerated tests at high temperatures and extrapolates the results to

FILM TECHNIQUES

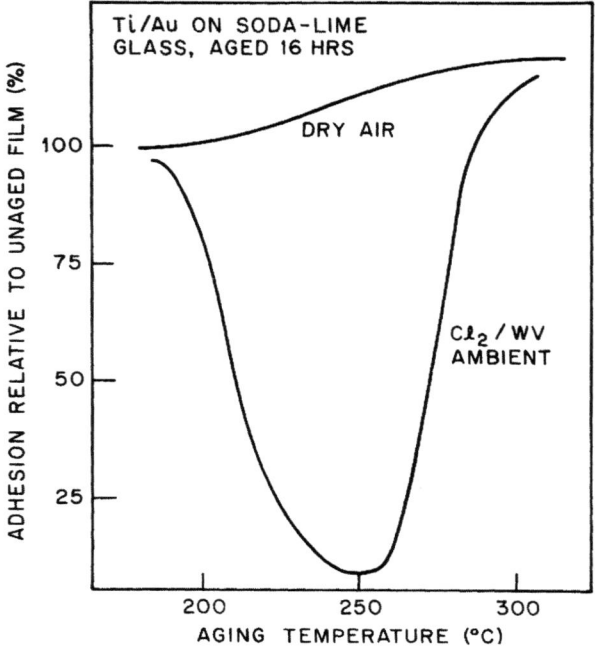

FIGURE 1. Influence of aging temperature on adhesion, as measured by the scratch test. Metallization was 500Å Ti + 10,000Å Au deposited on soda-lime glass. Aged 16 hours at 225°C in the Argon-H_2O-Cl_2 atomsphere. (After English and Turner [2]).

operating temperatures using an Arrhenius plot. This is a useful thing to do and it has been used satisfactorily in many situations - it does, however, have pitfalls and one of these is a failure rate which does not obey the Arrhenius equation.

Rather simple tests can be devised to determine the failure of a film after aging. One of these is a test for adhesion of the film to the substrate. This is done by pressing down adhesive tape and then removing the tape; the percentage of the film which adheres to the tape is a measure of the failure of adhesion to the substrate.

As an example of the use of accelerated aging to examine the reliability of the adhesion of a film to its substrate, evaporated films were aged for 16 hours in a flowing gas atmosphere at temperatures between 150 and 400°C. Typically there is 500Å of Ti, 100 to 500Å of an intermediate layer if present, and 10,000Å of gold.

Figure 1 shows the results for Ti-Au films in dry air between 200 and 300°C - clearly there is no degradation of adhesion under

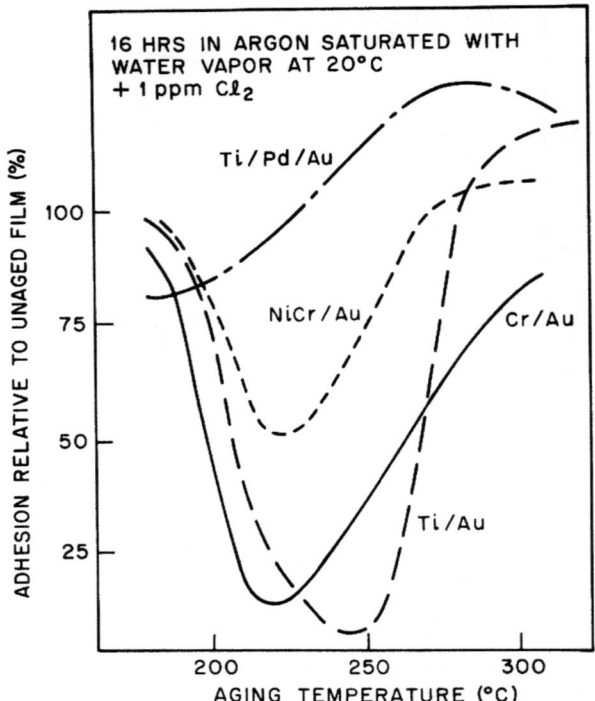

FIGURE 2. Influence of aging temperature on adhesion, as measured by scratch test, for several metallizations. (After English and Turner [2]).

these conditions. However, if argon saturated with water at 20°C and containing 1 p.p.m. chlorine is used, then there is attack on the film, and this attack shows a remarkable maximum in rate between 200 and 250°C (quite unlike an Arrhenius plot). Since 1/10 to 1/2 p.p.m. level of chlorine has been reported in the atmospheres of large cities, this choice of atmosphere is a realistic one. This effect is not confined to Ti-Au. Figure 2 shows it for Cr/Au, NiCr/Au; but note that if an intermediate layer of Pd is used, the trouble is eliminated. Pt and Rh are also effective. This is the benign effect of the intermediate layer.

There are several interesting observations which have been made concerning the mechanism of this failure:

1. There is no doubt that the corrosion which leads to loss of adhesion is corrosion of the Ti glue layer, and that this occurs at pinholes in the gold. This can be confirmed by visual observation of a film deposited on a glass slide.

2. It is particularly impressive that corrosion of Ti does not take place in the absence of gold. Gold circles may be deposited on a field of titanium, and corrosion of the titanium only takes place under the gold circles.

3. Very thin films of Pt or Pd may be used with good effect. Even 50 to 100A of Pt will have an appreciable influence. And, in fact, the protecting Pt layer does not have to be between the Ti and Au. It can be above the gold and still be reasonably effective.

4. The corrosion rate increases in the presence of water vapor and impurities such as halogens which form ions in water, and which produce an acid pH.

All these observations show that we are not dealing with a simple interdiffusion effect between components of the films, but that rather electrochemical corrosion is involved.

To go beyond this statement and to attempt to give a detailed explanation of the corrosion processes is not easy. However, D. M. MacArthur [1] has provided one outline of what may be going on, at least in the presence of bulk aqueous solutions. This outline is as follows.

It is well known that titanium can be anodically passivated. To produce the passivating layer, a sufficiently large anodic current must flow during the initial formation of the layer, and the titanium must subsequently be maintained at a suitable potential. Such current can be supplied externally, or sufficient electrochemical current can flow from a cathode material if the cathodic reaction rate is fast enough. Essentially then, a passive oxide layer is built up on the Ti surface, providing a critical anodic current density can be reached during film formation. It turns out that, in the case we are discussing, what determines the anodic current is the current which can flow at the corresponding cathode, and this cathodic reaction rate is controlled by the catalytic properties of the cathode material. A suitable potential will then maintain the passivating layer with the passage of only a very small electrochemical current.

Experiments by MacArthur have shown that in an electrolyte exposed to air, Au-Ti and Pt-Ti couples maintain a potential in the passive range for Ti. In other words, in the presence of a plentiful supply of oxygen, even the Ti-Au system appears to be relatively stable.

If, however, oxygen is excluded, then anodic attack can occur in an acid environment. In these experiments potential measurements of Ti either alone, or connected to other metals, at 60°C in 10% H_2SO_4 are made in the presence of different atmospheres.

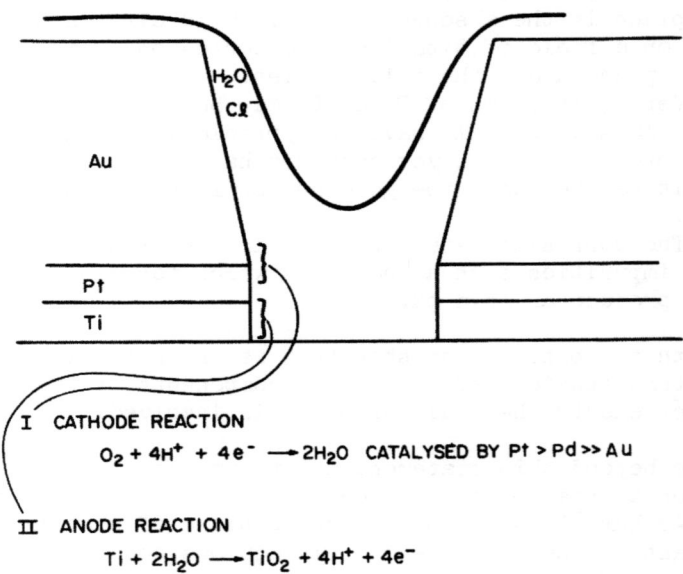

FIGURE 3. Schematic diagram of a pinhole in the metallization with suggested corrosion reactions as indicated.

The interesting point is that for Ti alone and for Ti-Au, in the absence of oxygen, the potential of the Ti moves into the region of active corrosion. Addition of oxygen restores passivity. However, Ti-Pd and Ti-Pt couples maintain a potential in the passive region. One has to be careful in carrying over these results in quantitative detail to other situations because relative areas of exposed metal may be important, but the importance of a changing concentration of oxygen is real, and paradoxically, the less oxygen is present the worse is the corrosion.

Figure 3 schematically represents a pinhole and shows the proposed electrochemical reactions occurring at different locations within the pinhole. Overall, the reactions lead to the oxidation of titanium. These reactions are subject to the following constraints:

1. We know corrosion occurs at pinholes, and at the bottom of a pinhole an oxygen deficiency can develop. We have seen that this actually favors corrosion, and this is because a low oxygen pressure does not allow reaction I to proceed fast enough to

produce a passivating layer on the Ti. At places where there is a plentiful supply of oxygen, a passivated layer is produced without trouble since reaction I can proceed rapidly.

2. Reaction I, the reduction of oxygen, is known to occur many orders of magnitude faster on Pt or Pd than on Au as a result of the catalytic action of these metals. It is the ability of Pt or Pd to allow reaction I to proceed rapidly, even in low oxygen pressures, that leads to passivation of Ti by these metals. Providing the Pt is close by, it clearly does not matter if it is placed above or below the gold layer, and this is consistent with experiment.

3. The presence of water vapor and halogens allows the corrosion to occur, supporting the view that electrochemical processes are occurring by ionic conduction in a solution. It is believed that this solution is usually present in the form of a thin adsorbed aqueous layer. This layer will not be present at high temperatures so corrosion will cease. On the other hand, at low temperatures the reactions rates are very slow. In this way we can understand, at least qualitatively, why the corrosion rate shows a maximum at intermediate temperatures. Evidently near 250° there is the optimum combination of water film thickness and reaction rates, which leads to maximum corrosion.

A model can therefore be constructed which appears to fit the observations. It emphasizes the complexity of the situation and shows why prediction of life by extrapolating high temperature data can be dangerous. It must be added however that English and Turner [2] have worked with a wider variety of metals using vapor phase corrosion tests, and their conclusion is that an important contribution of the intermediate layer may be simply to provide physical separation between the Ti and Au layers. This separation increases the electrolytic path between the two active metals, (the Pt or Pd are assumed to be passive), with the result that the electrical resistance is increased so that the corrosion current is reduced to a negligible value. Further experiments are required to elucidate these matters completely.

One moral to be drawn is that it is always potentially dangerous to use multilevel metal films as there can be insidious reliability and manufacturing problems. However, there do appear to be certain combinations of metals which can give satisfactory results, and two of these are Ti-Pt-Au and Ti-Pd-Au.

Liquid Phase Epitaxy

In the last few years, a novel method of growing thin crystalline films has come into rather remarkable prominence. It is termed Liquid Phase Epitaxy and it has been used to grow both epitaxial films and heteroepitaxial films. As its name implies, it is a process in which there is precipitation from a liquid phase of a crystalline layer onto a parent substrate, in which the crystallographic orientation of the layer is determined by that of the parent substrate. The difference from more familiar epitaxy is simply that precipitation occurs from a liquid phase rather than from the vapor phase. This technique has received wide application in the semiconductor field, particularly for the growth of III-V compounds. In addition, it has recently had remarkable success in growing heteroepitaxial films for the magnetic bubble devices. Both these topics will be discussed. Several review articles have been published and they may be referred to for further details [3,4].

Consider first the semiconductor applications with emphasis on GaAs and GaP, for both of which materials growth takes place from gallium rich solution.

While it is possible to use a "traveling solvent method," or steady state method, in which a layer of solvent travels between a cooler growing crystal and hotter source material, the more widely used method is the cooling method.

In Fig. 4, if a system in equilibrium at point A is cooled to point B, then the solution becomes supersaturated and arsenic is removed from solution in the form of GaAs and the liquid reaches a new composition at point C. Growth occurs more easily on a GaAs substrate than in the solution, with the result that under favorable conditions a uniform epitaxial film forms on the substrate. The thickness of the film grown will depend upon the melt volume, the cooling range and the area of the seed on which growth occurs.

Nelson [5] used this method first for GaAs. He used what has come to be called the "tipping" method illustrated in Fig. 5. When the Ga has come into equilibrium with the GaAs nutrient at 800-900°C, it is flowed onto the GaAs seed wafer by tipping the oven. The system is then cooled by several hundred degrees, and at an appropriate point the residue is tipped off. Ideally, all Ga flows away from the seed at this point, but in practice this does not always happen.

Before discussing further details of this process, it is appropriate to point out that as for any system in which there is a change in composition between melt and solid, there are opportunities for constitutional supercooling. It is particularly important here where the control of surface planarity and layer thickness are frequently critical matters. Figure 6 illustrates the effect. In it, temperature is plotted schematically as a

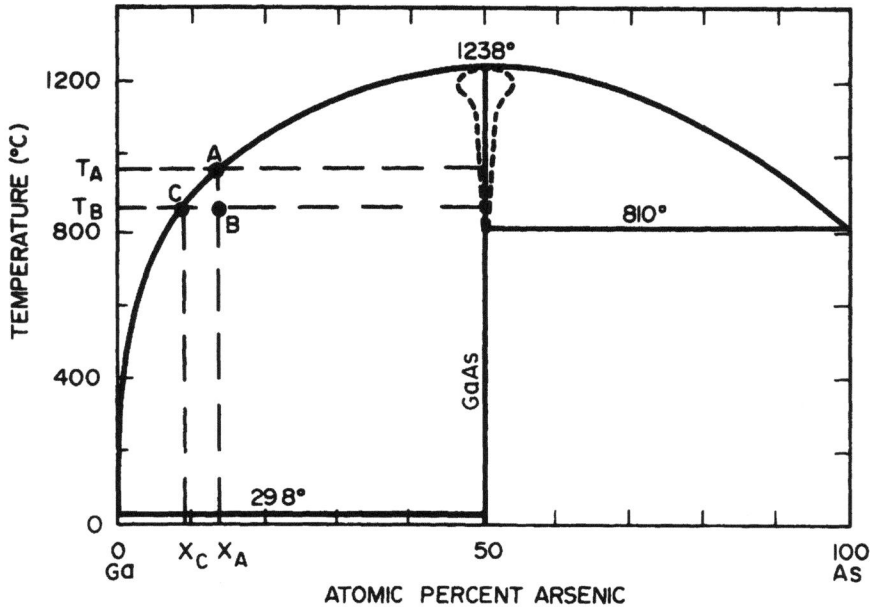

FIGURE 4. GaAs phase diagram. (After Hansen [21]). Schematic solidus curves have been added as dotted lines.

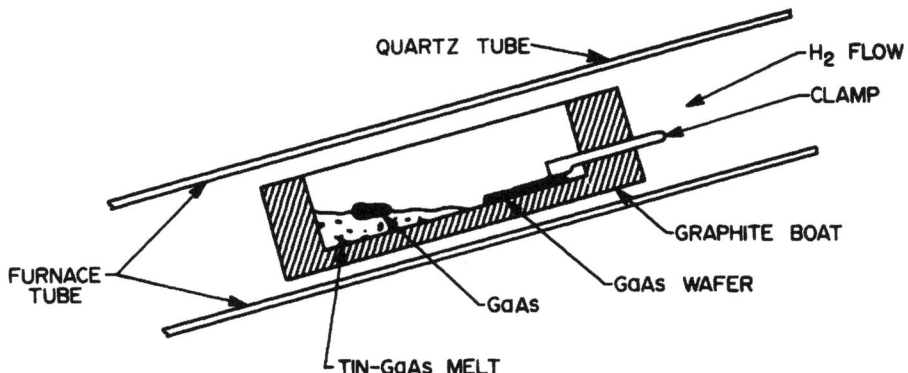

FIGURE 5. Apparatus for LPE growth of GaAs from Ga solution. (After Nelson [22]).

function of distance from the growing crystal surface. Consider the case of GaAs. If growth takes place at 800°C, the solution contains 20 times less arsenic than does the solid. As growth proceeds, the solution ahead of the crystal surface becomes depleted of arsenic, as a result of the limited diffusion of arsenic from deeper in the solution. This depleted solution will

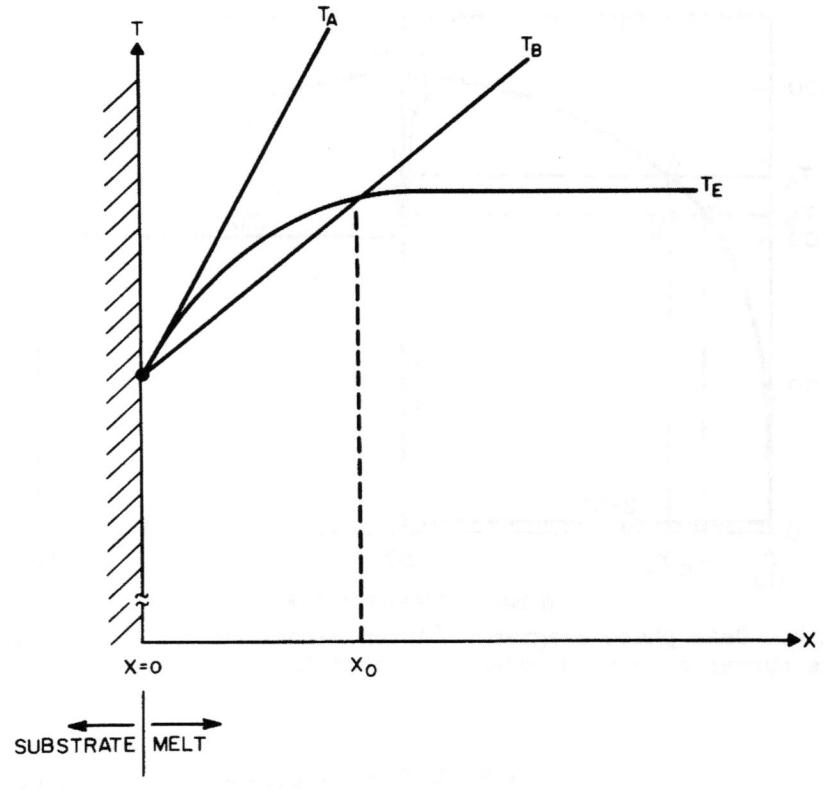

FIGURE 6. Temperature distributions illustrating constitutional supercooling. (After Dawson [3]).

exist in equilibrium with GaAs at a lower temperature than corresponds to that for the bulk solution. (We remember that the crystal is growing because it is cooler than the bulk solution.) The equilibrium temperature Te, therefore, has the general shape shown in the figure. If at any point the actual temperature falls below Te, a supersaturated solution occurs. This is called constitutional supercooling. The temperature gradient labeled T_A is the smallest one which avoids constitutional supercooling. For the shallower gradient T_B, the entire region from the growth interface to a point X_O into the melt is supercooled. Such supercooling can be harmful. If protrusions extend into these supercooled regions, rapid and uneven growth may occur. Or indeed spontaneous nucleation may take place in the supersaturated region, and the small crystallites may be grown into the layer as polycrystalline regions.

Attempts [6] have been made to calculate what temperature gradients are required to avoid constitutional supercooling. The problem is complex, involving diffusion coefficients in the liquid, and diffusion boundary conditions, and assumptions have to be made concerning any stirring or convective effects. Some of the results appear to be quite pessimistic, indicating a required temperature gradient of 100°C/cm. However, more favorable geometries can be designed so that gradients of 5°C/cm are likely to be sufficient, and using thin melts contained between material of high thermal conductivity such as graphite, these gradients may be achieved. The likelihood of some degree of supercooling not actually resulting in spontaneous nucleation will alleviate the problems. Constitutional supercooling is therefore a troublesome matter, but understanding the problem does seem to have produced a situation in which it is not a serious impediment to progress.

Advantages of Liquid Phase Epitaxy

LPE has a number of advantages which lead to its use.

1. The first major point is that growth can often take place several hundred degrees below the compound's melting point - perhaps 400°C for GaAs. Lower temperatures can lead to increased chemical purity. There are several documented cases [7,8] in which contamination of GaAs by Si from the walls of the containing vessel has been markedly decreased, with resultant mobility improvements, by growing crystals at lower temperatures. For instance LPE growth of GaAs from Ga solution at 800°C produced in one experiment an ionized impurity content of $\sim 10^{16}$/cc, whereas at 700°C, a concentration of $\sim 6 \times 10^{14}$/cc was achieved. There was an associated increase of electron mobility. It is likely, too, that stoichiometric and crystalline defects such as vacancies are also reduced in concentration.

2. Another advantage of lower temperatures is that the vapor pressure of volatile species may be greatly reduced. This is spectacularly illustrated by the case of gallium phosphide. This compound melts at 1465°C at which temperature it develops 35 atmospheres of phosphorus pressure - in fact elaborate and expensive crystal pullers are required to grow crystals from which slices may be cut to provide substrates for LPE growth. This subsequent LPE growth is done from Ga solution at 1000°C and the phosphorus pressure is now near 10^{-4} atmospheres, at which point phosphorus loss is negligible.

3. An additional important advantage of LPE seems to be that the epitaxial layer can grow with a reduced dislocation density compared to the parent substrate. One presumes that this is connected with the near equilibrium conditions which can obtain for slow growth from solution. At any rate, in GaP a threefold

reduction in the dislocation density, as judged by etch pit count, has been observed within 5μ of the substrate surface. Similar results have been obtained with GaAs and GaSb.

APPLICATIONS OF LIQUID PHASE EPITAXY

GaAs

There have been several applications of LPE to grow very pure GaAs for such applications as Gunn devices. Using pure containing vessels ionized impurity levels below $10^{14}/cc$ have been obtained, with mobilities at 77°K as high as 175,000 cm^2/Vsec.

One of the most spectacular applications of LPE has been to grow complex heteroepitaxy structures used as semiconductor junction lasers which operate at room temperature [9,10]. An important part of this device is the two layers on either side of the active GaAs, composed of a Ga, Al As alloy. It is also important that the interfaces should be extremely flat and parallel. In this case the problem is made more difficult by the tendency for Al contained in the liquid alloy to react with oxygen to form a scum of Al_2O_3. The problems were solved by using a slider illustrated in Fig. 7. Here all layers are grown in a single growth cycle, and the substrate is delivered to each melt, each of which has a controlled composition, by the slider which first wipes the melt surface free of any solid material. After bringing the system to equilibrium, a cooling rate is established and the solution holder is positioned to bring each melt successively into contact with the substrate. Systems such as this, which have been used with great success, all depend on the fact that the gallium solutions do not wet the graphite surfaces, so that leaks do not occur around the sliding contacts.

GaP

The commercial importance of this material is that appropriately doped p,n junctions, when forward-biased, emit visible light. This light may be red or green and quite high efficiencies can be achieved.

While junctions can be made in GaP by diffusion, these generally do not have high electroluminescent efficiency. Instead, junctions are made by LPE, often using two LPE layers. It is necessary to control the doping level of several impurities in forming these junctions - often quite high impurity concentrations are required to maximize the light output.

FILM TECHNIQUES 547

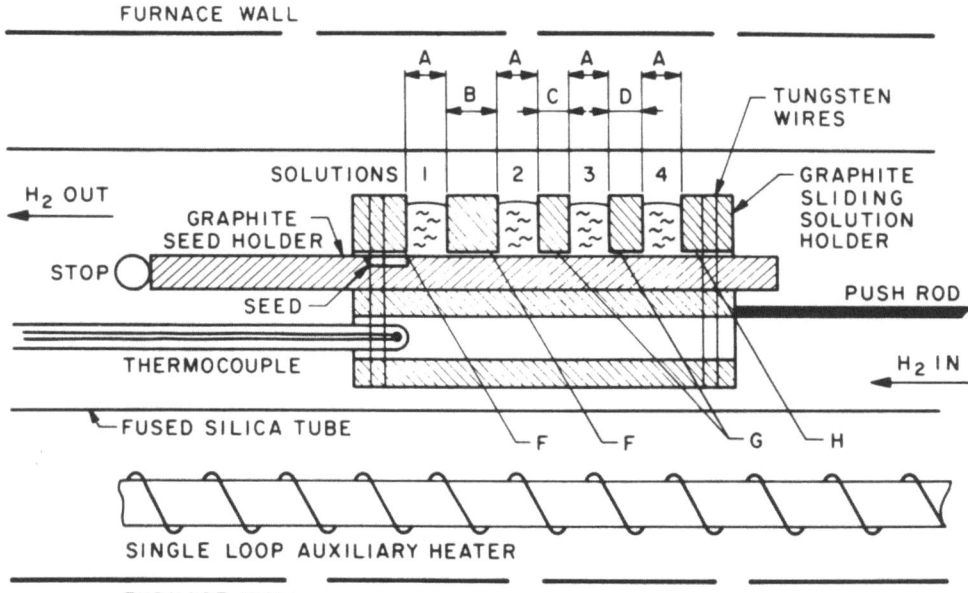

FIGURE 7. Apparatus used for the growth of four-layer heterostructure lasers. (After Panish, et al [23]).

Exactly why LPE gives more efficient diodes than does diffusion is not known. It is most likely associated with the high perfection of material grown this way with a flexible doping system which easily allows the simultaneous incorporation of several impurities, with the favorable stoichiometry provided by excess gallium, and perhaps with an efficient "gettering" action by the gallium solution for unwanted impurities.

The situation today in the industry making electroluminescent diodes is that most people are choosing to use GaAs,P alloys made by vapor phase graded epitaxy and then diffusing in impurities to make p,n junctions. Since these are direct band gap alloys, doping procedures are easy, and the diffusion process is more familiar than LPE. However, these diodes are less efficient than those made by LPE from GaP, and they cannot emit green light. At present, sections of the industry are addressing the problems of large scale, economic production of devices using LPE, and there is every indication that these efforts will be successful.

Several other semiconductor materials have been grown by LPE. These include InP, InSb, GaSb, Ge, PbS, and ZnTe.

LIQUID PHASE EPITAXY FOR MAGNETIC BUBBLE DEVICES

Magnetic bubble devices have the potential for performing great feats of permanent memory and for the logical handling of digital information.

The basic idea is simple. One applies a permanent magnetic field perpendicular to an appropriate sheet of magnetic material. If the field falls in a certain range of values cylindrical magnetic domains are formed which are called magnetic bubbles [11]. These domains can move quite easily, for their motion only requires the movement of the domain walls.

One can think of the bubbles as little magnets floating in a sea. They can be moved by a magnetic field. This can be done in a controlled and convenient way by patterning on the surface of the magnetic sheet a T-bar pattern of a soft magnetic material, and then having a horizontal rotating magnetic field. This horizontal field induces magnetic poles in the T-bar pattern, these poles will repel or attract the bubbles and as the field rotates the bubbles will propagate along the T-bar pattern. So we have the elements of a shift register. Bubbles can be made at will and can be detected. An important point is that because they are little magnets the bubbles also interact with each other. Thus if a bubble is present, a second bubble will go one way, but if the first bubble is absent, the second will go another way. Logic operations may therefore be performed.

There are however serious materials problems. There are minimum requirements which the material must have [12]:

1. It must be available in nearly defect-free single crystal plates, or films, with a unique magnetic axis of easy magnetization perpendicular to the plate or film. It has been established that domains do not propagate in polycrystalline films.
2. The material must be optically transparent. While there is nothing in the operation of a memory which requires it, there is presently no way of setting up a memory device without watching the domains. Future developments may waive this requirement, but for the present it is necessary.
3. The domains must have a useful size. A diameter of 0.3 mils (about 8 microns) will allow a storage density of about 10^6 bits per square inch, a very attractive density. A diameter of 3 mils, however, will reduce this by a factor of 100. On the other hand, domains cannot be made arbitrarily small. As the size is reduced circuit elements become increasingly difficult to prepare and detection problems rise formidably. Equilibrium domain diameter, d, depends upon the magnetic properties of the material in the following way:

$$d \approx \frac{(AK)^{\frac{1}{2}}}{M_s^2}$$

where A is the magnetic coupling constant, K is the anisotropy constant and M_s is the saturation magnetization. Since M_s and K depend on temperature it can be seen that for domain size to remain constant either the temperature must be controlled, or K, and especially the saturation magnetization must be independent of temperature near room temperature.

4. The domains must be mobile. For a given size, mobility determines the bit rate at which a memory may be operated. Although domain mobility can be traded to some extent for smaller domain size, a goal of a megacycle bit rate will require a mobility of about 100 cm/sec Oe for a 0.3 mil domain.

Given these requirements, the magnetic materials now available which fulfill them are few.

At present the most promising materials are based upon the garnet structure. It was less than two years ago that anyone suspected that the garnets would be at all suitable for this purpose. This was because everyone supposed that garnets were cubic crystals and so could not show the necessary magnetic anisotropy.

However simple garnets by no means provide satisfactory properties. The type of considerations used to achieve a useful composition may be described as follows.

It is necessary to achieve temperature insensitive characteristics. Garnets with flat magnetization versus temperature curves near room temperature can be prepared by using rare earths which form iron garnets with compensation points in the $4\pi M_s$ versus T curve. The magnetization curves of garnets possessing compensation points are shown in Fig. 8. The compensation points arise because the net magnetization of the iron containing sublattice opposes the magnetization of the rare earth sublattice. When the temperature dependence of the sublattice magnetizations is such that the moments exactly balance each other, the net moment becomes zero. Note that above the compensation point the magnetization curves are quite flat. Thus a garnet with a low temperature compensation point whose maximum in $4\pi M_s$ is at room temperature will have little temperature sensitivity.

Note that the magnetization of rare earth garnets with compensation points low enough to give us a flat magnetization curve at room temperature, also have a $4\pi M_s$ of several hundred, or

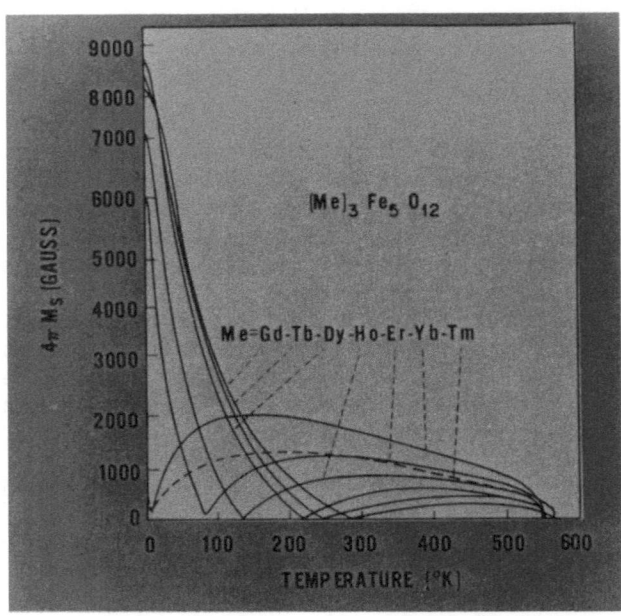

FIGURE 8. Spontaneous magnetization of a number of rare-earth iron garnets and yttrium-iron garnets, expressed in number of Bohr magnetons as a function of temperature (after Bertaut [24]).

even thousand, of gauss. The uniaxial anisotropy in garnets is low enough that domain diameters are too small to detect unless the magnetization is around 150. Thus nonmagnetic ions, gallium or aluminum, must also be added to the garnet to lower its moment. However, this moves the compensation point toward room temperature again. To achieve temperature insensitivity together with satisfactory magnetization values requires that a nice balance be struck between the species of rare earths on the dodecahedral sites, and the concentration of nonmagnetic ions on the tetrahedral and octahedral sites.

Finally there remains the question as to why the garnets show magnetic anisotropy at all since they are supposed to be cubic materials. One source of anisotropy which applies to epitaxial films grown on substrates is the strain induced in the film by the substrate. There is no doubt that this can be an important effect, and it has the characteristic of remaining despite high temperature annealing of the structure.

There is however another source of anisotropy which is not connected with strain [13]. It was discovered in single crystals.

FILM TECHNIQUES 551

COMPOSITION	COERC. H_C [Oe]	ANISO-TROPY	MOBILITY cm/sec-Oe
$Er_2Eu_1\ Ga_7Fe_{4.3}O_{12}$	0.5–1.0	HIGH	50–100
$Y_.9Gd_{1.5}Yb_.6Al_7Fe_{4.3}O_{12}$	≤ 0.1	LOW	~2000
$Y_1Gd_1Tm_1Ga_8Fe_{4.2}O_{12}$	≤ 0.1	MEDIUM	1000
$Y_{1.2}Eu_{1.6}Yb_2Al_1Fe_4O_{12}$	~0.5	HIGH	~500

FIGURE 9. Properties of various LPE garnet films. (After Varnerin [25]).

The single crystals were not however uniformly anisotropic - certain sections of the crystal were, and it was soon found that these sections occurred under certain facets of the growing crystal. It was clear therefore that the anisotropy arose from the direction imparted to the material as it grew under the influence of a particular crystal surface. The effect only occurred if the garnet had more than one rare earth, and it was concluded that as the crystal grew the rare earth atoms were not included in a random way along the various cube axes. It might be reasonable to suppose that the larger atoms were preferentially excluded from the axes in the growing plane. In $Nd_{0.1}Y_{2.9}Fe_5O_{12}$, Nd has a radius 10% greater than that of Y, and indeed it was found that for a crystal grown at 950°C there were 4-1/2 times as many Nd ions on the axis vertical to the growing surface as there were on the in plane axes [14]. This result was established by electron spin resonance experiments.

In fact this is an unusually high differential concentration. For more useful crystal compositions it has been calculated that a differential concentration of only a few percent can provide sufficient anisotropy for useful bubble devices.

As might be expected at high temperatures the atoms will randomize in a bulk crystal and the anisotropy will gradually disappear, never to be reinstated in that particular crystal. In fact this property of an adjustable anisotropy can be useful since it provides a means of reducing bubble size in a controlled way. Ultimately bubble size controls the information packing density in a crystal.

So finally we may arrive at such compositions as shown in Fig. 9. Much work has been done with the first of these

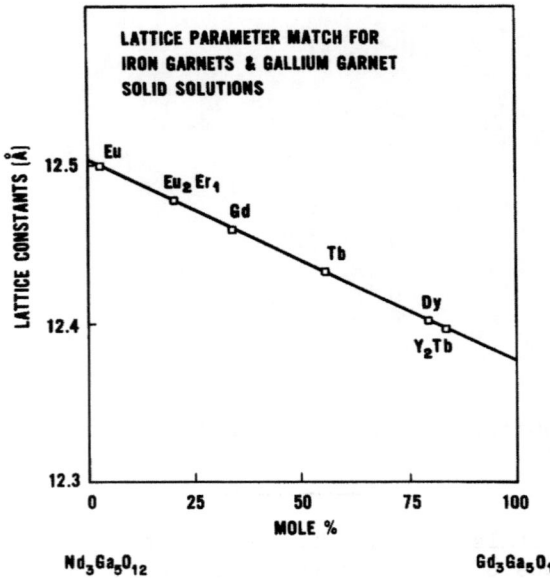

FIGURE 10. Lattice parameter match for iron garnets and gallium garnet solid solutions.

materials - this was the composition first grown with low defect density. However the mobility was rather low, and the coercivity rather high. Introducing a third rare earth in the second and third materials raised the mobility and lowered the coercivity. The fourth composition had superior temperature stability; however its coercivity is rather high. Development work continues in attempts to find superior compositions.

Now the question is, how can one obtain single crystal films roughly 5μ thick and uniform in thickness to within a few percent, of compositions such as these with 6 elements all in the correct proportions, with controlled anisotropy and defect free to a degree that one can make chips containing perhaps 10^4 bits (~160 mils square) with reasonable yield. It is a remarkable fact that liquid phase heteroepitaxy appears to put these goals within reach [15]. Note that this must be heteroepitaxy - the substrate must be nonmagnetic.

The first step is to obtain a suitable substrate. It was found that a variety of rare earth gallium garnets could be conveniently grown by Czochralski pulling. Furthermore it was found that mixed crystals could also be pulled. Figure 10 shows that in this way a variety of substrates would be produced by this means which had a corresponding variety of lattice constants.

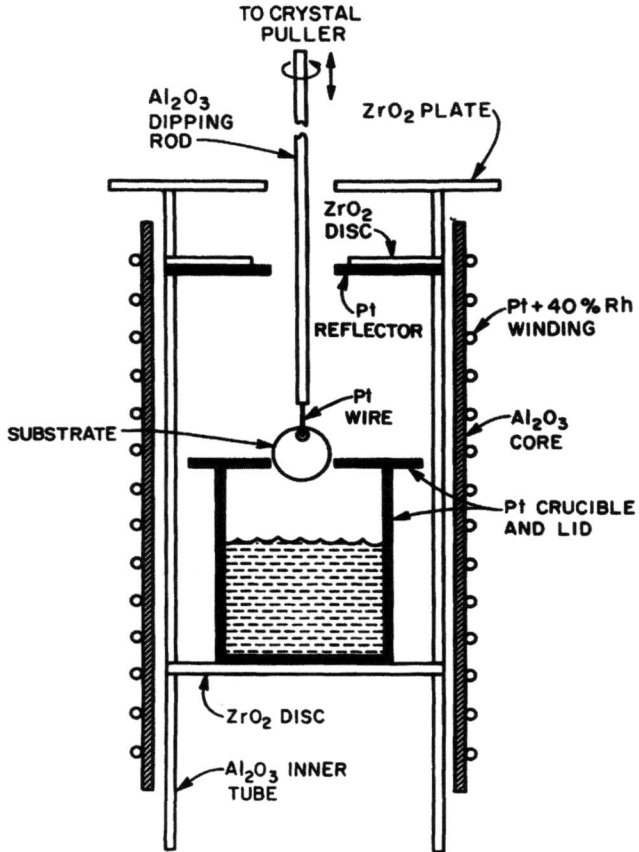

FIGURE 11. Cross section of crystal growth apparatus. (After Blank and Nielsen [26]).

This is important since it has been found necessary to match the substrate lattice constant to that of the magnetic film if good films are to be grown. In fact gadolinium gallium garnet, $Gd_3 Ga_5 O_{12}$, has been found to be a very useful substrate. Lattice matching to it has been no great problem and it can be grown with high perfection, providing one grows it in the presence of an inert atmosphere containing 2% oxygen. It is now even available from commerical sources.

The crystalline film is then grown on the carefully prepared substrate surface by a remarkably simple process illustrated in Fig. 11. It has been found that garnets may be deposited from a flux of oxides of the rare earths, Fe, Ga and Al dissolved in PbO and B_2O_3. The pot is equilibrated near 1100°C and then cooled to near 900°C [16]. The flux has the very desirable quality of being readily supercooled over a large temperature range. This means that a slice can be dipped into the flux and growth will take place under isothermal conditions. This has the advantage of avoiding temperature gradients, with consequent nonuniform growth, and it also avoids, of course, constitutional supercooling. Growth rate can be controlled by controlling the degree of supercooling. In a particular case 30°C of supercooling produces a 4μ thick film in 10 minutes. Maximum supercooling of 120°C has been observed. For supercooling of 10 to 40°C melts will remain without precipitation for 48 hours. The melt does not wet the substrate, so that after growth the substrate with its epitaxial film is simply withdrawn from the melt and it is ready for inspection and metallization.

It is extremely important that films grown in this way must be free of defects. In particular they must be free of dislocations. Dislocations can be detected through their birefringence, and they are known to pin dislocations.

A central fact of the garnet system appears to be that dislocations do not form easily. It is believed that this is because of the large lattice constant of garnets (12.40 Å compared to 5.42 Å for silicon). The energy of formation of a dislocation depends on the square of the lattice constant, and so it turns out that for garnets no interfacial dislocations will form so long as the lattice mismatch between film and substrate is no greater than 0.015 Å.

However dirt or mechanical damage on the substrate surface will cause trouble and in addition if there are any dislocations in the substrate these will grow in the epitaxial layer. Fortunately dislocations in the substrate may be readily detected by viewing between crossed polaroids.

The result is that dislocation-free substrates are not hard to obtain, and if these substrates are carefully polished and well cleaned a film can be grown heteroepitaxially with no misfit dislocations. Even where tensile stresses develop large enough to cause films to crack, no dislocations are observed. The garnet system must surely be unusual in this respect.

With sufficient care, results such as those illustrated in Fig. 12 can be obtained. The results pertain to the best 1 cm^2 area found in 2 cm diameter slices. They provide great encouragement that materials will be routinely available which will allow the fabrication of chips of useful size with reasonable yield.

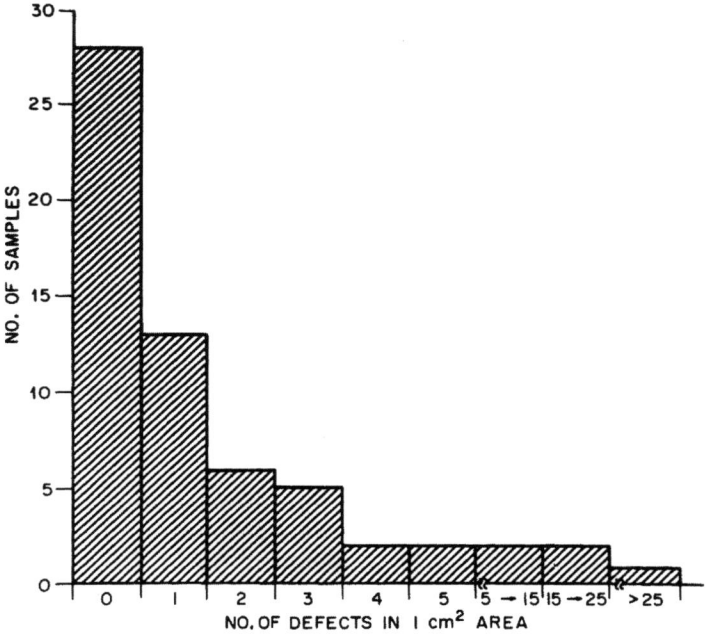

FIGURE 12. Hewitt's yield of the "best" 1 cm² garnet area obtained from a 2 cm diameter LPE sample. (After Varnerin [25]).

THIN FILMS AND THEIR POSSIBLE USE AS COMPONENTS IN OPTICAL COMMUNICATION SYSTEMS

The third topic is to consider how thin films could be important in a technology which has so far been the subject of a great deal of discussion and experimentation but which has not yet reached the stage of commercial exploitation. This is the possibility of using light waves as a medium for communication. As everyone knows the discovery of the laser operating at frequencies in the range of 10^{16} cycles/second makes possible, in principle, the transmission of enormous quantities of information over a single light beam. Someone has characterized this field of optical communication as being, today, one which has generated much optics but little communication. However, progress is being made, but it is quite certain that communication will not result until better components have been invented than exist today.

As far as the transmission medium itself is concerned, the light beams may travel in gas-filled pipes and be periodically refocussed by lenses - perhaps gas lenses - or possibly the light may be transmitted along low glass fibers which are clad with a

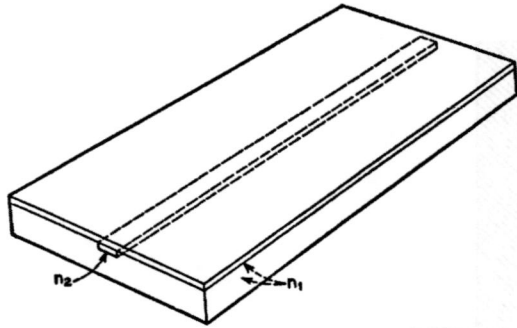

FIGURE 13. A schematic diagram of a planar waveguide which could be fabricated in glass using photolithographic techniques. (After Miller [17]).

lower refractive index medium so that surface scattering is minimized. There will be no further discussion of the transmission medium (reference may be made to Chapter 13 by N. B. Hannay).

Now of course the light that is transmitted must be modulated to begin with, and repeater stations will be needed at which signals can be amplified, demodulated, mixed with other signals, and so forth. It is here, in this area of components, that thin films may enter the picture.

The reason is that it is advantageous that the dimensions of the components which go to make up the modulator and repeater units, should be comparable with the wavelengths of the signals being handled. We are familiar with this in the microwave field where waveguides and components become smaller the higher the frequency of operation. Such small dimensions lead to economy, compactness, dimensional stability, stability of mode propagation, and to energy concentration.

Waveguides for light can conveniently be made by surrounding one transparent medium by another of lower refractive index. The difference of indices does not have to be large.

A simple form of waveguide can be made as shown in Fig. 13. Here there is a channel with a width of a few microns consisting of material with index n_2 surrounded by a medium with index n_1. This structure might be made using photolithographic procedures similar to those currently used in making silicon integrated circuits. A window in a mask or in photoresist, could be produced on the surface of a piece of glass, and diffusion, ion implantation or ionic replacement procedures could be used to

FILM TECHNIQUES 557

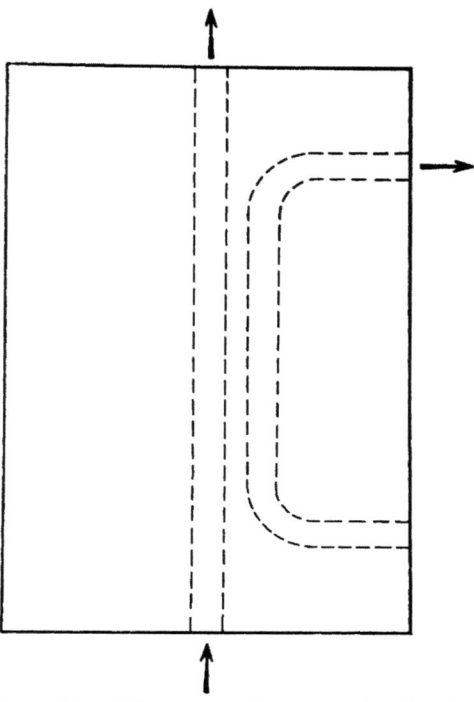

FIGURE 14. A schematic diagram of an optical microcircuit device which behaves as a directional coupler type hybrid. (After Miller [17]).

change the index of the glass. A covering might be sputtered over the pattern. It is not hard to imagine related structures which perform various circuit functions [17]. In Fig. 14 we see a directional coupler form of hybrid. In it there are two closely spaced guides, and the exponentially decaying field in the n_1 region provides continuous distributed coupling between the guides.

At present the greatest challenge is the problem of making suitable films and fabricating guides with required edge delineation to minimize loss. For instance edge waviness of about 500Å gives a loss of 1 db/cm. It is important that the films should not absorb the light. Some results concerning absorption are shown in Fig. 15. The films first studied [18] were ZnO which were sputtered onto hot glass substrates. It was disappointing to find that these films had a very high loss of some 60 db/cm at 6328Å, despite the fact that they were oriented with respect to the substrate. This loss was traced to the fact that the average grain size of the crystallites was about 1/2 micron, comparable to

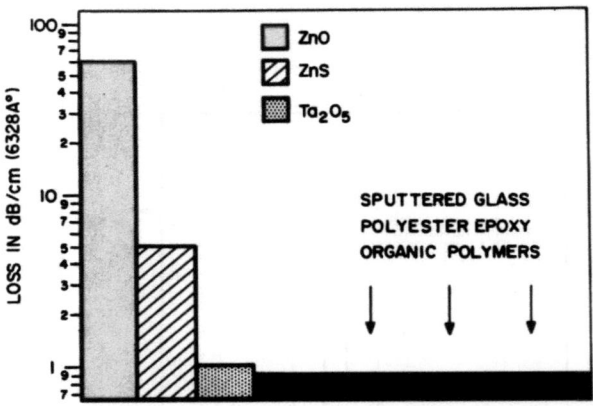

MATERIALS FOR LIGHT GUIDES

FIGURE 15. Losses in decibels per cm measured at 6328Å light wavelength for several semiconductor and organic films. (After Tien [18]).

---

the light wavelength, with the result that there was excessive scattering of the light in the film. Incidentally scattering losses will have to be thought about very carefully. Even if the intergranular effect can be avoided, the surface at the air interface will have to be very smooth. Calculations show that going down 1 cm of film, the light is equivalently internally reflected from the air-film interface 1000 times. There must be very little scattering at each reflection if much light is to be left after 1 cm of transmission.

Zinc sulphide evaporated onto glass at room temperature gave a polycrystalline film with very small crystals and indeed scattering losses were much reduced. A film with a loss of 5 db/cm was produced. Now, however, the losses were dominated by the tail of the intrinsic band gap absorption. To avoid this, films of Ta_2O_5 were made. This material has a large band gap and in amorphous sputtered films a loss of 0.9 db/cm was seen for red light.

At present all the low loss films are amorphous. These include sputtered glass, various epoxys and organic polymer films.

Such films as these may well be useful. However for certain purposes such as electrooptic modulation, frequency conversion or parametric oscillators involving signal, idler and pump frequencies, it is necessary to work in the realm of the nonlinear optics. For this purpose a medium is required which lacks a center of symmetry, so that it must be crystalline. However several other conditions are necessary for strong nonlinear interactions between waves of different frequencies. The crystal should have a high nonlinear coefficient, the energy density should be high, and the waves must travel together and be phase matched.

In principle thin films offer many advantages for such interactions. A thin film can concentrate laser energy for a long distance. The phase velocity of a light wave in a waveguide mode depends on the thickness of the film and the mode of propagation. This can lead to flexibility, and the possibility of phase matching without relying upon the birefringence of the crystal. Thus crystals such as GaP, ZnS, ZnTe, which have large nonlinear coefficients, but which have little birefringence can in principle be used for nonlinear experiments. In addition the nonlinear interaction can take place in the film, or in the substrate or in both.

In fact the only such experiment conducted to date does involve nonlinear interactions in the substrate [18]. An infrared beam is coupled into a polycrystalline ZnS film deposited upon a single crystal of ZnO. This fundamental wave produces a nonlinear polarization wave in the substrate. This wave in turn generates second harmonic radiation in the form of Cerenkov radiation propagating at an angle below the interface.

The most recent advances [19] in this field have come from the use of the heteroepitaxial garnet films grown in connection with the magnetic bubble devices. These films as we saw have been grown with a high degree of perfection, they are reasonably transparent to light in the visible and near infrared, and since they are heteroepitaxial films they may be expected to have a different refractive index from that of the substrate. It is fortunate that this difference has a sign which results in the trapping of light in the film. The reason is that the iron garnets have a larger refractive index than gallium or aluminum garnets. For instance at 1.523μ refractive index values are:

$$\begin{array}{ll} \text{Fe garnets} & 2.22 \pm 0.02 \\ \text{Ga garnet} & 1.94 \pm 0.02 \\ \text{Al garnet} & 1.82 \pm 0.02 \end{array}$$

Experiments have shown that indeed these films do act as waveguides for the light. (e.g., $Y_3Sc_{0.7}Fe_{4.3}O_{12}$ on $Gd_3Ga_5O_{12}$). TE and TM waveguide modes have been transmitted in films roughly 2.4μ thick, at several laser wavelengths in the visible and near infrared. Measurements indicated that the energy loss in the films were about 8 db/cm at 1.152μ and less than 3 db/cm at 1.523μ.

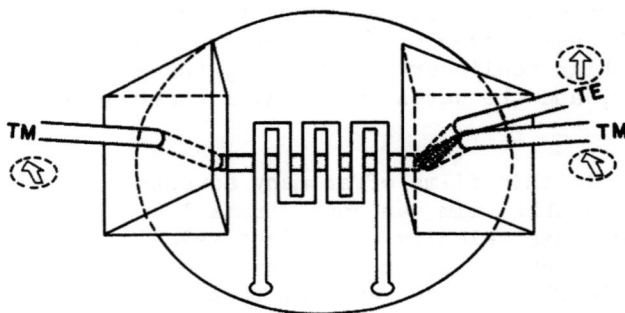

FIGURE 16. Experimental arrangement of a magneto-optic switch or modulator. (After Tien, et al [20]).

A quite intriguing device has been built using such films. It uses their magnetic properties to modulate and switch a guided light beam [20]. It is illustrated in Fig. 16. A prism couples light (1.152μ) into a $Y_3Sc_{0.4}Ga_{1.1}Fe_{3.5}O_{12}$ film grown on a $Gd_3Ga_5O_{12}$ substrate. These materials were chosen so that the stresses in the film resulted in a magnetic anisotropy in which the easy axis of magnetization was in the plane of the film, not vertical to it. Consequently, because of the high degree of symmetry in the plane, this axis can be rotated rather freely in the plane by a small magnetizing field. A small serpentine metallization pattern is fabricated on the surface of the film. Through this, electric current flows producing a small magnetic field. Faraday rotation within the film will convert one waveguide mode into another. This is controlled by the component of the crystalline magnetization vector which is parallel to the direction of light propagation. For the modulation experiment a D.C. magnetic field is held at $45°$ to the direction of light propagation. This field is then modulated by current flowing in the circuit. The two beams emerge separated by $20°\ 11'$. The same principle using different circuitry will provide switching between the two modes.

The important point is that with the very easy change of magnetization vector, very small fields are effective. Thus switching between modes can be produced by a field of 0.2 Oersteds, which is several times smaller than the earth's magnetic field of 0.6 Oersteds. Because of this the serpentine pattern need have only a very low inductance so that wide bandwidths are possible with low modulation powers - operation has been demonstrated at 80 Mc/s at which point the experiment was limited by the speed of the detector.

There are, therefore, exciting days ahead of us in this area of devices for manipulating light beams. It must be remembered, however, that these advances are largely made possible by quite remarkable advances in the materials science and engineering of thin epitaxial films. It is also important to remember that these advances have come from an intensive effort to solve rather specific problems set by particular devices. There is much nourishment to be gained by tackling problems thoroughly, so that materials can be produced which have high perfection, by methods which are truly reproducible. One may think one understands the problems when one has done something just once. But this is often not the case. Only by repeating it many times will one really discover all the problems, and having solved these, new opportunities may become apparent.

REFERENCES

[1] D. M. MacArthur, Bell Telephone Laboratories, private communication.
[2] A. T. English and P. A. Turner, "Stability of Conductor Metallizations in Corrosive Environments," Journal of Electronic Materials, Vol. 1, No. 1, 1972 (A Publication of the Metallurgical Society of AIME.)
[3] L. R. Dawson, "Liquid Phase Epitaxy," Progress in Solid State Chemistry, Vol. VII, ed. by H. Reiss and J. O. McCaldin (Pergamon, Oxford, 1972), p. 117.
[4] H. C. Casey, Jr. and F. A. Trumbore, "Single Crystal Electroluminescent Materials," Mater. Sci. Eng. 6, pp. 69-109 (1970).
[5] H. Nelson, RCA Rev., 24, p. 603 (1963).
[6] H. T. Minden, "Constitutional Supercooling in GaAs Liquid Phase Epitaxy," J. Crystal Growth 6, pp. 228-236 (1970).
[7] M. E. Weiner, "Si Contamination in Open Flow Quartz Systems for the Growth of GaAs and GaP," J. Electrochem. Soc. 119, p. 496 (1972).
[8] H. G. B. Hicks and P. D. Greene, "Control of Silicon Contamination in Solution Growth of Gallium Arsenide in Silica," Proc. 3rd Int. Symp. on GaAs, Aachen, 1970 (Inst. Phys. Phys. Soc., London, 1971), p. 92.
[9] I. Hayashi, M. B. Panish, P. W. Foy, and S. Sumski, Appl. Phys. Lett. 17, 109 (1970).
[10] L. R. Dawson, "Near Equilibrium LPE Growth of GaAs-Ga$_{1-x}$Al$_x$As Double Heterostructures," to be published.
[11] A. H. Bobeck, "Properties and Device Applications of Magnetic Domains in Orthoferrites," BSTJ 46, Oct. 1967.
[12] A. H. Bobeck, R. F. Fischer and J. L. Smith, "An Overview of Magnetic Bubble Domains - Material-Device Interface," AIP Conf. Proc. 5, 45-55 (1971).

[13] A. H. Bobeck, E. G. Spencer, L. G. Van Uitert, S. C. Abrahams, R. L. Barns, W. H. Grodkiewicz, R. C. Sherwood, P. H. Schmidt, D. H. Smith and E. M. Walters, "Uniaxial Magnetic Garnets for Domain Wall "Bubble" Devices," Appl. Phys. Letters 17, 131-134 (1 August 1970).

[14] R. Wolfe, M. D. Sturge, F. R. Merritt, and L. G. Van Uitert, "Facet-Related Site Selectivity for Rare-Earth Ions in Yttrium Aluminum Garnet," Phys. Rev. Letters 26, 1570-1573, (21 June 1971); M. D. Sturge, S. L. Blank and R. Wolfe, "Site Preferences for Nd^{3+} in YAG Films Grown by LPE," Mat. Res. Bull. 7, 989-998, (1972).

[15] L. J. Varnerin, "Approaches for Making Bubble-Domain Materials," IEEE Trans. on Magnetics, MAG-7, 404-409 (1971).

[16] H. J. Levinstein, S. J. Licht, R. W. Landorf, and S. L. Blank, "The Growth of High Quality Garnet Thin Films for Supercooled Melts," Appl. Phys. Lett., 19, 486-488 (1971).

[17] S. E. Miller, "Integrated Optics: An Introduction," BSTJ, Vol. 48 No. 7, pp. 2059-2069, September 1969.

[18] P. K. Tien, "Light Waves in Thin Films and Integrated Optics," Applied Optics, Vol. 10, No. 11, pp. 2395-2413, November 1971.

[19] P. K. Tien, R. J. Martin, S. L. Blank, S. H. Wemple, and L. J. Varnerin, "Optical Waveguides of Single-Crystal Garnet Films," Appl. Phys. Lett. 21, 207-209, (1972).

[20] P. K. Tien, R. J. Martin, R. Wolfe, R. C. LeCraw and S. L. Blank, "Switching and Modulation of Light in Magneto-optic Waveguides of Garnet Films," Appl. Phys. Lett., Vol. 21, No. 8, pp. 394-396, 15 October 1972.

[21] M. Hansen, Constitution of Binary Alloys, McGraw-Hill, N.Y. (1958), p. 165.

[22] H. Nelson, RCA Review 24, 603 (1963).

[23] M. B. Panish, S. Sumski and I. Hayashi, Met. Trans. 2, 795 (1971).

[24] F. Bertaut and R. Pauthenet, Proc. I. E. E. B104, 261-164 (1957). "Crystalline structure and Magnetic Properties of Ferrites Having the General Formula $5Fe_2O_3 \cdot 3M_2O_3$".

[25] L. J. Varnerin, "A Perspective on Magnetic Materials for Bubble Mass Memories," IEEE Trans. on Magnetics, MAG-8, 329-333 (September 1972).

[26] S. L. Blank and J. W. Nielsen, "The Growth of Magnetic Garnets by Liquid Phase Epitaxy," Journal of Crystal Growth 17 (1972) 302-311, North-Holland Publishing Co.

CHAPTER 21

SEMICONDUCTOR INTEGRATED CIRCUIT TECHNOLOGY

D. G. Thomas

Bell Laboratories, Murray Hill, New Jersey 07974

INTRODUCTION

To all intents and purposes the only semiconductor which is used to make integrated circuits today is silicon. Germanium is still widely used to make discrete devices such as transistors - it was the first material to be developed and products made with it will take a long time to pass from the scene - and there are many semiconductors from the III-V, II-VI and IV-VI groups of the periodic table which have significant uses, but not as integrated circuits.

The reasons for this are not hard to find. Silicon is an astonishing material. In common with many materials it can be made both p- and n-type, and stable p,n junctions can exist in a single crystal, with the consequence that transistor action can be obtained, and this of course is essential to making useful devices. What is apparently unique to silicon is the remarkable fact that by simply heating it in air, a crystal of silicon may be covered by an amorphous film of silicon oxide, which is an excellent insulator, which may be readily patterned using photo-lithographic techniques, and which will, at high temperatures act as a diffusion barrier against impurities or dopants which one wishes to introduce into the silicon through windows in the patterned silicon oxide. These circumstances allow integrated circuits to be made along the following lines. Useful general accounts of this subject and further references may be found in references 1 and 2.

FABRICATION OF BIPOLAR SILICON INTEGRATED CIRCUITS

We will consider first the fabrication of an epitaxial-diffused structure, sometimes called the standard buried collector process, to produce a bipolar integrated circuit. Figure 1 illustrates several of the steps involved. (The term bipolar refers to a device in which transistor action is achieved with charge carriers of both polarity, as opposed to unipolar devices which use only one.) A p-type slice of silicon serves as the substrate. It may be 6 to 15 mils thick, depending on its diameter, and it must be high resistance, typically 10 ohm cm, so that it forms a junction with a high breakdown voltage and low capacity. By heating it in wet or dry oxygen at 1000°C or higher an oxide film is grown 3000 to 7000 Å thick. The surface is now coated with a photoresist material. On exposure to light such a material may be polymerized and rendered resistant to removal by solvents which will dissolve the unpolymerized material, (this is called a negative working photoresist). The resist is in fact exposed through a previously prepared mask. Usually the mask is in direct contact with the slice, although today projection printing is being used which results in less mask wear and fewer defects produced by scratches on the photoresist. It goes without saying that the mask contains many repeated patterns so that many identical circuits are made together. It must be noted that mask making is itself a major undertaking. For a single circuit many master mask levels have to be made, and then working copies are fabricated from these. The standard way of making masks is to cut out a rubylith pattern and then to reduce its size photographically; an image of the reduction is then step and repeated many times to form the final array. Today more sophisticated systems are in use, designed especially to eliminate the manual cutting of rubyliths.

Diffusion and Cleaning

Using windows in the exposed and developed photoresist, windows are cut in the oxide layer. Impurities are now introduced into the silicon usually by diffusion at temperatures from 900 to 1100°C. The details of this diffusion process are critical to the processing. Furnaces must have a uniform temperature zone controlled to within a few degrees C, which is fairly straightforward. A more difficult matter is the diffusant source. The source must provide a constant and known concentration of the impurity at the surface of the silicon - this must be done for several slices in a run, and uniformly over each slice, and also must be reproducible from run to run. The source must also provide only one electrically active impurity. Very importantly the source should not leave on the surface compounds difficult or

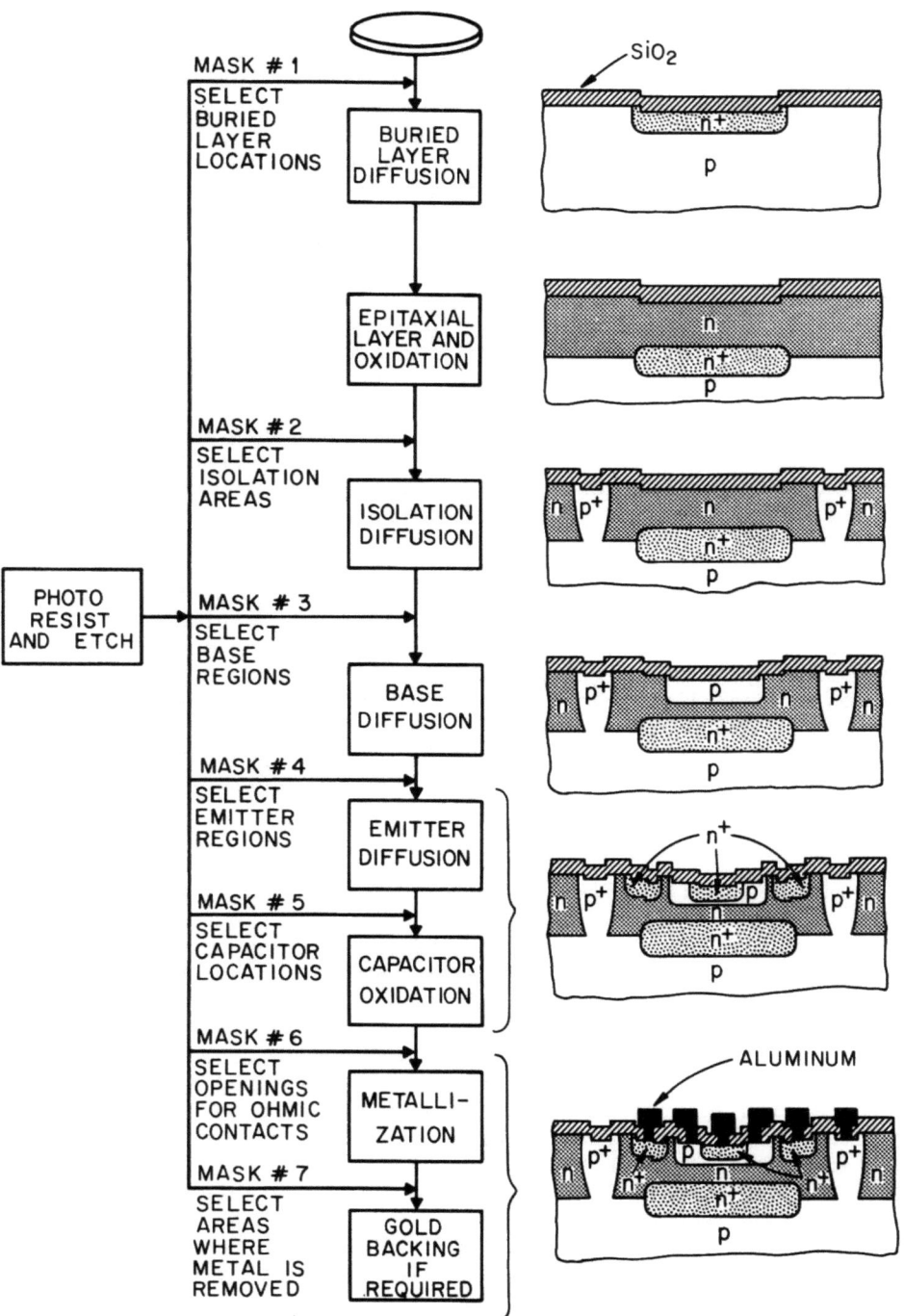

FIGURE 1. Flow chart of the sequence of processes used in the fabrication of a single-crystal monolithic circuit [1].

impossible to remove. This really emphasizes the importance of cleaning in making integrated circuits. Before almost any step it is usually necessary to go through a rigorous cleaning procedure. If some of the diffusant source is left this may be a source of an unwanted impurity in a subsequent diffusion step, or if it remains in a window it may prevent contact being made in a subsequent metallization step, or residual impurities can affect adhesion of materials such as photoresists or metals, and poor adhesion means low yield or low reliability devices, and impurities can also cause much electrical trouble at surfaces for both bipolar and MOS devices. So cleaning is vital and requires a great deal of attention. In a particular process sequence containing 73 steps, 25 of these were cleaning steps, and this probably underestimates the relative amount of care and attention which cleaning demands.

To return to diffusion, the first diffusion step is often that for a high conductivity n-type buried layer. Arsenic or antimony may be used for this purpose. Both elements can be introduced from open tubes. Sb_2O_3 is often used as a source of antimony, As_2O_3 for arsenic. Volatile compounds may be introduced by means of a bubbler arrangement. Usually an oxidizing atmosphere is used so that a doped glass, based on SiO_2 formed by oxidation of the silicon, is usually the diffusant source. Arsenic has much to be said for it for a buried layer dopant - it diffuses slowly, it has high solubility and it causes little distortion of the silicon lattice so that good quality epitaxial silicon can be grown upon it. One does have to be careful however that in growing the epitaxial silicon it is not doped with arsenic which evaporates into the gas stream from other parts of the silicon slice.

Silicon Epitaxy

All the oxide is removed and an epitaxial layer of silicon typically 8 or 9 microns thick is next grown on the surface of the substrate. This layer is n-type and so of the opposite type from the substrate. Its function is first of all to enable the circuits which are to be made to be electrically isolated from each other. This isolation is achieved with a reverse biased p,n junction which has a high electrical impedance - it may also be noted however that there is a considerable parasitic capacitance associated with these back-biased junctions, and other isolation schemes are sometimes used which will be mentioned later. The epitaxial region is also used as the material in which transistors and resistors are made; it may be doped at about 0.5 ohm cm, representing a compromise between the high resistivity needed for high breakdown voltage and low junction capacitance at the collector-base junction, and the low resistivity desired for low

$V_{CE(SAT)}$ and high frequency response for the transistors to be fabricated in it.

Epitaxial silicon is thus an important part of many integrated circuits - although not all, for most MOS devices do not require epitaxial silicon since isolation is achieved by means of the p,n junction which exists between source, drain or channel on the one hand, and the substrate on the other. Further details of this topic are discussed below when MOS circuits are considered.

Isolation and Junction Formation

After the epitaxial layer has been grown standard processing usually calls for the silicon surface to be oxidized. Again this is done by simply heating in an oxygen atmosphere near 1000°C. The next diffusion step is for isolation. This in effect forms an isolated tub of n-type material in which devices are made - these devices will always be electrically isolated from others in other tubs since they are separated by two back-to-back p,n junctions, so that no matter what polarity voltage might appear between the two tubs, there is always a back-biased diode between them. Boron is often used for isolation.

During the diffusion cycle a new layer of silicon dioxide grows over the diffused p-region, and the pre-existing oxide over the n-region grows thicker.

The next masking step (#3), has as its primary purpose the opening of windows through which boron again is diffused in order to produce the transistor base regions. At the same time this rather light doping which does not penetrate through to the substrate, can be used to make resistors.

Again during the diffusion an oxide grows over the diffused p-region.

The next masking and selective etching step produces windows through which phosphorus (or arsenic) is diffused for the formation of transistor emitters. In addition the cathode regions for diodes and capacitors may be made.

At the same time windows are cut into the n-type regions - the collector regions - so that there may be a heavily doped n-type surface layer on these regions. This heavy doping is produced by the same phosphorus diffusion as produces the emitters. The reason for this is that Al is usually used as the contacting and interconnecting metallization. Al is a p-type impurity in Si with a maximum solubility of 2×10^{19} atoms/cc. Hence a large concentration of phosphorus in the n-region is required to prevent the formation of a p,n junction when the Al is alloyed in to form the contact. In fact to have a good ohmic contact it is desirable to have the phosphorus surface concentration under the Al contacts in excess of 2×10^{20} atoms/cc. At these high concentrations there is easy tunneling between the valence and conduction bands, and ohmic contacts result.

At this point, after the heavy donor diffusion, the junction formation in the integrated circuit is complete.

The fifth masking step, (or sixth if capacitors have been formed on the silicon surface), is used to open windows to the emitter, base and collector contacts of the transistors, and also to diodes, resistors, and other circuit elements.

Metallization

A thin, uniform coating of aluminum is now vacuum-deposited over the surface of the wafer. An enormous amount of work has gone into perfecting metallization techniques [3]. Al is usually evaporated from a tungsten filament or by e gun evaporation. Step coverage is very important and so multiple sources are often used in the evaporation equipment, or the slices may be moved continuously during evaporation. The deposited Al is patterned and etched by conventional means (masking step #7 in Fig. 1); typically 10 μ line widths are used, although 2.5 μ lines have been shown to be feasible. In addition it has also been shown that multilevel metallization can be achieved if a layer of SiO_2 is deposited above the first Al level. One has to be careful here to avoid the formation of hillocks in the first level Al layer, since these may cause problems if a continuous dielectric layer has to cover them. These hillocks may be caused by stress in the Al, and the problem is reduced if 1% Si is present in the Al. In addition one must be careful to slope the windows in the intermediate dielectric so that continuous metal films result without opens or near opens. Finally, care has to be taken that an oxide on the Al surface does not prevent contact between the two Al layers.

To achieve a low resistance contact to silicon the slice is heated, typically for a few minutes at 550°C. The Al and Si form an alloy, but so long as the temperature is kept below 577°C, no liquid phase is formed. (In fact reaction between Al and SiO_2 begins at 500°C.) This represents an upper limit to the temperature for any subsequent processing, and therefore limits any gettering steps which can be performed with metal present - this is particularly significant for MOS devices. One advantage which Al has over some other metals is that it will reduce any SiO$_2$ which happens to remain in the contact windows - in other words it is rather forgiving of improper cleaning out of these windows.

Electromigration [4] has affected the performance of integrated circuits, particularly those made with aluminum metallization, and this effect will now be considered briefly.

Electromigration in Thin Film Metal Conductors

Electromigration is a potential wear-out mechanism for thin film metal conductor stripes. Due to the passage of a high current density, matter is selectively transported from place to place, resulting in local accumulations or depletions of material and eventually, open circuit failure by coalescence of voids. The matter flow is believed to be coupled to the electric current by a momentum transfer process between moving charge carriers (electrons in most common metals) and diffusing metal atoms.

The matter flow is accompanied by a uniform counter flow of vacancies with drift velocity

$$V_D = Z^* q \rho j \frac{D}{kT}$$

where D is the uncorrelated diffusion coefficient. Z^* is an effective ionic valence, q is the electronic charge, ρ is the resistivity of the metal, and j is the current density. On a macroscopic scale the vacancy flux J is given by

$$J = -D\nabla n + \frac{nD}{kT} Z^* q \rho j$$

representing the sum of the diffusion and field drift terms respectively, and where n is the local vacancy concentration. Wherever the two terms are not equal, an accumulation or depletion of vacancies will occur. The rate of accumulation or depletion is given by,

$$\frac{\partial n}{\partial t} = \cdot \nabla \cdot J = D \nabla^2 n + \nabla \cdot \left(\frac{nD}{kT} Z^* q \rho j \right) + \frac{n_\infty - n}{\tau}$$

where n_∞ is the equilibrium vacancy concentration, and τ is the mean lifetime of vacancies. Generally dn/dt will be zero. However abrupt changes in D, or temperature gradients will cause vacancy depletion and accumulation. Structural discontinuities such as abrupt changes in grain size, lead to abrupt changes in D. Joule heating due to the passage of the current can produce temperature gradients. Vacancy supersaturations eventually lead to void nucleation and growth, and finally to an open circuit.

The practical significance of this phenomenon lies in the implied limitations on design of thin film and silicon integrated circuits. Specifically, susceptibility to this type of wearout is increased by increased current density, narrower line width, reduced film thickness, higher operating temperatures, and the presence of steep gradients in temperature, composition or metallurgical structure within the conductor. Many of these adverse factors are intensified as dimensions are reduced and component densities are increased in the continuing effort to secure greater economies of scale.

At the present time, current densities at critical sections of conductor stripes are in a few instances as high as 10^5 A/cm^2 in devices currently being designed. Line widths, historically not much below 5 microns due to limitations imposed by photographic pattern delineation methods, may soon approach a micron or even less as electron beam methods of pattern generation are introduced. Device temperatures, too, are rising, and design values of 50-80°C are not uncommon today.

These trends can be compared with experimental results which show that failure after perhaps a thousand hours of operation is characteristic of gold thin film stripes passing a few million A/cm^2 at 185°C. While the extrapolation of these short-time test results to the conditions of actual service cannot yet be made with precision, it appears that in devices intended to operate for 10-20 years the limits of prudent design lie not very far from current practice.

Recent experiments have clarified the process by which electromigration failures occur, and, in some cases, suggested methods to reduce susceptibility to this mode of wearout.

For example, there is evidence that the current-induced mass transport is chiefly along grain boundaries, owing to the higher mobility of the metal atoms in the disordered boundary region. This suggests that large-grained (or single-crystal) films would be preferable, and this appears to have been verified. Improvements obtained by alloying may well reflect the importance of adsorbed solute atoms or precipitate particles in reducing atom mobility in the boundaries.

The choice of metal for the film is, of course, an important factor. Much of the industry uses aluminum metallization, and the first electromigration failures in commercial products were observed in aluminum films. Gold is somewhat less susceptible than aluminum, based on its lower atomic mobility at the same temperature. Despite this advantage, other less favorable factors (high operating temperature, say, or greater divergencies in grain size) may well be present and produce failures in a given circuit. The advantage associated with gold is less an absolute than a relative one. Solutions can be found to these difficulties, and if high reliability circuits are to be built it is important that these solutions be incorporated into the structures.

Testing

The metal is now in place and the circuits may be probed and electrically tested while they are still joined together on a slice. Today many manufacturers before this probe test go through one more step in which SiO_2 glass is deposited over the whole slice, and then windows are opened where contact has to be made to the outside world. The purpose of this step is to provide mechanical scratch protection to the delicate metallization pattern. It also offers a degree of protection from outside contamination.

It is at probe test of course that one gets the information which indicates the yield and so the success of ones processing.

In conventional processing the slices may be thinned down and then a diamond tool scribes fine lines between the chips. The slices are then simply broken apart along these scribed lines, and the good chips are removed.

Bonding and Mounting

The chips can be eutectically bonded to a conducting substrate which therefore provides an electrical contact to the bulk silicon in the chip. Usually contact is made to the metallic pattern on the chip by thermocompression bonding of wires 0.001 inch in diameter. Today chips are usually mounted in "DIP'S", or dual-in-line packs, which have become an industry standard.

The bonding wires are usually made of gold. Problems can arise in the bonding of gold to aluminum in that intermetallic compounds can form - this is popularly known as the purple plague. For this reason ultrasonic techniques have been worked out for the bonding of Al wires to Al pads, and in fact highly reliable bonds can be made this way. There remains however the problem of bonding the other end of the Al wire often to a gold-plated land.

However, provided high temperatures are avoided the reliability of Au to Al bonds as made today appears to be satisfactory. It is difficult though to be sure of long term reliability without being able to do high temperature testing.

Another problem with this wire bonding is that it is a one-at-a-time hand operation which is delicate and expensive. For this reason many manufacturers do this assembly work in areas of low labor costs.

There are ways to avoid some of these problems. One of these uses gold beam leads [5]. In this technology the gold beam leads are fabricated when the chip is made. They are exposed when the chip is etched apart - which takes the place of scribing and breaking. This does mean that interconnections within the chip are accomplished with a different metal system than simply Al - a highly reliable Ti-Pt-Au system is chosen, the gold of course being continuous with the gold beam leads.

An additional feature of these chips is that they are sealed against outside contamination with a layer of silicon nitride. This means that they can be used in a comparatively unprotected environment whereas for high reliability, Al devices are usually packaged in a hermetically sealed can.

Review of Bipolar Structure

At this point it is interesting to look back for a moment at the critical parts of the structure. These are of course the bipolar transistors. Figure 2 shows a cross section drawn to correct scale.

The various dimensions for window openings, diffusion depths, epitaxial thickness and so forth are marked. Note also the diameter of human hair. This figure clearly shows that the base width is much smaller than any of the other device dimensions. Here it is shown to be $0.3\ \mu$, or 3000 Å. In fact the thickness and doping level in the base region control the so-called Gummel number [6]. This number is the total number of active impurities in the base region. It controls the base resistance which is an important quantity as far as the high frequency performance of the transistor is concerned.

Figure 2 shows the n^+ region used for collector contacts. The collector region adjacent to the base is made of high resistivity material in order to reduce the collector-base junction capacitance, so the n^+ contact regions are necessary for good contact to the metallization. This diagram shows too, how

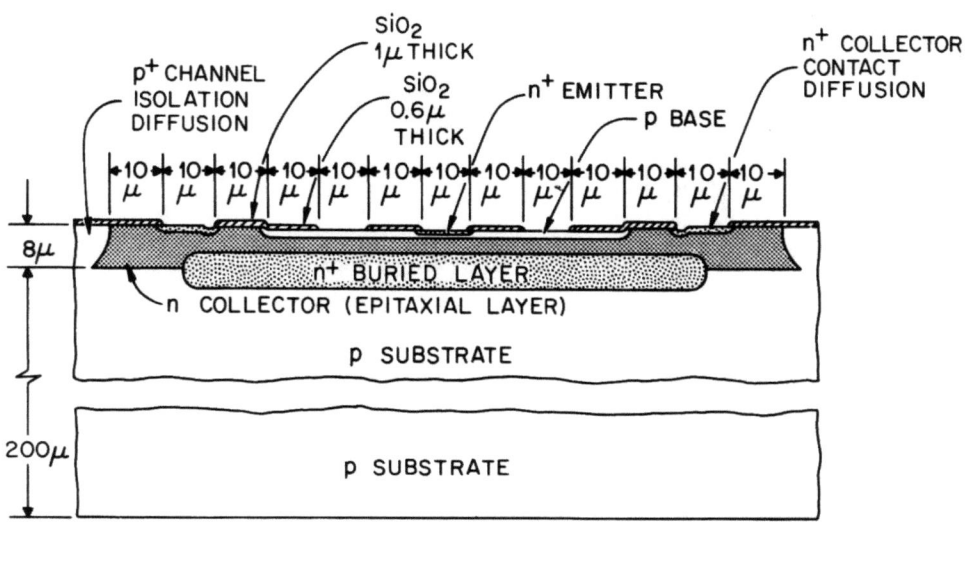

FIGURE 2. Cross section of an integrated circuit transistor shown to correct scale [1].

important the buried layer is in reducing the collector series resistance.

There are of course many variations possible to optimize performance. In the example shown here two base contacts, (which are usually in the form of stripes), are used to reduce base resistance. By eliminating one of them, one may reduce the base area thereby reducing the collector base capacitance, and also making easier the interconnection layout. In general narrow emitters improve practically all device parameters, reducing for instance current crowding effects. Small base widths generally improve intrinsic frequency response.

Finally, Fig. 3 shows a typical impurity distribution as a function of distance into the wafer for a transistor produced by diffusion into an epitaxial layer in the manner described previously. The emitter is very heavily n-doped so that high emitter efficiency is achieved - that is, most of the current across this forward-biased junction is carried by electrons.

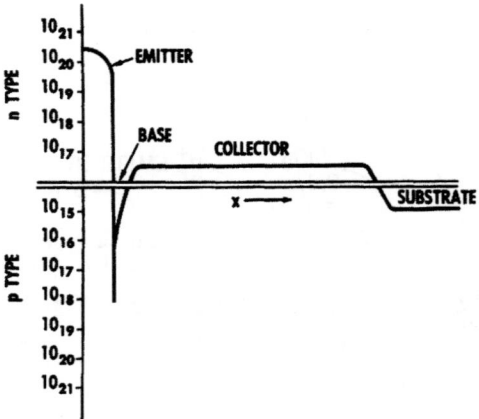

FIGURE 3. Approximate impurity distribution as a function of distance into wafer for a transistor produced by the epitaxial-diffused process [2].

There is a thin base p-type region shorter than the electron diffusion length, so that the collector to emitter current gain α is as near to unity as possible, with the result that the collector-to-base current gain, $\beta = \frac{\alpha}{1-\alpha}$, is as large as possible. As mentioned before the collector doping, which is of course the doping of the epitaxial layer, is chosen to be a compromise between the high resistivity needed for high breakdown voltage, and low collector-to-base junction capacitance, and the low resistance needed for low $V_{CE(SAT)}$, (the voltage across the transistor when it is driven into saturation), and good high frequency performance.

METAL-OXIDE-SEMICONDUCTOR (MOS) TRANSISTORS

So far a description has been attempted outlining how bipolar SIC's are made. In principle the process is straightforward, but it is a complex series of steps and so manufacture can be troublesome. One of the great attractions of MOS devices is that they are simpler to make. The basic structure is illustrated in Fig. 4.

The black regions are metal contacts to source, gate and drain. This is an n-channel device and so the source and drain are two heavily doped n-type diffused regions. The gate is metal deposited on an insulator. Conduction occurs between source and drain in an "inversion" n-type layer formed between source and drain under the influence of a potential applied to the gate electrode. Thus no special isolation steps are necessary. There

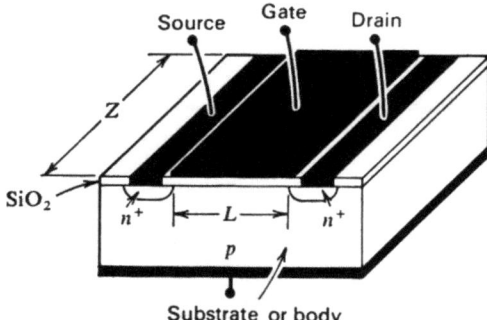

FIGURE 4. n-Channel surface field-effect transistor [20].

is in principle only one diffusion step necessary to make this transistor, for all parts of it are isolated from the substrate by a reverse-biased junction. As we saw earlier to make a standard bipolar transistor, one has to start with a carefully grown epitaxial layer, and then diffuse an isolation region, a base region and finally emitters.

Now things are by no means quite as cut and dried as this superficial comparison might suggest - there are for instance extra steps required to produce a satisfactory gate insulator - nevertheless conceptually, at least, there are perhaps half the number of photolithographic steps required to make an MOS as a bipolar circuit. In fact it is not only conceptually easier to make, it is also easier to invent - that is, it is easier to understand its operation - and it is an interesting historical fact that the general idea of using field effect devices as amplifiers was discussed in the early thirties. People worked on it at Bell Labs in the forties, and it was this work which more or less led to the accidental discovery of the bipolar transistor. Difficulties were encountered in making MOS devices, and it was not until the sixties that major interest in MOS developed - undoubtedly this came about partly as a result of the techniques worked out for the bipolar technology, particularly that for growing high quality thermal SiO_2 with excellent insulating properties. At any rate in 1972 in the U.S. MOS SIC sales are expected to be $180 million, compared to $440 million for bipolar SIC's. So the MOS IC business is large and it is growing.

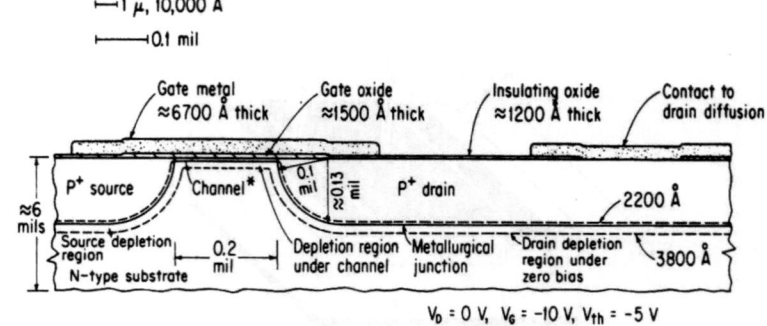

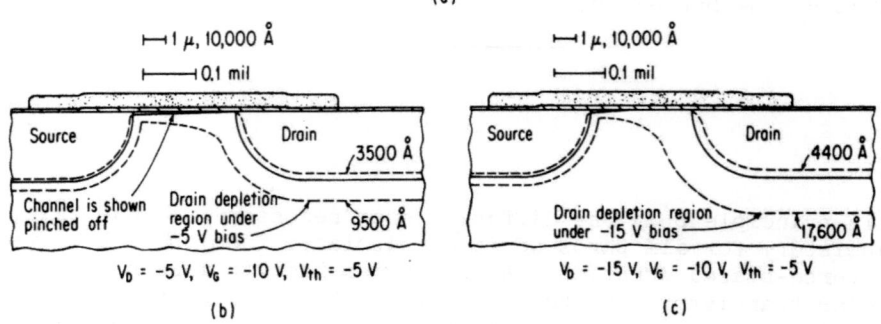

FIGURE 5. Scale drawing of a typical MOS structure in cross section: (a) $V_D = 0$ V; (b) $V_D = -5$ V; (c) $V_D = -15$ V; (d) change in scale showing the MOS in relation to the whole silicon wafer. The channel shown is exaggerated in the depth dimension. The source and substrate are both considered at grounded potential [21].

The workings of an MOS transistor can be examined in more detail in the scale drawing in Fig. 5.

This is now for a p-channel device, which up to the present time represents the bulk of the type of MOS devices which have been manufactured.

In these drawings it is assumed that the source and substrate are at ground potential. A negative potential of sufficient magnitude on the gate will produce a very thin p-type inversion layer at the silicon surface; in section (a) of this figure -10 volts has been applied to the gate, and a channel is produced 25 to 50 Å thick. Current will now flow from source to drain if a

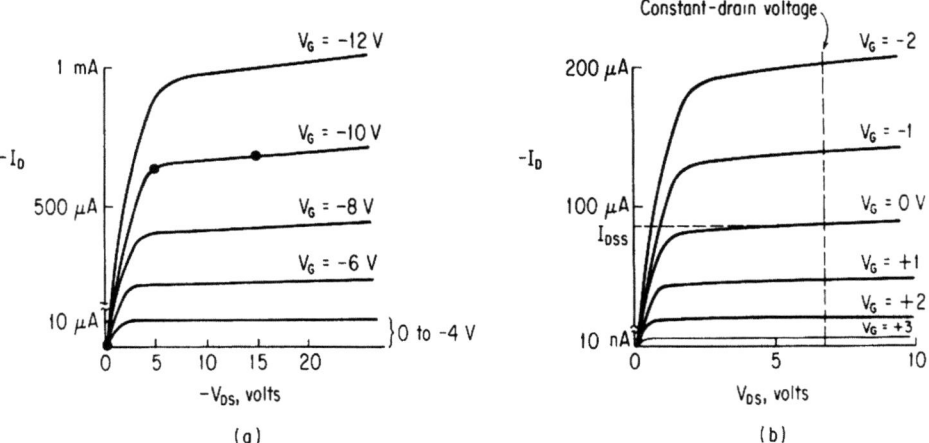

FIGURE 6. MOS characteristic curves, P channel: (a) Enhancement-mode device. (b) Depletion-mode device [21].

negative potential is applied to the drain, and this current can be modulated by the gate voltage. For low drain voltage the inversion layer extends across the entire channel, and drain current depends upon both drain and gate voltage.

For constant gate voltage, an increase in the drain voltage changes the situation in the gate region. Drain current produces an IR drop along the channel. This drop is of such a polarity as to oppose the field in the oxide produced by the gate bias. Consequently we can see how the inversion layer will almost "pinch off", and the current will saturate at the "pinch off" or "threshold" voltage (see sections (b) and (c)). Thus characteristic curves such as those shown in Fig. 6 are provided.

Section (a) of Fig. 6 shows the curves for the device just described, namely an enhancement type device in which zero current flows with zero gate bias. It is also possible to arrange matters so that current does flow at zero applied bias, and the current may then be increased or decreased by varying gate voltage. Characteristic curves for this situation are shown in section (b). This is called a depletion mode device, and is useful for fast switching applications.

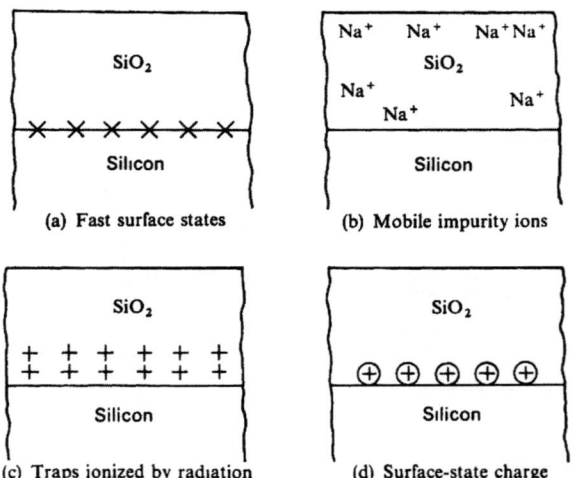

FIGURE 7. Charges and states associated with the Si/SiO$_2$ system [20].

The reason that MOS devices were rather slow to come, even after the discovery of the excellent insulating properties of thermal SiO$_2$, was primarily one of device stability. This is connected with the fact that extremely small numbers of atomic charges - much less than a monolayer - can cause quite unacceptable changes in threshold voltage. These changes are associated with several types of states which are depicted schematically in Fig. 7. Fast states occur at the silicon-SiO$_2$ interface; they have rapid communication with holes and/or electrons in the silicon. They are often thought to be connected with the states expected at a discontinuity in a periodic lattice such as occur at a surface. They often occur at densities near 10^{11} to $10^{12}/cm^2$ which is far too high for device purposes. However various annealing procedures, often in H$_2$, can reduce this to the vicinity of $10^{10}/cm^2$. Generally today this is not a major problem.

Ionizing radiation can produce holes and electrons in the SiO$_2$ and if a field is present these may separate and become trapped. Such things can happen during manufacture. Annealing near 300°C generally removes this damage.

A far more serious problem with early MOS devices was drift under bias of the flat band voltage. This can, and often is, caused by Na$^+$ ions which drift through the insulator. This subject will be discussed later when "gettering" is considered.

Finally another source of trouble has been what is called surface state charge, Q_{ss}. This charge may have values between 10^{10} and 10^{11} charges/cm^2. The charge is fixed, in that it is not altered as different potentials are applied, and it is located within 200 Å of the Si-SiO$_2$ surface. It does change with annealing conditions and there is a rather well known "Fairchild Triangle" which describes this [7]. Cooling slowly from 1100°C in dry oxygen increases the density of these states, whereas cooling in nitrogen decreases the density. This has led some people to speculate that these surface states are associated with excess ionic silicon in the oxide. Whatever the explanation, an important point is that if (100) silicon is used instead of the more usual <111> orientation then the Q_{ss} values will be roughly three times less in number. Since lower values of Q_{ss} mean that lower voltages are required to start bending the bands, this effect is sometimes used to produce MOS devices with low threshold voltages.

Gate Materials

So far in this discussion of MOS devices nothing has been said concerning the material which is used as the gate electrode. This is however an important matter and in fact has been largely responsible for modern developments in the MOS field.

The gate material first used, and still quite widely used today, is simply aluminum. This material while easy to handle and of good conductivity has two disadvantages:

1. It has a low work function, which results in rather high threshold voltages, usually in the range of 4-5 volts. This makes compatibility with bipolar circuits quite inconvenient to achieve. In addition the speed-power product of the device is inferior to one that has a threshold in the vicinity of 1-2 volts.

2. The second disadvantage of the aluminum gate is that it cannot be heated above about 550°C. This fact has two consequences. (a) As we will see when gettering is discussed the stability of the MOS devices is assisted by a high temperature step which serves to remove troublesome impurities. If this step can be performed after the gate is in place the gettering step can occur at a later stage in the processing, and this is not possible with Al gate. (b) If the source and drain doping can be performed with the gate in place then what is called a self-aligned structure results which avoids most of the parasitic overlap capacitance between gate and source and drain. This overlap, which can be seen in Fig. 5, is largely the result of alignment tolerances which must be observed in photolithography.

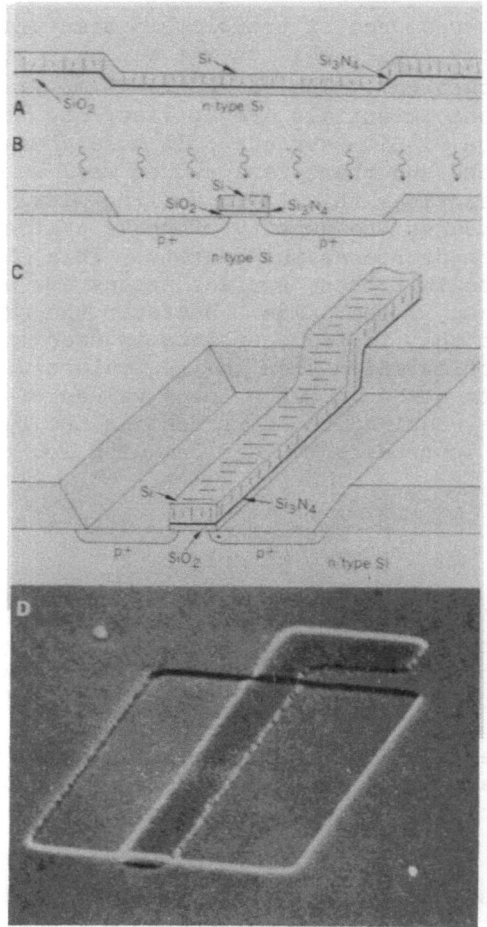

FIGURE 8. Silicon gate processing: A-Device after deposition of nitride and amorphous silicon sandwich. B-Definition of gate insulator and gate electrode dimensions and subsequent boron diffusion. C-Perspective view of structure of (B). D-SEM photograph of the structure at end of this fabrication stage. N.B. The Si_3N_4 layer is sometimes omitted [8].

Today many people are overcoming these disadvantages by using the Si gate process [8]. In this, doped polycrystalline silicon is used as the gate material, as illustrated in Fig. 8. A thin oxide is grown in a region where the source, gate and drain are to be made. A layer of polycrystalline silicon is deposited by the same reaction as is used to deposit epitaxial silicon. The polycrystalline silicon is patterned and the source

SEMICONDUCTOR INTEGRATED CIRCUIT TECHNOLOGY

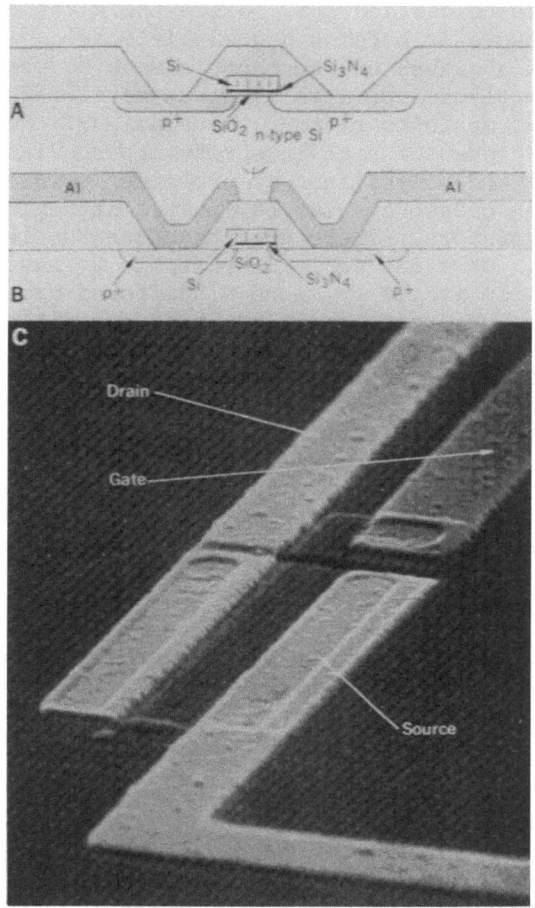

FIGURE 9. A-Device of Fig. 8 after deposition of silicon dioxide and definition of contact regions. B-Final structure after deposition and definition of metal interconnections. C-SEM photograph of the completed transistor structure [8].

and drain windows are opened. The source and drain regions are diffused, and at the same time the polysilicon is doped. The important point to note here is that the gate itself is now the mask against diffusion. The consequence is that with the exception of a small amount of lateral diffusion there is no overlap of gate with source and drain.

The next step in this process is also important, and is included in Fig. 9. It is to deposit a glass over the whole structure; it is at this stage at which a high temperature gettering step can be carried out. Since the glass can be heated to a high temperature it becomes a good dielectric. Windows are then opened in this glass, and contact is made to the source and drain and to the polysilicon with aluminum metal. The final critical masking step is to pattern the aluminum. Notice that because there is present a high integrity intermediate insulator, shorts between Al and polysilicon at crossover points are made less likely. Such crossovers can be located at regions remote from the MOS transistors. There are therefore two levels of interconnections. In addition it is possible to have a connection level in the silicon itself by means of diffused crossunders, although this form of interconnection has a parasitic capacity much larger than that of the other interconnection schemes. These three levels of interconnection provide great flexibility in circuit layout and lead to saving of silicon area. Notice that only four critical masking steps are required to make these devices. This is the same number as is used for conventional Al gate processing. (A refinement allows the polysilicon to make direct contact to the bulk silicon at the source or drain. This saves silicon area, but it comes at the cost of an extra masking step.)

Thus one of the advantages of this technology is multilevel interconnections. In addition these devices have low threshold voltages, typically near 2 volts, because of the high work function of the p-doped silicon. As we saw, this leads to better speed power product and to easier compatibility with bipolar integrated circuits.

An important advantage is also the self-aligned gate feature. This avoids the necessity of visual alignment to produce the gate with its attendant inaccuracies. As a result the Miller capacitance between gate and drain can be reduced by a factor of 3 to 5. In addition there can be a 50% reduction in gate capacitance because of a smaller thin gate oxide area, and a 30-40% reduction in junction capacitance mainly as the result of reduced junction area.

Probably today the best known application of the silicon gate technology is the p-channel Intel 1103 1024 bit random access memory. However n-channel Si gate is also in manufacture. In addition Si gate devices are used in logic applications.

Si gate is therefore very popular for good reason. There are those however who can see still further improvements in the materials system used to make MOS devices. One such proposed improvement is to use a refractory metal such as molydbenum [9] or tungsten in place of polysilicon. These metals can withstand high temperatures and do not react with SiO_2 so that gettering is possible; furthermore they can be used as a mask against diffusion just as can polysilicon. Their work functions are high, so that low threshold devices result just as for Si gate. Certain processing simplicities result since doping of the gate material is not required. However the chief advantage claimed for this technology is that the refractory metal gate has a conductivity at least 300 times better than the polysilicon. This should lead to faster MOS circuits, for there will be less delay in signal propagation, and noise margins should improve since there will be much lower I-R drops down the gate connection lines. It is also likely that for memory chips which have increasingly complex peripheral circuits around the main memory core, the higher conductivity of the first interconnection level will be advantageous.

"GETTERING"

"Gettering" has been referred to several times in connection with SIC fabrication, but so far it has not been described. What gettering amounts to is a particular treatment of the device at some stage in its fabrication, and this treatment seems to have the effect of removing certain impurities or imperfections from the structure. The impurities may cause device characteristics or device stability to be downgraded, and their removal is essential especially if large scale integration is to be practiced. The reason is that if there is a large number of devices on one chip, the yield for individual devices must be very high if the chip yield is to be acceptable. Often this can only be achieved if gettering is employed. For small scale integration, or ultimately for making discrete devices, the importance of individual device yield is diminished.

In fact there are two rather different gettering functions:

(a) The hardening of p,n junctions.
(b) The stabilizing of MOS devices.

Junction Hardening

The problem is that instead of a "hard" junction in which, under reverse bias, very little current flows until a high voltage breakdown occurs, there sometimes occurs a soft "leaky" junction in which current flows at small reverse voltages. Now this type of thing can be very sensitive to the device being made, it can be strongly affected by the materials used, and for a given device it can even be dependent on the location in which it is made, and the operator by whom it is made. The best thing to do therefore

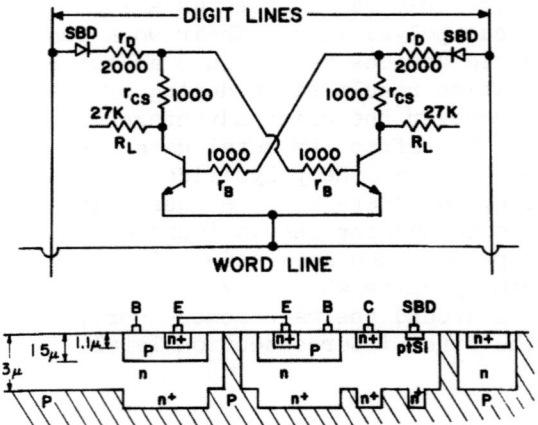

FIGURE 10. Circuit configuration of a fast bipolar memory [22].

is to attempt to outline the problem and its solution by considering a particular example [10].

Figure 10 shows the elements of a fast bipolar static memory chip. The chip has 64 bits on it, and so 128 transistors. Notice that in order to reduce power consumption some rather high value load resistors are included at 27 KΩ. Small leakage currents can therefore produce significant voltage drops and in fact one problem that this device presented was leaky base-to-collector junctions - notice that these junctions are rather close to the buried layer as the epitaxy is only 3 μ thick, (the thin epi means that high resistors can be fabricated in it without using excessive area). It is convenient to define breakdown voltage as the reverse voltage at which the base collector current reaches 10 μ amps. Calculation shows that this breakdown should be in the 25-30 volt range, and a value of 10 volts is acceptable for satisfactory device performance.

During a period of trouble a statistical count of devices gave the results shown in Fig. 11 for breakdown voltage at the collector base junction. This of course is not bad if one were making discrete transistors but is hopeless for satisfactory chip yield, which demands transistor yields close to 100%.

These results stimulated a diagnostic program. Diffusion was examined; different boron sources were tried, as well as experiments performed in different furnaces. Negative results were obtained. It was however observed that some epitaxial material gave much better results, indicating that the trouble

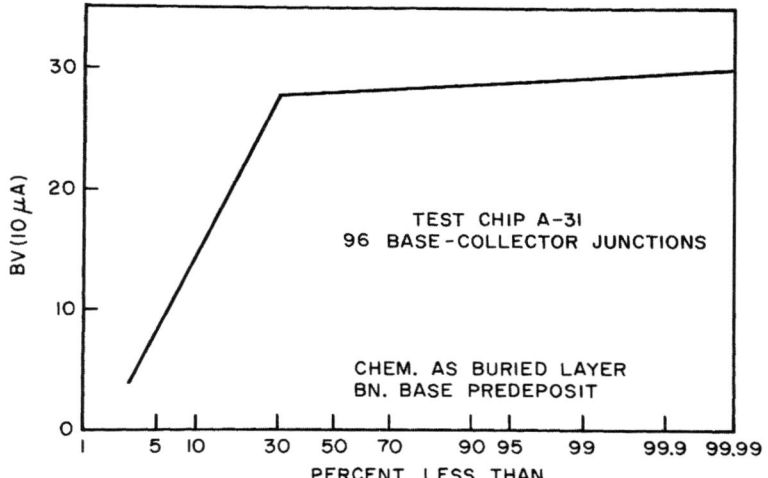

FIGURE 11. Breakdown voltages (i.e., voltage at which 10 μa flows) of base collector junctions [10].

might lie in certain epitaxial lots. Junctions were then formed in bulk silicon of the same doping as the epi (although without buried layers), and 100% yield was obtained for a sample of more than 500 diodes. This provides additional evidence of trouble in the epitaxy.

Experiments led to the suspicion - and such effects have been known since 1960 from the work of Shockley and Goetzberger [11] - that the trouble arose from defects or impurities, introduced from the epitaxial layer, present in the junction space charge region. These defects are hole-electron recombination-generation centers and so cause junction leakage. If impurities are the problem chemical gettering should help.

It is known that heat treatment in the presence of a phosphosilicate glass is a useful gettering process. Some slices were therefore reoxidized to protect the front surface, the back oxide was removed and they were given a 5-minute getter in a PBr_3 emitter diffusion furnace at 1000°C. Before and after measurements were taken on 96 diodes with the results shown in Fig. 12. One sees that in this case the low voltage sport population was reduced from 20 to <1%, and the number of junctions with less than 10 volt breakdown decreased to zero. In other words close to 100% yield was attained. This provides strong evidence that in this case at least, the low breakdown voltage problem is related to contamination or defects in the space charge region. It also shows that if this gettering step can be included in device processing it will

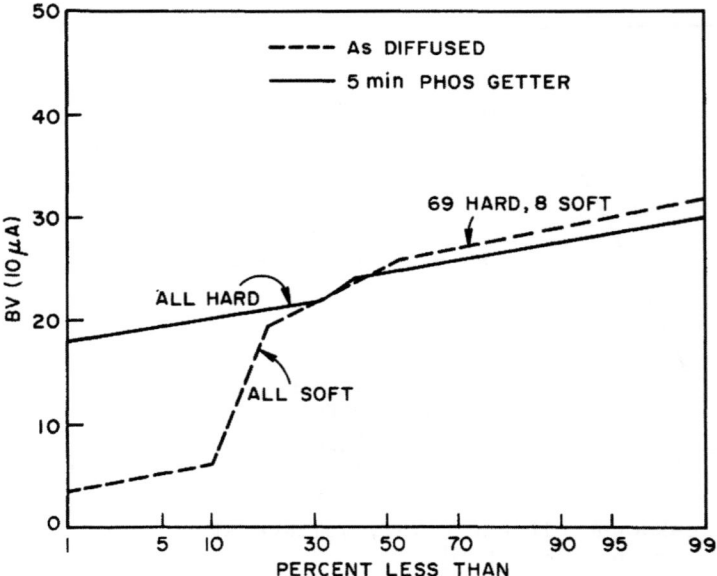

FIGURE 12. Similar breakdown voltage distribution for material from a more typical batch before and after a 5 min. 1000°C PBr_3 getter treatment [10].

provide a useful safeguard against the accidental inclusion of impurities.

The interesting question is what are these contaminants? Some measurements of the 2 MeV He^+ back-scattering spectrum before gettering on the backs of slices yielding a large number of soft junctions reveals a relatively small number of impurities there, a maximum of $2\times10^{14}/cm^2$ Cu and Fe and $10^{14}/cm^2$ atoms in the Pt-Pb mass range. After gettering, $2\times10^{15}/cm^2$ Cu and Fe have accumulated there, along with $4\times10^{14}/cm$ Pt-Pb. Thus, there is some evidence that elements such as Au, Fe, Co or Cu are indeed responsible for the poor junction leakage characteristics.

Before leaving the subject of junction gettering, it should be mentioned that there are other junction gettering methods. One of these is simply to damage the lattice in a region well away from the active parts of the devices, e.g., the back of the wafer; the damaged region evidently can provide a sink for the precipitation of troublesome impurities. Damage can be introduced into the back surface of the slice by simple sand blasting, but a more elegant method is to use ion implantation, for instance with 50 KeV argon ions. A heat treatment near 900°C for 30 minutes completes the treatment. Incidentally, the normal emitter

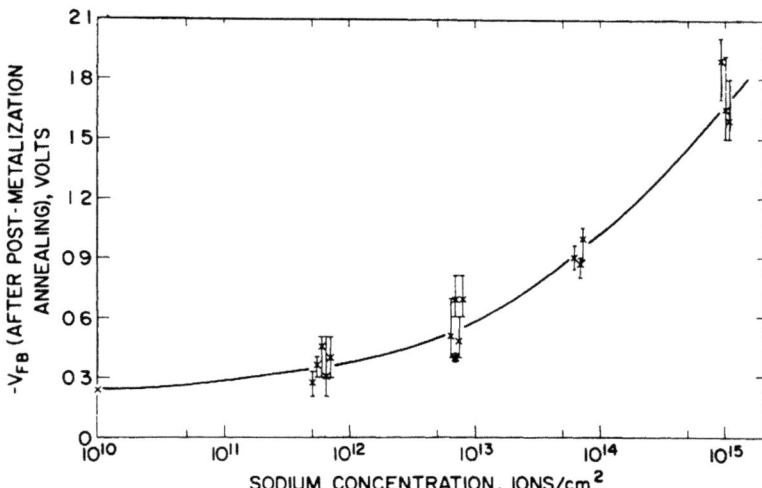

FIGURE 13. Dependence of the flat band voltage on sodium contamination level in Al/NaCl/PSG (4% P_2O_5, 125 Å)/SiO_2 (900 Å)/Si structures after a typical post-metallization annealing treatment (5 min. in N_2 at 500°C) [12].

diffusion with phosphorus undoubtedly provides some gettering action but this may not give optimum results. It may for instance occur at the wrong point in the process sequence. Thus there is some evidence that base collector junctions should be gettered before emitter deposition.

Stabilizing of MOS Devices

In MOS circuits there is another type of impurity removal, or at least of impurity prevention, which is of great significance. It is usually referred to as a gettering action of mobile charge. The problem is that certain ions, notably Na^+, can move comparatively rapidly through thermal SiO_2 which is used as the insulator under the gate of MOS devices. Even the movement of very small quantities of these ions can strongly affect device characteristics such as flatband voltage or threshold voltage. The effect is accelerated by an applied electric field. Since the oxide is usually 1000 Å thick, 10 volts applied to the gate produces a field of 10^6 volts/cm. Radio tracer work has shown that deposited Na is entirely transported as Na^+ ions through SiO_2 by moderate electrothermal stressing (e.g., a few hundred seconds at 200°C under a field of 10^6 V/cm).

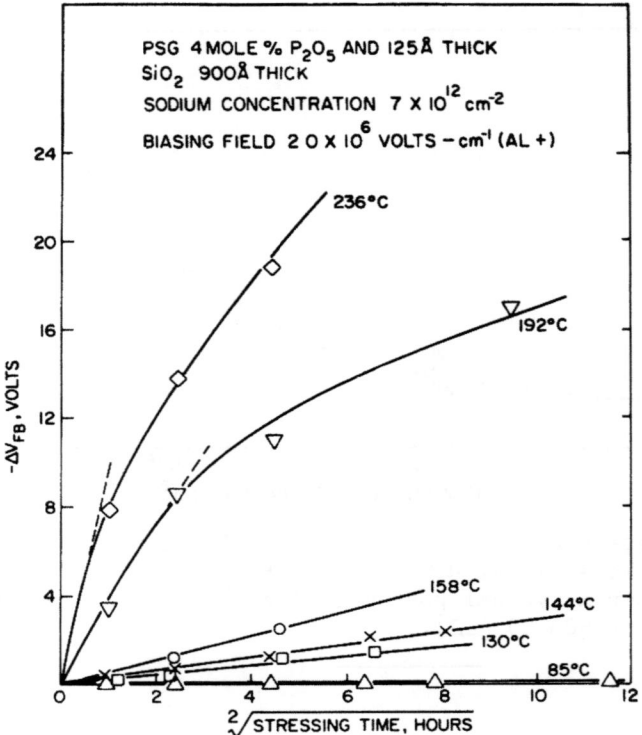

FIGURE 14. Effects of time and temperature on the extent of Na^+ ion drift for the indicated sodium-PSG-SiO$_2$ system [12].

J. M. Eldridge and D. R. Kerr [12] have studied this effect, and particularly how it can be minimized by the same type of phosphorus doped SiO$_2$ glass that was used for junction gettering. It was found that the P glass certainly greatly reduced the motion of the Na$^+$ ions, but it must be recognized that it is not a complete panacea. To emphasize this last point, even without an applied electric field there can be significant migration of Na through a P glass layer, as is shown in Fig. 13. In this experiment various surface quantities of Na have been deposited on 125 Å of P glass over 900 Å of SiO$_2$ on Si. An Al electrode completes the structure. Significant changes in the flatband voltage are produced by a brief high temperature unbiased anneal such as usually occurs after metallization. In fact one can calculate from these results that 5×10^{12} ions/cm^2 penetrated halfway into the glass.

Furthermore under the influence of a field the drift is much faster as shown in Fig. 14. An initial surface concentration of

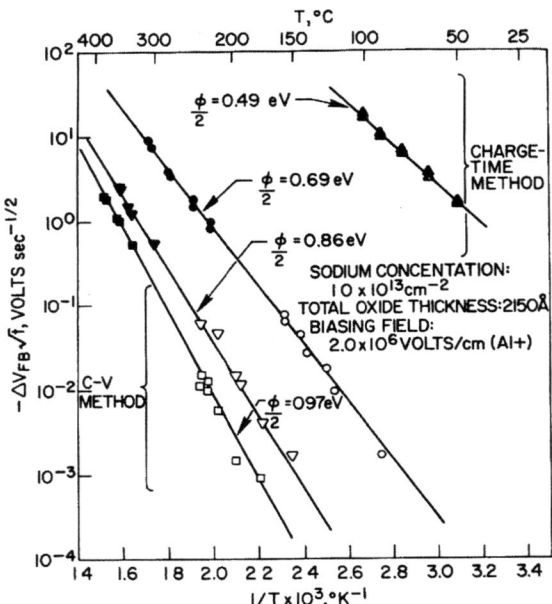

FIGURE 15. Temperature-dependence of $-\Delta V_{FB} \cdot t^{-1/2}$ values for various PSG barriers: ▲ pure SiO_2; ●, ○ 3.5% $P_2O_5 \times 125$ Å thick; ▼, ▽ 6% $P_2O_5 \times 230$ Å thick; and ■, □ 8% $P_2O_5 \times 190$ Å thick. Note that the sodium concentration is constant [12].

only 7×10^{12} Na atoms/cm^2 was chosen for this experiment. Notice that very large changes in flatband voltage can be produced, even in the presence of the layer of phosphorus doped glass. The initial slopes of these $\sqrt{t}$ plots are strongly dependent on temperature, and they can be plotted in the usual way against $\frac{1}{T}$, as is shown in Fig. 15.

In this figure the effect is examined as a function of temperature for different concentrations of phosphorus in the doped layer. This figure shows that while Na can indeed penetrate the P glass, its rate is greatly reduced from the rate which is achieved in pure SiO_2. Thus eight more orders of magnitude of time may be required for a given amount of sodium to drift through the gate dielectric region in P glass stabilized films than in pure silica.

No one knows exactly why P glass has this effect on the diffusion of Na. It has been suggested by Eldridge, et al., that introducing P_2O_5 in place of two SiO_2 groups will result in a structure in which the two P atoms are reasonably close together,

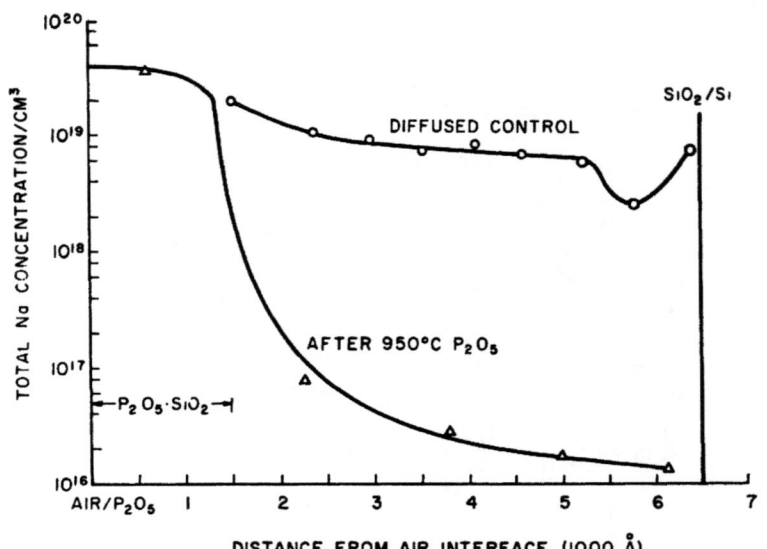

FIGURE 16. P_2O_5 effect on NaCl contaminated oxide. Control sample is typical of diffusion from NaCl contamination at 800°C for 1 hour, $\Delta Q_{MOS} = 1.2 \times 10^{12}$ cm^{-2}. After P_2O_5, $Q_{MOS} \approx 1 \times 10^{11}$ cm^{-2} for this sample [13].

so that the extra oxygen atom, which will be negatively charged, can jump from one P atom to another; (this hypothesis is supported by the observation that the bulk electric polarization of pure P glass is proportional to the square of the P_2O_5 concentration). The negatively charged oxygen atoms may then coulombically attract the Na$^+$ ions, and so will impede their motion, and if used as a layer under the gate metal it will therefore help to stabilize MOS devices.

It is clear however that the P glass does more than just decrease the diffusion coefficient of the Na. Since traps are produced for the Na, the P glass acts as a sink for Na and so can getter it from pure SiO_2 used as gate insulator. This property is illustrated by work of Yon, Ko and Kuper [13] from which Fig. 16 is taken.

This figure shows the result of an experiment in which the concentration of radio tracer Na has been measured as a function of depth in SiO_2 before and after a phosphorus gettering step. It can be seen that the Na concentration in the contaminated oxide is reduced by nearly three orders of magnitude by the phosphorus gettering step.

In fact modern MOS devices are usually gettered by depositing P glass over, not under, the refractory gate material - this is generally the case for Si gate and for refractory metal gates. The effectiveness of these treatments is testament to the fact that the P glass removes contaminants from the device structure. It must be emphasized again, however, that P glass is no universal cure-all and high standards of cleanliness must be maintained throughout manufacture.

RECENT DEVELOPMENTS

Up to this point an outline has been given of some of the electronic material considerations used in making more or less standard bipolar and MOS integrated circuits. Some of the more recent activities in this field will now be discussed, and at the same time a few specific examples of devices which these developments make possible will be described.

Bipolar Developments

Bipolar circuits are faster than MOS circuits. Most MOS circuits do not operate above 2 mc/s whereas bipolar circuits can operate at 100 mc/s or more. The fundamental reason for this is that the transconductance, $\left(\frac{dI}{dV}\right)$, is much higher for bipolars than for MOS devices. Crudely the current goes exponentially with voltage at a p,n junction in a bipolar device, whereas it goes more like V^2 in an MOS device. Thus stray capacity can be charged and discharged with much less power in a bipolar than in an MOS device.

While this is true, MOS devices with features such as n-channel and self-aligned gates have been made which have steadily improving performance. Thus 1024 bit MOS RAM's have been advertised with access times below 100 ns. In addition MOS circuits are generally regarded as simpler to make.

All of this has stimulated activity in the bipolar camp.

Two of the major factors which limit the performance and economics of bipolar devices are:

1. Parasitic Capacitance
2. Excessive area of silicon used to achieve a device structure.

On the first point the trouble is the inherent parasitic capacitance between each transistor's collector and ground. This capacitance is a result of the p,n junction isolation, with its capacitance associated with the reverse-biased junction. The magnitude of this capacitance depends on such things as the resistivity of the material used, the thickness of the epi layer and the area of the collector region.

Clearly, the transistor can switch only as fast as this stray capacitance can be charged and discharged, and this fixes the speed-power product of the device for any given current and voltage.

There are at the moment two general changes which are being introduced to mitigate both problems just mentioned. One is to use thin epitaxial silicon - instead of epitaxial material 7 to 10 microns thick, layers 1 to 3 microns thick are coming into use. The other is to use different isolation techniques between devices; the most popular of these is oxide isolation.

Thin epitaxy has several advantages. Since diffusion goes sideways as well as vertically it is clear that the isolation diffusion in bipolar circuits will spread out around the diffusion window a distance equal to the epitaxial thickness. This lateral diffusion must be allowed for in the circuit layout, and for thick epitaxy there are serious area penalties. Thinner epitaxy reduces these penalties. It also will reduce the collector-substrate capacitance so improving performance. In addition, for thin enough epi the inverse transistor gain can be increased. This inverse gain is the current gain when collector and emitter are interchanged in function - that is the collector junction is forward biased and the emitter base junction is reverse biased. The increase in gain occurs because of the proximity of the heavily doped buried collector to the base region, so that the injection efficiency of the buried collector used as an emitter is improved. This high inverse gain is useful in certain circuits; it can result in a decreased transistor storage time, and it can increase the performance of certain TTL circuits.

Of course such advances do not come without effort. High tolerances have to be maintained in the growth of the epitaxy and in the diffusion parameters. In particular junction breakdown voltages tend to decrease with shallower diffusions.

The use of thin epitaxy may be taken a step further to both reduce device area, and to reduce the number of processing steps. This comes about with the so-called, Collector Diffused Isolation or CDI circuits [14].

The CDI structure is different from the standard buried collector structure, (Fig. 1), as shown in Fig. 17. Instead of n epi, a thin layer of p epi is used which is of the same

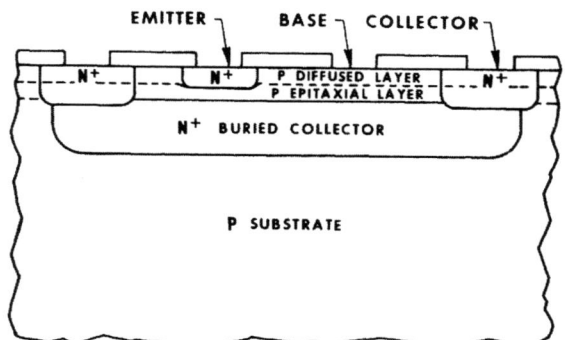

FIGURE 17. Diagram of a collector diffusion isolation structure [14].

conductivity type as the substrate. The p,n junction needed to provide electrical isolation is formed at the same time that the collector contact diffusion is performed. No separate isolation diffusion is required. Note also that no base diffusion is required since a portion of the epi material is used as the base - in other words we have a grown base rather than a diffused base. Only one critical diffusion is needed - the emitter diffusion, although an additional p^+ diffusion is usually added to make resistors and allow good contact to be made to the base. This usually intersects the collector diffusion at the surface and gives additional capacitance.

So provided tolerances can be maintained CDI offers as much as a factor of four in reduced area over SBC, and fewer processing steps.

Another scheme to reduce area still further and to improve performance uses oxide isolations [15]. It goes under various names such as Isoplanar, OXIM, Locos or Planox. It is being applied to MOS devices as well as bipolar. The important point in this process is that silicon nitride is used as a mask against oxidation. Silicon nitride is laid over the whole surface and is patterned with phosphoric acid. Oxidation of silicon takes place in the windows of this nitride pattern.

An isoplanar, or OXIM transistor is illustrated in Fig. 18. As with CDI thin epi is used, 1-2 μ thick, which is of the same electrical type as the substrate. Isolation comes from the p,n junction at the buried layer and the regions of SiO_2 which essentially reach down completely through the epitaxial layer. Again there is a grown base and a diffused emitter.

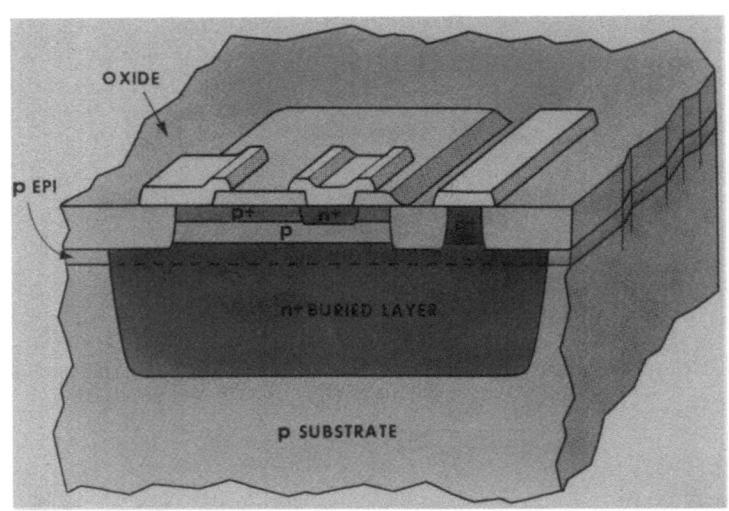

FIGURE 18. A diagram of the OXIM structure.

It is claimed that this structure requires roughly half the area for a given function than does SBC. It is also advantageous to have oxide isolation in that there is a reduction of sidewall capacitance. There are mask simplifications involved as the oxide regions are not sensitive to pinholes in photoresist, and the essentially planar nature of the surface makes metallization, particularly two level metallization, easier to accomplish.

Fully decoded static random access memory chips are being made with this planar technology. Memory cell size is approaching 10 sq. mils/bit, which compares favorably with MOS cell size, and access time is in the range of 60 ns, with 0.5 milliwatt/bit power dissipation.

So we see that there are many possibilities for improving bipolar SIC's, and the people who manage these activities have to choose their paths very carefully between the temptation of adopting the most recent technical advance, and the necessity of having what the customer wants at the right time and at the right price.

MOS Developments

The disadvantages of the standard Al gate MOS technology, namely excessive gate to drain capacitance and high threshold voltage, have already been alluded to, together with a discussion of the Si gate and RMOS technologies which go far toward alleviating the situation. Useful inventions and new processes are still being described which enhance still further the virtues of MOS devices, and three of these will now be briefly discussed.

(i) <u>n-Channel</u>. n-Channel MOS devices are now becoming available from commercial sources [16]. The attraction of these devices over the more usual p-channel structures is the higher mobility of electrons compared to holes. The electron mobility is a factor of 2-3 times greater than that of the holes, with a corresponding improvement in the speed-power product. n-channel devices can be made using conventional Al gate, or RMOS, but today the use of Si gate seems to be the most rapidly expanding technique in use.

It is interesting that some designers have chosen to exploit the improved speed-power product of n-channel devices, not to achieve high speeds, but rather to achieve reasonable speed at very low operating voltages, and in this way to produce circuits that are TTL compatible. Thus Intel has built dynamic recirculating shift registers which operate at 2 mc/s, and a static 1024 bit random access memory which operates at 1 mc/s. Both these devices operate from a single +5 volt power supply, as do TTL circuits.

The processing of these circuits is basically similar to that used for conventional p-channel Si gate. However now the starting material is boron-doped p-type silicon, with n^+ sources and drains. By control of the doping level in the gate region, and by adjusting the gate oxide thickness, threshold voltages of less than 1 volt are achieved, and this means that a low voltage power supply is sufficient to run the whole circuit.

There are other advantages to having low voltages running these devices. The width of depletion regions in reverse-biased junctions increases with the applied voltage, and if these junctions are too close together punch-through can result. This happens when the doping in the silicon at the surface is low, so that the depletion layer spreads far out, and there is a potential difference between the two diffused regions which are close together. Space charge limited current will flow, particularly under the influence of a suitable potential on overlying metal interconnections. With the +5 volt power supply diffusions may be placed close together (0.4 mil or 10 μ), and channels may be short (0.25 mil, or 6.4 μ). Thus tighter and smaller layouts are possible without the expense of tight alignment tolerances. In the example quoted by Intel for the static RAM, a cell size of 17.2 mil^2 is required for the p-channel device, but only 7.9 mil^2 for the n-channel layout.

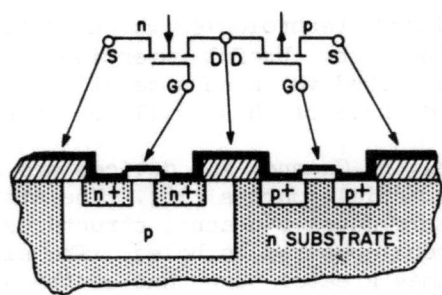

FIGURE 19.   Cross section of COS/MOS inverter-circuit chip [17].

In addition lower voltages reduce parasitic device interactions caused by large voltages appearing in the interconnection metal which can cause high capacitance and leakage from field inversion. Furthermore dynamic memory circuits, which are very useful for saving power and for saving silicon area, depend for their success on having p,n junctions with very low reverse leakage currents - if leakage currents are high then retention time is short and the advantage of the dynamic circuit is lost. Clearly if other things are equal, lower voltages lead to lower reverse-bias currents, and so to less critical processing steps to make adequate junctions.

Thus n-channel devices are being made today and their advantages are likely to lead to their steadily increasing use.

(ii) COS/MOS. One of the advanced forms of MOS circuits which is available today from commercial sources is the complementary MOS device, or COS/MOS, (for complementary symmetry) [17]. The characteristic of these structures is that they integrate both p-channel and n-channel enhancement type MOS transistors on the same monolithic substrate. This is in contrast to the simpler MOS structures which have just p-channel, or just n-channel transistors. For this added circuit flexibility one pays the usual price of increased processing complexity.

In its barest outline a C MOS circuit has the cross section illustrated in Fig. 19. In this structure n-channel and p-channel MOS transistors are connected together. In principle the processing is straightforward. As described by RCA the substrate is n-type with (100) orientation. The first step is to diffuse the lightly doped p-type well in which the n-channel devices will be made. This may be followed by p^+ diffusions for source and drain regions of the p-channel devices, and finally n^+ diffusions for analogous n-channel sources and drains. In addition, as for all MOS devices, there are usually protective diodes which prevent

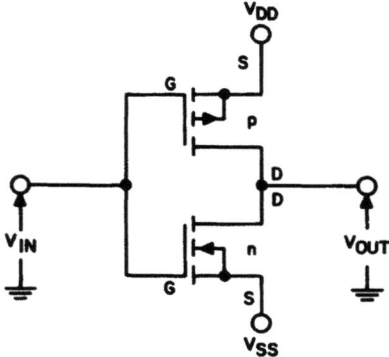

FIGURE 20. Schematic diagram of a COS/MOS inverter circuit [17].

breakdown of the thin gate oxides. Today the gate metal is aluminum. n-channel MOS devices are traditionally more demanding of processing techniques than are p-channel devices. This is because the sign of the voltage applied to the gate drives Na^+ ions to the more critical Si - SiO_2 interface, so that these ions must be rigorously excluded if stable, reliable n-channel devices are to be made. In fact the manufacturers state that a "clean oxide" technology has been developed to eliminate the contamination.

The n-channel units exhibit the higher carrier mobility associated with electrons. They have approximately twice the transconductance of p-channel units with identical geometry. Therefore the matching of a p-channel with an n-channel device requires that a p-channel unit with a given channel length have approximately twice the channel width of the n-channel unit with which it is to be matched. (The channel width is the dimension measured at right angles to the source-to-drain direction.)

A major advantage in certain applications of the C MOS circuits is their very low power consumption when in the quiescent or holding mode. This might, for instance, be very useful for SIC's intended for wrist watch applications. The reason for the low power consumption can be illustrated by considering the basic inverter circuit shown in Fig. 20. When the voltage at the input of the inverter is zero, (logic 0), the gate-to-source voltage of the p-type transistor is equal to the positive supply voltage V_{DD}. Since this is a p-type device and the gate is negative with respect to the source, this device is ON. Now there is a low impedance between V_{DD} and the output, and a high impedance to ground, and the output is therefore close to V_{DD}, (i.e., is logic 1).

When the input is $+V_{DD}$, the source-to-gate voltage of the p-channel transistor is zero so this transistor is off, however at the n-channel transistor the voltage is V_{DD}, but now with the gate positive with respect to source. Since this is an n-channel device it is turned on, with the consequence that the output approaches zero, (logic zero).

The point is that because of the complementary nature of the devices in either logic state, one MOS transistor is on while the other is off. Therefore the quiescent power consumption is very low, being largely determined by the leakage currents of the various back-biased p,n junctions.

Other advantages are claimed for this technology. By paralleling n- and p-channel MOS transistors a transmission gate, or switch, can be made with superior switching characteristics and with improved speed over a single channel MOS device. High noise immunity and wide temperature operating margins are also important.

In fact combining the inverter with the transmission gate allows a full family of digital circuits to be made including NOR and NAND gates, D type flip flops, counters, shift registers, arithmetic blocks, memories and so forth.

It is likely that other more advanced MOS techniques such as Si gate or ion implantation will be applied to this technology, if they have not already been, and it may be expected that despite its increased complexity, the improved performance of C MOS will ensure it a significant part of the SIC market.

(iii) <u>Charge transfer devices</u>. These devices are rather recent in concept and use, but are closely related to MOS devices. They include the "Bucket Brigade" devices described in 1969 [18] and the "Charge Coupled" devices described in 1970 [19]. Both concepts were motivated, at least in part, by a desire to provide fewer and simpler processing steps, and so to allow higher levels of integration. To a degree this expectation has already been brought about.

Most of the discussion will be of CCD's. In these devices information is handled in the form of packets of charge consisting of minority carriers. These packets are moved physically near the surface of the semiconductor under the influence of potential wells created by electrodes on the crystal surface. Figure 21 illustrates the basic idea. The Si - SiO_2 interface is used, and the SiO_2 is used as an insulator. Charge may be stored as minority carriers in one potential well, and it may be passed along as shown in Fig. 21, using a 3-phase electrode array. It is in fact possible to pass charge along with a 2-phase structure provided asymmetrical electrodes are employed.

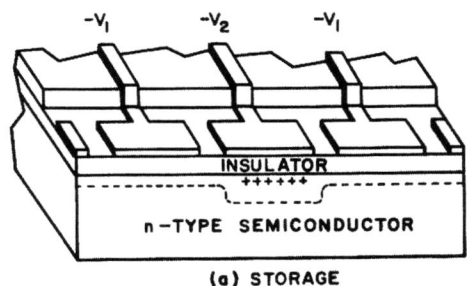

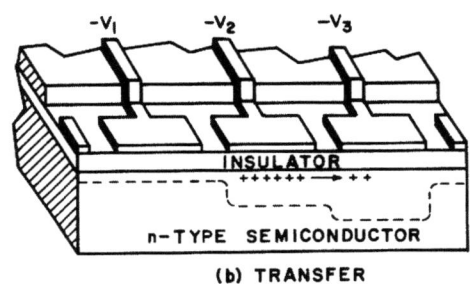

FIGURE 21. Cutaway of a charge-coupled device in (a) the storage condition and (b) the transfer condition [22].

There are several points to make about these devices. First, there is no inherent gain in the charge coupled device, so that degradation of the charge packet may be expected as it passes beneath the electrodes. Second, the minority charge packets are formed in the presence of a reverse-biased electrode, with the consequence that their existence depends on low reverse leakage. In other words the device is a dynamic memory, or dynamic shift register. Finally the device is not just a digital device, but rather analog since the charge packets can vary in magnitude.

In these devices it is important that there should be efficient transfer of charge between the electrodes, which may result in the interelectrode gaps having to be small, perhaps 3 μ. However the essential part of the device is very simple to make - there are no diffusions into the silicon, (although there may be channel stops around the periphery), and one simply has to define a metal pattern on a plane oxide surface. Charge can be introduced into a CCD from an input diode which is connected to the first electrode by means of an inversion region which can be switched on and off by an MOS control gate. Similarly at the output the electrons are gated to an output collector diode.

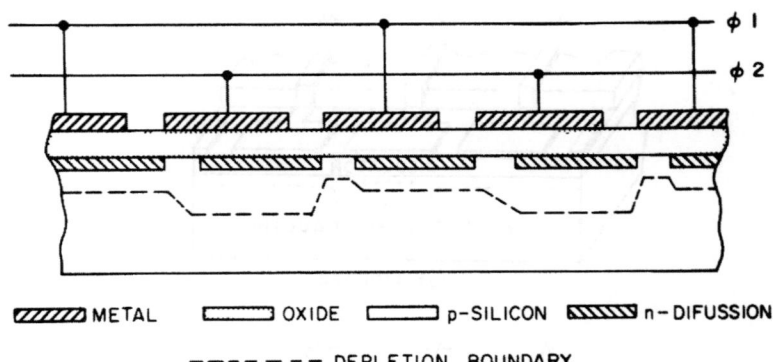

FIGURE 22. MOS Bucket Brigade device [23].

The limitation to these devices is the efficiency with which the charge packet may be transferred from one electrode to the next. At very low speeds the charge may decay by recombination from reverse current, but generally a more serious matter is incomplete transfer at high frequencies. The trouble arises partly because it takes a finite time for the particles to move under the combined effects of drift and diffusion, and partly because charge becomes trapped at surface states and is released from these traps only slowly. There is a smearing of the charge pulses. The effects are minimized by proper electrode design and by the usual care associated with control of surface states. As an example at 10 mc/s, transfer efficiencies of 99.9% per cycle have been achieved. Where necessary it is not hard to include simple MOS transistors to regenerate charge packets in an extended shift register.

Figure 22 illustrates the Bucket Brigade device. The point here is that every cell has a diffused region, although it is not necessary to make contact to these regions. The information is again stored as charge, but now as majority carriers in the diffused regions. Transfer between the diffused regions takes place in more or less normal depletion regions under the metal electrodes. A two-phase system is illustrated. If one phase is turned on and the other off, half the diffusions act as effective sources and the other half as drains, so that excess charge transfers from one to the other. At the next half cycle, each source becomes a drain and vice versa and the excess charges are once more passed along the line. It is sometimes pointed out that these devices could be assembled from discrete devices, whereas the CCD is an entity unto itself and has no equivalent circuit using discrete components.

One of the most spectacular, and possibly most important, uses of these charge transfer devices is as all solid state imaging devices. For this purpose minority charges are generated by light absorbed by the silicon. The quantity generated of course depends on the intensity of the light. Because of the simplicity of fabrication it becomes feasible to think of making an array with sufficient detail that an image can be sensed with reasonable resolution. In fact it seems certain that all solid state TV cameras will be made using CCD's. These will be stable, they will be free of lag and they will require very simple driving circuitry.

ACKNOWLEDGMENTS

The author wishes to acknowledge helpful contributions from Messrs. E. N. Fuls, C. F. Gibbon, J. Simpson, G. E. Smith and P. A. Turner.

REFERENCES

[1] Motorola Series in Solid-State Electronics, prepared by the Engineering Staff, Motorola Inc., Semiconductor Products Division, <u>Analysis and Design of Integrated Circuits,</u> Editors: D. K. Lynn, C. S. Meyer, D. J. Hamilton, McGraw-Hill Book Company, New York (1967).

[2] Motorola Series in Solid State Electronics, prepared by the Engineering Staff, Motorola Inc., Semiconductor Products Division, <u>Integrated Circuits Design Principles and Fabrication,</u> Editors: R. M. Warner, Jr., J. N. Fordemwalt, McGraw-Hill Book Company, New York (1965).

[3] G. L. Schnable and R. S. Keen, "Aluminum Metallization - Advantages and Limitations for Integrated Circuit Applications," Proceedings of the IEEE, Vol. 57, No. 9 (September 1969), pp. 1570-1580; Ibid., L. E. Terry and R. W. Wilson, Metallization Systems for Silicon Integrated Circuits," pp. 1580-1586.

[4] F. M. D'Heurle, "Electromigration and Failure in Electronics: An Introduction," Proceedings of the IEEE, Vol. 59 (1971), p. 1409; J. R. Black, "Electromigration - A Brief Survey and Some Recent Results," IEEE Trans. Electron Devices, Vol. ED-16 (1969) p. 338.

[5] S. S. Hause and R. A. Whitner, "Manufacturing Beam-Lead, Sealed-Junction Monolithic Integrated Circuits," The Western Electric Engineer (December 1967), p. 3-15; M. P. Lepselter, "Beam-Lead Technology," The Bell System Technical Journal, Vol. XLV, No. 2 (February 1966), p. 233.

[6] H. K. Gummel, "Measurement of the Number of Impurities in the Base Layer of a Transistor," Proceedings of the IRE, Vol. 49, No. 4 (April 1961).
[7] B. E. Deal, M. Sklar, A. S. Grove, E. H. Snow, J. Electrochem. Soc. 114, March 1967.
[8] L. L. Vadasz, A. S. Grove, T. A. Rowe, G. E. Moore, "Silicon Gate Technology," IEEE Spectrum (October 1969), p. 28-35.
[9] W. J. Laughton, Electronics, p. 68, April 12, 1971.
[10] Private communications, E. N. Fuls, C. F. Gibbon and R. A. Moline, Bell Laboratories.
[11] W. Shockley and A. Goetzberger, J. Appl. Phys. 31, 1821 (1960).
[12] J. M. Eldridge and D. R. Kerr, "Sodium Ion Drift through Phosphosilicate Glass-SiO_2 Films," J. Electrochem. Soc., Vol. 118, No. 6 (June 1971), p. 986-991.
[13] E. Yon, W. H. Ko and A. B. Kuper, "Sodium Distribution in Thermal Oxide on Silicon by Radiochemical and MOS Analysis," IEEE Transactions on Electron Devices, Vol. ED-13, No. 12, (February 1966) pp. 276-280.
[14] B. T. Murphy, V. J. Glinski, P. A. Gary and R. A. Pedersen, "Collector Diffusion Isolated Integrated Circuits," Proceedings of the IEEE, Vol. 57, No. 9 (September 1969), pp. 1523-1527.
[15] D. Peltzer and B. Herndon, "Isolation Method Shrinks Bipolar Cells for Fast, Dense Memories," Electronics (March 1, 1971), pp. 53-55.
[16] Electronics, May 8, 1972, pp. 106-114.
[17] RCA COS/MOS Integrated Circuits Manual (Technical Series CMS-270) RCA/Solid State Division, Somerville, N. J. (1971).
[18] F. L. J. Sangster and K. Teer, "Bucket-Brigade Electronics - New Possibilities for Delay, Time-Axis Conversion, and Scanning," IEEE J. Solid-State Circuits, Vol. SC-4 (June 1969), pp. 131-136.
[19] W. S. Boyle and G. E. Smith, "Charge-Coupled Semiconductor Devices," Bell System Technical Journal, Vol. 49 (1970) pp. 487-493.
[20] A. S. Grove, Physics and Technology of Semiconductor Devices, John Wiley and Sons, Inc., New York (1967).
[21] Robert H. Crawford, MOSFET in Circuit Design (Metal-Oxide-Semiconductor Field-Effect Transistors for Discrete and Integrated-Circuit Technology), (Texas Instruments Electronics Series), McGraw-Hill Book Company, New York, (1967).
[22] W. S. Boyle, G. E. Smith, "Charge-coupled devices - A New Approach to MIS Device Structures," IEEE Spectrum, Vol. 8, No. 7, (July 1971) pp. 18-27.
[23] Private communication with M. F. Tompsett, Bell Telephone Laboratories.

CHAPTER 22

THE USE OF ELECTRONIC MATERIALS IN COMPUTER MEMORIES

Andrew H. Eschenfelder

IBM Research Laboratory, San Jose, California

PART A. MEMORY TECHNOLOGY CANDIDATES

INTRODUCTION

In this chapter we will discuss materials as used in memory applications. Many different technological candidates have been proposed for this application, and for each technological candidate there are classes of materials that could be used. The properties of materials utilized include magnetic, ferroelectric, semiconducting, superconducting, optical and structural. In fact, the possibilities are too numerous to cover here and it seems more appropriate to discuss the prime candidates in order to develop an understanding of the way in which materials are used for this application, the characteristics that are the most important, and what needs to be done to make further progress. It is appropriate to begin the discussion by considering the function that memory provides in a typical computing system.

THE MEMORY FUNCTION

Figure 1 depicts a typical medium performance computer configuration. Not shown are the input/output devices such as card readers and printers. There is a central processing unit (CPU) wherein the processing of data is accomplished. In addition to the CPU there are five memory units, including the buffer. The data to be processed is entered from cards via the reader, or may be in residence on the system in a disc file facility or may

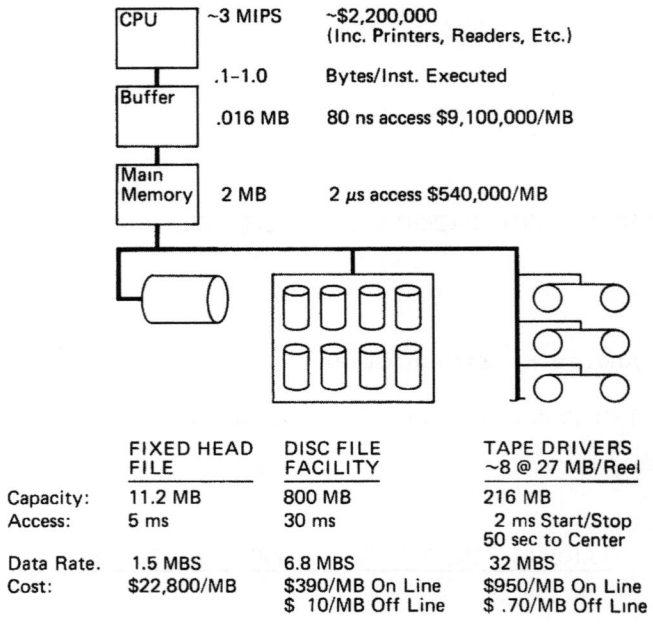

FIGURE 1. Typical computing facility.

be fed in from a terminal or from reels of tape that are mounted on the drives by the operator when called for by the system. The programming system by which the machine is operating is usually stored in the fixed head file. Located there also are the specific instructions for the current job and one or more other jobs that have been accepted by the machine and are ready to go as soon as the facilities are available.

In the figure, for each of the memory units the following relevant information is shown: the data capacity for the device, the access time, the data rate, and the price normalized to a price per megabyte (MB) of data capacity. Of course, these values are constantly changing with time as improvements are made. However, these one-point-in-time values, given in the figure, enable us to understand some points that remain true even though the numbers change. In the case of the disc and tape files, the disc pacs and tape reels can be removed and stored on a shelf. Therefore, there is an on-line price which includes the price of the drive unit and there is an off-line price which is the shelf-storage price of just the disc pac or tape reel. The access time is the delay between a request for data from the device and the receipt of data from the device. Once access has been made to a particular data set, the data can be read from the device at the data rate.

THE USE OF ELECTRONIC MATERIALS IN COMPUTER MEMORIES

Most CPUs are capable of executing from 0.1 to 10 million instructions per second (MIPS). (Some operate slower or faster than this range.) It is clear that data must be transferred to the CPU from memory very rapidly to keep the CPU operating at this rate. Since typical jobs will execute 1-8 instructions per byte of data, the data requirement is of the order of one million bytes per second (MBS). If the specific data needed could be anticipated and lined up ahead of time, the memory devices could keep up with this. However, real problems generally involve selection of blocks of data from sequentially different parts of the file. This is especially true for information systems (e.g. airline reservations) where there is random access to small blocks within a very large data file or for interactive computing via terminals where the data sets may not be large but there are many customers interleaved in time. As a result the access time is generally more important than data rate.

Semiconductor memories can be accessed electronically in less than 100 ns. Core memories are also electronic, but their access time is an order of magnitude slower. However, the price is commensurately less. Even though the core memory costs have been reduced substantially over the years, it still costs over a million dollars to put 2 MB of core on the system. Semiconductor memory has been so expensive that only a small amount could be used. However, by a clever use of 2 MB of core as a main memory and only 16,000 bytes of expensive but fast semiconductor buffer memory it is possible to gain an average high speed, yet modest average cost. This is accomplished by continuously replacing 32 byte blocks of data in the buffer from the core memory in such a way that the CPU finds what it wants in the fast access device 99% of the time. Mattson [1] gives a description of this approach and the methods used to design such a two-level memory hierarchy.

The only way to get sufficient memory on a system (some installations need over a billion bytes on-line) at a reasonable price is to make extensive use of memory devices that utilize mechanical access. In this case the access mechanism is amortized over millions of bytes of data and so the cost per MB is much lower. On the other hand, the mechanical access introduces an access delay in excess of a millisecond. This is the case for the fixed head files, the discs and the tapes. In the case of these devices we see from the Figure there is also a tradeoff of price and access time. In a fixed head file there is a magnetic head for each track of data so that the only delay is due to the rotation time of the drum. However, so many heads are expensive. As a result, the use of fixed head files is limited to data that must be frequently accessed, such as the programs for operating the system. A disc file facility, on the other hand, has one head per recorded surface and this is mechanically moved to the proper track. Thus the cost is lower, but there is an additional delay. The average cost can be further decreased because the disc pacs can be substituted and the data retained on them. This disc pac cost is only $10 per MB. The same principle is carried much further with tape devices where the price of stored reels is only $0.70 per MB. On the other hand, the drives are expensive so that it is not feasible to keep tapes continuously mounted in residence on the system as done with some discs. Tapes are changed for each job, whereas most discs are not. Access to the tapes, even once mounted, is also slower than discs since there is a 2 ms start time and a sequential pass through half the tape on average to get to the desired data.

Thus, on a real system there is a struggle to get enough memory with fast enough access at a reasonable price. Each memory unit has a different tradeoff of these parameters and, in fact, it requires a combination of different memory units to get the best result. The job of the technologist and materials scientist is to find ways of improving the cost/performance beyond that of the memory devices shown either by advances in the magnetic technologies or by developing other technologies which are judged to have potential advantages. We shall list some of the suggested candidates and go on to give some indication of the way in which their technological characteristics are influenced by the physics of the relevant phenomena and the configuration of the materials used. Finally we will try to provide a framework for the relative comparison of the most popular memory configurations.

First, however, we might make some observations:

1. Data storage is a key element of an information processing facility, and the memory technologies must keep pace with the development of the other elements, such as the logic. Just as semiconductor logic is making rapid strides in cost, performance, and miniaturization, so also should memory technology if it is not to hold back the total system.

2. The access method is very dominant in determining performance and cost. High physical density of data storage is a key factor in getting the best performance and cost for a given access method. Therefore, candidates must possess an inherent storage phenomena, a convenient and rapid access method and very high storage density.

3. Magnetics has been the workhorse for twenty years. Many other technologies have challenged magnetics, but none have been successful until the recent introduction of semiconductor memories in the high performance area and the use of photo-images in a few specialized, high-capacity, slow access files. It is not

THE USE OF ELECTRONIC MATERIALS IN COMPUTER MEMORIES

sufficient that a technology work well, but it must be good enough to replace magnetics.

4. The magnetic technology has been continuously improved so that it now represents a complex, highly developed technology with a large investment in design understanding, process technique, and manufacturing plant. Thus magnetics presents a moving target and the challengers will not suceed without a substantial investment. Any new technology must be flexible enough to cover a broad range of application characteristics so as to amortize the required investment. Special purpose candidates are not likely to be successful. Other variations on magnetic and semiconductor technology would appear favored by virtue of the already great technical investment in them.

MEMORY CANDIDATES

In the following several sections we will list the technological candidates which will be discussed more fully in later sections. Here we want to identify storage phenomena, the configurations used, and the access method. For convenience we will categorize them according to access method. These categories will be Magnetic, Optical, and Electronic. In the Magnetic category, we will consider those where the data is sensed by magnetic flux linkage induced signals. Optical technologies will include those where an optical beam is used for access. Of course, both of these ultimately are sensed by the magnetic or optical creation of a voltage or current state in a sensor, but we will separate those where the voltage or current state is intrinsic to the device (e.g. switchable resistors) into the Electronic category.

Magnetic

(i) Cores. The first magnetic configuration used for computer memories and in the past champion for fast, reasonable-cost main memory is the core array. These cores are tiny toroids of ferrite ceramic. They are produced by powder metallurgy and then strung on wires in a two dimensional array so that each toroid is linked by two orthogonal wires. Electrical current passed down the two wires will create a magnetic field that will magnetize the core in either of the two directions around the core. The core has a square hysteresis loop with a firm threshold field so that the field from one wire will not change the state of the core, but the two operating together will. Thus, individual cores in the array can be switched by selecting the appropriate two wires. A third wire is passed through all the cores to act as a sense line. Thus, when an attempt is made to switch a core in a given direction, if the core were already in that state it would not switch and no induced signal would be detected in the sense line. If the core were not in that state, however, it would switch with a resulting sense signal. Thus the wires can be used to write a magnetic state into a pre-selected core and also to read the state. If cores are formed into an N by N array, 2N current drivers are necessary. Therefore, the larger the array, the lower the cost of drivers per bit stored. On the other hand, there are losses in the lines, some spurious signals from unselected cores and electromagnetic delays in propagation down the wires in addition to the loading on the drivers, so there are practical limits to the size of the arrays. Core planes are stacked into three-dimensional arrays, and complete units contain of the order of a million cores.

(ii) <u>Planar films</u>. Planar film arrays were extensively explored and, to some extent, commercially used in an attempt to eliminate the cost of handling the individual core elements and improve the speed of magnetic memories. Thin films of a magnetic alloy, such as FeNi, were deposited on a substrate, and then the orthogonal array of conductors deposited with suitable films of electrical insulation. The operation was analogous to cores, but the magnetic path now passed out of the magnetic material so very thin films were necessary.

(iii) <u>Plated wire</u>. Another variation on this same theme is wire plated with a magnetic coating. In this case, the wire acts as one drive current conductor, and in this case the magnetic path is contained totally within the magnetic material. In order to provide selection of bits, wires are wrapped around a layer of the memory wires. These orthogonal wires can be activated to bias particular sections of the memory wires so they can be switched by the drive current.

(iv) <u>Magnetic head</u>. Costs of magnetic memories can be reduced considerably by eliminating the need for electronic access to each bit location in the memory. One approach to this involves using mechanical motion to provide part of the access and thereby letting one access mechanism serve thousands of bits. Conventional magnetic recording by means of a magnetic head exploits this approach. The magnetic head is used to both write and read magnetic information stored on a continuous surface of magnetic material. Variation in the current in the head winding causes changes in the fringing field seen by the surface, which is moving with respect to the head. The trailing edge of the head field will then leave a pattern recorded in the surface that reflects the pattern of head current changes. Conversely, for reading, as the pattern in the surface passes the reading head, fringing flux changes will be intercepted by the head structure and will induce in the head winding a corresponding pattern of voltage signals which can be detected. The recording surface can be of many different materials and deposited in many different ways. For instance, magnetic alloy surfaces can be deposited by evaporation, plating, or sputtering. On the other hand, the most widely used surfaces have been "painted" surfaces where the paint has fine magnetic γ-Fe_2O_3 particles suspended in a non-magnetic, polymeric base. This process has been well controlled and inexpensive and yields properties similar to an all magnetic surface, provided the particles are much finer than the stored magnetic pattern. Several configurations have been used, including tapes, drums and discs. For tape recording, the magnetic paint is applied to a reel of flexible substrate, such as mylar. A compound head will be used that has one recording gap for each track on the tape. In this case it may be necessary to pass through an entire reel of tape to get to the information desired. The drum removes that problem by using many more tracks that are much shorter. In this case the substrate drum is rigid and can be revolved at high speed for rapid mechanical access. On the other hand, in this configuration so many heads must be used that the device becomes expensive. The disc represents a compromise. Here, too, the substrate is rigid and can be rotated rapidly. However, there is usually only one head per surface and this is mounted on a movable arm so that it can be moved from one track to another as needed. Thus the head cost is greatly reduced compared to the drum, but the time for the head to be moved to the track must be added to the rotational delay before sought information can be read. These several configurations, then, use the same magnetic head recording mechanisms, but provide a tradeoff of cost for access delay.

(v) <u>Bubbles</u>. Other configurations have been proposed to circumvent the delays inherent in the mechanical motion of magnetic head recording while at the same time making use of the concept of shared access mechanisms so as to avoid the high cost of the purely electronic access. One particularly attractive configuration utilizes magnetic bubbles. This was made possible by the discovery that in certain types of magnetic surfaces small domains of reverse magnetization (or magnetic bubbles) can be nucleated, propagated along given tracks and collapsed. In this case, then, instead of stationary magnetic patterns that are passed by a head through the mechanical rotation of the substrate, the magnetic patterns could be propagated along a stationary substrate and passed by a detecting station. Very small bubbles and very rapid propagation are possible.

Optical

Optical storage uses an optical beam for access. That optical beam can be generated in a number of different ways. Furthermore, for access, either the recording medium must be movable so as to position the desired information in the beam or the beam must itself be deflected. The beam must cause some stable, persistent change in the recorded medium, and also must be able to detect that different state for reading of the information stored. The information can be stored discretely in small regions of the medium, or it can be stored in a superimposed fashion in a hologram. We will first mention several different forms of beam sources, then list some of the properties of media that are used for optical storage, and finally say a few words about holographic storage.

(i) _Sources_. If the medium is going to be moved, ordinary optical sources and conventional optics can be used as long as the storage mechanism does not require some special aspect of the light (such as coherence, high intensity, etc.) which would necessitate lasers. On the other hand, it is very desirable to minimize the mechanical motion. Therefore some proposed optical configurations utilize the CRT as a source. The CRT allows a raster of selectable spots on the order of 1000×1000. Such a raster has also been further multiplexed by using an array of lenses so that each CRT raster spot is simultaneously imaged on over 100 blocks of storage which can be separately controlled. Thus, of the order of 10^8 bits could be associated with each CRT without mechanical motion. The light intensity from a CRT passing through one lens is low enough that very sensitive media are required. If a laser source is used to increase the intensity and relax the requirements on the medium, deflection can also be accomplished, but only by a separate, rather expensive deflector. The parts cost of both of these approaches is of the order of $10,000 for 10 MB. The access speed would be largely determined by the deflection time and this time could be less than 50 μs. We can see that this approach might compete with fixed head file technology. However, it would be just as expensive for lower capacities and so will not compete with semiconductor storage for the very small, fast buffers. On the other hand, capacities in the neighborhood of a billion bytes would require hundreds of such units, and so some form of mechanical motion is also involved with large capacity optical stores in order to achieve reasonable costs.

As soon as we introduce mechanical motion similar to that used in magnetic recording, we recognize that similar delays in accessing are implied. If optical stores do not provide faster access than magnetic stores, what possible advantages do they have? The access mechanisms are not intrinsically cheaper, nor are the recording media costs per square inch. On the other hand, (i) optical techniques should allow higher storage densities on the medium and therefore lower cost per bit, (ii) optical deflection within a block not requiring mechanical motion is simple, fast and can be accurately servoed, and (iii) the transducers can be kept away from the storage medium in contrast to the magnetic case where the required proximity for high densities introduces contact and wear problems.

Once we allow mechanical motion, we can consider a variety of configurations, including coated discs and drums, flexible strips or chips that can be stored in cartridges which, in turn, can be moved, and tape which may be in cassettes or on reels. Tradeoffs can be made between the relative amount of optical deflection and mechanical motion. And with a lesser requirement on optical deflection, very inexpensive laser sources can be used that are fabricated in integrated arrays using the techniques of integrated circuitry.

(ii) <u>Media</u>. There are many phenomena which interact with light and which persist so that they can be used for information storage. Some of these phenomena are essentially electronic in which the storage is in reality a small volume of material in which the electrons have been induced to a stable state different from that of the surrounding material. On the other hand, the information can be represented by a structural change in a small region of the recording material. The phenomena may interact with the light quanta directly (such as in the case of photochromics or photoconductivity) or it may be influenced by thermal energy imparted by the light (such as thermomagnetic writing). In addition, the phenomenon may be reversible and therefore admit of easy erasure and rewriting, or it may be reversible within limits and restricted to applications where rewrite is seldom necessary, or it may be irreversible and used for read-only applications such as archives.

The two most popular irreversible phenomena are silver halide and hole-burning in metal films. The great stability of these phenomena enhances their value for read-only archives. Commercial memory systems have been produced using each technology. The IBM 1360 Photodigital Storage System was first installed in 1967 and is described in the Proceedings of the AFIPS Conference [2]. Information is photographically stored on rectangular film chips which are stored in cartridges that could be accessed in five seconds. After the cartridge is retrieved, a particular film chip is automatically picked from the cartridge and the information which is stored at 2×10^6 bits/in^2 is read optically. 2×10^{11} bytes of data are included in the store at the low cost of 10^{-4}¢ per bit. The Unicon System by Precision Instruments has been delivered and operated with the ILLIAC computer. In this case, strips of a polyester 3.5 inches by 31 inches are coated with 200 Å of a metal film and stored in a carrousel from which they can be mechanically retrieved in 6-8 seconds. Writing is accomplished by a laser vaporizing small holes ($3\mu \times 4\mu$) in the metal film. Each strip contains about 200 MB and after being automatically mounted on a rotating drum can be scanned and the pattern of holes read at a rate of .4 MBs per second. Errors are partially corrected through redundancy. With 20% of the capacity used for redundancy, errors have been as low as one in 10^8 bytes. The writing energy of the laser is 300 mw with only 10 mw being used to read. The strip cost is very low - $40 for 200 MB, and the complete system of 10^{11} bytes sells for under $2 M [3].

The partially reversible systems are less attractive. In addition to their limited reversibility they usually have other drawbacks. For example, sensitivity of photochromics is only gained through compromising the stability of the storage. We will not discuss such systems here because of the limitation of time.

There are a number of candidates for reversible optical storage. Magnetooptic storage is electronic in that the information is represented by a small volume where the magnetization has a particular orientation but the material is physically as uniform as possible. Chalcogenide storage is structural since the information is represented by small regions where the crystallographic phase has been transformed. In the magnetooptic case writing is accomplished by irradiating a small area of the recording surface by a laser in the presence of a biasing magnetic field. The bias field is not sufficient to change the magnetic state of the material at the ambient temperature. The laser heats the material so that the coercive force drops below the bias field and a change in magnetization occurs locally. The thermodynamics of the structure is such that this change is quickly frozen in place when the laser radiation is removed. Reading is also accomplished by the laser using either the Faraday or Kerr Effects. We will discuss the magnetooptic case further later. In the chalcogenide case a phase transformation is locally induced by thermal heating from a laser. If the right material is chosen the transformation can be very rapid in one direction. Erasing the material, however, is necessarily slower. The transformed spot has a different reflection characteristic and can be detected optically with good signal to noise. Time does not permit a discussion of this approach, but additional insight can be gained from the paper by Feinleib et al [4].

(iii) <u>Holographic storage</u>. It is possible to utilize the coherency of the laser radiation to achieve advantages in storage. The basic concept involves storing an array of information over a given area so that instead of each bit of information occupying a correspondingly small part of the storage area, it is spread over the whole storage area. Consider for instance an array of 100×100 opaque and transparent areas in some random arrangement that represents stored bits. A photographic recording of the two-dimensional diffraction pattern produced by the coherent illumination of this grid is a hologram. The hologram consists of the superposition over the entire area of the diffraction pattern produced by each of the bits. Now if this hologram is itself illuminated coherently, a replica of the original 100×100 array is produced and this can be imaged on a 100×100 array of detectors. Thus if the hologram is stored rather than the array, several important advantages are gained. In the first place, a small defect in the storage medium that might otherwise represent the loss of one bit now only decreases the signal for each bit slightly and no bit is lost. Secondly, slight displacements of the hologram will not cause the pattern of information to miss the corresponding detector array. This system thus offers tolerances that can be important. Hill [5] presents an analysis of the use of optical holographs in memories. Lin and Beauchamp [6] have used the techniques with thermoplastic recording materials and Rajchman [7] with magnetooptic materials. None of these approaches has so far been commercially successful.

Electronic

(i) **Bipolar transistor**. Bipolar transistor circuits would be a natural candidate for memories in view of their extensive development for logic applications. Indeed the use of bipolar transistor circuits for relatively small memories has been well established and documented in the literature. On the other hand, these circuits are relatively expensive and not economical for memories of large capacities. This has led to the pursuit of several other electronic devices as alternatives.

(ii) **FET**. The Field Effect Transistor circuits (FET) offer advantages over the bipolar transistor circuits in that they can be fabricated with a much simpler process and therefore can be made with a higher yield and at substantially lower cost. At the same time they perform with a reasonable speed. Memories using FETs larger than 100,000 bytes have been operated at one microsecond. For true random access as in core memories it is necessary to use multiple transistors per bit. Two of the devices comprise a conventional flip-flop, and the others are for reading, writing, noise isolation, etc. In order to reduce the associated circuitry, it is possible to use FETs in a shift register configuration. In a shift register the FETs are arranged in a loop, and the information is inserted at one point of the loop. After that it is circulated around the loop and withdrawn at the end. Such a configuration greatly decreases the input and output circuitry and also reduces the number of interdevice connections that have to be fabricated on the chip. These factors have the result that the cost per bit is decreased. It is still true, however, as in the case of bipolar transistor circuits, the information is volatile. That means that when the power is turned off the information is lost.

(iii) <u>Charge coupled</u>. Charge coupled devices are fabricated with an even simpler process than FET circuits. On the other hand, they are utilized in a shift register configuration and the information is still volatile. Silicon (Si) is coated with about a one thousand angstrom layer of oxide and then a pattern of electrodes superimposed. Potentials on the electrodes cause potential wells to be created at the surface of the semi-conductor in which charge can be stored. By sequencing the applied voltage on successive electrodes, the stored charge can be shifted along the string. The processing is very simple requiring no diffusions and only one oxidation plus two levels of metalization. It is still necessary to use three electrodes per bit since the charge can spill in either direction.

(iv) <u>MNOS</u>. MNOS devices have been designed in an effort to eliminate the volatility of the other electronic devices. An MNOS device is an FET with an additional layer of silicon nitride (Si_3N_4) on top of the oxide. There are traps at or near the oxide-nitride interface which will store charge for over 1000 hours and thereby perpetuate the state of the device. The state of the MNOS device is set or reset by voltage pulses of 20-25 volts of appropriate polarity. While these devices have memory, writing speeds are much slower than regular FETs.

(v) <u>Switchable resistances</u>. Switchable resistances are two terminal devices that have two electrodes separated by either a thin film or a sandwich of materials. These devices have two stable resistance states that are switched when either current or voltage exceeds a given threshold. They have memory in that the given resistance states will persist even when the voltage is removed from the device.

There are three particular switchable resistance devices that have received a considerable amount of attention. One involves chalcogenide glasses which are based primarily on tellurium compounds. These materials are ordinarily fairly high resistivity. However, when the electric field across a film of such material exceeds a certain value, a phase transition will occur in narrow (1μ) filaments of that material so that the effective resistance of the device drops by several orders of magnitude. The high resistance state can be reestablished by pulsing the device with a sufficiently high current. Thus the fundamental mechanism of this device is a thermal electric metallurgical instability and this presents certain problems. In the first place, because it depends on atomic rearrangements, it is not a fast device. In addition, there are potential aging mechanisms because of these atomic rearrangements. Thus in spite of the fact that this is a very simple device, it would be preferable to have such devices where the two resistance states are due to electronic rearrangements rather than atomic.

Heterojunctions between zinc selenide (ZnSe) and germanium (Ge) or between gallium phosphide (GaP) and silicon (Si) provide such devices. In this case a sufficient voltage applied in the proper direction will empty electronic states associated with the heterojunction and reduce the resistance. On the other hand, if a sufficiently high current pulse is applied in the opposite direction, this will saturate the electronic states, reestablishing a high resistance which will persist. Another device which operates in a similar fashion is a niobium oxide device where a layer of Nb_2O_5 is sandwiched between niobium (Nb) and bismuth (Bi). One complication of these devices is that pulses of opposite polarity are required to switch between the states. This requires more complicated associated circuitry.

All of these switchable resistance devices are used in memories by locating them at the intersections of an array of x lines and y lines down which the appropriate switching and sensing signals are transmitted. Since the devices have a nonzero conductivity, even in their high resistance state, each one must be coupled by some current limiting device such as a diode in the case of the chalcogenide devices, and a more complicated Schottky diode with Zener breakdown in the case of the niobium oxide and heterojunction devices that require pulses of opposite polarity for set and reset. Thus these arrays of devices are not as simple as might be first supposed.

PART B. PHYSICS OF MEMORY TECHNOLOGIES

INTRODUCTION

Now that we have recognized many technological candidates for the memory function, we want to consider their relative merits. The cost of the devices depends on the simplicity of fabrication and, in the cases of integrated devices, on the density with which they can be crowded in one fabricated unit. The performance characteristics depend on the physics of their

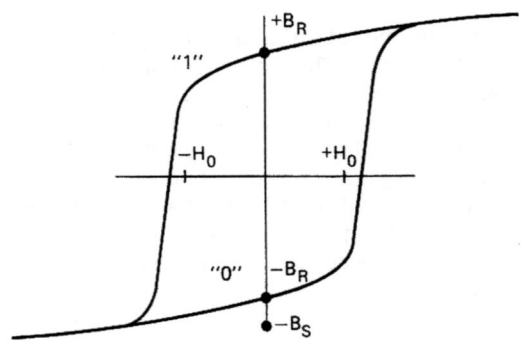

FIGURE 2. Hysteris loop.

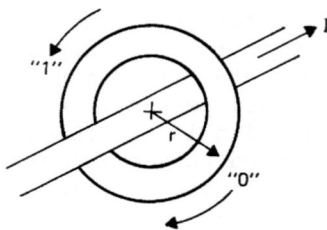

FIGURE 3. Magnetic core.

operation. We could spend a complete lecture on the physics of any one device. In order to get some feeling for the major ones, therefore, we will bring out only some of the key physical concepts that relate to performance and cite literature where the students can find more adequate treatments.

CORES, FILMS, AND PLATED WIRE

The same physics underlies the operation of cores, films, and plated wire. Let us first consider a small toroid made from a material with a magnetic hysterisis loop as in Fig. 2. This is a so-called square-loop material where $B_R \simeq B_S$ as long as the demagnetizing field, H, does not exceed H_0. When H does exceed H_0 the material switches almost completely from $-B_R$ to B_S and relaxes to B_R when the applied field is removed. Thus, the material is essentially bistable with only two magnetic states. For a toroid of this material the two states are for essentially complete magnetization in either direction around the toroid as shown in Fig. 3. One sense of magnetization can be used to designate an information bit "1" and the other a "0".

Now, if a current, I, is passed along a wire intersecting the core, it will generate a field in the core of

$$H = \frac{I}{5r} \qquad (1)$$

where r is the radius of the core. If this field is large enough to switch the core, a sense wire passed through the core will have a voltage generated in it of

$$V = 10^{-8} \frac{d\varphi}{dt} \qquad (2)$$

where $\frac{d\varphi}{dt}$ is the time rate of change of flux which is proportional to the cross-sectional area and B_R.

When the core switches, it has been found experimentally that the larger the field exceeds the threshold field H_0, the faster the switching. The following empirical relationship has been found:

$$\frac{1}{\tau} = \alpha_s (H - H_0) \qquad (3)$$

where τ is the observed time to reverse magnetization and α is a constant depending on the particular material.

There are two switching mechanisms in magnetic materials. In "rotational" switching the magnetization of large regions of the material rotates into the new direction uniformly. This is a very fast process but also requires high energy and therefore relatively high threshold fields. In "domain-wall" switching only a small region initially reverses with a thin "domain-wall" separating this region from the unswitched portion. Then the domain-wall quickly propagates to complete the switching. For this mode, since only the material within the domain-wall has its magnetization in high energy states, the energy needed to nucleate switching is less than for "rotational" switching but the speed is slower. Both of these switching mechanisms have been understood and in both cases the switching speed would be proportional to the applied field as in Eq. (3). However, for each mechanism a different α_s and different threshold field H_0 would apply. Domain walls can be inhibited and rotation enhanced by a variety of techniques, especially in thin films, if the most rapid switching is desired.

Since each of the drive lines in a core array intersects many cores besides the one to be switched, it cannot carry a

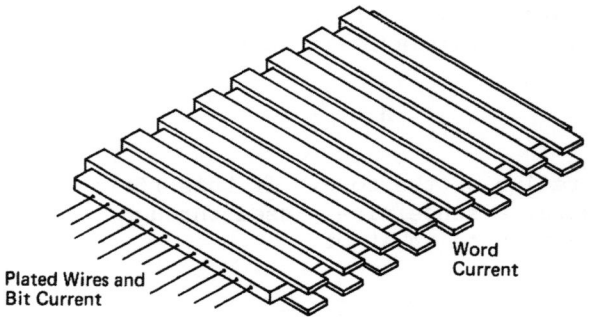

FIGURE 4. Plated wire configuration.

current greater than I_0 corresponding to the threshold field H_0. Then the core to be switched is selected by an x wire that carries I_0 and a y wire that carries I_0. The net switching current is therefore $2I_0$. The applied field is twice the threshold field and the switching speed would be $\frac{1}{\tau} = \alpha H_0 = \frac{\alpha I_0}{5r}$.

We can now make several observation: To minimize the requirements on the current drivers a small I_0 is desired. Decreasing I_0 yields slower switching times and this can only be compensated by making the cores as small as possible and by using materials that have the largest possible switching coefficient, α_s. As the cores are made smaller the read signals would also suffer since the cross-sectional area is smaller and so would be φ. On the other hand, the concomitant increase in switching speed sustains $d\varphi/dt$ and helps to preserve the read signal. The read signal should not be too high, anyway, because it appears as a back-voltage on the drive lines and cannot be allowed to seriously perturb the drivers.

Core technology has progressively developed new compositions to optimize α, H_0 and loop squareness as well as the techniques to fabricate, test, and automatically wire extremely small cores. Greifer [8] has written an extensive article on ferrite core memories with many references.

Planar films and plated wire are variants on the same theme and also require small geometries for high performance. In plated wire memories, the central conductor acts as one drive conductor and the other is passed orthogonally around a plane of wires as in Fig. 4. This configuration admits of simple assembly and yet a very small and compact configuration can be achieved. The switching is slightly different in that the word lines are used to rotate the magnetization slightly out of a pure circumferential direction around the wire so that it will reverse under the influence of the bit current. If the bit current is not on, the magnetization must not switch but the deflection from circumferential will cause a sufficient voltage in the bit line that it can be used to detect the sense of the magnetization under the word line. Thus, by pulsing the word line and sensing each bit line, all of the bits in one word can be read non-destructively. This contrasts with cores where the reading process reverses the bit and the information must be restored. On the other hand, in plated wires it is more difficult to achieve the needed properties of the wire and to compensate for the "open-flux" nature of the word lines. Since these word lines are not enclosed by magnetic material higher currents are necessary and adjacent bits are not isolated. Of course, wires involve platable magnetic alloys rather than the insulating oxides used for cores. However, the same basic physical phenomena are used to explain both technologies. Plated wire technology is reviewed in an article by Mathias and Fedde [9]. England [10] discusses the application of plated wire for memories with NDRO (non-destructive readout).

Planar films represent a third attempt to gain ease of fabrication by sequentially depositing in a planar configuration a ground plane, magnetic film, and then two orthogonal layers of strip lines with the required intervening insulating layers. Very rapid switching has been achieved by designing the films in such a way that domain-wall motion is suppressed and the high-speed rotational process operates. Once again, the same basic physics is involved with a different emphasis. While some planar film memories have been shipped with high performance machines they have not been able to compete on a cost/performance basis with cores and plated wires. The article by Pugh et al [11] illustrates the factors that must be considered in designing a magnetic film memory. A special variation that enhances flux closure is the use of coupled films. Lee [12] describes a coupled magnetic film array and gives additional references.

In the case of alloys used in wires and planar arrays, the switching conditions are controlled by imparting special magnetic anisotropies to the materials during fabrication. Thus, the magnetization is more tightly bound to particularly selected directions. One example is the stress-induced anisotropy used in plated wires which is described in a paper by Lutes [13].

MAGNETIC HEAD

We have already discussed how variations in magnetic head current will leave a pattern of magnetization reversals in a magnetic medium on a moving surface and also how that head can pick up the fringing fields from such a pattern and hence the

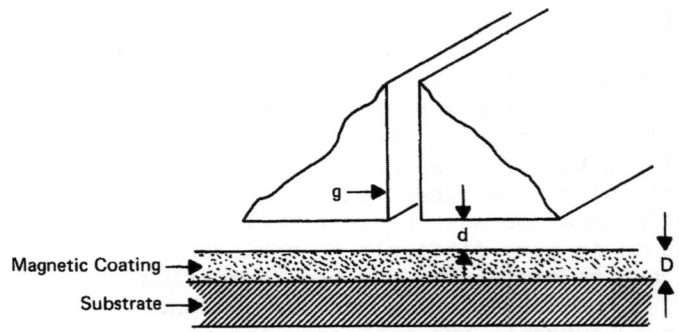

FIGURE 5. Bit size limitations.

pattern read by sensing the voltage signals in the windings of the head.

Very rapid oscillations in the head current compared to the relative motion of the head and surface should result in a very fine pattern of small bits. Two things keep those bits from being arbitrarily small. In the first place, if the bits are smaller than the head gap, then they will not be differentiable on reading. More than one bit would be in the gap at the same time and the head would see the total field from several bits. In addition, each bit sees a demagnetizing field from the adjoining material that is polarized in a different sense. Thus, neighboring regions of opposite polarity will wash each other out unless the geometry and magnetic properties of the medium are in the correct relationship.

There have been many complex calculations of the exact magnetic patterns and read signals resulting therefrom. One example is the paper by Potter and Schmulian [14] and they give additional references. Middleton [15] gives a very simple expression that conveys the essential tradeoffs. He gives for the smallest bits a formula

$$\ell = 2\sqrt{g^2 + (d+a)(d+a+D)} \qquad (4)$$

where the small letters represent dimensions shown in Fig. 5. $a = \frac{2MD}{H_c}$ where M is the magnetization of the medium and H_c its coercive force. It can be seen that if g and d are very small and $M \gg H_c$, then a is half the limiting bit length. In this case, however, the bit is appreciably larger than the medium thickness because the coercive force is not high enough to withstand the demagnetizing field unless the bit is long and thin.

To get smaller stable bit patterns, the coercive force is increased by going to intrinsically high coercive force materials or by increasing the coercive force by structure, such as small acicular particles in an organic matrix. $2M/H_c$ cannot be made too small however. If H_c is too large it becomes too hard to write and if M becomes too small there will be no appreciable read signal. Typical values of H_c are 200 oersted and $2M/H_c \simeq .3-1.0$.

To get the smallest bits, it is seen that all three geometrical factors have to be kept small. In practice they have been approximately equal, that is $g \simeq d \simeq D$. In 1956, these spacings were about 20 μ. By 1962, they had been reduced to 5 μ and now they are in the neighborhood of 1 μ. With 1 μ, 5 μ bits are typical.

It has become increasingly difficult to reduce the characteristic dimensions. The medium must be not only very thin but uniformly so and smooth so that a head can be flown steadily at a 1 μ spacing. Any small variations will cause fluctuations in response and may also lead to intermittent contact and wear of the surfaces. In tapes contact recording is used and since relative velocities are fairly low, wear is not a serious problem. In the high speed discs, however, the heads are mounted in aerodynamic sliders which fly over the surface balanced between the aerodynamic forces of the air film and spring loading. At atmospheric pressure, 1 μ spacing is only about 16 times the mean free path of air. It is therefore clear that as closer spacings are sought the physics of the slider bearing will become more complicated.

As bit densities go up, the data rate increases and the cost/bit decreases since more bits are stored on a comparable mechanism. Achievable track densities do not depend on magnetics but on mechanics. It is true that the bit must be large enough to give a signal many times the electronic noise from the associated circuits and the noise from non-uniformities in the surface. As the bits get shorter this puts a constraint on the width of the bit, the head, and the track. The factors that have to be taken into account in considering signal to noise rates are illustrated in the paper by Mallinson [16]. These will be of increasing importance as yet higher densities are achieved. However, so far, the track densities have been limited by problems of mechanical registration between the head and the track.

Thus, we see that advances in magnetic recording using heads require achievement of smaller dimensions in surfaces, heads and relative spacings as well as uniformity and smoothness of the surface. For thinner, smoother surfaces, evaporation, plating and sputtering fabrication processes will be more seriously considered for the high performance devices although they could

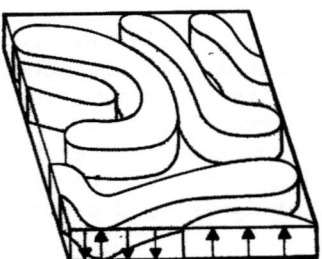

FIGURE 6(a). Snake pattern.

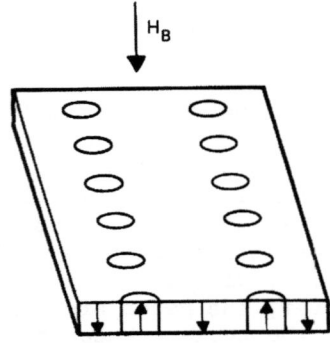

FIGURE 6(b). Bubble pattern.

not equal the low cost of conventional γ-Fe_2O_3 coated surfaces. Plating and evaporation will shift attention to the magnetic alloys with a different set of corrosion and wear characteristics. As in the core and film memories, the problems of achieving advances are not as much involved with the discovery of new magnetic materials or new physics as with fabricating demanding configurations required by the physics.

BUBBLES

There are many materials that have a uniaxial anisotropy. Most often this is due to a uniaxial symmetry in the crystal structure. These materials can also be deposited in films so that their easy axis of magnetization is normal to the surface. The magnetization of such films with no applied field will break up into "snake" patterns of opposite polarity as in Fig. 6(a). The domains of opposite polarity are separated by conventional 180° domain walls. The snake pattern represents a balance between the magnetostatic energy due to the exposed magnetization and the energy of the domain walls. If a bias field of .3 to .6 of $4\pi M_S$ is applied (this can be 20-50 oersteds) the snake pattern will collapse to individual, unconnected, cylindrical domains or

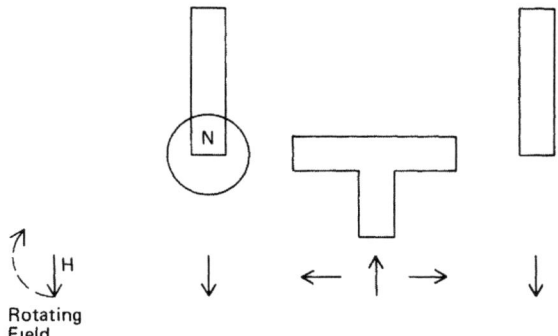

FIGURE 7. T-Bar bubble circuit.

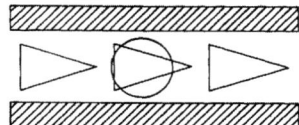

FIGURE 8. Angel fish bubble circuit.

"bubbles" in a sea of reverse magnetization as depicted in Fig. 6(b). These bubbles will have a spontaneous diameter of from 1 to 300 μ depending on the material. Their diameter will vary slightly with the applied bias field and they are stable over a diameter range of about 3:1. The bubbles naturally repel each other with a spacing of approximately four diameters. The bubbles are mobile and can be moved at a speed in the neighborhood of one bubble diameter per 100 ns.

One method of channeling bubbles has been the T-bar pattern and rotating field as illustrated in Fig. 7. If a magnetic field of about 10 oersteds is rotated in a clockwise direction the bubble will proceed along the T-bar pattern from left to right.

Its position on the T-bar pattern will be at an extremity of a bar segment to which the applied field would repel it. The field orientation corresponding to each bubble position is indicated.

Another channeling method is the "angelfish" pattern of Fig. 8. In this case, the bias field is alternated in magnitude. As the field causes the bubble to expand the bubble will overlap the next arrow head and as the bubble contracts, with a decrease in the field, it will slide off the tip of the trailing head. Thus, propagation occurs to the right in the diagram.

Bubble generators have been invented such that with each rotation of the field a new bubble peels off the generator and enters the T-bar pattern. Bubble splitters, that double bubbles, and bubble diverters have also been invented. Bubbles can be sensed at the end or branch of a shift register pattern by a number of effects: magnetoresistance, Hall Effect, magnetooptic, or inductively. For instance, a magnetoresistance line placed at a point of the T-bar will experience an oscillating resistance as the driving field rotates. When a bubble passes that point on the T-bar there will be an additional increase in resistance due to the magnetic bubble presence.

The optimum thickness of the film is given by Thiele [17]:

$$T = \frac{4\sigma_w}{4\pi M_S^2} = \frac{32\sqrt{AK}}{4\pi M_S^2} \quad (5)$$

where σ_w is the wall energy per unit area, A is the exchange energy constant, and K the anisotropy energy coefficient. The bubble diameter is then about twice this optimum thickness.

A variety of materials have been explored. Magnetoplumbite ($PbFe_{12}O_{19}$) has a high $4\pi M_S \simeq 4000$ oe and gives very small bubbles ($\sim 1\,\mu$). It, however, is slow. Rare earth orthoferrites ($RE\text{-}FeO_3$) have smaller M_S and larger K so the bubbles are larger. They operate twenty times as fast as magnetoplumbite but the bubbles are very large ($\sim 25\,\mu$) and thus the density would be much lower. Garnets ($RE_3Fe_5O_{12}$) represent a happy medium in that small bubbles ($\sim 3\,\mu$) can be sustained and moved with good mobility. Epitaxial films of these materials have been grown with defect densities of $10/cm^2$ or less. With small bubbles this should allow usable bubble densities of .4 Mbits/cm^2. The data rate should approach 1 MB for the best materials.

Rapid progress is being made in materials and device invention. However, many uncertainties exist including the effect of inhomogeneities (such as strains and defects) in the films that might be introduced during the array fabrication. Temperature variations could be troublesome because of thermal sensitivities (e.g., differential expansion) built into the array or because of intrinsic thermal variations in the bubble phenomenon (e.g., bubble diameters have been observed to vary by up to 13%/degree C). The arrays certainly will be susceptible to stray magnetic fields.

There are many articles on the various aspects of bubble devices in the literature, especially the IEEE Transactions on Magnetics. The article by Bobeck, et al [18] and several articles that immediately follow it in the same volume discuss material

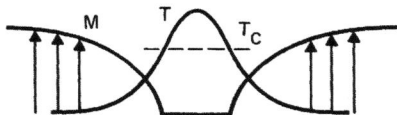

FIGURE 9(a). Impressed thermal and magnetic pattern.

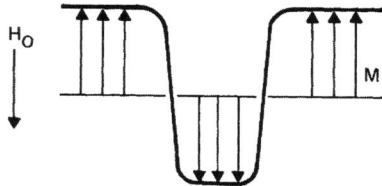

FIGURE 9(b). Final magnetic orientation.

---

requirements, theory of bubble domains, bubble generation and propagation.

MAGNETOOPTICS

We will now consider how a laser beam can be used to write and read magnetic domains in a film of uniaxial magnetic material with the easy axis normal to the surface. This can also be done with films with in-plane magnetization but it is slightly more complicated and does not allow as high a bit density.

The writing depends on thermal heating from the laser beam either close to or above the Curie temperature. A focused laser will produce an energy absorption which, in competition with thermal conduction losses (both in the film and to the substrate), will yield a temperature profile such as shown in Fig. 9(a). Wherever the temperature is above ambient the magnetization will be reduced and it will go to zero where the temperature exceeds the Curie temperature. The resulting magnetization pattern during irradiation is also depicted in Fig. 9(a). If a bias field tending to reverse the magnetization is present, when the laser beam is turned off, the magnetization will be reversed wherever the heating had been enough to decrease the coercive force below the bias field, H_0. Now the magnetization is distributed as in Fig. 9(b). Thus, a small region of reverse magnetization is written onto the film.

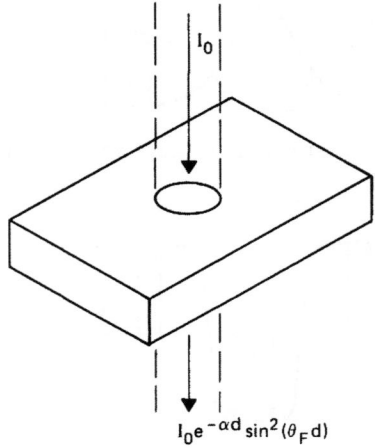

FIGURE 10. Faraday rotational sensing.

If the laser is turned on again with reduced intensity and the beam is polarized, the direction of polarization will be rotated by the magnetization according to the Faraday effect. If the analyzer is set so there is a minimum output for the unwritten magnetization state, the light intensity seen by the analyzer where the beam traverses the written spot will be

$$I = I_o \, e^{-\alpha d} \, \sin^2(\theta_F d) \qquad (6)$$

where I_0 is the initial intensity, α the absorption coefficient, d the thickness of the film, and θ_F the Faraday coefficient. This is shown in Fig. 10. If the film is too thin there will not be much rotation and so no distinguishable signal. If the film is too thick the beam will be so absorbed that there is hardly any signal. The optimum thickness is

$$d_{opt} = \frac{1}{\theta_F} \tan^{-1} \frac{2\theta_F}{\alpha} \qquad (7)$$

which for $2\theta_F \ll \alpha$ yields $d_{opt} = \frac{2}{\alpha}$. \qquad (8)

Using this thickness of material $\frac{I}{I_0} = \eta = \left[\frac{2\theta_F}{e\alpha}\right]^2$.

An appropriate figure of merit for reading is then

$$F_{read} = \frac{\theta_F}{\alpha} \tag{9}$$

Mn Bi has a large $\theta_F \simeq 10^4$ cm^{-1} and $\alpha \simeq 4\times10^5$ cm^{-1} at 850 nm. So that $d_{opt} = 500$ Å, $F_{read} = .03$ and $\eta \simeq 3\times10^{-4}$. Thus the detector sees .03% of the original laser beam and the rotation of polarization is $\sim 2.5°$.

The writing energy is given by

$$E_W = 4.186 \, M\bar{C}_v \Delta T = 4.186 \, \rho A \bar{C}_v d \Delta T \tag{10}$$

where ρ is the density, A the area of the spot, $\bar{C}_v$ the specific heat averaged over the temperature excursion, and ΔT the temperature rise. This must be corrected for losses due to reflection, conduction away, etc.

Neglecting conduction

$$E_W = (I_0 - I - I_0 R)t = I_0 t(1 - R - e^{-\alpha d}) \tag{11}$$

where R is the reflection coefficient and t is the time duration of the laser pulse. For

$$d_{opt} = \frac{2}{\alpha} \text{ and } R = 0; \; I_0 = \frac{9.67 \, \rho A C_v \Delta T}{\alpha t}$$

an appropriate figure of merit for writing is

$$F_{write} = \frac{\alpha}{\rho C_v \Delta T} \tag{12}$$

(where ΔT now is the temperature rise necessary to exceed the Curie temperature) since this factor for the material gives an inverse measure of the incident energy density pulse necessary to write. Once again, for MnBi $C_v = .046$ cal/gm deg, $\rho = 9$ g/cm^3 $\Delta T = 340°$ and $F_{write} = 2480$ which corresponds to 40 mW to write a 10 μ^2 bit in 10 ns neglecting reflection. The experimental values are higher because of losses due to reflection and conduction.

These considerations show that the following are criteria for good materials:

    a. Can be deposited with a uniaxial anisotropy normal to the surface with K/M large enough that it won't spontaneously form domains of reverse magnetization.

b. A Curie temperature, T_c, above room temperature but not too far above so that the temperature doesn't have to be raised too high for writing. On the other hand, T_c must be high enough that the magnetization state is not disturbed by the small energies used for reading.

c. Large absorption, α, so that thin films can be used and writing energy minimized.

d. Large θ_F/α so a good reading signal results.

e. Large enough heat conductivity that the film will cool rapidly when the writing laser is turned off, but not any higher than necessary so that writing energy is not dissipated. The thermal recovery time should be approximately equal to the write pulse duration.

f. The film structure must be stable within the temperature excursion.

MnBi is a very good material with respect to all criteria but b. and f. The Curie temperature of 360°C is higher than desired and there is an accompanying phase transition at 360° and a decomposition that sets in around 445°. Chen and Aagard [19] have shown that repeated thermal writing gradually converts the written spot to the high temperature phase. Feldtkeller [20] reports that this occurs when the integrated time above 360° approaches 10^4 seconds. Fortunately, the high temperature phase has good properties and Feldtkeller [20] has reported that when titanium is added to MnBi to form $Mn_{.8}Ti_{.2}Bi$ the high temperature phase can be quenched in and is stable. The α still is high (3×10^5 cm^{-1}) as in MnBi. θ_F is only slightly diminished (3×10^3 cm^{-1}) and the Curie temperature is reduced to 125°C. Thus, $\theta_F/\alpha = .01, \eta = .5 \times 10^{-4}$ and the rotation is 1°. In addition, there are many other materials which could have properties suitable for thermal writing and magnetooptic reading.

Of course, lasers must be available to match the needs of the materials. GaAs lasers have been batch-fabricated into arrays by Marinace [21] with 10-20 lasers per unit. Wieder and Werlich [22] have studied the characteristics of these lasers and they each emit 250 mw with 750 mw input. These lasers are operated at low temperatures to gain these characteristics but before long it is expected adequate lasers operating at room temperature will be developed. Lasing at room temperature has already been demonstrated [23].

ELECTRONIC

Integrated circuit devices are covered in other lectures, but some remarks are appropriate with respect to their use in memory.

In order to compete with the other technologies for extensive memory applications, other than fast buffers, the electronic technologies must be low in cost. This implies the simplest fabrication process to reduce the cost of the chip and maximum density of bits on the chip to get lowest possible cost/bit. Unless the fabrication is much more complicated a technology that requires fewer devices per bit will be preferable. Also, for this reason, shift register configurations may be chosen to reduce read/write/noise isolation etc. circuitry even though speed is sacrificed.

MOSFET circuits offer a fairly simple fabrication process, requiring no built in junctions but only an oxide layer and electrodes. Charge coupled devices are even simpler. They involve no diffusions, no penetrations of the oxide layer, and no complementary devices but merely repetitive electrodes on top of the oxide. However, they are very slow and have not yet been extensively explored. In order to be successful, the material must be such that the charges have a sufficiently long lifetime (>1 sec have been reported) and the charge transfer efficiency from electrode to electrode must be very high (99.9% for 1/3 of the charge to survive 1000 shifts). This charge transfer efficiency will depend on charge recombination processes in the potential wells and close proximity of adjacent wells.

In both MOSFET and charge coupled devices, bits (which contain more than one device) could be spaced on about 25 μ centers in a 25 mm square active chip area to get $\sim 10^4$ bits/chip. Such densities could bring the cost per bit down to 0.1¢, over 2 orders of magnitude reduction from recent prices. This would also be lower than projected core and plated wire costs. There is obviously a real opportunity and challenge to reach these goals. The review by Matick [24] contains a substantial discussion of the electronic memory candidates, including Charge Coupled and Switchable Resistance.

A different sort of electronic technology, which should be mentioned even though it's possible practical use is questionable and would occur only after extensive further investigation and progress, is Josephson junction technology. Josephson predicted unusual properties for thin junctions between superconducting films, and most of these properties have been experimentally confirmed. The physics is very interesting and the devices could be used for both computer logic and memory as well as other applications. The advantages of the technology are extremely high speed and low power consumption. Matisoo [26] provides an excellent introduction to this subject.

SUMMARY

We have discussed a number of technologies. All of them have very interesting physics, both of the underlying phenomena and also of the particular device configurations. This is particularly true of the newer technologies like bubbles, magnetooptics, and unconventional electronic devices. A wide variety of materials are available for each of the technologies and the properties peculiar to each of these materials need to be understood. Perhaps the greatest challenge for all of them is the fabrication of uniform, defect-free multilayer film structures with high production yield to get low cost. Thus, greater attention is needed to the physics and chemistry of the deposition processes including: evaporation, sputtering, plating, painting, vapor growth, and photolithography.

PART C. MEMORY TECHNOLOGY COMPARISONS

COMPARISON FRAMEWORK

Let us begin our discussion of the comparison of memory technologies by considering Fig. 11, which gives a framework against which technologies can be placed for comparison.

Figure 11 is a plot of the access time of some commercially available storage products against the total storage capacity in bytes of those products. Representative selling prices and price per kilobyte are also indicated. It is seen that, in general, the technologies allow a range of products of increasing capacity but it is necessary to sacrifice access time to get the larger capacities at a reasonable price. The faster technologies are also the most expensive so that mere replication of the faster technologies to get increased capacity is not satisfactory.

Semiconductor memories are presently the fastest but also the most expensive. Access times of less than 100 ns can readily be obtained, but the cost has been in the neighborhood of $3,000 per KB. Thus, a relatively small amount of semiconductor memory has been used in order to keep the total price reasonable. This semiconductor memory has been used to interface the high speed CPU with the rest of the lower cost memory hierarchy. As costs continue to improve, the amount used in a given system is increasing.

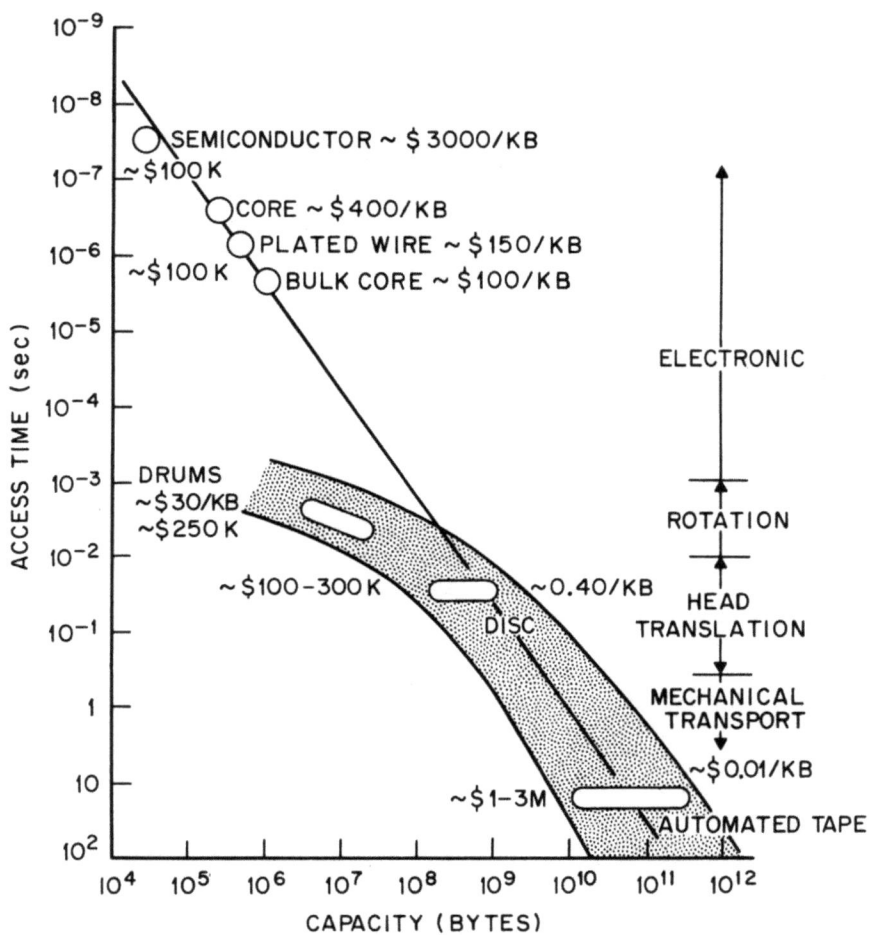

FIGURE 11. Random access memory.

Cores and films with access completely by electronic means are the fastest magnetic devices. Fast core memories have access times of several hundred ns and cost about $400 per KB, in sizes of several hundred KBs. Core memories have been made larger as shown by the bulk core point. Since the required electronics increase less than linearly with the number of cores, the price per KB drops somewhat to around $100 per KB. But in order to have a memory of reasonable cost, it is still necessary to limit the size to several MBs and this size implies added circuit delays so that the access time increases to several microseconds. Plated wire memories of several hundred KBs have also been shipped and compete with cores at a price in the neighborhood of $150 per KB.

To get increased capacity at a reasonable cost, one uses magnetic recording which involves some mechanical access and this necessarily slows down the access. The rotating medium of disc and drum products implies access times of milliseconds since the carrier of the medium cannot be rotated faster than 100 RPS unless it is very small and, therefore, of limited capacity. In order to keep the cost of disc products reasonable, it is necessary to have a minimum number of magnetic heads since these are a major cost item. Thus, in a disc store, it is usual to use only one head per surface and to move that rapidly from track to track. This head translation implies a further access delay of tens of milliseconds. By so doing, however, the cost can be kept well under $1 per KB for capacities of several hundred MBs. These capacities can be achieved for about the price of a 1 MB bulk core memory. Going to cartridges of magnetic strips or tapes or some other configuration that involves an additional physical transport introduces access dealys of the order of one or more seconds depending on the total capacity and the distance over which the transport takes place. On the other hand, the price is once again reduced by a couple orders of magnitude so that now memories of several hundred billion bytes are economically feasible. Thus, we see that of the several forms of memory technology that are available all fall close to the same access time/capacity tradeoff line. There is presently a gap between core and disc products. As we try to speed up the magnetic recording products by using more heads in order to reduce the head translation time, the price increases very rapidly. Thus, in the case of drums, where all head motion has been eliminated by going to one head per track, the price per KB has increased by one hundredfold and it is necessary to cut the capacity substantially in order to keep the price of the product reasonable. Only by finding new techniques to make cheap multiple batch fabricated heads will we be able to keep to the line in the several millisecond access range with inductive magnetic recording.

As time progresses, all of these technologies are being improved so that the access time/capacity tradeoff line has gradually moved to the right. That movement has clearly been slowing down but nevertheless a new technology must make a substantial penetration to the right to justify the development and production tooling investment.

THE USE OF ELECTRONIC MATERIALS IN COMPUTER MEMORIES

OBSERVATIONS ON STATUS AND PROSPECTS

With regard to the status and prospects of the several technologies, we can make the following observations:

1. Semiconductors have already invaded the main memory market. Just one example is the shipment by IBM of bipolar main memories in System 370 in the Models 135 and 145. Increases in level of integration beyond 128 bits per chip can be expected with concomitant reductions in price while still keeping access times in the 100 ns range.

2. MOSFET memories have a substantial advantage over the bipolar in simplicity of fabrication and density of the chip. This, in turn, yields cost advantages. The level of integration seems to be going up by about a factor of 2 each year. Present chips have over 1000 bits and operate in a 300 to 400 ns range. They are approaching $100 per KB which makes them competitive with cores and it is probable that prices under $10 per KB will be achieved later in the decade.

3. Ferrite cores have been the dominant technology except for small capacity high performance memories. With the onrush of semiconductors, core prices are being dropped in order to remain competitive, but they do not have the potential to reach the price goals that have been forecast for MOSFET.

4. Plated wire is available and being used currently. The plated wire has not had enough advantages to supplant cores except where its read - only capability has been significant. This technology has about the same price/performance in the larger sizes as core memories.

5. Semiconductors can also be used for read only memories. ROM chips of greater than 8000 bits have been announced in both bipolar and MOS. The MOS chips provide a 500 ns access performance at about $100 per KB which is cost competitive with plated wire and is improving. The bipolar ROM can be ten times faster at only twice the cost but it also consumes ten times the power.

6. MOSFET will apparently gradually replace cores and plated films as the levels of integration continue to increase and the costs decrease.

7. Magnetic recording is still the winner for large capacity random access storage. The IBM 3330 provides 4000 BPI, 200 TPI, 30 ms seek time, and this is achieved through reduction of the head-surface spacing to about one micron. It is reasonable to expect a factor of two technological improvement in each of these

parameters with further development of the technology. A factor of three is possible, but represents a very challenging target. Thus, the density would improve by a factor of four to ten with commensurate reductions in cost. A storage bit density of 4×10^6 bits per square inch certainly seems like a reasonable potential. The major limitations are as follows:

    a. Close spacing, which must be accomplished for high density but at the same time must be managed so that wear of the head and recording media is avoided.

    b. Track density. The tracks must be wide enough for an adequate signal to noise ratio. But even to reach those limits will require the fabrication of narrower heads and sophisticated servo techniques in order to achieve precise positioning of the heads on the tracks.

    c. Seek time. Improvements in seek time involve optimization of both the rotational latency and also the positioning of the head including settling time after the rapid track access. Both of these factors could be improved by using multiple heads, but this will only be economically possible if new techniques are found to fabricate inexpensive batch fabricated heads.

    8. Beam addressable file technologies offer some real advantages, but do not avoid all the problems limiting magnetic recording. While it is possible to design fairly small memories that would have complete optical access, for the larger capacities mechanical motion will still be necessary. There need be no mechanical motion of the head since this can be avoided by batch-fabricated injection lasers with auxiliary beam deflection, but rotation of the disk or some other mechanical motion of the medium appears to be required. Thus, the latency can be reduced, but not obviously eliminated.

    The major advantage in beam addressable file is in track spacing. Because of the integrated lasers, beam deflection and optical servoing, two orders of magnitude in track spacing beyond what can be achieved in magnetic recording appear reasonable. This would raise the storage densities to greater than 10^8 BPI^2. The increased storage densities would be accompanied by a concomitant reduction in cost per MB and raise appreciably the storage capacities of these fast memories. A second advantage of the laser is the potential low cost of arrays which would allow multiplexing of access paths and thereby faster access times when the extra cost is warranted.

    Another substantial advantage of the laser technology is that it is not necessary to fly the lasers as close to the surface as in magnetic recording in order to get the high densities and thus the wear problem is eliminated.

At present, the magnetooptic approach to beam addressable file has a lot more momentum than the amorphous semiconductor. The MnBi with titanium additions has very good properties and there are prospects of finding even better materials. The amorphous semiconductors require somewhat higher energies but give better signals. It is not at all clear at this point, however, whether appropriate properties will be achieved and if the material and phenomenon will be stable. Rapid progress is being made on the lasers which are necessary whichever material succeeds. Low temperature lasers are already available and operation at room temperature has been announced. Development in lasers is proceeding rapidly.

9. Low-cost, large capacity automated tape files will become more generally available. Progress is being stimulated by early models already available such as the Ampex Terrabit Memory (TBM). It was delivered to the first customer in September of 1971. It is made up as follows:

a. There are modules that have two 10.5 inch reels of 2 inch tape, each reel containing 5.6×10^9 bytes. Storage density is $.6 \times 10^6$ BPI made up of 7,500 BPI and 80 TPI. Each reel is then equivalent to approximately 125 standard reels of magnetic tape recorded at 1,600 BPI.

b. The total file has from 1 to 32 such modules and ranges in price from 1 to 3 million dollars. A full 32 modules is equal to 3.6×10^{11} bytes or 10^4 reels of standard tape.

c. The tape can be moved at 1,000 inches per second for rapid search and provides an average access time of 15 seconds. When the record is reached, a rotary head sweeps across the tape at 800 inches per second while the tape slowly moves by at 5 inches per second. Experience has shown that the uncorrectable errors in reading this file are less than three for every 10^{11} bytes. Thus, it is apparent that very large capacity stores can be provided at prices of one cent per KB. Grumman has also shipped a mass tape memory and we can expect to see many others.

10. The future of bubbles is not yet clear. Very rapid progress is being made on both materials and novel configurations. If 100 ns per shift can be obtained, this would give a data rate of 10 MBs and an average access delay for a 512 bit loop of 25 μs. A price potential of 10^{-3} cents per bit has been projected and if achieved, this would make a very attractive technology. On the other hand, it is unlikely that memories of any significant size will be available before 1975.

SUMMARY

We can summarize the situation with respect to memory technologies as follows:

1. The data base requirements are continuing to grow. There is a demand for ever more on-line storage.

2. In any given file, some information must be accessed frequently while other parts of the file are referred to in larger blocks but less frequently.

3. Processors are getting faster and, therefore, need data available at higher rates.

4. No memory technology is projected to satisfy the total requirement for speed, capacity and reasonable cost.

5. To meet the requirements, it will be necessary to continue to use a hierarchy of storage devices. Semiconductors will be used for high speed and will increasingly take over main memory from cores, plated films, etc. Disc memories will continue and, for this purpose, improvements in cost performance of magnetic recording will occur. It is yet unclear to what extent beam addressable file technologies will be successful and have sufficient advantages to replace magnetic recording. Many new varieties of automated, high capacity tape files can be expected. Beam addressable technologies may well contribute in this area also.

6. The most rapid technological strides are foreseen for the semiconductor memory technologies. With the dramatic decrease in costs, it will be possible to economically include much more of this type of memory in systems. This will decrease the frequency with which the slower parts of the memory hierarchy have to be accessed. Nevertheless, it is not anticipated that it would decrease so much that it could interface directly to the large capacity automated tape files and squeeze out the intervening technologies like discs.

7. The major dark horse in the memory technology area is bubbles. Rapid progress is being made and the full potential of that technology remains to be discovered.

This section is necessarily very brief. The literature contains more extensive comparisons of memory technologies. One example is the recent article by Matick [24], although he emphasizes technologies for the larger capacity memories. The IEEE Transactions on Magnetics contain an abundance of good articles on all phases of memory technologies, including some nonmagnetic. The interested student is urged to scan the Tables of Contents of those volumes in order to comprehend the variety and extent of information available. The IEEE Transactions on Magnetics are published quarterly by the Institute of Electrical and Electronics Engineers, Inc., 345 East 47 Street, New York, New York, 10017.

REFERENCES

[1] R. L. Mattson, "Evaluation of Multilevel Memories," IEEE Trans, MAG-7, 814 (1971).
[2] J. D. Kuehler and H. R. Kerby, "A Photodigital Mass Storage," AFIPS Conf. Proc. 29, 735 (1966) Spartan, Washington, D.C.
[3] H. Tate, "Trillion Bit Laser Memory System," (Precision Instrument Co.). Material presented at IEEE Magnetics Meeting, March 23, 1972, Stanford University, Stanford, California.
[4] J. Feinleib, J. deNeufville, S. C. Moss, and S. R. Ovshinsky, "Rapid Reversible Light-Induced Crystallization of Amorphous Semiconductors," App. Phys. Ltrs. 18, 254 (1971).
[5] B. Hill, "Some Aspects of a Large Capacity Holographic Memory," Applied Optics 11, 182 (1972).
[6] L. H. Lin and H. L. Beauchamp, "Write-Read Erase in Situ Optical Memory Using Thermoplastic Holograms," Applied Optics 9, 2088 (1970).
[7] J. A. Rajchman, "Promise of Optical Memories," Jrnl. App. Phys. 41, 1376, (1970).
[8] A. P. Greifer, "Ferrite Memory Materials," IEEE Trans, MAG-5, 774 (1969).
[9] J. S. Mathias and G. A. Fedde, "Plated Wire Technology; A Critical Review," IEEE Trans MAG-5, 728 (1969).
[10] W. A. England, "Plated Wire Memory for Military and Space Application," Computer Design, 83 (1972).
[11] E. W. Pugh, V. T. Shahan, and W. T. Siegle, "Device and Array Design for a 120 Nanosecond Magnetic Film Main Memory," IBM Jrnl. R&D, 11, 169 (1967).
[12] F. S. Lee, "A High Density Coupled Magnetic Film Memory Array," IEEE Trans, MAG-7, 868 (1971).
[13] O. S. Lutes, "Magnetic Stress Anisotropy Field in Plated Cylindrical Permalloy Fields," IEEE Trans, MAG-7, 861 (1971).
[14] R. I. Potter and R. J. Schmulian, "Self-Consistently Computed Magnetization Patterns in Thin Magnetic Recording Media," IEEE Trans, MAG-7, 873 (1971).
[15] B. K. Middleton, "The Dependence of Recording Characteristics of Thin Metal Tapes on their Magnetic Properties and on the Replay Head," IEEE Trans, MAG-2, 225 (1966).

[16] J. C. Mallinson, "Maximum S/N Ratio of a Tape Recorder," IEEE Trans, MAG-5, 182 (1969).

[17] A. A. Thiele, "Theory of the Static Stability of Cylindrical Domains in Uniaxial Platelets," Jrnl. App. Phys. $\underline{44}$, 1139 (1970).

[18] A. H. Bobeck, R. F. Fischer, A. J. Perneski, J. P. Remeika, and L. G. vanVitert, "Application of Orthoferrites to Domain-Wall Devices," IEEE Trans, MAG-5, 544 (1969).

[19] D. Chen and R. L. Aagard, "MnBi Films: High-Temperature Phase Properties and Curie-Point Writing Characteristics," Jrnl. App. Phy. $\underline{41}$, 2530 (1970).

[20] E. Feldtkeller, "MnBi Films and their Suitability for Magnetooptic Memories," IEEE Trans, MAG-8 (1972).

[21] J. C. Marinace, "Experimental Fabrication of One-Dimensional GaAs Laser Arrays," IBM Jrnl. R&D $\underline{15}$, 258 (1971).

[22] H. Wieder and H. Werlich, "Characteristics of GaAs Laser Arrays Designed for Beam Addressable Memories," IBM Jrnl. R&D $\underline{15}$, 272 (1971).

[23] I. Hayashi, M. B. Parrish, P. W. Foy, and S. Sumski, "Junction Lasers which Operate Continuously at Room Temperature", App. Phys. Ltrs. $\underline{17}$, 109 (1970).

[24] R. E. Matick, "Review of Current Proposed Technologies for Mass Storage Systems," Proc of IEEE $\underline{60}$, 266 (1972).

[25] J. Matisoo, "Josephson-Type Superconductive Tunnel Junctions and Applications," IEEE Trans MAG-5, 848 (1969).

INDEX

Absorption, optical, 18, 202, 231, 310
  coefficient, 199
  direct transitions, 206
  edge, 17, 210
  excitons, 56-58, 214-217
  free electrons, 224-226
  impurity, 217-224
  indirect transitions, 206
  interband transitions, 203-212
  intraband transitions, 214
  lattice, 226-229
  phonon processes, 218
  reststrahlen, 227
  transition probability, 201
Acoustooptic effect, 272-273
Ag_3AsS_3, 271
Al_2O_3:Cr, 261
AlAs, 71, 509, 513
Alkali halides, 221, 227, 323
Alloys, magnetic, 407-446
Alnico, 438-441
AlP, 71, 509, 513
AlSb, 71, 509, 513
Amorphous semiconductors, 149-165
Anisotropy, magnetic, 387-390
Antiferromagnetic interaction, 372, 397
Antiferromagnetism, 393-399
AsS_3, 151
Auger processes, 53, 233
Avalanche effects (in p-n junctions), 49, 51, 87

Barium sodium niobate, 270, 272, 273, 281
BeO, 509, 513
Bipolar silicon integrated circuits, 564-575, 591-594
Birefringence, 249, 267, 279, 327, 357
Bloch walls, 390, 410
BN, 509, 513
Boltzmann transport equation, 96
Bonding in solids, 505-515
Bragg-Williams model, 483
Bridgman method, 495-496
Brillouin zone, 4
Bubble memory materials, 548-555, 554-557

CaF_2:Er, 264-265
Carrier concentration, 26-34, 43
Carrier mobility, 34-38, 71-73
Carrier transport, 34-38
Cathodochromic displays, 323
$CaWO_4$:Nd, 263
Cd chalcogenides, diffusion, 135
$CdCr_2Se_4$, 171
CdS, 509, 513
CdSe, 509, 513
CdTe, 509, 513
  energy gap, 16
Center of inversion, 241
Charge compensation, 263
Chemical bonds in solids, 505-515

Chemical potential, 43
Chemical vapor deposition, 498-500
Circulator, 472
Coherence length, 249
Cohesive energy, 512-514
Compounds (see Semiconductor Compounds)
Computer memories (see Memories)
Conduction band, 16,24,27,70
Configurational entropy, 516, 528-529
Constitutional supercooling, 491,542-545
CoO, 398
Coordination, 506-511
Covalent bonds, 506-510
Crystal growth, 281,479-502, 541-555,566-567
  Bridgman method, 495-496
  Czochralski method, 281-283, 495-498
  floating-zone, 498
  flux growth, 500-501,553-554
  impurity control, 490-492
  interface structure, 482-487
  liquid phase epitaxy, 541-555
  morphology, 487-490
  methods, 492-502
  rate of, 479-487
  Stockbarger, 495-496
  theory of, 479-487
  vapor phase epitaxy, 62,500, 566-567
  Verneuil method, 498-499
Crystal momentum, 5
Crystal structures, 1
CuBr, 509,513
CuCl, 509,513
$CuCr_2O_4$, 403
CuI, 509,513
Cu_2O, 216
Curie temperature, 278,384, 397,401
Curie-Weiss law, 385
Cyclotron resonance, 20
Czochralski method, 281-283, 495-498

Defects, 515-526
  concentration, 516-520
  interactions, 520-526
Density of states, 27
Dielectric constant, 200,507
Diffusion, 127-143,524
Diffusion-hardening alloys, 437-441
Direct-gap semiconductors, 15, 17,19,54,70
Disorder,
  Frenkel, 517
  Schottky, 516
Dispersion, 200,248,268
Displays (see Optical displays)
Distribution coefficient, 490
Domains, 390-393
Dye lasers, 285-303

Effective mass, 15,71
Effective mass approximation, 19
Einstein relation, 38
Electroluminescence, 54-63, 336-339
Electroluminescent phosphors, 335
Electromigration, 569-571
Electronegativity, 505
Electrooptic effect, 245, 268-270,272
Electrooptic materials, 272
Energy bands, 5-26,70,155
Energy gap, 16,70,507-512
  homopolar, 508
  ionic, 508
Epitaxy,
  heteroepitaxy, 66,542, 552-554,559
  liquid phase, 541-555
  vapor phase, 62,500,566-567
EuO, 180,398
EuS, 398
EuSe, 398
EuTe, 398
Exchange interaction, 371-382
Excitons, 56-58,214-217
Extrinsic semiconductors, 30

Index

F-centers, 221,323
Fe, 417
Fe-Co alloys, 431
Fe-Ni alloys, 422-431
FeO, 398
Fermi level, 27,32,43
Ferrimagnetism, 399-403,452-461
Ferrites, 403,408,451-475,461, 462-469
Ferromagnetic interactions, 372,383
Ferromagnetic resonance, 387-390, 463
Ferromagnetism, 383-393
Fibers (see Optical fiber waveguides)
Films,
  bubble memory, 548-555
  metallic, 536-541,568-571
  optical, 555-561
  semiconductor, 542-547, 566-567
  techniques, 535-561,566-571
Floating-zone method, 498
Fluorescence, 285
Flux growth, 500-501,553-554
Franck-Condon principle, 203
Free carriers, 55
Free electron energy, 8
Free electrons, absorption, 224-226
Free energy, 516,527-531
Frenkel disorder, 517

GaAs,
  absorption, 227
  diffusion in, 128-133
  effective mass, 19,71
  electrical applications, 83-87,91
  energy bands, 7
  energy gap, 16,30,69,509
  epitaxial films, 542-546,62
  excitons, 216
  heat of formation, 513
  "hot" electrons, 101
  impurities in, 521-523,60, 78,23
  light emission, 60
  mobility, 69
  relaxation times, 114,123
  scattering, 94,71
GaAs-GaP, 73,61
GaN, 60
GaP,
  acoustooptic effect, 273
  effective mass, 71
  electroluminescence, 57-63, 337,524-526
  energy gap, 71,509
  epitaxial films, 546-547
  excitons, 216
  heat of formation, 513
  light emission, 57-63,337, 524-526
  mobility, 72
Garnets, 262,403,459-461, 549-555,624
  bubble memory devices, 549-555,622-625
  liquid phase epitaxial growth, 549-555
  magnetic properties, 459-461
  optical circuits, 559-561
  optically pumped lasers, 262
  structure, 459-461
GaSb, 71,111,113,125,509,513
  "hot" electrons, 113,125
  relaxation times, 123,111
Gas discharge displays, 339-343
Gate materials, 579-583
Ge,
  absorption, 205,210,212,222, 225,226
  donors, 24
  energy bands, 6,10,15,16
  energy gap, 30,69,509
  impurities, 524-525
  mobility, 69
  relaxation times, 110,113, 119,122
Gettering, 583
  junction hardening, 583-587
  stabilizing of MOS devices, 587-591
Glasses, 149-165,307-316
  AsS_3, 151
  electrical properties, 157-165

Glasses (Cont'd.)
  energy bands, 155-157
  formation of, 149-150,313
  Ge-As-Te, 151
  memory, 158-164,614-615
  mobilities in, 154
  optical, 310-314,316
  optical imaging, 165
  optical losses in, 310-312,316
  optical properties, 154,165
  oxide, 150,151
  Se, 151
  semiconducting properties, 150-155
  Si-As-Te, 151
  switching, 158-164
Gunn effect, 86,91,109
Gyrator, 472

Heat of formation, 505,512-513
Heteroepitaxy, 66,552,559,552-554
Heterojunctions, 615
HIO_3, 270,272,273
"Hot" electrons, 84,89-104, 107-126
Hysteresis, 409

Image light valve projection displays, 348-366
Imperfections, 515-526
InAs, 7,16,71,123,509,513
  energy bands, 7
  energy gap, 16,71
Indirect-gap semiconductors, 15,17
Injection lasers, 41,63-67
InP, 71,91,101,509,513
InSb, 509,513
InSb,
  diffusion, 128
  effective mass, 19,71
  energy bands, 8
  energy gap, 16,71
  "hot" electrons, 99
  photoconductivity, 231
Integrated circuits, 536,563-601
  bipolar, 564-575,591-594
  fabrication, 564-601
Internal reflection, 308
Interstials, 517

Intervalley scattering, 109
Intrinsic semiconductor, 27
Invar, 429-431
Iodic acid, 270
Ionic bonds, 506-510
Ionic character, 505-512
Ionicity, 505-512
Ionization energy, 31,55
Ion-pairing, 524-525
Isoelectronic impurities, 57
Isolator, 471

Josephson junction, 629
Junction hardening, 583-587

KH_2PO_4 (KDP), 242,251,253,272, 357-364
Kramers-Kronig relation, 200, 507

Laser displays, 346
Lasers,
  dye, 285-303
  injection, 41,63-67
  mode-locking, 291-295, 302-303
  optically pumped, 261-266
  Raman, 257
  semiconductor, 19,63-67
  tunable, 254,287-291
Law of Mass Action, 518-526
Lead chalcogenides, 82
Lead molybdate, 270,272,273
Light emitting diodes (LED's), 60-63,336-339
$LiIO_3$, 270
$LiNbO_3$, 253,255,256,270,272, 273,277-281
  optical damage, 281
  stoichiometry, 277-281
Liquid crystals, 322,326-334
Liquid phase epitaxy, 541-555
$LiTaO_3$, 270,272,273
Lithium formate, 272

Magnetic alloys, 407-446
  hard magnetic materials, 432-445
  losses, 409

Index

Magnetic alloys (Cont'd.)
  soft magnetic materials, 408-432
Magnetic bubble devices, 548-555,609-610,622-625, 635-636
Magnetic bubble materials, 548-555,622-625
Magnetic compounds, 398, 402-405
Magnetic ions, 371-382
Magnetic memories, 607-609, 630-637
  bubbles, 609-610,622-625, 635-636
  cores, 607-608,616-619, 631,633
  magnetic head, 608-609, 619-622,632-634
  physics of, 616-625
  planar films, 608,616-619, 631
  plated wire, 608,616-619, 633
Magnetic moments, 371
Magnetic semiconductors, 169-196
  donors and acceptors in, 176-182
  electronic energy levels in, 170-185
  ferromagnetic, 170-176
  transport properties, 185
Magnetic spirals, 405
Magnetism, 169-196,371-405, 407-446,462-469
Magnetooptic materials, 612, 625-628
Magnetooptic memories, 625-628,635
Magnons, 385-387
Memories, 603-637
  comparison of, 630-637
  glasses, 158-164,614-615
  Josephson junction, 629
  magnetic, 607-610,616-625, 630-637
  magnetic bubble devices, 548-555,622-625,635-636
  magnetooptic, 625-628

  optical, 610-613,634-635
  semiconductor, 582,584,594, 629-637,613-615
  superconducting, 629
  systems requirements, 603-607
Metal films, 536-541,568-571
Metal-oxide-semiconductor (MOS) transistors, 575-591, 594-601
  charge-transfer devices, 598-601
  COS/MOS, 596-598
  gate materials, 579-583
  n-channel, 595-596
  stabilizing of, 587-591
Microwave devices, 84-87,469-475
Miller's rule, 244,267,275
MnBi, 628
MnO, 398
MnS, 398
MnS_2, 398
MnSe, 398
$MnSe_2$, 398
$MnTe_2$, 398
Mobility, 34-38,71-73
Mode-locking, 291-295,302-303
Mode in optical fiber waveguides, 308
Morphology, crystal, 487-490

Néel temperature, 394,397
Ni-Fe alloys, 422-431
NiO, 398
Nonlinear optical materials, 265-283
Nonlinear optics, 239-258, 265-283
  acoustooptic effect, 272-273
  electrooptic effect, 245, 265,268-270,272
  parametric oscillation, 253-256
  Pockels effect, 245,357
  second harmonic generation, 247-252,265,267-268,271, 280
  stimulated Raman effect, 256-258
  susceptibility, 239-243,265, 268,275

Nonradiative processes, 52
Nonstoichiometry, 277-281, 517-520

Optical absorption (see Absorption)
Optical circuits, 556-561
Optical communications systems, 307, 555
Optical damage, 281
Optical display materials, 317-366
Optical displays, 317-366
  active, 330
  cathodochromic, 323
  electroluminescent phosphor, 335
  gas discharge, 339
  laser, 346
  light emitting diode (LED), 336
  liquid crystal, 327, 331
  particle orientation, 325
  passive, 322
  photochromic, 323
  projection, 343
  semiconductor glasses, 165
  upconversion, 337
  Xerographic, 324
Optical fiber waveguides, 307-316
  fiber drawing, 314
  losses in, 316
  physical principles, 308-310
Optical films, 555-561
Optical memories, 610-613, 634-635
  holographic storage, 612-613
  materials, 611-612
  sources, 610-611
Optical properties of solids, 54-67, 79-83, 199-235, 239-258, 261-283, 507
  absorption, 18, 202-231, 310
  electroluminescence, 54-63, 336-339
  injection lasers, 63-67
  nonlinear, 239-258, 265-283
  optical constants, 199-202
  optically-pumped lasers, 261, 266
  photoconduction, 229-232
  photoemission, 231-235
  semiconductors, 79-83, 54-67
Optically-pumped lasers, 261-266
Order-disorder hardening alloys, 441
Orthogonalized plane waves, 13

Pair spectra, 58
Parametric oscillation (optical), 253-256
Particle orientation display, 325
PbS, 82
PbSe, 82
PbTe, 82
Permalloys, 423-429
Phase matching, 249, 267, 279
Phonon scattering of electrons, 35, 71, 92, 108
Phosphorescence, 285
Phosphors, 319, 335
Photocells, 80-83
Photochromic displays, 323
Photoconductivity, 229-231
Photoemission, 231-235
Photoluminescence, 56, 58
Picosecond pulses, 291-302
Piezoelectricity, 273
p-n junctions, 41-52
Pockels effect, 245, 357
Poisson equation, 44
Polarization, 239-247
Precipitation hardening alloys, 441-443
Projection displays, 343-366
Proustite, 271
Pseudopotentials, 8, 14, 513
Pyroelectricity, 273

Quartz, 272
Quasi-Fermi level, 46

Radiative recombination, 54-59

Index

Raman lasers, 257
Rayleigh scattering, 311
RbI, 216
Reciprocal lattice, 3
Recombination, 18,50,52-59
  Auger processes, 53
  excitons, 56-58
  non-radiative processes, 52
  radiative processes, 54-59
Rectifier, ideal, 48,50
Refractive index, 200,248,267
Relaxation times in semiconductors, 107-126
Resonance, ferromagnetic, 387-390,463
Retrograde solid solubility, 491
Ruby, 261

Schottky disorder, 516
Se, 151
Second harmonic generation, 247-252,265,267-268, 271,280
Semiconductor compounds,
  II-VI compounds, 2,74,79,133
  III-V compounds, 2,55-67, 69-74,78,84-87,128
  IV-VI compounds, 74,143
  ternaries, 76
Semiconductor lasers, 19, 63-67
Semiconductor memories, 582, 584,594,613-615,629-637
  bipolar transistor, 613
  charge-coupled, 614-629
  MNOS, 614
  MOSFET, 613-614,629-633
  switchable resistance, 614-615
Semiconductor (see also Semiconductor compounds)
  amorphous, 149-165
  bonding, 2
  carrier concentration, 26-34
  crystal structures, 1-4
  density of states, 27
  diffusion, 127-147
  diodes, 50
  direct-gap, 15,17,19,54,70
  donors in, 22-26,31,54
  doping of, 26-34
  electroluminescence, 54-63
  energy bands, 5-26,70
  extrinsic, 30
  films, 542-547,566-567
  Group IV, 2
  hole concentration, 28
  indirect gap, 15
  injection lasers, 53-67
  intrinsic, 27
  lasers, 53-67
  magnetic, 169-196
  memories, 582,584,594,613-615,629-637
  microwave devices, 84-87
  mobility, 34-38,71-73
  optical properties, 17,54-67, 79-83,199-235
  photocells, 79
  photoluminescence, 56,58
  p-n junctions, 41-52
  recombination, 50,52
  relaxation times, 107-126
  structure, 506-511
  transistors, 51
Si,
  absorption, 222,226,229
  bipolar integrated circuits, 564-574,591-594
  energy bands, 6,10,15,16
  energy gap, 30,69
  epitaxy, 500,566-567
  impurities, 22,520-522
  integrated circuits, 563-601
  metal-oxide-semiconductor (see MOS)
  mobility, 69
  MOS transistors, 575-591, 594-601
  photoemission, 234
  relaxation times, 120,121
SiC, 76
Silicon-iron, 418-422
Silver gallium sulfide/selenide, 270
Snell's Law, 308
Solar cell, 51,82

Solid solutions, 526-533
  free energy, 528-529
  stability, 530-533
Solubility of impurities, 491, 520-523
Solutions (see Solid Solutions)
Space charge, 44
Spectroscopic splitting factor, 372
Spinels, 402, 453-461
Spinodal decomposition, 529
Spin-orbit coupling, 387
Spin-orbit effects, 12, 15
Spin waves, 385-387
Steel,
  low-carbon, 417
  martensitic, 436-437
Stimulated Raman effect, 256-258
Stockbarger method, 495-496
Superexchange, 379-382
Susceptibility (magnetic), 384
Susceptibility (optical), 239-243, 268, 275
Switches, semiconductor glasses, 158-164

TeO_2, 270, 272, 273
Te, relaxation times, 122
Ternary compounds, 76
Tight binding approximation, 11
Tipping method, 542
Transistors, 51
  bipolar, 564-574, 591-594
  metal-oxide-semiconductor (see MOS)
  MOS, 575-591, 594-601
Transitions, optical (see Absorption)
Traps, 49

Upconversion, 337

Vacancies, 516-520
Valence band, 10
Vapor phase epitaxy, 62, 500, 566-567
Verneuil method, 498-499
Vicalloy, 442

Waveguides (see Optical fiber waveguides)
Wurtzite structure, 3

Xerographic display, 324

Yttrium aluminum garnet (YAG):
  Nd, 261-263, 266

Zinc-blende structure, 3, 70
Zn chalcogenides, 16, 272
  diffusion, 134
ZnO, 509, 513
ZnS, 509, 513
ZnSe, 16
  energy gap, 509
  heat of formation, 513
ZnTe, 272, 509, 513
Zone-melting, 498
Zone-refining, 492, 494

If you have any concerns about our products,
you can contact us on
ProductSafety@springernature.com

In case Publisher is established outside the EU,
the EU authorized representative is:
**Springer Nature Customer Service Center GmbH
Europaplatz 3, 69115 Heidelberg, Germany**

Printed by Libri Plureos GmbH
in Hamburg, Germany